The Scientific Enterprise in Antiquity and the Middle Ages

READINGS FROM ISIS

The Scientific Enterprise in Antiquity and the Middle Ages

EDITED BY
MICHAEL H. SHANK

THE UNIVERSITY OF CHICAGO PRESS
Chicago and London

The essays in this volume originally appeared in *Isis.* Acknowledgment of the original publication date can be found on the first page of each article.

The University of Chicago Press, Chicago, 60637
The University of Chicago Press, Ltd., London
 Published in 2000
Printed in the United States of America
ISBN: (cl) 0-226-74950-9
ISBN: (pa) 0-226-74951-7

04 03 02 01 00 5 4 3 2 1

Library of Congress Cataloging-in-Publication Data

The scientific enterprise in antiquity and the middle ages : readings from Isis / edited by Michael H. Shank.
p. cm.
Includes bibliographical references and index.
ISBN 0-226-74950-9 (cloth) — ISBN 0-226-74951-7 (pbk.)
1. Science, Ancient. 2. Science, Medieval. I. Shank, Michael H.

Q124.95 .S37 2000
509.01—dc21 99-088364
CIP

The paper used in the publication meets the minimum requirements of American National Standard for Information Sciences—Permanence of Paper for Printed Library Materials, ANSI Z39.48.1984.

CONTENTS

CONTENTS

Introduction

*Michael H. Shank**

AT THE END of his undergraduate career at M.I.T., physicist Richard Feynman reportedly told his advisor, John C. Slater, that he hoped to stay on for graduate school because "[M.I.T.] is the best school for science in the country."

"You think that?"

"Yeah."

"That's why you should go to some other school. You should find out how the rest of the world is."[1]

Slater's advice is sound in general. At its best, exposure to different outlooks and practices broadens one's understanding, shakes up expectations, exhumes and challenges buried assumptions, and sharpens the critical faculties. Readers who appreciate Slater's wisdom will approach this volume in a congenial frame of mind. Indeed, the scholarship from three-and-a-half millennia of ancient and medieval science in this collection attempts to "find out how the rest of the world [was]."

How did people in earlier cultures make sense of what we now call the "natural world?" Answering this question can be unnerving, since it requires immersion in one or more different cultures and alienation from the security of one's own. It is also exciting, and for the persistent the rewards are many. Few experiences are more exhilarating than gaining even modest access to the creative thoughts and practices of the long-dead, making headway in cracking the cultural codes of distant societies, or puzzling out the internal logic of claims and behaviors that initially strike us as odd, if not irrational or absurd. Such encounters can be deeply transformative. As Thomas Kuhn famously reported in the opening paragraph of his epoch-making *The Structure of Scientific Revolutions*, "exposure to out-of-date scientific theory and practice radically undermined some of my basic assumptions about the nature of science and the reasons for its special success."[2] A fair warning to the unwary reader!

THE PRACTITIONERS AND THEIR SOURCES

How do we learn about the scientific enterprise in the remote past? How do we gain access to its sources? Finally, how does scholarship of the caliber represented in this volume come to be produced? Budding scholars who find questions about early science enticing enough to make a long-range commitment to it must invest much

* I would like to thank David C. Lindberg, Ronald L. Numbers, Robert S. Westman, and the participants in the University of Wisconsin Memorial Union Terrace Summer Discussion Group for comments and constructive criticisms.

[1] Richard Feynman, *"Surely You're Joking, Mr. Feynman!" Adventures of a Curious Character* (New York: Norton, 1985), p. 47.

[2] Thomas S. Kuhn, *The Structure of Scientific Revolutions* (Chicago: University of Chicago Press, 1962, rev. 1970), p. v.

time before they can make any headway in their field. Typically, they first learn as much as possible about the intricate webs of signs that our predecessors used to communicate—not only their languages, but also the cultures in which these languages functioned and the contexts that gave them meaning.

These prerequisites are merely a beginning, however. Historians in some fields of early science can pull their sources off the library shelf and start to read, but many cannot. They often need to reconstruct and "establish" the texts on which they work. In the worst cases, they must literally reassemble, like a puzzle, the physical fragments of clay tablets, papyri, or manuscripts scattered in various libraries.[3] In other instances, the assemblage is more metaphorical. Scholars confronted with multiple copies of a text usually try to reconstruct the original version based on existing copies that are often separated by several centuries and significantly different from one another. Text editors compare the variant readings in these copies of the "same" text, transcribe them, and must decide which differences are scribal omissions, additions, amplifications, simplifications, glosses, or misreadings. From this work they eventually assemble a "critical edition" of the text in the original language and often translate the result into a modern one. (An example appears on page 291, in the appendix to the William Newman article). The reliable establishment and translation of a text may seem a trivial and boring activity (as I too once thought, before I knew better). Some aspects of this task are indeed work-a-day, but many are anything but routine. They confront the editor with frequent puzzles and difficult judgment calls, and fill the translator with trepidation about problems of meaning, understanding, and expression as well as interpretation.[4]

Once the critical edition is complete, scholars attempt to diffuse their work by publication. When successful, their colleagues can gain access to the new work and contribute to the ongoing task of figuring out what the text says and means and what it might have meant to its author(s) and readers. Such texts often embody our best hope of appreciating the thoughts, circumstances, goals, questions, and values of past authors, their audiences, and their place in these early societies.[5] The value of studies such as these is measured not by high-volume sales but by staying power: against the background of fleeting interpretations, few scholarly endeavors make a more enduring contribution to knowledge than an excellent textual edition.[6]

Scholarship in early science proceeds slowly and tends to be sharply focused—not surprising, since it takes years for apprentice-scholars to get up to speed, while the study of languages and of cultural contexts remains an ever-present goal for everyone. This scholarship uses evidence from artifacts, images, sculptures, build-

[3] Most of the material on Greek science is a notable exception, but even here important textual work remains (see the Alexander Jones selection below). For a classic statement of the challenges presented by Babylonian sources, see Otto Neugebauer, *The Exact Sciences in Antiquity* (Princeton, N. J.: Princeton University Press, 1952, rev. ed. New York: Dover, 1969), ch. 3.

[4] See the remarks by G. E. R. Lloyd in the first section of his article, below. For one perspective on the process, see M. L. West, *Textual Criticism and Editorial Technique* (Stuttgart, Germany: B. G. Teubner, 1973).

[5] For allusions to these processes, see, for example, the article by Jones, "The Adaptation of Babylonian Methods in Greek Numerical Astronomy," and the edition from scattered manuscripts and the translation appended to Newman's article on page 291.

[6] There is, of course, nothing magically transformative about a new edition that no one reads or uses to challenge fond conclusions in the existing literature.

ings, city plans, etc., whenever they are available and relevant but texts continue to form its core. This orientation is not simply a preference, but often a necessity. Texts devoid of context are sometimes all that we have. And when the author of a document, its date, place of composition, and provenance are unknown, textual techniques—careful analyses of the language, the argument, the meaning, the script, and the source as physical object—are singularly more promising tools than sociological methods. In small details of language, usage, and content, the historian will sometimes detect unsuspected clues about the world from which the text emerged, giving it the chance to find a place among larger patterns and historical generalizations. The challenge of the hunt is not the least pleasure of this type of work.[7] Eventually, as the body of specialized studies develops, historians increasingly become able to answer broader questions and criticize or modify existing generalizations. How and why did a once-favored practice fall into disuse? How did another, at one time unusual and perhaps treated with suspicion or hostility, spread like wildfire and become characteristic of an institution, era, or culture? How did a specific import (a concept, technique, approach, text, theory, or instrument) come to transform the importing culture? Answers to such questions require attention to larger cultural patterns, to political, social, and educational institutions, and to modes of communicating, teaching, and diffusing knowledge about nature.

Scholars who teach and learn the history of early science often work in an atmosphere of craft and apprenticeship. Conversations within the "guild" are as likely to focus on the handling of sources and on tools of the trade as on historical interpretations. Practitioners of early science on the whole tend to be a critical lot, an outlook perhaps related to an existential awareness of the many opportunities for going astray as one proceeds from ancient document to modern interpretation, usually for reasons of craft rather than interpretation.

Professional formation, however, has a reverse side—"professional deformation," as the French expression snidely puts it. Some scholars tend to focus on fascinating trees to the neglect of the forest and to write for the colleagues in their own field who already share a deep commitment to the enterprise rather than for the general public, which needs to be convinced. With some notable exceptions, scholars of early science have not grasped the importance of communicating either their findings or their enthusiasm to their nonspecialist colleagues or the larger public.

Similar tasks, approaches, and problems are not unknown in the history of more recent science. But they occupy such a prominent place in the history of early science that they have shaped the way in which much ancient and medieval scholarship presents itself and the way it reads. Upon approaching early science for the first time, readers may feel as if they are entering a different world in terms of both content and the way in which it is presented. Well they might: "The past is a foreign country; they do things differently there"[8]—and the remote past is more foreign yet. To be effective, its ambassadors need to have one foot in each world, and the

[7] See, for example, Carlo Ginzburg, "Clues: Roots of an Evidential Paradigm," in his *Clues, Myths, and the Historical Method*, trans. John and Anne Tedeschi (Baltimore, Maryland: The Johns Hopkins University Press, 1989), pp. 96–125; and Robin W. Winks, ed., *The Historian as Detective: Essays on Evidence* (New York: Harper and Row, 1969).

[8] L. P. Hartley, *The Go-Between*; cited in David Lowenthal, *The Past is a Foreign Country* (Cambridge, England: Cambridge University Press, 1985), p. xvi.

language that they speak will often contain curious terms and betray the literary equivalent of a strong accent. Acting as a go-between is not always easy. At the heart of the predicament lies the challenge of presenting ancient and foreign concepts, institutions, rationalities, or practices in ways that preserve their otherness and yet allow modern readers to grasp them. A reader aware of the challenge may be in a better position to appreciate, if not the idiosyncrasies of this work, then at least a few of the reasons for them.

At early stages in their careers, scholars in these early fields are sometimes urged by well-meaning friends not to waste their time on such distant centuries and foreign cultures. Almost a century ago, as a young graduate student, the budding historian of medieval science, Lynn Thorndike, received gentle words of caution from his older brother Ashley, a scholar of Elizabethan literature.[9] Historians of early science will sympathize with the mildly xenophobic suspicions that travelers confront when they return to their homeland after extensive immersion in another culture. Benign prejudices about the irrelevance of studying early science sometimes shade into imputations of pernicious escapism.[10] Such attitudes, combined with many prerequisites and low external rewards, also have a silver lining: careerists eager to make a "quick splash" go elsewhere. Attrition thus creates small fields populated by individuals whom the sympathetic describe as determined, careful, and dedicated to their distant fields; the uncharitable as stubborn, anal-retentive, and antiquarian. William Paley would have marveled at this wonderful adaptation to the scarcity of research support and academic positions for specialists in early science.

LOOKING BACKWARD

In the third millennium, historiographers of premodern science will surely see the twentieth century as a watershed. The last hundred years have utterly transformed both the study and our understanding of ancient and medieval science, even as new worlds have emerged. The news, however, has not yet reached many quarters. The disjunction between scholarly achievements and public consciousness stands out starkly in the scientific timeline from the astronomer Carl Sagan's *Cosmos*, published in 1980. This companion book to Sagan's popular television series by the same name offers a timeline that begins with Thales, the sixth-century B. C. Greek philosopher, runs through other pre-Socratics, Plato, Aristarchus, Ptolemy, down to the death of Hypatia and the burning of the library at Alexandria in the fifth century of our era. From this date until the late fifteenth century, there is a large, ten-century gap until Columbus, Leonardo, and Copernicus reopen the flow of significant contributors to the scientific stream. The timeline caption concludes: "The millennium gap in the middle of the diagram represents a poignant lost opportu-

[9] Columbia University, Special Collections, Lynn Thorndike Papers, File "Letters to 1907 or so" under "A", Sept 9 (no date, but soon after 1902): From Ashley Thorndike writing from Evanston, Ill.: "I shouldn't go too deep into Mediaeval lore till I saw my way out . . ."

[10] "Joining a general antirational retreat from the unhappy reality of defeat [after World War I], German historians immersed themselves in Paracelsus, Arabic alchemy, Egyptian mathematics, and Babylonian astronomy." Paul Forman, "Independence, Not Transcendence for the Historian of Science," *Isis* 82 (1991) p. 81. Clearly, these bizarre subjects would intrigue no normal person. Politico- or socio-pathological explanations must therefore be invoked to make sense of such aberrant behavior.

nity for the human species."[11] Now Sagan was a well-educated Cornell professor, a Pulitzer Prize winner, and a popular lecturer with access to an excellent library collection in the history of science. Yet his timeline is out of date by half a century. Being in or out of fashion is not the issue; rather, the problem is Sagan's failure to take into account massive amounts of new empirical knowledge generated since the late nineteenth century, some of it available in syntheses that are almost 50 years old.

Since late antiquity, the expression "ancient science" in the Near East, North Africa, and Europe, has largely meant "Greek science," precisely as in Sagan's timeline. This is not surprising if one remembers that, until rather recently, no one could understand the writings of the most ancient cultures of the world, in contrast with the writings of China or Greece. With the decipherment of Egyptian hieroglyphics in the early nineteenth century and of Mesopotamian cuneiform scripts in the late nineteenth century, a small group of pioneering historians finally gained access to the writings of these civilizations. These texts offered glimpses of unsuspected worlds, particularly in medicine, mathematics, and astronomy. The twentieth-century dividends of this work for the history of ancient science have been enormous, and none is more significant than the work of Otto Neugebauer (d. 1990) on mathematics and astronomy in Egyptian and especially Babylonian civilizations.[12] The results, however, have been slow to diffuse and a massive amount of work still remains. Two examples must suffice. A full millennium before Pythagoras, Babylonian mathematicians were not only acquainted with the "Pythagorean" theorem, but were exploring ways of generating arithmetically groups of three numbers such that the square of the first equals the sum of the squares of the other two.[13] By the beginning of our era, the Babylonian science of divination had developed into a quantitative science of omens so successfully predictive that we call it astronomy, and so important historically that one cannot understand Ptolemy's *Almagest* without it. These and many others results have turned the standard story of the Greek origins of scientific thought into an untenable proposition (see especially the Rochberg and Pingree selections). But the task has scarcely begun.

At the end of the nineteenth century, "ancient science" may have had an excessively narrow meaning, but at least it meant something. The expression "medieval science," by contrast, was an oxymoron. Unfortunately, it has remained one well into the twentieth century and, in many quarters, to this day. The message of the caption below Sagan's timeline is clear: nothing happened for a millennium. One of the most enduring legacies of the early sixteenth century is the concept of a "middle age" (*medium aevum*) that functioned effectively as a long, place-holding zero separating its own day from classical antiquity. Despite the growing body of scholarship in this area, the picture of a scientifically vacuous medieval millennium remains canonical. Like the old canard about the flat earth of the Middle Ages, it makes a good story fully consistent with deeply entrenched prejudices about the period.

[11] Carl Sagan, *Cosmos* (New York: Random House, 1980) p. 335.

[12] When it is not ignored on account of its difficulty, Otto Neugebauer's work in this area is underappreciated and misunderstood (e.g., in the Bernal selection). For a survey of Neugebauer's contributions, see Noel Swerdlow, "Otto E. Neugebauer (26 May 1899–19 February 1990)," *Proceedings of the American Philosophical Society* 137, #1 (1993), pp. 139–165.

[13] Otto Neugebauer, *The Exact Sciences in Antiquity*, 2nd edition (New York: Dover, 1969), pp. 34–44;

The man most responsible for changing the situation was the French physicist, philosopher of science, and historian of science Pierre Duhem (d. 1916). Ironically, Duhem for most of his career would not have objected to the Sagan timeline. Only toward the end of his life did he discover late-medieval natural philosophers whose discussions seemed to foreshadow such features of early modern science as the rotation of the earth and inertia. Duhem's eventual argument for the conceptual continuity between the fourteenth and seventeenth centuries was both premature and overly optimistic. But he did establish beyond a doubt that the history of science had a medieval chapter of far greater interest than anyone had suspected.[14]

Some objections to its name notwithstanding, "medieval science" is now a well-established field of endeavor, with growing bodies of primary and secondary literature covering several different cultures. Although much fundamental work remains to be done on late-fourteenth- and fifteenth-century Latin material, the most exciting frontiers of medieval science have shifted eastward, and earlier. Tens of thousands of scientific sources in Arabic and Sanskrit remain unstudied, many scarcely catalogued. More seriously, numerous Sanskrit sources are in imminent danger of disappearing altogether, as the fragile leaves on which they are written decay. But it is Arabic civilization between the eighth and fifteenth centuries that presents the widest prospect for new discoveries and sweeping reinterpretations of science in the post-Hellenistic period.[15] Joseph Needham's massive *Science and Civilization in China*, which focuses heavily on the "medieval" period, now gives the nonspecialist interested in comparative questions a place to start, a wealth of information, and critical questions. Who knows what may yet turn up in Central Asia, where all these civilizations traded when they did not clash?

The medieval period was a particularly fruitful one for, among other things, the birth and development of institutions crucial for scientific pursuits. It is ironic that Sagan spent his entire scientific career associated with universities and observatories, two enduring institutions of great significance for science that emerged in the Latin and Arabic civilizations, respectively, during that desolate medieval millennium of "lost opportunity for mankind."[16] Since these and others institutions often join intellectual and pedagogical concerns with social, economic, and political

[14] For three very different perspectives on Duhem, see John E. Murdoch, "Pierre Duhem (1861–1916)," in Helen Damico, ed., *Medieval Scholarship: Biographical Studies on the Formation of a Discipline.* Volume 3: Philosophy and the Arts. (New York: Garland Publishing, Inc., 1999), pp. 23–42; and on the continuity debate generally, Roger Ariew and Peter Barker, eds., *Revolution and Continuity: Essays in the History and Philosophy of Early Modern Science* (Washington, D.C.: Catholic University of America Press, 1991), esp. pp. 1–18; and H. Floris Cohen, *The Scientific Revolution: A Historiographical Inquiry* (Chicago and London: The University of Chicago Press, 1994), pp. 45–53.

[15] For a glimpse at the extent of these untapped resources, see David A. King, *A Survey of the Scientific Manuscripts in the Egyptian National Library* (Winona Lake, Indiana: Eisenbrauns, 1986); Ekmeleddin Ihsanoglu, ed., *Osmanli Astronomi Literatürü Tarihi* [History of Astronomical Literature During the Ottoman Period], 2 vol. (Istanbul: IRCICA, 1997); Ekmeleddin Ihsanoglu, ed., *Osmanli matematik edebiyati tarihi* [History of Ottoman Mathematical Literature], 2 vol. (Istanbul: IRCICA, 1999).

[16] See Edward Grant, "Science in the Medieval University," in James Kittelson and Pamela Transue, eds., *Rebirth, Resilience and Reform: Universities in Transition, 1300–1700* (Columbus: The Ohio State University Press, 1984); and Aydin Sayili, *The Observatory in Islam* (Ankara, Turkey, 1960; reprinted New York: Arno Press, 1981). The Astronomical Bureau in China also deserves mention as a powerful force during this period; see Joseph Needham, *Science and Civilization in China*, vol. 3 (Cambridge: Cambridge University Press, 1970), pp. 186–194.

parameters, they offer excellent opportunities for examining the scientific enterprise in a well-rounded way. In this realm, too, much fundamental work remains to be done in several cultures.

By comparison with the surprising new worlds recovered by scholars of Babylonian astronomy and mathematics, of Arabic, Latin, and Chinese science, and of cross-cultural transmission, the impressive twentieth-century advances in the history of European science since 1500 look like a minor reshuffling of interpretative categories.

WHAT DOES "SCIENCE" MEAN IN ANTIQUITY AND THE MIDDLE AGES?

So far we have treated the title of this volume as unproblematic. But the reader may already have wondered: Can we properly designate anything more than 500 years old as "scientific?" And even if we can, is it helpful to do so? These questions are difficult and hotly contested. Scholars have thought hard about them, and several have lost patience with the answers of their colleagues (see, for example, Neugebauer's "The Study of Wretched Subjects," on page 21). It is a question with which each article in this collection grapples implicitly, and sometimes explicitly (most notably in the contributions by Pingree, Lloyd, von Staden, and Copenhaver on pages 30–71, 373).

Finding answers to the preceding two questions is no mean feat. One can quickly reply "no," simply by defining science as a post-Copernican, or a post-Newtonian, or a post-1800 enterprise. Before the terminus of choice, so the argument goes, the premodern inquiry into nature differed so strikingly from its modern form, however defined, that it falls altogether outside the category "science." Perhaps this early thinking about nature is deemed, by our lights, overly philosophical or metaphysical or mythical or magical, or insufficiently secular or mathematical or experimental or observational, or perhaps its practitioners are thought insufficiently numerous or institutionalized or professionalized—that is to say, in every case, insufficiently modern.

No matter how clean as a definition, such a narrow demarcation of "science" is fundamentally absurd and ahistorical. It is absurd, for it amounts to stating that since ancient science is not modern science, it does not count as science. (An analogous argument would duly eliminate personal identity between the newborn and the adult). The absurdity of this position comes from its being ahistorical. Historians study—and seek to understand—change: to expect that premodern science should look like modern science in order to qualify as a proper object of study is to expect precisely that little should have changed during an interval of sweeping transformations. While some prefer to use the term "natural philosophy" to designate early science, this approach runs into similar problems of anachronism and misapplication, for this expression is too specific and too Greek to encompass all of the premodern scientific traditions and their many cultures (Babylonian mathematical astronomy, for example, was most emphatically *not* natural philosophy). In fact, we can communicate very well with our own contemporaries if we persist in pointing out that the English word "science" is a generic term, just as its etymological ancestors once were. It encompasses many species, of which some are recent (molecular biology) and some extinct (medical astrology), while yet others have survived a long history (planetary theory). Within each species are individual (hence dissimilar) specimens that come into being and die even as they share family resemblances and a common name in spite of many notable differences.[17] Freed from excessively parochial definitions of science, twentieth-century scholarship has found fascinating new ancestors of the scientific tradition, which belong in even the most simplified timeline. It has also turned up some intriguing extinct species, which belong among the twists, turns, and dead-ends of the more intricate versions of the story. The common ground and the tensions between the selections by Neugebauer, Pingree, Lloyd, Sabra, and Copenhaver in particular will help the reader think critically about the scope and utility of the category "science."

SCIENTIFIC TRADITIONS IN CROSS-CULTURAL PERSPECTIVE

The emphasis, in the historiography of science during the last two generations, on scientific revolutions has led to many fruitful questions. But it has diverted attention away from the ways in which *one* aspect of a scientific tradition can undergo significant changes while remaining largely within that tradition. Typically overlooked in these discussions of revolutionary scientific change is an appreciation of the fact that traditions are almost never static.[18] Indeed the most enduring traditions survive not because they are immutable, but precisely because they manage to adapt to new circumstances. They succeed in doing so because they are not changing in every respect at once. So it is that remnants of multiple traditions of ancient and medieval science survive not only in texts and artifacts, but also in fundamental aspects of our modern vocabulary, frames of reference, and institutions. For example, the Egyptian 24-hour division of the day; the Babylonian sexagesimal divisions and subdivisions of the hour and the degree; the Byzantine hospital; the Latin university; and the Arabic astronomical observatory continue to function in modern contexts that are foreign to their birthplaces. This history of usage and survival serves as a two-fold symptom of the longevity of scientific traditions and of indisputable changes, and it raises fascinating questions about both diffusion, survival, and appropriation.

The history of ancient and medieval science offers case studies of cross-cultural diffusion that are considerably more complex than the obvious model of propagation by force, through conquest and colonization. The modern pattern of scientific diffusion associated with the colonists—in fact a well-tested Greek model—stands to be enriched by the relatively frequent case of conquering peoples that are seduced by, and adopt specific elements of, the cultures of the conquered, including their sciences, medicine, and natural philosophy. Why were Hipparchus and Ptolemy so impressed by "barbarian" Babylonian mathematical astronomy? Why did Muslim and Jewish scholars in the Arabic and Persian worlds in turn take up the admiring (and critical) study of such pagan foreigners as Euclid and Ptolemy, Aristotle and Galen? Why did Christian and Jewish scholars in medieval Europe

[17] This discussion owes much to my colleague Dave Lindberg.

[18] Although Thomas Kuhn's concept of "normal science" did point to the significance of traditional elements in both scientific research and pedagogy, his work has been received largely as a statement about revolutions. Kuhn, *The Structure of Scientific Revolutions*, esp. ch. 2 and 11. His notion that "[scientific revolutions] are the tradition-shattering complements to the tradition-bound activity of normal science" (p. 6) points to a perception of traditions as fossil-like—static and brittle. Some are—but others are malleable, elastic, even alive.

come to be fascinated by the same pagans and by original works by Muslim infidels, to boot? Why did the thirteenth-century nomadic Mongol conqueror Hulagu Khan set up an astronomical observatory in Northwest Persia run by his Muslim former prisoner, Nasir al-Dīn-al-Ṭūsī? What did these people think they were doing when they reached out to such aliens? And how did the fragile and foreign transplants that resulted from these improbable contacts not only survive, but also put down roots and become defining elements of their new environments?[19] Each of these cases involves long-lived foreign scientific traditions improbably finding admirers and a home in new cultural contexts.

Even unanswered, the larger questions that these cases raise make abundantly clear the great cultural diversity of early scientific activity and the significance of cross-cultural diffusion. It is therefore odd that, in the debate about multiculturalism in the 1990s, science in the period before 1500 in Near-Eastern and European history suffers from a two-fold marginalization. On the one hand, early science (typically understood as Greek science) is easily dismissed for its genetic connections with "Western civilization" today, and as constituting, therefore, a scholarly manifestation of European hegemony. In this framework, the "affinities" between Greek and modern science (see von Staden's "Affinities and Elisions. Helen and Hellenocentrism" on page 59), which were once an occasion for gloating, have become objects of opprobrium. On the other hand, all of these early periods and inquiries, not just the Greek, are easy to dismiss for their irrelevance to the present—the thoughts of long-dead men recorded in very dead languages, a characterization to which late-twentieth-century students can "relate," as they say.

One cannot, of course, have it both ways. An undue emphasis on similarities between the past and the present of science is as misleading as a complete denial of all kinship. One of the historian's many challenges is to make sense of both the similarities *and* the differences, the connections *and* the disjunctions, the continuities *and* the breaks—and to keep them together, in tension, in the same narratives.

Its impact on later European civilization notwithstanding, much of the Greek scientific material, properly taught in context ought to strike our students and us not as cozily familiar, but as curiously alien. Conversely, the material that seems familiar (e.g., a theorem from Euclid's *Elements*) does so not because it is timeless, but because our schools continue to teach it as a basic part of mathematical education, unencumbered by its Hellenistic context. Von Staden's article suggests that our accounts of Greek scientific material need to restore the older "elisions," elements that were quietly dropped because they undercut a triumphalist story. If we can, in addition, take an equanimous view of the affinities between some aspects of Greek and modern science, foregoing excessive praise or blame, the resulting picture becomes more complicated and culturally pluralistic in issues large and small (see Von Staden's fascinating footnote about the word "barbarian").

[19] A. I. Sabra, "The Appropriation and Subsequent Naturalization of Greek Science in Medieval Islam: A Preliminary Statement," *History of Science* 25 (1987), pp. 223–243; reprinted and discussed in F. Jamil Ragep and Sally Ragep, with Steven Livesey, eds., *Tradition, Transmission, Transformation: Proceedings of Two Conferences on Pre-modern Science held at the University of Oklahoma* (Leiden, The Netherlands: Brill, 1996), pp. 3–27, and passim.

It should by now be obvious that the premodern period fairly bristles with case studies of various forms of interaction between cultures. The past of early science, it would seem, is not a foreign country, but several. For travelers who plan to tarry there, anthropological sensitivity is at a premium and comparative work is inevitable. They must be patient enough to wrestle with ways of thinking alien to their own and to learn to communicate with past individuals in their own languages and frameworks. This is multiculturalism with teeth. Its aim is not to make anyone of us feel good, but to unsettle all of us by challenging our intellects, our assumptions, and our narratives.[20] This task also demands cross-cultural comparisons of specific contributions to the scientific enterprise according to specific, historically sensitive criteria. It takes but a moment's thought to realize that an evaluation in which the activities of all cultures and geographical regions emerge on a par in every respect is a useless exercise. The serious comparison of cultures and civilizations in specific areas of endeavor requires attention to strengths and weaknesses, to differences as well as similarities. But differences cannot be discussed if we banish from our vocabulary the words that express them, modifiers like "more" and "less," or predicates like "present" and "absent." Historians of science may take comfort from the fact that they are far from the first to engage in the analysis of differences. The individuals who made it their business to appropriate some foreign cultural imports, but not others, did so on the basis of comparisons.

For the history of science, the period from antiquity through the Middle Ages is one of extraordinary importance. If one takes the long view, even the dramatic changes within the scientific enterprise of the last several centuries pale by comparison with those made between the invention of writing and 1500. In the aggregate, the latter contributed a large body of material to early modern science—concepts and procedures, ways of thinking and techniques, ideals of argumentation and criteria of proof, patterns of support and institutions, means of teaching, recording, and presenting knowledge—and some of these have survived well into the modern period. The Latin university, the Islamic observatory, the Byzantine hospital; the impressive predictions that used rough empirical data cleaned up with mathematical techniques in Babylonian planetary theory; the mathematization of physical problems in Archimedes and of natural philosophy in late medieval Europe; the patronage of astronomy/astrology and medicine by rulers in every major civilization; the astonishing cross-cultural diffusion of works, instruments, techniques, and data; and, perhaps most fascinating of all, the appropriation of the latter in contexts very different from those in which they originated throughout the entire premodern period—this cursory list of issues barely scratches the surface of these fascinating centuries. We truly owe it to ourselves to "find out how the rest of the world is."

[20] "There are ways of reading Shakespeare and Dante that can be quite illuminating and profound, and ways of reading Toni Morrison that are flat. I want to know how people are reading these texts and what kind of historical sense they have. Are they just occasions for cathartic expressions of identity? Just moments for the building of a self-confidence that has been undermined by white supremacist and male supremacist forces? This may be morally justifiable, but it is not intellectually justifiable. It's intellectually debilitating." Cornel West, *Beyond Eurocentrism and Multiculturalism*, vol. 2: *Prophetic Reflections: Notes on Race and Power in America* (Monroe, Maine: Common Courage Press, 1993), pp. 120–121.

THE COLLECTION: AN OVERVIEW

Unlike the other "Scientific Enterprise" readers in this series, this one benefits from no unity of time, place, or theme.[21] The articles in this collection treat material that extends over three millennia and more than half a dozen cultures, providing a rich sampling of themes, patterns, comparisons, debates, and interactions. Although the articles are drawn from recent and classic work in *Isis*, the leading journal for the history of science, they do not, indeed cannot, cover the full chronological, disciplinary, geographical, and historiographical range of early science, or even be representative of it. Since the 1960s, *Isis* has, for example, published far more articles on early astronomy than on early medicine.[22] In recent years, it has competed with a growing number of journals that specialize by field of science, as well as with journals devoted to specific areas of history, philosophy, language area, or culture.

Scholarship does not stand still. Accordingly, the selected articles in this volume do not represent the final word on any given topic. Rather, they illustrate stimulating, controversial, readable, and/or now-classic treatments of historical problems in the history of science before 1500. They offer the reader many insights into the new worlds of the scientific past, and the teacher convenient springboards into classroom discussion. With much regret, I have left several excellent and many good articles out of the collection for reasons of accessibility and space. The essays are generally organized in chronological fashion, with some attention to grouping by content.

Topics and Themes

The volume opens appropriately with "The Study of Wretched Subjects" (1951), a one-page rebuke from Otto Neugebauer, the dean of historians of ancient mathematical astronomy. His target was a review in which George Sarton, founder-editor of *Isis* and Harvard professor devoted to early science, had used the adjective "wretched " in reference to astrological material. Sarton proved fair-minded enough—and perhaps chastised enough—to publish it. At issue was the value of studying historical materials that were once taken seriously, but now look like nonsense to many (in this case, Sarton), a matter of central concern to the history of ancient and medieval science, which Neugebauer places here in bold relief.

What, then, counts as science? The question surfaces in the collection, most sharply perhaps in "The Cultures of Ancient Science: Some Historical Reflections" (1992), a four-part discussion, with an introduction by Francesca Rochberg, devoted to the contested place of Greece in ancient science, particularly in relation to Egyptian and Babylonian civilizations. Rochberg's introduction critically sets the four pieces in a larger context, beginning with changing trends in historiography

[21] Peter Dear, ed., *The Scientific Enterprise in Early Modern Europe: Readings from* Isis (Chicago, 1997); Ronald Numbers and Charles Rosenberg, eds., *The Scientific Enterprise in America: Readings from* Isis (Chicago, 1996); Sally Gregory Kohlstedt, ed., *History of Women in the Sciences: Readings from* Isis (Chicago, 2000).

[22] Happily, an excellent remedy for this defect in the later period is Michael R. McVaugh and Nancy G. Siraisi, eds., *Renaissance Medical Learning: Evolution of a Tradition* [*Osiris*, 2nd series, 6 (1990).

and philosophy of science since the 1960s. She offers examples of the familiar, well-entrenched outlook that reads Greek approaches to nature as both radically different from those of neighboring cultures and fundamentally akin to modern science, hence at the origins of modern science.

The four contributions to the discussion all grapple with (1) the meaning of science in its premodern and modern incarnations (or alternatively the proper objects of study for the historian of science); (2) the related thorny question of the origins of scientific thought and practice (to which the historian's definition or usage of the word "science" is very germane); and finally (3) the diffusion of this knowledge. David Pingree's spirited "Hellenophilia vs. the History of Science" makes a strong case for reconfiguring the map of one specific subset, the exact sciences in antiquity. The standard picture of scientific origins tautologically traces science, defined in Greek terms, to Greece (one of the symptoms of the disease that Pingree identifies as "Hellenophilia.") Instead, Pingree offers the following: "science is a systematic explanation of perceived or imaginary phenomena, or else is based on such an explanation." This inclusive definition—too inclusive for some—finds room for Sarton's "wretched subjects" and leads to a very different map of scientific origins, one in which a mathematically fortified Babylonian science of divination radiates outward from Mesopotamia toward Egypt, Greece, and India.

G. E. R. Lloyd's discussion of the "Methods and Problems in the History of Ancient Science: The Greek Case" reflects the new-found tentativeness of historians of Greek science in light of insights from the philosophy of science and cultural anthropology. Lloyd shows nicely how the usual difficulties of historical inquiry are exacerbated for more ancient (and hence more alien) sources. While willing to entertain questions of origins in specific areas of inquiry, he points to the futility of seeking a single origin for "science as such," emphasizing the pluralism of, and controversies among, the activities that fall under this broad rubric. Lloyd is a physician who has healed himself with solid doses of anthropology: the contrast between these new approaches and Lloyd's triumphalist views in his earlier textbooks (cited in Bernal's article) is a measure of the changes that have swept the field, not least under Lloyd's own leadership.

Heinrich von Staden's "Affinities and Elisions: Helen and Hellenocentrism" gets the award for the subtlest title in the collection. His article addresses the charge of Hellenocentrism, a milder cognate of Pingree's Hellenophilia, this time from the vantage point of the Greek world. The "affinities" refer to the congenial aspects of Greek scientific thought, that is, those that are highlighted, lifted out of their contexts, and praised because they resonate with moderns sensibilities and provide a ready-made account of scientific origins (von Staden critically scrutinizes in particular the conflation of alphabetic literacy with Athenian democracy and with Pre-Socratic—non-Athenian—natural philosophy). The "elisions" are the suppressed, less palatable surroundings of these affinities, such embarrassing—if not wretched—endeavors as astrology and magic, which are often valued and practiced by the very individuals who appear in the affinity list. Von Staden reminds the reader of forgotten associations of traits in Greek culture—associations that defy simplistic modernization (e.g., the prephilosophical and prescientific association of both criticism and authorial voice with the myth-making environment). This complex, pluralistic picture of Greek culture even turns up some new affinities: since the Greeks too had their orientalizing phases and Egyptomanias, caution and critical judgment remain in order when reading the Greeks' own origin myths.

A scholar of nineteenth-century China, Martin Bernal wrote "Animadversions on the Origins of Western Science" as an outsider to the study of antiquity, but one whose views on the subject were drawing widespread attention in the news media. His essay illustrates for the reader both the provocative questions and the weaknesses of his approach in *Black Athena*, but now on the specific terrain of Egyptian and Greek science. His essay, which uses probabilistic language, calls into question Otto Neugebauer's evaluation of the Egyptian "exact sciences" as "primitive." Issues of tact aside, Neugebauer was the foremost editor, translator, and interpreter of texts in both Egyptian and Mesopotamian mathematics and astronomy. His judgment was a comparative one, based on an exhaustive immersion in Egyptian, Babylonian, Greek, and many other sources. Bernal's arguments for Egyptian priority in various areas rely, curiously, on a standard of science that is very Greek and sometimes later (the less supernatural and magic and religion, the more scientific), while avoiding detailed comparisons with the Mesopotamian mathematics or astronomy that were the grounds for Neugebauer's comparative judgment. Robert Palter has criticized this essay extensively, on both evidentiary and interpretative grounds, in his "Black Athena, Afrocentrism, and the History of Science," which I urge the reader to study with care.[23]

The four contributions to this forum present a fascinating study in contrasts with respect to the formulation of historical problems, assumptions, historiography, definitions of science, evidence, and—not least—tone and scholarly rhetoric. Two are measured and diplomatic; the other two are highly charged, even angry; and of these, the one without footnotes does not need them whereas the one with footnotes might be better without them. All in all, this is a thought-provoking introduction to important and controversial issues in the historiography of early science, as the authors offer fruitfully different assumptions and approaches, with insights that can help sharpen our understanding of the medieval era as well.

The next group of articles touches upon three prominent areas of the Greek scientific enterprise—mathematics, astronomy, and medicine. Eudoxus (d. ca. 337 B.C.), the renowned mathematician responsible for the "Euclidean" theory of proportion that some consider the crown jewel of Greek mathematics, plays a pivotal role in "A New View of Early Greek Astronomy" (1972). In this now-classic article, Bernard Goldstein and Alan Bowen call into question the standard picture of early Greek astronomy as geometrical and oriented to planetary problems. Instead, they argue that the organization of stars into constellations, attention to their risings and settings, weather, and calendaric problems were primary objects of inquiry.[24] On their account, Eudoxus became the key agent of a momentous change. His most important contribution was not, famous though it is, the concentric-sphere

[23] Robert Palter, "Black Athena, Afrocentrism, and the History of Science," in Mary Lefkovitz and Guy MacLean Rogers, eds., *Black Athena Revisited* (Chapel Hill: University of North Carolina Press, 1996), pp. 209–266. This is an updated version of Palter's response to Bernal's article in the journal *History of Science* 31 (1993), pp. 227–287, to which Bernal in turn responded, and Palter replied in the same journal 32 (1994), pp. 445–468.

[24] For a more radical and controversial statement about the lack of evidence for planetary concerns in early Greek astronomy, see Bernard Goldstein, "Saving the Phenomena: The Background to Ptolemy's Planetary Theory," *Journal for the History of Astronomy* 28 (1997), pp. 1–12.

models of planetary motion on which Aristotle embroidered. Rather, Goldstein and Bowen argue, it was a two-sphere model of the universe according to which the heavens and the earth are both conceived as spheres with a mathematically analogous grid ("spider web") of reference lines and axes (equator, tropics, poles, etc). Here lies the beginning of (the nonquantitative phase of) geometrical modeling in Greek astronomy, to the basic features of which the article offers a clear introduction.

In "The Adaptation of Babylonian Methods in Greek Numerical Astronomy" (1991), Alexander Jones addresses a second crucial turning-point in Greek astronomy—the association of numerical values (with predictive capabilities) with geometrical models in Hellenistic astronomy between Hipparchus and Ptolemy. Jones thus effectively instantiates the general theme of cross-cultural transmission raised especially in the Pingree and von Staden contributions to "The Cultures of Ancient Science." Drawing on fresh studies of papyrus material, the article shows a fascinating nonlinearity in the interaction between Greek geometrical and Babylonian arithmetical methods in astronomy between Hipparchus (second century B.C.) and Ptolemy (second century) and provides clear evidence of Babylonian influence on Greek astronomy in Hellenistic Egypt. Jones's article also gives the lie to the widespread notion that all important Greek materials have been edited.

Owsei Temkin's "Greek Medicine as Science and Craft" is a classic article by one of the leading historians of medicine of the last generation. It introduces the reader to key elements, terminology, interpretative problems, and earlier historiographical debates in Greek medicine to the time of Galen, from the problem of the Hippocratic corpus to the divisions between the medical sects. At several junctures, one can see Temkin thinking his way through such problems as the pluralistic relations of medicine and philosophy and of medical practice or craft to the later dominance of Galenic theory.

The articles by David Lindberg and Stephen McCluskey explicitly tackle the question of science and religion, an issue that, broadly understood, implicitly underlies every selection in this volume. Indeed, whereas many historians of more recent science think of this question as an aspect of the humanistic relations of science, historians of early science are as likely to think of it as an issue in the relations of science and society. Lindberg's "Science and the Early Christian Church" (1983) makes a fitting transition from antiquity to the Latin Middle Ages but, more importantly, contributes a much-needed interpretative overview of science and religion in the Patristic period. Like many other topics in ancient and medieval science, the vast domain of science and religion is still bedeviled by untenable, yet tenacious nineteenth-century interpretations. After summarizing and criticizing the "Draper-White thesis" about the warfare of science and religion, Lindberg turns to the Church Fathers themselves, whose quotable quips (e.g., Tertullian's "What has Athens to do with Jerusalem?") he replaces in their various contexts. The results yield a broad spectrum of complex views about the proper place of various Greek natural philosophies in a Christian worldview, from diffidence to enthusiasm. The simple polarities of the latter-day defenders of Christianity and of Draper and White do not do justice to the evident pluralism at the end of the classical world.

McCluskey's "Gregory of Tours, Monastic Timekeeping, and Early Christian Attitudes to Astronomy" (1990) offers a specific case study in a later phase of the preceding problem as he examines the practice of astronomy in the early medieval period. He demonstrates the indebtedness of Latin monastic culture to Near-Eastern computational methods, and the transformation of its Greco-Latin pagan roots. Such pre-Benedictine monastic practices of astronomy had made their way from John Cassian in fourth-century Lower Egypt to Gregory of Tours in sixth-century Gaul. Particularly noteworthy are McCluskey's case for the role of very basic observations in sixth-century Gaul, and his argument for the impetus of monastic timekeeping in "baptizing" the study of astronomy in the Latin Middle Ages. The regulation of daily life in Benedictine monasteries effectively gave religious and institutional sanction to elementary nighttime observations to determine the time of prayers. Interestingly, a similar religious sanction for astronomy will appear in a more complicated guise in Islam (see A. I. Sabra's "Situating Arabic Science").

Linda Voigts's "Anglo-Saxon Plant Remedies and the Anglo-Saxons" (1979) goes well beyond her title to question earlier generalizations about not only herbal remedies, but also changing climates and cultivated vs. indigenous plant species. Her examination of the four oldest Old English medieval medical texts challenges standard interpretations of these works as mere academic exercises without practical use. Voigts shows instead that the revisions of these pre-Conquest works and the marginalia in them reflect concerns about identification, differentiation, and medical uses that are incompatible with strictly bookish interests. Looming in the background are issues of readership and practice, of geographical diffusion and classification. The article also gives the reader a glimpse of the complex textual and source problems with which the scholar who uses unpublished sources must grapple in order to make interpretative headway. Voigts's article won the Zeitlin-Verbrugge prize of the History of Science Society in 1981.

In "Celestial Orbs in the Latin Middle Ages," Edward Grant surveys the late-medieval natural philosophical and astronomical literature in the Latin world for clues about the physical characteristics of the celestial spheres. The stimulus for this inquiry is a pair of sixteenth-century problems, namely Noel Swerdlow's claim that Copernicus (d. 1543) believed in solid spheres and the demise of these solid (i.e., hard) spheres in the wake of the comet of 1577. But did late medieval natural philosophers also hold this view? Grant argues that, curiously, they did not address the question directly. For most of them, the late-medieval heavens consisted of material spheres that were probably fluid rather than hard, a result that will keep historians of astronomy and cosmology on the look-out for unambiguous evidence.

Bernard Goldstein's "Theory and Observation in Medieval Astronomy" examines crucial topics that had received surprisingly little joint attention and shows how observational arguments led to the modification of theories. The article draws particular attention to physically oriented arguments involving planetary sizes, distances, and illuminations (such as the failure to observe expected phases for Venus or to see expected variations in the size of the Moon as predicted by Ptolemaic theory) in Arabic and Hebrew sources. Goldstein shows how the fourteenth-century Jewish polymath from Provence, Levi ben Gerson, used such physical considerations to choose between mathematically equivalent models or to construct alternatives to Ptolemy's models when he found the latter wanting.

A. I. Sabra's "Situating Arabic Science: Locality versus Essence," the Distinguished Lecture of the History of Science Society for 1995, offers a concise overview of the state of scholarship in Arabic science and a vigorous manifesto for future research in the field. Sabra makes a strong case for the dangers involved in approaching historical and cross-cultural problems in terms of the "essences" of peoples, cultures, and religions, all of which call for historical analysis from the point of view of *local* circumstances. The reader unfamiliar with Sabra's earlier work will find a concise updating of his groundbreaking thesis about the significance of the "appropriation" and eventual "naturalization" of Greek science in Arabic civilization (a thesis with which the present article should ideally be read).[25] Together, these articles shed a new light on an area of scholarship that has long been neglected, and that in coming decades will continue to grow in significance. In contrast with the traditional view that Islam acted as a hostile, or at best passive, container of Greek science until the Latin world (eventually) made something of it, Sabra documents the initial enthusiasm for this material and demonstrates the symbiotic transformations of this new material in a new context. He emphasizes here three prominent institutional loci of these new scientific developments, most notably court patronage, the endowed madrasa (school/college), and the mosque.

Victor Roberts's "The Solar and Lunar Theory of Ibn ash-Shāṭir: A Pre-Copernican Copernican Model" (1957), while difficult for the uninitiated, presents the early evidence for one of the most exciting cross-cultural questions in medieval and early modern science of the past generation. His analysis notes the identity between these models, invented by Ibn ash-Shātir (d. 1375/6) in Damascus, and several "novel" ones used by Copernicus. How did Copernicus learn about these and the other Arabic models that he used? This tantalizing question still has no satisfactory answer, but it is easy to see why this discovery of probable cross-cultural transmission generated much excitement. In most accounts, both popular and scholarly, Copernicus's astronomy still appears to be a bolt out of the blue. Now one surprising link connects him not merely to medieval astronomy, but to Islamic civilization.[26]

Behind the quaint title of Lynn Thorndike's "The True Place of Astrology in the History of Science" (1955) lies an important but oft-overlooked theme—the pervasiveness and rationality of the pre-Newtonian view that the heavens exerted widespread influence on the natural world, from the tides and seasons to the deaths of rulers, through the flow of bodily humors. Thorndike was making his case at a time when the study of astrology as a serious historical phenomenon still found little sympathy among his colleagues (Neugebauer's "The Study of Wretched Subjects," which mentions Thorndike as a fellow-laborer, appeared only four years

[25] A. I. Sabra, "The Appropriation and Subsequent Naturalization of Greek Science in Medieval Islam: A Preliminary Statement," *History of Science*; also reprinted and discussed in F. Jamil Ragep and Sally Ragep, with Steven Livesey, eds., *Tradition, Transmission, Transformation: Proceedings of Two Conferences on Pre-modern Science held at the University of Oklahoma* (Leiden, The Netherlands: Brill, 1996).

[26] Since Roberts's article, Noel Swerdlow has found a new link between Copernicus's heliocentric universe and the fifteenth-century Latin astronomy that preceded him. In the *Epitome of the Almagest* that he coauthored with Georg Peurbach, Johannes Regiomontanus (1436–1476) proved that an alternative planetary model, which Ptolemy held to be impossible, was indeed possible. Swerdlow has argued persuasively that Copernicus was working with Regiomontanus's alternative eccentric model of the inferior planets, a model that transforms directly into one centered on the Sun. See Noel Swerdlow, "The Derivation and First Draft of Copernicus's Planetary Theory: A Translation of the Commentariolus with Commentary," *Proceedings of the American Philosophical Society*, 117 (1973), pp. 423–512, esp. 471–78.

earlier). While Thorndike's characterization of celestial influence as a "law" presents problems, the concise point that his language is trying to make remains timely for many readers and can serve as an excellent basis for discussion of both content and historiography. Astrology did indeed provide the stimulus for most of the work that we call "astronomical" from Ptolemy into the early modern period.

The astrological assumptions that Thorndike sees as a dominant feature of pre-Newtonian science pervade Helen Lemay's "The Stars and Human Sexuality: Some Medieval Scientific Views" (1980). While highlighting the impact of Islamic astrological writings on Latin learning through the translation movement of the twelfth century, her article also communicates the *differences* in astrological assumptions among various practitioners and theoreticians. Astrologers proper, physicians, and natural philosophers all deserve attention when considering the extent of celestial influence on various aspects of sexual life, from conception to marital fidelity, through sexual proclivities and attestations of virginity.

Nathan Sivin's "Chinese Alchemy and the Manipulation of Time" (1976) presents a concise introduction to fundamental concepts of Chinese alchemy and cosmology. He shows that the larger framework of this endeavor was not chemistry in any modern sense, but a schema of cosmic perfectibility guided by numerological relationships. Specific laboratory techniques (such as "fire phasing") allowed the alchemist to create an analogy to the cosmos in the laboratory on a much more limited scale of time. Beyond its ostensible orientation to a product with health benefits, the alchemist sought a kind of ecstasy—a very different goal than that of (most!) modern chemists. The article concludes with a tantalizing comparison between Chinese and Hellenistic alchemy.[27]

A complement to Sivin's article, William Newman's "Technology and Alchemical Debate in the Late Middle Ages" (1989) sketches the changes in attitude toward alchemy from the late twelfth century to the early fourteenth century in the Latin world. Newman shows how, in their defenses of alchemy, Roger Bacon and Paul of Taranto presented the manipulations of human "art" (or technique) as matching or improving upon the productions of nature—a striking departure from the traditional superiority accorded to the operations of nature. Newman argues compellingly that this thirteenth-century Latin alchemical literature, despite opposition to it, survived and marks the birth of the technological optimism that we associate with a later era of European history. His essay, to which he has appended a short, but very important text, won the History of Science Society's Schuman Prize for 1986.

Bacon the thirteenth-century Franciscan reappears in David Lindberg's "Science as Handmaiden: Roger Bacon and the Patristic Tradition" (1987), which is fighting overly simplistic characterizations of the medieval scientific enterprise. Like von Staden, Lindberg is concerned that modern affinities to Bacon be tempered by restoring their elided context. Against an older but still prevalent characterization of Bacon as an early-modern experimental natural philosopher ahead of his time and lost in the thirteenth century, Lindberg argues that Bacon's approach to the relation of religious faith and learning was deeply traditional (largely

[27] Sivin edited an earlier collection of *Isis* articles devoted to science and technology in East Asia, a number of which deal with antiquity and the medieval period; Nathan Sivin, ed., *Science and Technology in East Asia* (New York: Science History Publications, 1977).

Augustinian). It was within this framework that Bacon enthusiastically promoted the new Greek and Arabic natural philosophical material that was flooding the universities in the thirteenth century. He argued that all of this learning could be useful to the Church, not as a subterfuge, but because he deeply believed that all wisdom was one. This article won the History of Science Society's Derek Price Award in 1989.

Bacon was also a master of optics or *perspectiva*, a prominent area of medieval scientific activity in Islam and later in the Latin world. A. Mark Smith's revisionist "Getting the Big Picture in Perspectivist Optics" (1981) argues for greater attention to the philosophical framework of the analysis of light and vision. His controversial argument seeks to place questions of cognition at the heart of *perspectiva*, one of the best-developed areas in which mathematics interacted successfully with natural philosophy.

Marshall Clagett wrote "The Impact of Archimedes on Medieval Science" (1959) as a prelude to his five-volume *Archimedes in the Middle Ages* (1964–1984), which was devoted to the impact of this "physico-mathematical" corpus between the twelfth and fifteenth centuries. The article illustrates the narrow thread by which the Archimedean tradition hung in both the Latin and Arabic worlds and hints at the significance of this medieval Latin tradition on the sixteenth-century Archimedean revival. The strong emphasis on manuscripts, routes of transmission, and methods of translation is typical of the preliminary stage of a textual investigation.

In "Taddeo Alderotti and Bartolomeo da Varignana on the Nature of Medical Learning" (1977), Nancy Siraisi uses two academic physicians from the late thirteenth and early fourteenth centuries to illuminate the interactions and tensions between medicine and natural philosophy in the institutional context of the University of Bologna's new faculty of medicine and arts. Whereas Alderotti (d. 1295) saw medicine as subordinated to natural philosophy, his student (and later opponent) da Varignana (d. after 1321) argued for the independence of medicine. Despite its affinities to natural philosophy, the practical purpose of medicine marked it off as distinct from, and not subordinated to, natural philosophy. This debate thus sheds light on the contested place of medicine in the larger schema of learning and on the relation of theory to practice.

Michael McVaugh's article, "Arnald of Villanova and Bradwardine's Law," is one of many pieces of scholarship on quantification in late medieval science, a theme that the more recent historiography of the Scientific Revolution has largely ignored. It illustrates the general importance attached to the theoretical quantification of natural phenomena in the early fourteenth century. In the absence of a smoking gun, McVaugh makes a compelling case for the continuity between Arnald of Villanova's quantification of the virtues of pharmaceutical compounds in late thirteenth-century Montpellier, and the quantification of motions associated with "Bradwardine's Law" in fourteenth-century Merton College, Oxford. His argument highlights the value of casting one's net broadly, unencumbered by such latter-day disciplinary categories as "pharmacy" and "dynamics."

The two concluding contributions open onto one of the large problems in the periodization of the history of science, namely the relations between the late Middle Ages, the Renaissance, and the Scientific Revolution, as these cultural developments persist in being called. In "Did Science Have a Renaissance?" Brian P.

Copenhaver helps set the record straight on several topics of central concern to the Renaissance, that paradoxical period that looks at once forward and backward. He treats the question he raises by highlighting positions that are counterintuitive to the widely diffused picture of the period, but characteristic of the period itself, notably the persistence of Aristotelian natural philosophy and the development of magic (two trends that will seem retrograde to a "progress"-oriented reader). Copenhaver answers his question by offering pointed criticisms of such influential interpretations of the period as those of Alexandre Koyré and Michel Foucault. Copenhaver's added warning against the easy, but misleading, association of magic and irrationality appropriately harkens back to the important theme raised by Neugebauer and expanded by Thorndike. Pamela Long's "Power, Patronage, and the Authorship of *Ars*" offers a wide-ranging and stimulating study of fifteenth-century works in the mechanical arts in Italy and South Germany. She argues that the act of writing such works, which proliferated in fifteenth-century manuscripts just before the printing press, and the act of dedicating them to patrons eased their transition into the realm of literate, discursive knowledge. This phenomenon, which has powerful implications for the interpretation of the Scientific Revolution, belies the widespread notion of the mechanical arts as having low status before the seventeenth century. Long's work is particularly exciting because it cuts across and challenges several major historiographical divisions—theoretical and practical knowledge; orality and literacy; manuscript and print; medieval, Renaissance, and early modern era—even as she moves freely across the major historiographical obstacle of the Alps.

The exciting sequel to these amazing developments appears in Peter Dear, ed., *The Scientific Enterprise in Early Modern Europe: Readings from Isis.*

The Study of Wretched Subjects

By O. Neugebauer

IN the last issue of *Isis* (vol. 41, 125–126, p. 374) there is a short review by Professor Sarton of a recent publication by E. S. Drower of the Mandean "Book of the Zodiac" which is characterized by the reviewer as "a wretched collection of omens, debased astrology and miscellaneous nonsense." Because this factually correct statement [1] does not tell the whole story, I want to amplify it by a few remarks to explain to the reader why a serious scholar might spend years on the study of wretched subjects like ancient astrology.

The great Belgian historian and philologist, Franz Cumont, in August 1898 signed the preface to the first volume of the *Catalogus Codicum Astrologorum Graecorum*, the twelfth volume of which is now in preparation. The often literal agreement between the Greek texts and the Mandean treatises requires the extension of Professor Sarton's characterization to an enterprise which has enjoyed the wholehearted and enthusiastic support of a great number of scholars of the very first rank. They all labored to recover countless wretched collections of astrological treatises from European libraries, and they succeeded in giving us an insight into the daily life, religion and superstition, and astronomical methods and cosmogonic ideas of generations of men who had to live without the higher blessings of our own scientific era.

To the historian of science the transmission of ideas is rightly one of his most important problems. Astrological lore furnishes us one of the most convincing proofs for the transmission of Hellenistic astronomy to India; astrological manuscripts help us to estimate more accurately the combination of Syriac, Arabic, Hindu, and Greek sources in the building up of Islamic science. No Arabic astronomer can be fully understood without a thorough knowledge of astrological concepts. The only hope of obtaining a few glimpses of the astronomical methods of the time of Hipparchus rests in the painstaking investigation of wretched writers like Vettius Valens or Paulus Alexandrinus. Six large volumes of miscellaneous nonsense were published by Professor Thorndike and have become a treasured tool for the study of Mediaeval scientific literature. And the history of the art and philosophy of the Renaissance has gained immensely from the researches carried out by the Warburg Institute on the astrology of preceding periods.

The book by Mrs Drower is only one modest contribution within the larger task of research in the history of the civilization of the Near East. The difficulties of the problems involved are by far greater than in the case of Greek or Latin texts with their so much better known terminology and background. We have to thank Mrs Drower for exposing a new source which may one day furnish the missing link in the transmission of doctrines which have left their imprint in almost all phases of Mediaeval learning, Medicine, Botany, Chemistry, etc.

All these facts are, of course, well known to Professor Sarton. But when the recognized dean of the History of Science disposes of a whole field with the words "the superstitious flotsam of the Near East," he perhaps does not fully realize how much he is contributing to the destruction of the very foundations of our studies: the recovery and study of the texts as they are, regardless of our own tastes and prejudices.

[1] The classification under "Babylonia and Assyria" is misleading because Mandean astrology belongs to the Islamic, and thus ultimately Hellenistic, civilization.

An astrological cuneiform tablet from second-century B.C. Uruk showing the Babylonian representations of Leo and Hydra and Jupiter—the eight-pointed star (reprinted from Weidner, Gestirn-Darstellungen auf babylonischen Tontafeln, *1967).*

*THE CULTURES OF ANCIENT SCIENCE: SOME HISTORICAL REFLECTIONS**

Introduction

By F. Rochberg

FROM THE EIGHTEENTH CENTURY until nearly the present, the origins of science were associated with works from classical Greek and Greco-Roman antiquity, particularly the "exact" or mathematical sciences, such as astronomy, harmonics, and optics, and the pure mathematical fields of geometry, arithmetic, and trigonometry. These products of classical antiquity were held to be prized possessions of the Western tradition and regarded as the highest intellectual watermark of Western civilization. The role of Indian and Arabic science did not go unrecognized, but these could be seen as genetically related to classical science, from which they derived and for which they functioned as cultural transmitters. As our understanding of science as a cultural, intellectual, social, and historical phenomenon changes, so does the manner by which we investigate its history and the question of its origins. As historians consider new kinds of sources appropriate for the reconstruction of ancient science, they confront the disparities between the modern concept of "science" and what seems to characterize the sciences of other periods and other cultures.

The early nineteenth century saw the emergence of general histories of science, such as William Whewell's *History of the Inductive Sciences* of 1837, which traced the growth and development of the various but related parts of civilization classified as "science."[1] Characteristic of Whewell's time were the assumptions that these special parts of civilization shared certain methods (observation, quantification, experimentation, hypothesis testing) and manners of thought (abstract, inductive, mathematical) identified as "scientific" and that the history of the sciences demonstrated an inherent tendency to progress in prescribed stages, for example, from observation to theorization. Inquiry into the origins of science was simultaneously an inquiry about ancient Greek science, the reconstruction of which was wholly dependent upon continuity with the scholastic traditions of European medieval and Renaissance science. But the reconstruction of ancient science is no longer exclusively a matter for classical philologists,

* This cluster of essays owes its existence to the promptings and suggestions of Michael H. Shank and F. Jamil Ragep.—THE EDITOR

[1] See also William Whewell, *Philosophy of the Inductive Sciences* (1840).

and the continuities from antiquity to the European Renaissance in the history of astronomy in particular are now known to reach back to the ancient Near East.

The decipherment of cuneiform at the end of the last century yielded an unexpected wealth of texts dealing with mathematics and mathematical as well as observational astronomy, not to mention astrology, medicine, and various types of scholarly divination relating to a wide range of natural phenomena. The nature of this Near Eastern material and its relation to that of other ancient traditions—Egyptian, Indian, and Greek—place the history of science and the question of its origins on entirely different foundations. The effects of this change go beyond the establishment, by mere translation of texts, of the content of just another form of ancient science. The non- and pre-Greek, non- and pre-Western material fundamentally challenges our old definitions of science, our understanding of its development, and consequently the way in which we should reconstruct and write its history.

That a special section of *Isis* should be devoted to the origins of science signals a readiness on the part of the profession at large to come to grips with the antecedents of Western science, both in the light of their impact on Western tradition and, wholly apart from that influential role, as topics worth consideration in their own right. Indeed, the selection of contributors and the essays they offer mark, albeit in different ways, a turning point in the historiography of science. It is furthermore interesting to note that philosophers and historians of science alike argue increasingly strenuously against the position that the definitions, goals, and methods of modern science are paradigmatic for all science. Significant in this regard are developments in the philosophy of science since Thomas S. Kuhn broke down some of the reconstructions proffered by inductivist historians of science who maintained the inexorable, gradual advancement of science, as though science were an organic, evolving entity independent of human influence. In Kuhn's view, the nature of change in science would be viewed as an integral and dynamic part of human scientific communities; then science itself would be viewed, in Stephen Jay Gould's words, as "a social construction, uninterpretable as nature speaking directly to us through bits of fact in a logic divorced from human contexts."[2]

Since the 1960s, anthropologists concerned with "commensurability" and the translation of cultures have probed the issue of "mentalities," modes of thought, and the difficulties of expressing in our modern Western languages any understanding or analysis of non-Western cultures. These anthropological inquiries are not unconnected to those in the philosophy of science issuing in, for example, Kuhn's claim of incommensurability between an older and a newer scientific paradigm. While such a notion as incommensurability enables, or actually results in, a nonlinear view of developments in the history of science, there is a danger in applying the concept so broadly as to encompass not just different theories with their terms, but different cultures with their languages. By this account, says Hilary Putnam, "we could not translate other languages—or even past stages of

[2] Stephen Jay Gould, review of Martin Rudwick, *The Great Devonian Controversy* (Chicago: Univ. Chicago Press, 1986), in *New York Review of Books,* 27 Feb. 1986, p. 12.

our own language at all. . . . To tell us that Galileo had 'incommensurable' notions and then to go on to describe them at length is totally incoherent."[3]

Historicism is another attempt to do away with the assumption of a linear progress of science demonstrable by objective historical criteria. When pushed to extremes, however, it has only the purest and most uninteresting kind of relativism to offer: namely, that historical phenomena and historians alike are "historically conditioned." This approach achieves the same result as the incommensurability thesis or the anthropology of "difference": the gulf between us and the subject of our inquiry becomes unbridgeable, and the chances of reaching any understanding become nil.

In the papers that follow, a middle ground emerges in which the historian remains sensitive to the variable historical circumstances and cultural contexts of science but attempts, as Arnaldo Momigliano would have it, to "discriminate between truth and falsehood."[4] Each author, explicitly or implicitly, and in his own way, confronts the question of the validity of the category "science" as epistemologically demarcated from "nonscience." And whereas none articulates a position with respect to the universality of human thought—that is, against the notion of discontinuous mentalities, either between historical eras or between different cultures—something bordering on this position must be allowed lest our nonmodern, non-Western sources remain forever inaccessible and incomprehensible, at least to scholars from the West.

The rising interest in ancient and non-Western science and growing awareness of its broader implications have coincided with the explosive and complex effect of multiculturalism on academia. A consequence of the consideration of ancient Near Eastern cultures as an integral part of the origins of science is a kind of "decolonization" of the history of ancient science, much like that Momigliano advocated for classical history as early as 1967. In a comment on Werner Jaeger's *Paideia,* he defined German Hellenism as that which drew too sharp a line between the Greek and oriental cultures, carrying with it "the conviction that 'to understand the Greeks it is necessary to think in Greek or, alternatively, in German.' " It is well to note a distinction between the meaning of the word "culture" as it is used in the history of science and its usual usage in social history or anthropology. Otto Neugebauer pointed out that in the exact sciences the meaning of the term "culture" approaches that of "tradition," making the assignment of ethnicity extremely complex, even beyond the particular problems involved in identifying ancient ethnic groups in any way other than through ethno-linguistic affiliation.[5] At the same time, however, the relation between culture and tradition

[3] See Hilary Putnam, *Reason, Truth, and History* (Cambridge: Cambridge Univ. Press, 1981), p. 115. For an example of recent anthropological inquiry see S. J. Tambiah, *Magic, Science, Religion, and the Scope of Rationality* (Cambridge: Cambridge Univ. Press, 1990). Influenced by this kind of anthropological discourse is G. E. R. Lloyd, *Demystifying Mentalities* (Cambridge: Cambridge Univ. Press, 1990). Earlier, see, e.g., George W. Stocking, Jr., *Race, Culture, and Evolution* (New York: Free Press, 1971); and Dan Sperber, "Apparently Irrational Beliefs," in *Rationality and Relativism,* ed. Martin Hollis and Steven Lukes (Cambridge, Mass.: MIT Press, 1984).

[4] See Arnaldo Momigliano, "Historicism Revisited," in *Sesto contributo alla storia degli studi classici e del mondo antico* (Rome: Edizioni di Storia e Letteratura, 1980), pp. 31–32.

[5] Quoted in M. I. Finley, "The Historical Tradition: The *Contributi* of Arnaldo Momigliano," in *The Use and Abuse of History* (New York: Penguin, 1987), p. 79; and Otto Neugebauer, "Exact

mitigates the problem of translation between cultures, since the transmission of certain traditions was possible because at some level an understanding was achieved that allowed one culture to receive knowledge or methods from another one foreign to it, adapt and transform that knowledge or those methods, and produce a new stage in the life of that "tradition." In these terms, "tradition" (scientific or otherwise) transcends culture and language.

This bears on the question embedded in Martin Bernal's essay of the identity of the Egyptians associated with Egyptian science; because much of it was Hellenistic, the Egyptian tradition would seem to be Egyptian only because it was carried on in the cosmopolis of Alexandria. We should not forget the wonderful Greek forgeries under the names Hermes Trismegistus and Nechepso-Petosiris, the existence of which, Momigliano said, "underscores the Greeks' belief that they had learned great things from oriental masters, and [suggests that] a closer look at this material would foil any attempt to identify actual ethnic elements."[6] The Egyptians were not uniquely privileged as objects of historical-cultural invention; Greeks also produced forgeries under the name of Zarathustra. These orientalia show that the Greeks went in for their own form of "philia," crediting generic orientals with traditions that the cultures of ancient Egypt and Persia never knew.

Serious work on ancient Near Eastern science, particularly from Babylonian and Egyptian texts, has been largely the contribution of Neugebauer and his colleagues, especially the Assyriologist Abraham Sachs, the historian of mathematics and astronomy Asger Aaboe, and the Egyptologist R. A. Parker. For editions of and commentaries on Sanskrit scientific texts, David Pingree is practically singlehandedly responsible. Without the editions of sources prepared by these scholars, there would be no historiography of science for these civilizations.[7] Yet despite their prodigious output, the field as a whole still has not taken full account of the ancient Near Eastern and South Asian sources now available, with the consequence that, by and large, works of "universal" history of science do not go beyond acknowledging a certain technical sophistication in ancient oriental science.

Virtually all the general histories of science produced during the 1950s and 1960s (and reprinted continuously to the present), even those by authors evidently aware of the work of Neugebauer *et al.*, make the same false claims. I selected the following quotation because it does attempt to take account of the debt of Western science to the ancient Near East. In Chapter 1 of *The Growth of Scientific Ideas,* William P. D. Wightman acknowledges that "the heritage [from the Near East] of natural knowledge upon which the Greeks were able to draw is

---

Science in Antiquity," in *Astronomy and History: Selected Essays* (New York/Berlin: Springer, 1983), p. 31.

[6] Arnaldo Momigliano, *Alien Wisdom: The Limits of Hellenization* (London/New York: Cambridge Univ. Press, 1975), p. 146.

[7] In particular, Otto Neugebauer and Abraham Sachs, *Mathematical Cuneiform Texts* (New Haven, Conn.: American Oriental Society and American Schools of Oriental Research, 1945); Neugebauer, *Astronomical Cuneiform Texts,* 3 vols. (London: Lund Humphries, 1955); Neugebauer and R. A. Parker, *Egyptian Astronomical Texts,* 3 vols. (Providence, R.I.: Brown Univ. Press, 1960); and the many collaborative papers of Aaboe, Neugebauer, and Sachs in the *Journal of Cuneiform Studies,* spanning a good twenty-five years from the 1950s to the 1980s. See also David Pingree, *Census of the Exact Sciences in Sanskrit,* 4 vols. (Philadelphia: American Philosophical Society, 1970–1981).

proving year by year to be far richer than first supposed." Nonetheless, he adheres faithfully to the doctrine that it was the Greeks who really gave birth to science and that the Near Eastern contribution was just so much technical practical skill. "In a word," he says,

> Greek thought differed from all that had gone before in respect both of *generality* and *rigour*. In *general* thinking, which opens up possibilities undreamed of by the severely "practical man," we pass from *percepts*—things to which we can point, like triangular fields—to *concepts*—creations of the process known as *abstraction*. . . . Science is, among other things, the generalisation of perceptual experience by means of adequate concepts. . . . We shall therefore take it as proved that *so far as the existing evidence goes* the birth of scientific IDEAS took place in Asia Minor in the seventh century B.C., on the ground that at that time and place the idea of nature as a whole first took shape, was explicitly stated, and systematically expounded.[8]

The fallacy in this thinking is even clearer in the conclusion drawn from the misunderstanding of ancient Near Eastern science as merely technical. To view this material in terms of technical feats that had no impact on the kind of "thought" associated with science in the West is an ideological distortion not only of ancient Near Eastern science, but of science as a whole.

Much has been made of abstract thought as a necessary quality of true scientific thought—and as yet another thing that was first achieved by classical Greeks. Note, however, Salomon Bochner's opinion that "the Greeks, for all their cleverness, were not able, or not yet able, to make abstractions that were more than idealizations from immediate actuality and 'external' reality; their abstractions were rarely more than one-step abstractions, and as such they remained within the purview of what is called 'intuitive' in an obvious and direct sense." Despite rare insights such as this, general historians of science have almost invariably compared the ancient Near Eastern achievements unfavorably to Greek developments, which constituted "real" science.[9] In other words, despite the viewpoints reflected in the three-way mirror representing the philosophy, history, and anthropology of science, the orientation that David Pingree calls "Hellenophilia" perpetuates inductivist and continuist historical methods that proceed from a definition of science as "that which scientists happen to be doing now, following a method or methods adumbrated by the Greeks, but never fully understood or utilized by them."

From very different points of view, the papers that follow implicitly criticize an earlier historiography of science that presupposed progress as a characteristic of science and objectivity as a characteristic of its historians. All four authors con-

[8] William P. D. Wightman, *The Growth of Scientific Ideas* (New Haven, Conn.: Yale Univ. Press, 1951), pp. 4–6.

[9] Salomon Bochner, *The Role of Mathematics in the Rise of Science* (Princeton, N.J.: Princeton Univ. Press, 1966), p. 51. For illustrations of historians' tendencies to valorize Greek achievements see Stephen Toulmin and June Goodfield, *The Fabric of the Heavens: The Development of Astronomy and Dynamics* (New York: Harper Torchbooks, 1961); F. Sherwood Taylor, *A Short History of Science and Scientific Thought* (New York: Norton Library, 1963), pp. 2–3; R. J. Forbes and E. J. Dijksterhuis, *A History of Science and Technology: Nature Obeyed and Conquered* (Baltimore: Pelican, 1963), pp. 15–16; and S. F. Mason, *A History of the Sciences* (New York: Collier, 1971), p. 11. Even David C. Lindberg, *The Beginnings of Western Science* (Chicago/London: Univ. Chicago Press, 1992), fails to understand the character of Babylonian science and does not take full account of the relation between the Near Eastern and Hellenistic traditions.

demn histories of science that define science in such a way as to allow the Greeks to be its sole inventors, and they caution that the representation of Greek science as an anticipation of modern science in its goals and methods both misunderstands and distorts. An example of the object of condemnation can be seen in A. C. Crombie's claim, made in 1957 at Marshall Clagett's famous conference on the history of science, that

> it was the Greeks who invented science as we know it, by their assumption of a permanent, uniform, abstract order and laws by means of which the regular changes observed in the world could be explained by deduction, and by their brilliant idea of the generalized use of scientific theory tailored according to the principles of non-contradiction and the empirical test. It is this essential Greek idea of scientific explanation, "Euclidean" in logical form, that has introduced the main problems of scientific method and philosophy of science with which the Western scientific tradition has been concerned.[10]

Heinrich von Staden points out that *if* the purpose is indeed "to establish continuities between ancient and modern science, or to search for the 'origins' of the forms of discourse, of the terminological categories, and of the methodological distinctions with which it has become customary to identify modern Western 'science,' " then emphasizing continuities (that is, those elements having affinities to our own concept of science) and letting the discontinuities (that is, elements of magic and folklore) remain insignificantly in the background will indeed achieve its continuist goal. "Continuism" has been charged with the unfortunate weakness, stemming from its underlying premise that each historical event or development is some kind of reaction, *pro* or *contra,* to a preceding one, of reconstructing history in the absence of documentary evidence, producing series of must-have-beens, retrojected into the past and assigned historical reality. The continuist scheme is rife in the history of Hellenistic science, notably in the oft-repeated reconstruction of the kinematic models of planetary and lunar motion as the historical consequence of continuous linear development from the homocentric spheres of Eudoxus (a model itself wholly unsubstantiated in ancient texts) to Apollonius to Hipparchus to Ptolemy.

The goal implied in this collection of papers is to recognize the complexities and (to our eyes) contradictions found in the culture of ancient science, and therefore to give full consideration to all those aspects of ancient science once deemed "primitive," such as the irrational, the nonempirical, and the inexact, and to eschew, in von Staden's words, "using cultural comparison only to reaffirm . . . the roots or 'anticipations' of key features of our scientific selves in any given ancient culture, be it an Aegean, African, Asian, or early New World culture." Pingree, in full accord, points out further that in widening our historical perspective we are better able to account for the origins of the Western tradition and obtain a deeper understanding of "the cultural factors that impel sciences and scientists to follow one path rather than another."

Geoffrey Lloyd comments on fact and objectivity in the history of science,

[10] A. C. Crombie, "The Significance of Medieval Discussions of Scientific Method for the Scientific Revolution," in *Critical Problems in the History of Science,* ed. Marshall Clagett (Madison: Univ. Wisconsin Press, 1959), p. 81.

seeing scholars not only as observers of ancient science but also as limited in their ability to come to terms with ancient sources in accordance with the degree of incommensurability between their notions and ours. He applies to the historian of science the philosopher of science's claim that there is no such thing as raw sense-data independent of interpretation on the part of the scientist-observer of nature, arguing that we, as moderns, come to the science of any period "burdened" with our modern assumptions. From this point of view pure description becomes as impossible in historical inquiry as it is in practical natural inquiry, and Lloyd urges the redirection of the history of science toward recovering the contexts, goals, and methods of the ancients in their investigations of nature. Lloyd's concern is to escape the objectivist position, associated with such nineteenth-century historians as Leopold von Ranke and Lord Acton, that claims that history is built on philosophically unmediated "ascertained facts." This concern, derived from the extreme reaction to "ultimate history" and the concomitant claim that there is no objective historical truth, is tempered here by the view that, in the words of E. H. Carr, "the facts of history never come to us 'pure,' " as "they are always refracted through the mind of the recorder."[11] Although it now appears naive to present facts as right and true and opinions as wrong and false, the question of objectivity and truth is not so easily dismissed, and it seems all the more exigent when the subject matter is science.

While not necessarily directly influenced by one particular philosophy or anthropology of science, and certainly not unified by any such, this cluster of essays represents a direction in the historiography of science consistent with some of the "newer" ideas generated by the philosophy of science and social anthropology of the last thirty years. They present a less ideologically limited, less category-driven approach to sources identified as belonging to science and therefore to its historical development, and adopt methods in general history that attempt to find a balance between the recognition of the historian's own cultural and philosophical contribution—hence the limits on objectivity—and the desire to construct a working model consistent with available knowledge and textual evidence. The change of direction seems to be a consequence of the study of the enormous corpus of primary texts of Near Eastern and South Asian provenance in the broader context of the history of Western science. These sources have not only changed the way the story of the history of Western science is reconstructed, but have affected basic historical methods in our field. Finally, although an argument can probably be made for advancement in science, in that an ever-greater range of phenomena are understood and explained, that overall advancement neither correlates with a necessary historical pattern of unidirectional linear progress nor should it determine retrospectively our understanding of the cultural or substantive dimensions of science. An awareness of these epistemological and historiographical problems should have an increasing impact on the historiography of science, particularly as it bears on the question of origins.

[11] E. H. Carr, "The Historian and His Facts," *What Is History?* (New York: Vintage, 1961), p. 24.

Hellenophilia versus the History of Science

By David Pingree

THE GENESIS OF THIS PAPER lies in a conversation that I had with A. I. Sabra of Harvard on the perennial problem of the definition of science appropriate to a historian of science; its corruption (including the deliberately extreme mode of its expression) is entirely a result of my own labors. For the piece represents the attitudes toward the subject that I have developed over some three and a half decades of studying the history of the "exact" sciences (as I will persist in calling them despite the lack of exactitude in some of them), as practiced in ancient Mesopotamia, in ancient and medieval Greece, India, and the Latin-speaking West, and in medieval Islam. It is this experience, then, and the desire to reconstruct a complex history as accurately as possible, that motivates me—these two, and the wish to provide an apologia for my claim to be a historian of science rather than of quackery. For the sciences I study are those related to the stars, and they include not only various astronomies and the different mathematical theories they employ, but also astral omens, astrology, magic, medicine, and law (*dharmaśāstra*). All of these subjects, I would argue, were or are sciences within the contexts of the cultures in which they once flourished or now are practiced. As such they deserve to be studied by historians of science with as serious and thorough a purpose as are the topics that we usually find discussed in history of science classrooms or in the pages of *Isis*. This means that their intellectual content must be probed deeply, and not simply dismissed as rubbish or interpreted in the light of modern historical mythology; and that the intellectual content must be related to the culture that produced and nourished each, and to the social context within which each arose and developed.

In stating these opinions I may appear to have set myself up as a relativist, but I would deny the applicability of that epithet to my position since my interest lies not in judging the truth or falsehood of these or any other sciences, nor in discovering in them some part that might be useful or relevant to the present world, but simply in understanding how, why, where, and when they worked as functioning systems of thought and interacted with each other and with other systems of thought.

It is with these considerations in mind, then, that I have embraced the word employed in the title of this article, "Hellenophilia," as it is a most convenient description of a set of attitudes that I perceive to be of increasing prevalence

This paper was originally delivered as a lecture at the Department of History of Science, Harvard University, 14 November 1990.

within the profession of the history of science, and which I believe to be thoroughly pernicious. I like "Hellenophilia" as a word because it brings to mind such other terms as "necrophilia," a barbaric excess that erupts as a disease from the passionate rather than from the rational soul; whereas the true love of the Greeks, Philhellenism, though also an attribute of barbarians such as are we—the epithet "Philhellene" was proudly borne by ancient Parthians, Semites, and Romans—arises preeminently from well-deserved admiration. A Philhellene is one who shares in what used to be, when children in the West still were taught the classics, a virtually universal awe of Greek literature, art, philosophy, and science; a Hellenophile suffers from a form of madness that blinds him or her to historical truth and creates in the imagination the idea that one of several false propositions is true. The first of these is that the Greeks invented science; the second is that they discovered a way to truth, the scientific method, that we are now successfully following; the third is that the only real sciences are those that began in Greece; and the fourth (and last?) is that the true definition of science is just that which scientists happen to be doing now, following a method or methods adumbrated by the Greeks, but never fully understood or utilized by them.

Hellenophiles, it might be observed, are overwhelmingly Westerners, displaying the cultural myopia common in all cultures of the world but, as well, the arrogance that characterized the medieval Christian's recognition of his own infallibility and that has now been inherited by our modern priests of science. Intellectually these Western Hellenophiles are still living in the miasma that permeated Europe until the nineteenth century, before the discovery of Sanskrit and the cracking of cuneiform destroyed such ethnocentric rubbish; such persons have simply not been exposed to the knowledge they would need to arrive at a more balanced judgment. But, sadly, I must report that many non-Westerners have caught a form of the disease Hellenophilia; they are deluded into believing that the greatest glory an Indian, a Chinese, an Arab, or an African scientist can have acquired is that gained by having anticipated either a Greek or a modern Westerner. So some Indians, for instance, busily reinterpret their divinely inspired *Ṛgveda* so that it teaches such modern hypothetical theories as that of relativity or the latest attempt to explain black holes, as if these transitory ideas were eternal complete truths. In doing this they are behaving as did those Christians who once believed it important to demonstrate that Genesis agrees with Greek science. These attempts do not enhance the brilliance of the authors or the reinterpreters of their sacred or scientific texts, but rather reveal a severe sense of cultural inferiority.

Parallel to this form of cultural denigration, practiced by the culture itself or by historians of science, is, say, the false claim that medieval Islam only preserved Greek science and transmitted it as Muslims had received it to the eager West. In fact Arab scientists, using Indian, Iranian, and Syrian sources as well as their own genius, revised the Greek sciences, transforming them into the Islamic sciences that, historically, served as the main basis for what little science there was in Western Europe in the twelfth and following centuries and for the amazing developments that happened three and four centuries later in Italy and Central Europe.

Another form that this Western arrogance takes is the naive assumption that other peoples in the world not only should be like us, but actually are or were—

"were" because this particular fallacy usually affects those who study Stone Age and other preliterate cultures that have been left defenseless in the face of modern reconstructions of their thoughts by their inability to record them in permanent form. In the history of the exact sciences the scholars who perpetrate wild theories of prehistoric science call themselves archaeoastronomers. The basic premise of some archaeoastronomers is that megalithic and other cultures in which writing was not known built stone monuments, some quite massive, in order to record their insights into the periodicity of celestial motions. This seems to me a trivial purpose to motivate such monumental communal efforts as the building of Stonehenge or the pyramids. There are many strong arguments to be raised against many of these interpretations. At this point, however, I wish only to point out that they go against the strong evidence from early literate societies that early man had little interest in the stars before the end of the third millennium B.C.; the cataloguing of stars and the recording of stellar and planetary phenomena are not a natural, but a learned activity that needs a motivation such as that which inspired the Babylonians, who believed that the gods send messages to mankind through the celestial bodies. The realization that some of these ominous phenomena are periodic can be dated securely in Mesopotamia to a time no earlier than the late second millennium B.C.; mathematical control of the relation between solar and lunar motion came only in about 500 B.C. The Egyptians also first began using selected stars as a sort of crude clock only in about 2000 B.C. and progressed no further in mathematical astronomy till they came under Babylonian influence. The earliest traces of a knowledge of astronomy in Greece and India seem also to be derived, in the early first millennium B.C., from Mesopotamia. I cannot speak of the astronomy of the early Chinese with authority because I am ignorant of their language, but I gather from what I have read that not even the beginnings of the system of the *hsiu* or lunar lodges can be dated before the late second millennium B.C. From the written evidence, then, it appears that an interest in the stars as omens arose in Mesopotamia after 2000 B.C. and started to develop toward mathematical astronomy in about 1200 B.C., but that the Babylonians began to invent mathematical models useful for the prediction of celestial phenomena with some degree of accuracy only in about 500 B.C. From Mesopotamia these astronomical ideas rapidly radiated to Egypt, later to Greece and India, and finally, perhaps, to China; in each of these cultures they were molded by the recipient scientists into something new, though still having recognizable Mesopotamian origins. The astral sciences spread from one civilization to another like a highly infectious disease. It is within the context of this documented history that I find implausible the suggestion that less advanced civilizations, without any known systems of writing or accurate record keeping, independently discovered complex lunar theories, or precession, or even an accurate intercalation cycle. The example of the Babylonians, with their need for a specific motive for observing stars and the fact that it took them a millennium and a half to arrive at a workable mathematical astronomy, and the examples of the Egyptians, Greeks, and Indians, if not the Chinese, who initially borrowed their astronomies from the Babylonians before each developed its science in its own way, seem to me to invalidate the theoretical basis for much of archaeoastronomy.

I return now to the four variants of Hellenophilia that I mentioned earlier.

Each, I would claim, distorts the history of science in two ways: passively, it limits the phenomena that the historian is willing or able to examine; actively, it perverts understanding both of Western sciences, from the Greeks till now, and of non-Western sciences. Thus those who still believe that the Greeks invented science either are altogether ignorant of, say, Babylonian mathematics and astronomy or else, though aware of them, fall into my second category and refuse to recognize them as sciences. The ignorance of the first group, of course, can and should be remedied through education; the obstinacy of the second in not acknowledging that Old Babylonian investigations of irrational numbers like $\sqrt{2}$, of arithmetical and geometrical series, or of Pythagorean triplets are science even though they are mathematically correct, or in asserting that the arithmetical schemes that they successfully used to control the many variables involved in the prediction of the time of the first visibility of the lunar crescent cannot bear the august name of *scientia* even though the predictions were essentially correct—this obstinacy is hard to deal with. It leaves the obstinate, however, in the awkward position of denying the status of science to one of the main contributors to the Greek astronomy that is the forebear of our positional astronomy. Such a person, of course, can name an arbitrary date at which positional astronomy comes to fit into his or her definition of science; but this cannot be accepted by a historian, as it is the historian's task to seek out the origins of the ideas that he or she is dealing with, and these manifestly lie, for astronomy, in the wedges impressed on clay tablets as well as in the observed motions of the celestial bodies. It is certainly possible to be a modern scientist without knowing history, or even with a firm belief in historical mythology; but can a historian of science function effectively under such disabilities?

While ignorance of Babylonian astronomy destroys the historian's ability to understand the origins and development of Greek and other astronomies together with their more modern descendants, it also tempts him or her to imagine that there is no other way to do astronomy than through the Greek and modern way of making observations and building geometric models. But Babylonian astronomy reveals how few observations are needed and how imprecise they may be if the astronomers are clever enough; and it also demonstrates that simple arithmetical models suffice for predicting the times and longitudes of periodic celestial phenomena. The Babylonian solutions are brilliant applications of mathematical structures to rather crude data, made purely to provide the possibility of prediction without any concern for theories of cosmological structure or celestial mechanics. The Greeks added the concern both for the geometrical structure of the universe and for the cinematics of the heavens, with a strong prejudice in favor of circles or spheres rotating with uniform motion; but they also, to a large extent, simply expressed the Babylonian period relations and arithmetical zigzag and step functions in a geometrical language, using observations to modify Babylonian parameters and to fine-tune their own geometrical models. A third variety of astronomy emerged from the synthesis of Babylonian arithmetical models, Hellenistic geometrical models, and local mathematical traditions that occurred in India in the fifth century A.D. In this astronomy questions of celestial cinematics receded into trivial mechanisms while computational finesse harnessed a broad range of mathematical techniques to the solution of astronomical problems, with the role of observations being limited to the confirmation, if possible, of accepted theory. There were historically many more astronomies, manifesting a variety of ways in which the same phenomena might be made predictable by mathematical means. These different astronomies reflect the different intellectual traditions of the various cultures as well as the specific problems that each society wished its astronomers to address—for example, the Babylonians were interested in certain horizon phenomena that they regarded as omens; the Greeks brought philosophical and physical problems into a science that had been purely mathematical, and as well introduced the more social aim of casting horoscopes; and the Indians devoted their efforts to the purely pragmatic goals of casting horoscopes, of predicting eclipses because of their significance as omens, and of evolving and regulating an extraordinarily complex calendar. I believe that those historians who limit themselves to the study of only one of these approaches to mathematical astronomy will blunder, as indeed many have, by not fully understanding the range of possibilities or the shaping force of purely cultural factors on the course that any science takes.

If it is evident that for a historian the proposition that the Greeks invented science must be rejected, it necessarily follows that they did not discover a unique scientific method. Indeed, they—and we—have not one, but many scientific methods; biologists, physicists, and astronomers went, and go, their separate methodological ways. I choose, therefore, to focus on the pride of Greek science: Euclidean geometry—which, of course, is purely logical and nonexperimental. This is often and justly praised for the rigor and power of its axiomatic system and for its ability to offer logical deductive proofs. Indeed, Babylonian and Indian mathematics are frequently criticized for relying not on proofs but on demonstrations. But without axioms and without proofs Indian mathematicians solved indeterminate equations of the second degree and discovered the infinite power series for trigonometrical functions centuries before European mathematicians independently reached similar results. These achievements amply demonstrate, I believe, that the Euclidean approach is not necessary for discovery in mathematics. Those who deny the validity of alternative scientific methods must somehow explain how equivalent scientific "truths" can be arrived at without Greek methods. And in their denial they clearly deprive themselves of an opportunity to understand science more deeply.

To depart briefly from the exact sciences, I would like to draw attention to the fascinating work of Francis Zimmermann on *āyurveda,* the Indian science of longevity, as it was and is currently practiced in Kerala. While he strongly supports the idea of the historical dependence of *āyurveda* on the Galenic theory of humors, Zimmermann is also keenly aware of the many ways in which Indian *vaidyas* have altered foreign notions while incorporating them into their own cultural traditions to create a theory of harmony and mutual influence between the humans, animals, and plants inhabiting any region and that region's terrain and climate. In Kerala these "ecological" ideas produced a local interpretation of *āyurvedic* doctrine that, it would appear, maintained a generally healthy human population, at least to that population's satisfaction, as long as the people were not subjected to wars, famines, or epidemics. The *āyurvedic* approach to medicine does not inspire its practitioners to make discoveries in molecular biology, but it is the correct medical science for the cultural context within which it operates. And it attempts to address psychological, social, and environmental aspects of health that our mechanistic medicine tends to ignore. Western doctors have something to learn about medical care from *āyurveda,* and so do Western historians of medicine.

The third fallacious opinion that I have associated with the Hellenophiles is that the only sciences are those that accredited Greeks recognized as such. This opinion generally takes the form of allowing Aristotle to define science for us, so that it excludes even the genuinely Greek sciences of astrology, divination, magic, and other so-called superstitions. This brings us squarely to the fundamental question of this paper: What is the proper definition of science for a historian of science? I would offer this as the simplest, broadest, and most useful: science is a systematic explanation of perceived or imaginary phenomena, or else is based on such an explanation. Mathematics finds a place in science only as one of the symbolical languages in which scientific explanations may be expressed. This definition deliberately fails to distinguish between true and false science, for explanations of phenomena are never complete and can never be *proved* to be "true." Obviously, this shortfall is as true of modern scientific hypotheses as of ancient ones. It is, therefore, inappropriate to apply a standard of truthfulness to the sciences, at least viewed as historical phenomena, for the best that modern scientists can claim—I cannot judge whether justly or not—is that they are closer to some truth than were their predecessors; nor, for the reasons I have already stated, can the methodologies of science be limited to just those employed by present-day scientists.

If my definition of science as it must be viewed by a historian is accepted, it is easy to show that astrology and certain "learned" forms of divination, magic, alchemy, and so on are "sciences." Some may regard this procedure for elevating superstition to the rank of scientific theory as arbitrary and unfair, but remember that modern science is the initial culprit in that it arbitrarily sets up its own criteria by which it judges itself and all others. If I am a relativist, then, it is precisely at this point where, as a historian, I refuse to allow modern scientists who know little of history to define for me the bounds of what in the past—or in the present—I am allowed to consider to be science. It pains me to hear some scientists, who have not seriously considered the subject, denounce astrology as "unscientific" when all that they mean is that it does not agree with their ideas about the way the universe functions and does not adhere to their concept of a correct methodology. It pains me not because I believe that astrology is true; on the contrary, I believe it to be totally false. But the anathemas hurled at it by some scientists remind me more of the anathemas leveled by the medieval Church against those who disagreed with its dogmas than of rational argument. In its persecution of heretics as in its missionary zeal and its tendency to sermonize and to pontificate, our scientific establishment displays marked similarities to the Church, whose place in our society it has largely usurped.

That Church, like modern science, condemned divination, astrology, and magic, though on the grounds that they limit God's power and human free will rather than that they fail to conform to our current "laws of nature." Both of these arguments, to my way of thinking, are arbitrary and irrelevant to a historian, who should remain free of either the Church's or modern science's theology.

To turn to history: Babylonian divination is a systematic explanation of phenomena based on the theory that certain of them are signs sent by the gods to warn those expert in their interpretation of future events; there is no causal connection, but only one of prediction so that appropriate countermeasures may be undertaken. In this system the stars and the planets, which are regarded as the manifestations of the individual gods in the sky, indicate by their changes in quality and in location the events that will befall mankind at large, or a country in general, or a specific king and his family. Omens that appear on earth are contrived by Šamaš, the god of the sun, to warn individual men of the coming of good or evil. In themselves, texts describing the rules of the interpretation of omens—*Enūma Anu Enlil* for celestial omens and *Šumma ālu* for terrestrial ones—are scientific; they provide systematic explanations of phenomena. The former is in addition closely allied to the origins of mathematical astronomy, for the tablets of *Enūma Anu Enlil* contain the first recorded realization of the periodicity of certain celestial phenomena and the first attempts to provide mathematical models for predicting the occurrences of such phenomena. This close linkage between divination and mathematical astronomy, as well as a linkage between divination and the observations necessary for constructing and refining mathematical astronomy, persisted in the first millennium B.C. in the cuneiform *Letters, Reports,* and *Diaries*. Moreover, as the mathematical models became more sophisticated and the descriptions of the observed phenomena became more precise, the rules for predicting terrestrial events from the celestial phenomena became more complex. Babylonia, then, provides us our first example of the fruitful interplay between the theoretical and the applied aspects of a science.

Astrology grew out of a union of aspects of advanced Babylonian celestial divination with Aristotelian physics and Hellenistic astronomy; this union—illicit, some may think—occurred in Egypt in the second century B.C. The product was the supreme attempt made in antiquity to create in a rigorous form a causal model of the *kosmos,* one in which the eternally repeating rotations of the celestial bodies, together with their varying but periodically recurring interrelationships, produce all changes in the sublunar world of the four elements that, whether primary, secondary, or tertiary effects, constitute the generation and decay of material bodies and the modifications of the parts or functions of the rational and irrational souls of men, animals, and plants. In other words, ancient Greek astrology in its strictest interpretation was the most comprehensive scientific theory of antiquity, providing through the application of the mathematical models appropriate to it predictions of all changes that take place in a world of cause and effect; it is not surprising, then, that it was called simply *mathēsis* or "science" by Firmicus Maternus and others.

But even within Greek astrology there was a movement toward a relaxation of the rigidity of the theory, both because of the frequent failure of the predictions and, more important, because of people's desire to circumvent unpleasant predictions—a practical rather than a theological demand for a modicum of free choice and self-determination. This trend toward an astrology that indicates predispositions instead of concrete inevitabilities was accentuated when the Greek form of this science was transmitted to India and transformed into a system that rapidly increased the complexity of the mathematical models in order to diffuse and mollify the inescapability of a simpler predictive scheme. The Indians attempted to match the bewildering variousness of real lives by an equally bewildering multiplicity of mathematically computable variants in astrology.

If Greek astrology is based on the idea that the motions and interrelations of the celestial spheres are ultimately the causes of all terrestrial phenomena, astral magic, which was concocted out of Babylonian and Indian liturgies and iconographies mingled with Greek astrology, Ptolemaic astronomy, and Hellenistic philosophy by the self-styled Sabaeans of Ḥarrān in the ninth century, assumes that the magus's soul is free of inhibiting stellar influences, so that, by manipulating terrestrial objects, he can reverse the processes of astrology and change the wills of the planetary spirits. In this way the magus can employ the astral influences defined by astrology to effect the changes he wishes in the sublunar world. This was a dream still dreamt by two founding members of the Royal Society, Kenelm Digby and Elias Ashmole.

These same Sabaeans invented also a second type of learned magic based on Plato's and Aristotle's theories of animal and human souls. In this magic, which I have dubbed psychic, the magus artificially creates new animals by uniting either within a womb or within a womblike chamber animal or human parts representing the material body and the particular part or function of the soul that he wishes his creation to be endowed with. The magus can then employ his artificial animal to accomplish wonders. Astral and psychic magic we may not wish to test in order to determine their validity, but as historians we must regard them as scientific, if for no other reason than because many Western scientists in the sixteenth and seventeenth centuries took them to be genuine sciences. And, of course, they do fall under the aegis of my definition of science.

The same status must be accorded, then, to alchemy—be it Greek, Arabic, Chinese, or Indian—and to other systematic theories that explain phenomena, whether the lapidaries and physiognomics that spread from Mesopotamia to Greece, to Iran, and to India, or the science of determining sites suitable for different types of buildings—a science found in different forms in China and in India—or the purely Indian analysis of the processes of converting thought into sound in order to produce intelligible speech. These and other sciences cannot be dismissed simply because they do not fall into the intellectual system favored by some Greek philosophers.

I have already, if I have been at all successful, persuaded you that the fourth variety of Hellenophilia, in which one defines science as that which modern Western scientists believe in and the methodologies with which they operate, is inappropriate to a historian, though it may be useful to a modern Western scientist. And I have already mentioned that among the advantages provided to the historian by looking outside of the confines of such a restricted definition are a realization of the potential diversity of interpretations of phenomena and of the actual diversity of the origins of the ideas that have developed into modern Western science among other sciences, and an objectivity born of an understanding of the cultural factors that impel sciences and scientists to follow one path rather than another. The loss of all these advantages is the price paid for suffering the passive effects of this form of Hellenophilia.

Its active form is more pervasive in and pernicious to history. This results in the attitude that it is the task of the historian not to study the whole of a science within its cultural context, but to attempt to discover within the science elements similar to elements of modern Western science. One example I can give you relates to the Indian Mādhava's demonstration, in about 1400 A.D., of the infinite power series of trigonometrical functions using geometrical and algebraic arguments. When this was first described in English by Charles Whish, in the 1830s, it was heralded as the Indians' discovery of the calculus. This claim and Mādhava's achievements were ignored by Western historians, presumably at first because they could not admit that an Indian discovered the calculus, but later because no one read anymore the *Transactions of the Royal Asiatic Society,* in which Whish's article was published. The matter resurfaced in the 1950s, and now we have the Sanskrit texts properly edited, and we understand the clever way that Mādhava derived the series *without* the calculus; but many historians still find it impossible to conceive of the problem and its solution in terms of anything other than the calculus and proclaim that the calculus is what Mādhava found. In this case the elegance and brilliance of Mādhava's mathematics are being distorted as they are buried under the current mathematical solution to a problem to which he discovered an alternate and powerful solution.

Other examples of this dangerous tendency abound. For instance, since the 1850s historians ignorant of Mādhava's work have argued about whether Indian astronomers had the concept of the infinitesimal calculus on the basis of their use of the equivalent of the cosine function in a formula for finding the instantaneous velocity of the moon, a formula that occurs already in a sixth-century Sanskrit text, the *Pañcasiddhāntikā*. I cannot tell you how that formula was derived, since its author, Varāhamihira, has not told me; but I find it totally implausible that some Indian discovered the calculus—a discovery for which previous developments in Indian mathematics would not at all have prepared him—applied his discovery only to the problem of the instantaneous velocity of the moon, and then threw it away. The idea that he might have discovered the calculus arises only from the Hellenophilic attitude that what is valuable in the past is what we have in the present; this attitude makes historians become treasure hunters seeking pearls in the dung heap without any concern for where the oysters live and how they manufacture gems.

One particularly dangerous form of this aspect of Hellenophilia is the positivist position that is confident that mathematical logic provides the correct answers to questions in the history of the exact sciences. I, of course, am not denying the power of mathematics to provide insights into the character and structure of scientific theories; obviously, Otto Neugebauer's brilliant analysis of the astronomical tables written in cuneiform during the Seleucid period gives us a profound understanding of how this astronomy worked mathematically, and it tells us something about some stages in the development of the science as recorded on the hundreds of tablets that he investigated. But it does not and cannot, as Neugebauer well knew, answer a whole range of historical questions. We do not know by whom, when, or where any Babylonian lunar or planetary theory was invented; we do not know what observations were used, or where and why they were recorded; we do not know much about the stages by which Babylonian astronomers went from the crude planetary periods, derived from omen texts, found in MUL.APIN to the full-scale ephemerides of the last few centuries B.C. Historians need to be very careful in assessing the nature of the questions the material at hand will allow them to answer with a reasonable expectation of probability; and they must hope for and search out new evidence. But the positivists jump in to claim that mathematical models (and they usually use quite simple ones) suffice to describe the idiosyncratic behavior of people and to account for the perverse quirks in their personalities. I will not here name names, but the number of historians of the exact sciences who suffer from this malady is appallingly large; they can be easily recognized by the characteristic trait that, in general, the remoter the time and the scantier the evidence, the more precise their computations. Millennia of history are made to depend on the measurement of an arc of a few minutes or degrees when it has not even been convincingly demonstrated that any arc was being measured at all.

So far I have been attempting to discredit Hellenophilia on the grounds that it renders those affected by it unable to imagine many significant questions that legitimately should be addressed by historians of science and that it perverts their judgment. Much of my argument has been based on the anthropological perception that science is not the apprehension of an external set of truths that mankind is progressively acquiring a greater knowledge of, but that rather the sciences are the products of human culture. But this viewpoint must be modified by a further consideration, to which I have from time to time alluded since it strengthens the arguments in favor of the definition of science that I proposed. This consideration is that, as a simple historical fact, scientific ideas have been transmitted for millennia from culture to culture, and transformed by each recipient culture into something new. This is particularly noticeable in the astral sciences that I study—astronomy, astral omens, astrology, and astral magic—but can be readily discerned in many others. The taproot and trunk of the tree of the astral sciences are buried in the Mesopotamian desert, with subsidiary roots in Egypt and China (I have lopped the Mayas off this arboreal image, as they are self-rooted). From Babylonia the tree branched out to Egypt, to Greece, to Syria, to Iran, to India, and to China; grafted onto different cultural stocks in each of these civilizations, it developed variant leaves, shoots, and flowers. The process of the intertwining of these diverse varieties of astronomies throughout Eurasia and North Africa was amazingly complex, as ideas, mathematical models, parameters, and instruments circulated rapidly over the vast expanse of divergent traditions. Out of this process modern Western astronomy sprang from a rather late branch that grew from and was fed by an incredibly complicated undergrowth. For very complex reasons this modern Western astronomy has choked off all of its rivals and destroyed the intellectual diversity that mankind enjoyed before it moved from simple communication to Western domination. We cannot know what the Islamic, Indian, or Chinese astral sciences might have become had this not happened, except that they would not have become what our culture has produced. But unraveling the intertwined webbing of these sciences is a fascinating and a rewarding task for a historian, and one in which much remains to be done. I strongly recommend to those of you who have the opportunity thus to broaden your perspectives to grasp it.

Methods and Problems in the History of Ancient Science

The Greek Case

By G. E. R. Lloyd

THE AIM OF THIS PAPER is to set out, as briefly as possible, some of the chief methodological issues in the study of science in Greek antiquity. What are the main difficulties confronting such a study, what can it hope to achieve, what indeed is it a study of? It used to be assumed that one can straightforwardly pose problems to do with the "origins" of science and the meaning of "science" in antiquity. As will become apparent, I believe the questions have to be reformulated to be investigable.

I shall begin with two fundamental principles, one borrowed from the philosophy of science, the other from social anthropology. The first relates to the distinction between observation and theory, the second to that between actors' and observers' categories. Applying the latter to the Greek case, I shall then focus on one of its fundamental features, namely its pluralism. That in turn will provide a basis to tackle, finally, the problems of the validity of the question of origins and the usefulness of the category of science.

I

Modern philosophy of science has, for some time, insisted that the distinction between observation and theory is a relative, not an absolute one.[1] There is no set of data out there in reality to which we have unmediated access, that is to say data that can be described in theory-free terms and then used to check the theories suggested to explain them. The data the scientist has to work with are not "raw": their description itself presupposes a theoretical, that is a conceptual, framework. There is no such thing as a totally value-free, theory-free, account of the data.

This is not to abolish the distinction between observation and theory entirely, of course. Some descriptions are more, some less, theory-laden. To cite my favorite examples from Mary Hesse, we can contrast talk of "particle-pair annihilation" in a cloud chamber with talk of "two white streaks meeting and terminating at an angle."[2] Or again we can contrast the use of the term "epileptic fit"

[1] See, e.g., Ian Hacking, *Representing and Intervening* (Cambridge: Cambridge Univ. Press, 1983).
[2] Mary Hesse, *The Structure of Scientific Inference* (Berkeley: Univ. California Press, 1974), p. 24.

with the type of description of loss of voice, foaming at the mouth, clenching of the teeth, and convulsive movements of the hands that we have in the Hippocratic treatise *On Sacred Disease*.

We can, as these examples show, retreat from greater theory-ladenness to lesser, but that retreat does not lead to an ultimate refuge of an entirely theory-free vocabulary in which raw sense-data can be set down innocently of all assumptions.

These points, developed in the philosophy of science, have their relevance and applicability also to the history of science. Any attempt to write the history of scientific inquiry at any period involves selection, judgment, a conception of the subject under review, and methodological presuppositions about the historical inquiry itself. The history of science can never be *purely* descriptive any more than the practice of science can. There will always be elements of evaluation in any account offered. The methodology of the historian may or may not be set out explicitly: but there will be methodological assumptions implicit in the account nevertheless. Since methodology is inevitable, it is better to be self-conscious about it.

As in the scientific case, this does not mean that the distinction between greater and lesser theory-ladenness is useless. It must, however, be perceived as a matter of degree. The remark that all history of science is evaluative does not mean that there is no check on our evaluations. That would mean that free rein could be given to prejudice. No: the fact that the facts are not raw data here either does not mean that they cannot be used to revise and modify theories and explanations, even though the facts themselves never come free from all theoretical preconceptions. My own point, that no fact is just a raw fact, applies, of course, to the remark I have just made, to my own use of "fact," for it clearly reflects a point of view, the one I am putting forward in this section.

None of the above points leads to extreme skeptical, subjectivist, or relativist conclusions, though they do counsel caution, in general and with regard to one's own preconceptions in particular. We certainly have to be wary about claims to have understood, let alone claims to have reached satisfactory explanations, in the history of science. As the history of our subject abundantly illustrates, any understanding achieved is at best only provisional, always revisable, often radically so. The conclusions we reach are certainly subjective in the sense that they reflect our preconceptions. But that does not mean that there is no way in which we can become critical of those preconceptions. Dialogue between historians of science is possible, even though no two historians will share exactly the same ideas either on substantive issues in the material they study or on the nature of the study itself.

Understanding another's point of view is difficult and can never be complete, for we must always be prepared to revise what we had thought we had understood. Yet it is not impossible to reach at least an approximate understanding, as the very possibility of such revision presupposes. I can never be sure that what you mean by the terms "star" or "sun" or "heart" or "blood" is precisely what I think you mean by them, in general or in particular collocations. But the principle of charity of interpretation allows and dictates that I adopt some notion of your understanding as a working hypothesis, open to revision as the dialogue between us proceeds.[3] A similar point applies across natural languages, where, to be sure, the divergences in the networks of approximately corresponding senses and references will be far greater—not that there is a single reality being mapped in different ways that could even in principle be set down in a language neutral as between different points of view.

In the study of ancient scientific texts the problems are, of course, appreciably more severe than those of the attempted dialogue between contemporary historians. We cannot enter into dialogue with our authors. We cannot even directly check what they wrote. If the evidence in a dialogue between contemporaries is never fully in, the situation with regard to ancient evidence is immeasurably worse, not just in that it is incomplete, but also because what we have has been biased by the selective processes of transmission.

We must be clear, then, as a first principle, that the history of science is inevitably evaluative: it always presupposes a conceptual framework and a methodology. Our best hope, as historians, is to be as self-conscious and as self-critical as we can about these—not that we shall thereby escape being evaluative.

II

My second principle appeals to, and adapts, the social anthropologists' distinction between actors' and observers' categories.[4] However, unlike in some social anthropology, that distinction must be subordinated to the philosophical principle I have just adumbrated. The anthropologist studying a foreign society—or indeed the sociologist studying our own—aims, so far as possible, to express the ideas, beliefs, practices of the society in question in the terms used by members of the society themselves—the actors. At the limit this would involve describing those ideas entirely in the actors' own natural language. But short of doing that, it is often necessary to keep many important terms untranslated, allowing their significance to emerge from examples of their use, though these have inevitably to be accompanied by interpretative glosses.

The aim of keeping as close as possible to the actors' own categories is twofold. Negatively, first, it helps to minimize the risks of assimilating alien ideas to our own, of assuming that the subjects studied have the same conceptual framework in mind that is suggested by the interpreters' own (observer) categories. Positively, second, it allows an alien network of meanings to be built up in its own terms and be seen for what it is, as alien.

Of course the moral of my first philosophical principle is that no anthropological attempt to state the ideas and beliefs of a foreign society in what they, the anthropologists, represent as the actors' own categories can ultimately succeed, and this for two reasons. First, the anthropologist is, after all, an observer, not an actor, at least not qua anthropologist. Second, we said, no observation statement is entirely theory-free.

[3] Cf. Donald Davidson, *Essays on Actions and Events* (Oxford: Clarendon, 1980); and Davidson, *Inquiries into Truth and Interpretation* (Oxford: Clarendon, 1983).

[4] See, e.g., the collection of essays edited by Robin Horton and Ruth Finnegan, *Modes of Thought* (London: Faber & Faber, 1973).

Similarly, no historians of ancient science can put themselves precisely in the position of ancient scientists. We could not do so, with Greece, even if we chose to write our histories in (imitation) ancient Greek. Yet as with the social anthropologist, so with the historian of science, it is clearly of cardinal importance to avoid, as far as possible, the assimilation of our subjects' ideas to our own modern conceptual categories. We have to recover, as fully as possible, the original context in which the ideas were put forward, the goals, assumptions, and methodologies of those who put them forward, and again the social and cultural institutions within which they worked and with which they interacted—in every case in their very considerable complexity. Although students of ancient societies cannot, in the strict sense, be ethnographers—we cannot interview our subjects—that should not prevent us from emulating, as far as our materials permit, the kind of dense—or, as Clifford Geertz calls it, "thick"—account that considers the data concerning belief systems against the whole background of the geographical, economic, technological, social, and political situation of the society under investigation.[5] Doubtless it would be absurd to suppose that our aim should be to think like an ancient scientist. Yet we should and can attempt to investigate their reported ideas and beliefs in their contemporary context, in the light of what we can discover about what they thought they were doing, their motivations, and the social background of their work. We can thereby hope at least to avoid some of the grosser anachronisms that have bedeviled the subject from the uncritical application of modern scientific categories to ancient ideas. I shall have more to say on this topic in due course.

III

When we attempt to apply the two principles I have sketched out to the Greek case in particular, we must be fully aware, first, that the data we have to work with have already been extensively processed: by the selectivity of the evidence, mostly literary, that has survived, by the vagaries of transmission, by the pressures of constant reinterpretation. However, that does not prevent us from being able, second, to recover something of the original *pluralism* of their—actors'—views of the inquiries and activities they engaged in.

When we ask what categories and concepts the ancient Greeks themselves used of their own inquiries, we find quite a variety in play. The terms we conventionally translate "doctor" (*iatros*), "mathematician" (*mathēmatikos*), "philosopher" (*philosophos*), "physicist" or "natural philosopher" (*phusikos, phusiologos*), "musician" (*mousikos*), "architect" (*architektōn*), "engineer" (*mēchanikos*), all have points of contact with, but all diverge to a greater or lesser degree from, what those conventional renderings may suggest. Moreover, none of those terms picks out a single inquiry or activity about which the Greeks themselves had clear and unanimous views. Each was in antiquity already the subject of disagreement and dispute, with rival individuals or groups implicitly adopting or explicitly defending divergent conceptions.[6]

We may begin by taking two of what might seem to be among the least prob-

[5] Clifford Geertz, *The Interpretation of Cultures* (New York: Basic Books, 1973).
[6] See G. E. R. Lloyd, *The Revolutions of Wisdom* (Berkeley: Univ. California Press, 1987).

lematic terms, namely "doctor" and "mathematician." When we study Greek medicine we have first to rid ourselves of many standard modern assumptions, for example, concerning the differences between doctor and layperson. The institutions we associate with modern medicine—the hospitals, teaching schools, research laboratories—have no equivalents in Greek antiquity, where indeed there were no officially recognized qualifications that gave medical practitioners a legal right to engage in healing the sick.

Although we may talk loosely of "Greek medicine," we can never afford to forget that that term covers some five or six very different styles or traditions, and it follows from the point made in the last paragraph that *every* style or tradition competed with every other on more or less equal terms. The situation was quite different from that which obtains today between official practice and one or more varieties of so-called alternative medicine. There were, for instance, the root-cutters (*rizotomoi*) and the drug-sellers (*pharmakopōlai*), about whom we hear, for example, from Theophrastus. Then there were those called *maiai,* conventionally translated "midwives," though it is clear that the activities of the individuals in question were far from confined to what we call midwifery. Quite how far each of these three groups took for themselves or were given by others the title "doctor," *iatros,* is a delicate and disputed question. Thus none of the male writers represented in the extant Hippocratic treatises dating from the fifth and fourth centuries B.C. can bring himself to refer to his women associates as a doctor: the nearest we get to that is one text that speaks of "the healing woman," *hē iētreuousa,* using the participle from the cognate verb. However, in the inscriptional evidence from the fourth century onward we have cases of women healers who are referred to as doctors, *iatros* or *iatreinē,* using the noun in either its masculine or feminine form.[7]

But in addition to these three diverse traditions there were also those who practiced medicine in the shrines of healing gods or heroes, such as Asclepius, establishments that from the fourth century B.C. onward were increasingly well endowed and that remained popular down to Galen's day, in the second century A.D., and beyond. Those healers may be distinguished, in turn, from others who also invoked the discourse of the divine or the demonic, the itinerant sellers of charms and purifications about whom we hear from Plato, for instance, as well as from some of our extant Hippocratic writers.[8]

Then there are those Hippocratic writers themselves, not that they are anything like a homogeneous group.[9] They disagreed not just with the representatives of other traditions of healing, but also often among themselves, and not just about particular theories of disease or of treatment, but about the aims, status, and methods of medicine itself. Nor are these disagreements just a feature of the fifth and early fourth centuries B.C., for they too clearly continued down to Galen's day and beyond.

[7] See G. E. R. Lloyd, *Science, Folklore, and Ideology* (Cambridge: Cambridge Univ. Press, 1983), esp. Pt. 2, Ch. 2. Cf. Ann Ellis Hanson, "The Medical Writers' Woman," in *Before Sexuality,* ed. David M. Halperin, John J. Winkler, and Froma I. Zeitlin (Princeton, N.J.: Princeton Univ. Press, 1990), pp. 309–338.

[8] See, e.g., G. E. R. Lloyd, *Magic, Reason, and Experience* (Cambridge: Cambridge Univ. Press, 1979), Ch. 1.

[9] See, e.g., Wesley D. Smith, *The Hippocratic Tradition* (Ithaca, N.Y.: Cornell Univ. Press, 1979).

Thus one fundamental dispute related to whether the medical practitioner should or should not base his medical practice on general physical and physiological theories about the constituents of the human body, about the nature and causes of such processes as digestion and respiration, and so on. Some Hellenistic medical practitioners took the view that the job of the doctor was simply to heal and that it is either impossible or useless or both to inquire into such questions as the physical constitution of the body.[10] The two groups concerned, the so-called Empiricists and the Methodists, both rejected the use of dissection, that is of "anatomy" in the primary sense, that of "cutting open." To that others, including Galen, responded by insisting that dissection is important not just for theoretical reasons (for what it can teach you about physiological processes, for instance), but also for vital, practical ones. "If a man is ignorant of the position of a vital nerve, muscle, artery or important vein," he writes in *On Anatomical Procedures* 2.2, "he is more likely to be responsible for the death, than for the saving, of his patients."[11]

To study ancient Greek medicine, then, we have to try, among other things, to recapture the divergent views held and expressed about the nature and aims of medicine itself. We have, as noticed already, to be aware that our evidence exhibits a very distinct bias, not so much as between one literate group and another (although that is sometimes the case), as between the literate groups on the one hand and those traditions that did not privilege the written word. But having said that, we must make the best use we can of the sources available to us, and even if much of the detail escapes us, they certainly make the general point about pluralism abundantly clear.

It might be thought that some of the issues I have been referring to are specific to medicine, always a problematic and at best only partially scientific activity. One might suppose, for instance, that the nature and aims of mathematics, at least, were uncontroversial and agreed. Yet that is far from being the case.

There are two points here, one relating to the pluralism we can discern in what we recognize as mathematical inquiry in our terms, and the other to do with what the term *mathēmatikē* covered in Greek antiquity. Under the first head we should note that while mathematics in the Euclidean, that is to say axiomatic-deductive, style is well represented in our extant sources, that was far from being the only tradition of mathematical inquiry. There was also, for example, the so-called Heronic tradition of metrical geometry, characterized by its concern with the solution of problems of mensuration. Again there is the distinct tradition, recently explored by Wilbur Knorr in *The Ancient Tradition of Geometric Problems,* concerned with the investigation of such special problems as the squaring of the circle, the trisection of an angle, and the duplication of the cube.[12] Yet again there is a further distinct area of interest in the applications of mathematics to music theory.

Second, due attention has to be paid to the fact that the Greek term we con-

[10] See, e.g., Michael Frede, *Essays in Ancient Philosophy* (Minneapolis: Univ. Minnesota Press, 1987), Chs. 12–15.

[11] Translation based on Galen, *On Anatomical Procedures,* trans. Charles Singer (New York/London: Oxford Univ. Press for the Wellcome Historical Medical Museum, 1956).

[12] Wilbur R. Knorr, *The Ancient Tradition of Geometric Problems* (Boston: Birkhäuser, 1986); and Knorr, *The Evolution of the Euclidean Elements* (Dordrecht: Reidel, 1975).

ventionally translate "mathematics" is far more general than that translation might suggest. *Mathēmatikē* is connected with the term *mathēma,* which can be used of any branch of learning, being derived from the verb *manthanein,* an entirely general term for "to learn." What we nowadays call mathematics is included, certainly, in the corresponding ancient term, though some of the styles of mathematics practiced today would scarcely be recognizable to ancient Greek investigators. But while they regularly included arithmetic and geometry under "mathematics," that term covered a good deal more. Thus when Ptolemy contrasts *mathēmatikē* first with theology and then with physics in the opening chapter of his theoretical astronomical treatise, the *Syntaxis* or *Almagest,* he includes the contents of that treatise under that term.

But it is not just that ancient *mathēmatikē* included some of what we might think of as the exact sciences. *Mathēmatikos,* "mathematician," is the word used not just for the person who engages in theoretical astronomy, but also for the astrologer. The distinction between two types of study is drawn by Ptolemy himself, for instance, in terms of the different types of prediction they attempt, although he thinks of both of these as predictions achieved by means of *astronomia* (*Tetrabiblos* 1.1). Thus one type of investigation aims, among other things, to build theoretical models that will enable the movements of the sun, moon, and planets to be predicted. But the other attempts predictions concerning events on earth from the study of planetary configurations, horoscopes, and the like. There are two ancient Greek terms, *astronomia* and *astrologia,* that *can* be used to draw the distinction that Ptolemy makes. Thus Sextus Empiricus contrasts that part of *astrologia* or *mathēmatikē* that is also called *astronomia* with the part relating to the casting of nativities (*genethlialogia*) (*Against the Mathematicians* 5.1). Again there are two corresponding Latin terms, though as W. Hübner has recently shown their use diverges from the Greek.[13] But far more often, in Greek and in Latin, the terms in question are employed indifferently and with no sense that *astronomia* and *astrologia* deal with distinct areas of study. Moreover, neither term regularly carries any pejorative undertones.

As with medicine and the term for "doctor," the point is not merely a question of semantics. The status and validity of the various kinds of studies of the heavenly bodies were as much topics of dispute as the goals and methods of healing. How far *are* predictions concerning events on earth possible on the basis of the study of the heavens? Again, how far are the theoretical models, based on epicycles and eccentrics and their combinations, credible? The former question is at the center of a heated debate where not just Ptolemy's *Tetrabiblos* but also Cicero's *De divinatione,* for instance, provides rich materials for the analysis of the arguments used *pro* and *contra,* while the second question is pressed strongly by Proclus among others. While we must certainly acknowledge that astrology was well established and widespread, it would clearly be grotesque to claim that it was unproblematic to the ancients themselves.[14]

[13] W. Hübner, "Der Titel zum achten Buch des Martianus Capella," in *Vorträge des ersten Symposions des Bamberger Arbeitskreises "Antike Naturwissenschaft und ihre Rezeption"* (*AKAN*), ed. K. Döring and G. Wöhrle (Wiesbaden: Harrassowitz, 1990), pp. 65–86.

[14] See, e.g., A. A. Long, "Astrology: Arguments Pro and Contra," in *Science and Speculation,* ed. Jonathan Barnes, Jacques Brunschwig, Myles Burnyeat, and Malcolm Schofield (Cambridge: Cambridge Univ. Press, 1982), pp. 165–192; and Nicholas Denyer, "The Case against Divination: An

The upshot of my argument so far in this section has been that both "medicine" and "mathematics" in the ancient Greek world diverge from their modern counterparts and that neither was a unitary investigation with an agreed goal, subject matter, and method. But what about the term that might seem to correspond more closely to what we mean by natural science? What about the "inquiry concerning nature," *peri phuseōs historia*, the study associated especially, though not exclusively, with those known from the fourth century B.C. onward as *phusikoi* (physicists) or *phusiologoi* (in the primary sense of those who offer an account of nature)?

It is clear that from some time in the fifth century B.C. onward certain thinkers did begin explicitly to focus on the investigation of nature itself. Yet part of the context of this development is supplied by the concern shown by some of the individuals in question to mark out their style of inquiry from others belonging to traditional wisdom. The concept of *phusis*, nature, itself was used, among other things, for polemical purposes,[15] that is to say to establish a contrast between the natural and the supernatural, the former the proper subject of inquiry, the latter the domain of superstitious belief. We can see this at work in the Hippocratic treatise *On Sacred Disease*, for example. There that author appeals to the concepts of nature and causation in order to *rule out* the assumptions and practices of his main rivals (the purifiers) on the grounds that they ignore the fundamental point that the sacred disease, like all other diseases, has a natural cause and can be treated with natural therapies.

Again nature could be, and was, invoked to contrast the domain of objective truth with that of conventional or culturally relative norms. In the fifth century B.C. *phusis* came to be placed in opposition to *nomos*, covering custom, convention, and law. Yet those who used that contrast were anything but unanimous as to how it was to be interpreted, whether in moral or political philosophy or in the study of nature. For some nature is the only domain of objective truth; but others insisted also on objectivity in the sphere of *nomos*. For some nature itself is value-free: but many others maintained that it was not, but was rather charged with values and indeed with moral significance, even though what that moral significance was was intensely disputed. Some argued that nature taught that might is right, while for others nature demonstrated the benevolence of the craftsmanlike force at work in cosmic creation.

The implications of these disputes for our understanding of the position of those who insisted on the importance of "the inquiry concerning nature" are clear. Although much of the research that passed by that name is indeed research into one or other of the fields that, for us, make up natural science, we should not take it that what it was to study nature was unproblematic or uncontroversial. On the contrary, the very insistence that there is a domain of nature there to be studied was a key move in the bid by new-style claimants to wisdom to outdo

---

Examination of Cicero's *De Divinatione*," *Proceedings of the Cambridge Philological Society*, 1985, no. 211 (N.S., no. 31), pp. 1–10.

[15] See G. E. R. Lloyd, "The Invention of Nature," in *Methods and Problems in Greek Science* (Cambridge: Cambridge Univ. Press, 1991), Ch. 18.

their rivals from other traditions. However, just how that study is to be pursued, its goals and methodology, was hotly disputed among those who agreed, at least, that it was worth pursuing.[16]

IV

These points about the pluralism and controversiality of Greek inquiries have a moral for my fourth and final topic, the problems associated with any pursuit of the question of the origins of science. It is manifestly absurd to seek the origins of Greek medicine, or mathematics, as if they were homogeneous entities and as if those entities had historically datable beginnings. The same point can be extended well beyond Greece and generalized. A moment's reflection is enough to show how bizarre it would be to try to identify *an* origin of medicine or mathematics, as such, for every society employs some form of healing and has some use for number, even though the studies of social anthropologists as well as historians have made clear how diverse those forms and uses are and how varied are systems of numeration.

Similarly with whatever we adopt as our working definition of science. The idea that science *as such* could be said to have a single beginning, whenever or wherever we choose to place that, is equally extravagant. As I put it in a recent study, "The intellectual outputs of ancient Egypt, Babylonia, India and China especially all have their contributions to make to our understanding of the origins of the scientific enterprise, and all pose their special problems of interpretation—as does the exploration of the crises of successive 'scientific revolutions' in later European history." "The origins of science . . . should certainly not be construed as a matter of a single historical event. Rather a whole range of issues would have to be considered, in any full account, in connection with widely separated historical periods, both long before and long after classical antiquity. . . . The difficulties that philosophers of science have encountered in specifying items either of content or of method, either of theory or of practice, as constitutive of science as such as a whole, should just strengthen our resolve in this regard."[17]

As those quotations indicate, it is not my view that all talk of origins and developments has to be ruled out. On the contrary, so far as ancient Greece goes, there are certainly determinate questions that can be investigated, even if we have always to be on our guard both against the too-easy associations with sudden change that go with the genetic model—the talk of origin as birth—and against teleology, the idea that earlier work was targeted at later results.

Some examples will illustrate the different types of problems that different topics pose. We can study, for instance, the history of dissection and vivisection as research methods in ancient Greece, where we can investigate, among other questions, what the first attempts to generalize the method (in Aristotle) owed to

[16] See Lloyd, *Revolutions of Wisdom* (cit. n. 6).

[17] G. E. R. Lloyd, *Demystifying Mentalities* (Cambridge: Cambridge Univ. Press, 1990), pp. 15–16, 28–29.

a general epistemology and theory of causation, and where we can explore the ongoing controversies about its use—where, as already mentioned, there was no general agreement about its usefulness and relevance.[18]

Again, we can study the Greek use of the pulse in diagnosis, how that came to be recognized and elaborated. We do not have the original texts of two of the main early theorists, Praxagoras and Herophilus, though we have reports of their work from Galen, among others, and his extensive treatises on the subject have survived. Once again the evidence available to us demonstrates that the technique was and remained controversial. Galen's own use of it, so he tells us himself, was criticized by some of his contemporaries as magic.[19]

In those two cases our main sources are the extensive extant medical texts and ancient histories of medicine. But if we take as a third example the use of trepanning, that clearly antedated our first Greek medical writings by centuries. So questions of origins there have to be explored through the archaeological record—of the actual skulls on which the operation has been performed—and the most we can hope for is to establish very broad limits for how far back the practice can be traced. What the medical writings of the classical period add is that this technique too was controversial and some of its hazards recognized.

Similarly in physics or astronomy, we have extensive materials—in Aristotle's *On the Heavens,* in Archimedes' *Sand-Reckoner,* in Ptolemy's *Syntaxis,* among other works—that throw light on the pre- and post-Aristotelian debates on the shape of the earth, its position in space, and whether it is at rest and if so why. Clearly none of these sources is neutral on the issues, and none is exactly a reliable witness to the ideas and arguments of those who, for one reason or another, denied that the earth is at rest at the center of the universe.

Again we can go some way, even though not as far as we should like, to trace the history of the astronomical observations that Greeks carried out themselves, the systems of chronology used, and the gradual development of coordinate systems replacing very rough and ready techniques of location in relation to prominent stars or constellations. Here is an important case where the question of what the Greeks knew of Babylonian work, at different periods, is as important as the study of their own observational methods and results.

We can study, as well, one outcome of careful and sustained observational work: Hipparchus's discovery of precession. Here too we have to rely on a later source, the account we are given by Ptolemy of how Hipparchus tried various hypotheses before concluding that it was in relation to the sphere of the fixed stars as a whole, not just those near the ecliptic, that this motion occurred. Moreover, this is yet another example where what we might think of as a concrete result was the subject of ancient dispute. Although Ptolemy certainly accepted precession himself, it was thereafter often ignored and sometimes flatly denied, as, for example, by the fifth-century philosopher Proclus—and this despite the fact that the evidence for it increased with the passing centuries.

These examples are mostly ones that may seem familiar enough to anyone coming to the ancient world with conventional twentieth-century ideas about

[18] See Mario Vegetti, *Il coltello e lo stilo* (Milan: Il Saggiatore, 1979); cf. G. E. R. Lloyd, "Alcmaeon and the Early History of Dissection," in *Methods and Problems in Greek Science* (cit. n. 15), Ch. 8.

[19] See Lloyd, *Revolutions of Wisdom* (cit. n. 6).

science. But others help rather to remind us how alien some Greek theories and inquiries were. This is true, for instance, of the widespread idea that the exacerbations and remissions of diseases are governed by patterns of what were known as critical days. Yet here, too, there was no general agreement, let alone unanimity, about *which* are the critical days or how their sequences are arranged. Some favored combinations of either odd or even days, counting from the onset of the disease; others adopted other patterns; and many recognized the great variety of experience in observed exacerbations and remissions, and some of these commented explicitly on the difficulty of establishing exact periodicities.

A second example would be provided by the common, but also contested, medical and physiological idea that the womb wanders around all over the inside of the female body. This is one among many instances that lead the historian to examine deep-seated assumptions about gender difference in ancient Greece and the place of women in Greek society, in addition to the medical and physiological aspects of this nexus of beliefs.[20]

Yet a third example would be the belief in the harmony of the spheres, an idea at the intersection of acoustic theory, of astronomy, and of cosmology, where beliefs about the structure of the universe and about the place of humans in it are at stake, as also the whole question of the intelligibility of the cosmos.

These are just a few of the many questions that the historian of Greek antiquity can tackle. Any selection of themes or topics has, of course, to be justified in terms of the particular aspects of the overall subject that they can illuminate; but it must be acknowledged that a *completely* comprehensive account is an unattainable ideal. Scholars will have their own ideas as to the areas that deserve particular attention, and some that were scarcely explored at all a mere twenty years ago have begun to produce interesting results. There is every reason, in fact, for us to be as pluralist in *our* approaches as the ancient Greek theorists we are studying were in *their* views of their own inquiries.

One group of problems that seems to me particularly well worth investigating relates to the hard-hitting ancient epistemological and methodological debates that I have mentioned several times already. This is for three reasons especially. First, those debates are one of the chief contexts where the Greek pluralism I have spoken of manifests itself. The divergences of approach adopted by different groups of ancient Greek theorists are often expressed in these debates, the chief forum for the explicit justifications offered for rival ideas concerning the proper aims and methods of inquiry. Second, these debates may be rather a distinctive, or at least a distinctively prominent, feature of ancient Greek science, and on that score worthy of attention. Third, some of the ideas put forward, notably the competing models of science as demonstrative, deductive, and explanatory, or alternatively as empirical and conjectural, were particularly influential at different stages in the subsequent development of European science.

Let me elaborate those last two points and introduce some cautionary remarks, and first concerning influences. The added interest that some ancient Greek methodological disputes gain from the way that some of the ideas in question

[20] See the essays collected in Averil Cameron and Amelie Kuhrt, eds., *Images of Women in Antiquity* (Detroit: Wayne State Univ. Press, 1983); and Lesley Dean-Jones, "Menstrual Bleeding according to the Hippocratics and Aristotle," *Transactions of the American Philological Association*, 1989, *119*:177–192.

were later taken up is, it is true, a very mixed blessing. Of course, in the Renaissance battles between the Ancients and the Moderns, the Greeks themselves were often badly misrepresented on both sides, whether cast in the role of heroes or as villains. However, that does not mean that it is not worth investigating the ancient forms of those disputes and how they originated. On the contrary, the attempt to get the record straight concerning the ancient models themselves is all the more important as a necessary preliminary to the study of those later battles.

As regards statements of Greek distinctiveness, they always run the risk of being mistaken for, or confused with, claims to Greek superiority. As historians, however, we should be resolute in insisting both that *every* culture should be studied in its own terms and that, to understand *any* culture, we have, so far as possible, to adopt a comparativist, and diachronic, perspective. Certainly to try to understand any of the complex and varied products of Greek speculative inquiry we need to do more than just study the Greeks.

To be sure, plenty of explanatory hypotheses have been suggested that aim to identify the significant factors in Greek culture or society that may have contributed to, or even have been responsible for, some of those products. Economic, technological, geographical, political, social, and cultural considerations all need to be reviewed, as well as factors internal to the intellectual inquiries themselves. But in evaluating those hypotheses the comparativist approach seems to me to be essential. How else can we hope to assess what contribution the spread of literacy, notably the use of an alphabetic system of writing, may have made? How else evaluate what difference it made that the Greeks were aware of their non-Greek neighbors, from whom indeed they borrowed many of their ideas? Or again, how else to judge what differences it made that the Greeks developed the political and legal institutions of the city-state—where many citizens of the classical period (and not just in the democracies) gained extensive direct experience not just of political debate in general, but of certain styles of adversarial argument in particular? I have attempted some such evaluations myself in several publications, but it is not that I claim that they represent a definitive resolution of the issues.[21] Quite the reverse, since they were offered to further, not to close, discussion of the problems. Moreover, it is not that I hold that the issues I claim to be important are the *only* ones worth pursuing in Greek antiquity, just that they are *among* those that are.

So I turn, finally and in conclusion, to the question posed at the beginning, of the appropriateness or otherwise of the category of "science" itself in our historical studies of the ancient world. That this concept runs certain risks and suffers from certain drawbacks is easy to see. To start with, it is our category—an observer's, not an actor's one—and I dare say that is true not just so far as the ancient Greeks are concerned. Second, there are obvious dangers of anachronism as soon as we start judging ancient work by the criteria of modern science, let alone start looking, whiggishly, for anticipations of later results. Should we, then, abandon all such talk of science entirely, and limit ourselves to discussion of those broad ancient categories of "medicine," "mathematics," "the inquiry concerning nature," and so on that we have spoken about? Or should we be content

[21] See, most recently, Lloyd, *Demystifying Mentalities* (cit. n. 17).

with other, even broader, ancient categories, such as "wisdom" (*sophia*) or even "knowledge" (*epistēmē* or Latin *scientia*)?

Both of those solutions suffer, in turn, from their own disadvantages. The problem, if we limit ourselves to "medicine," "mathematics," and so on, is that studies so defined may not take sufficient account of the interactions between these fields. Yet the rivalries I have spoken of were a matter of conflict not just *within* medicine (for instance) but *between* doctors (of certain types) and philosophers, and even involved the attempts of some doctors to emulate mathematical methods. Three methodological developments that seem particularly important—the recognition of the value and importance of empirical research, the development of the notion of axiomatic-deductive proof, and the application of mathematics to physical explanations—all are inter- or cross-disciplinary, the first spanning "medicine" and "the inquiry concerning nature," the second both those two and mathematics, the third mathematics and "the inquiry concerning nature."[22]

Conversely, the use of such an all-embracing category as "wisdom" or "knowledge" runs the risk of including too much. We are not historians of poetry (often the field, par excellence, of *sophia* in ancient Greece), nor of history (in our sense of the study of historical events), nor yet of political philosophy. We are not dealing with ancient views of just *any* phenomena, nor with *any* ancient explanations, not even with just those explanations that can, in some sense, be described as systematic. The types of ancient knowledge we are concerned with are related directly or (in the case of mathematics) indirectly to natural phenomena, even though, as I have insisted, what counted as "nature" was itself a subject of ancient dispute.

Neither modern nor even ancient categories will do precisely for all our purposes. I have argued that in general ancient ones are preferable to modern, but also that it would be absurd to think that we can use the ancient ones entirely to replace our own. Any search for an entirely neutral language in which to report and discuss ancient ideas is, in any case, bound ultimately to fail. But that does not mean that we cannot discuss those ideas at all. Rather, it means that whatever terms we use must be treated as provisional and revisable.

In this sense the consideration of the convergences and the divergences between ancient Greek ideals, methods, and practices and later European ones can be used, judiciously, to throw light on the former—as well as on the latter. So too can comparisons with other ancient societies, where those between ancient Greek and classical Chinese inquiries seem to me particularly fruitful. It is not that we should try to judge either by the criteria of the other. It would be foolish to ask why the Greeks had no concept of *Qi* (氣) or the Chinese no exact equivalent to *stoicheia* (elements)—for they spoke rather of phases, *xing* (行). It is rather that the exploration of markedly different traditions of intellectual inquiries in contrasting social and cultural circumstances can serve as a reminder of the parochialism of each—or at least of their culturally specific characteristics.

Thus, as I have claimed, it is largely by the comparative method that we can

[22] These points are documented in Lloyd, *Magic, Reason, and Experience* (cit. n. 8); and Lloyd, *Revolutions of Wisdom* (cit. n. 6).

reach clearer ideas both of the explananda and of possible explanations. Our goal, as historians of Greek inquiries, is to reach as accurate an understanding of them as we can—in all their complexities and controversiality. It goes without saying that any hypotheses, on that score, that we offer have to be tested against the evidence available, and both the limitations of the latter and the tentativeness of the former have duly to be acknowledged. But the very fact that the complexities and the controversiality are now much more generally appreciated than used to be the case can be taken as an indication that our understanding of what it means to talk about the *science* of the ancient Greeks can indeed make some progress.

Affinities and Elisions

Helen and Hellenocentrism

By Heinrich von Staden

But Helen, born of Zeus, had other things in mind:
And now she dropped into the wine they were drinking
a drug—an anodyne, bile-allaying, causing one to forget all ills.
Whoever swallows it when it's mixed in the wine bowl
would not drop tears down his cheeks for a day,
not even if his mother has died and his father too,
nor if they cut down with bronze his brother or his own son
right in front of him and he saw it with his eyes.
Such cunningly good drugs the daughter of Zeus had;
drugs Polydamna, mistress of Thon, had provided her
in Egypt, where food-giving fields yield most
kinds of drugs: many good when mixed, many baneful.
And each healer there is knowledgeable beyond all
humans; for they are of the race of Paeon [the divine healer] . . .

Homer, *Odyssey* 4.219–232

I

Helen of Troy might have been intrigued by the recurrent though far from novel issue of Hel(l)enocentrism.[1] More often than not a term of indictment, "Hellenocentrism," in the case of the history of science, tends to entail at least a double charge. The first is that historians of science privilege Greek science over the science of other ancient cultures, often with significantly distorting consequences. The second is that Eurocentric historians tend to adopt a version of "science" that allows them to credit the Greeks with the invention of science and of "the" scientific method. The ancient Greeks might have responded to this modern controversy with a mixture of bafflement, understanding, and sly amusement.

I am grateful to J. T. Vallance and Ronald L. Numbers for valuable comments on an earlier version of this paper. Several ideas developed here were presented at a conference entitled "Critical Problems and Research Frontiers in the History of Science and Technology," held under the auspices of the History of Science Society and the Society for the History of Technology, Madison, Wisconsin, 30 October–3 November 1991. I am indebted to Frederick L. Holmes for the invitation to participate in the conference.

[1] For present purposes I prefer the term "Hellenocentrism" to "Philhellenism" (and certainly to the recently voguish "Hellenophilia" and "Hellenomania"), since "Hellenocentrism," though also not free from evaluative connotations, is not limited to *philia* or *mania*. "Philhellenism," moreover, often has specific epochal connotations (e.g., Romantic Philhellenism); see note 16.

Homer's Helen, with her "cunningly good drugs" from Egypt, is not the only ancient Greek, fictional or historical, to whom a crucial knowledge of other cultures is attributed by ancient sources. From Homer to later antiquity, Greek culture displays an explicit awareness not only of its own agonal heterogeneity but also of the complex yet often permeable boundaries it constructed between the "Greek" and the non-Greek. The Greeks, while often deploying "barbarian" to fence off "Greek" from non-Greek, and while insisting on their own innovativeness, also admiringly marked the palpable, physical presence of the non-Greek in their midst.[2] Thus they explicitly—even if at times spuriously—labeled treasured parts of their materia medica "Ethiopian," "Indian," "Egyptian," and so on.[3] Moreover, some ancient Greeks attributed the invention of the mathematical sciences, notably of astronomy, to the Egyptians.[4] Writing, too, is said by some Greeks to have been invented by Phoenicians or Assyrians or Egyptians.[5] Resonances between Greek art and Near Eastern art, including Egyptian art, can be traced as early as the Bronze Age, and some archaic vase painters in Greece explicitly play on their foreign origins, conspicuously signing their works "the Lydian," "Amasis," and so on.[6]

[2] The word "barbarian" (Greek *barbaros,* a reduplicative onomatopoeic formation initially playing on the incomprehensibility of a stranger's speech sounds: "bar-bar- . . .") is, however, neither a Greek invention nor the property of the Greeks alone. Cf. Sumerian *barbar,* "foreigner"; Babylonian *barbaru,* "the stranger"; Indic *barbaraḥ.* See A. Walde and J. Pokorny, *Vergleichendes Wörterbuch der indogermanischen Sprachen,* Vol. 2 (Berlin/Leipzig: De Gruyter, 1927), pp. 105–106; Edith Hall, *Inventing the Barbarian: Greek Self-Definition through Tragedy* (Oxford: Clarendon, 1989); Timothy Long, *Barbarians in Greek Comedy* (Carbondale: Southern Illinois Univ. Press, 1986); Julius Jüthner, *Hellenen und Barbaren* (Leipzig: Dietrich, 1923); T. J. Haarhoff, *The Stranger at the Gate* (London/New York: Longmans, Green, 1938); B. Laurot, "Idéaux grecs et barbarie chez Hérodote," *Ktèma,* 1981, *6*:39–48; François Hartog, *The Mirror of Herodotus: The Representation of the Other in the Writing of History,* trans. Janet Lloyd (Berkeley/Los Angeles: Univ. California Press, 1988); Steven W. Hirsch, *The Friendship of the Barbarians: Xenophon and the Persian Empire* (Hanover, N.H.: Univ. Press New England, 1985); and Edmund Leach, "The Idea of the Barbarian Others," in *Storia delle idee: Problemi e prospettivi,* ed. M. L. Bianchi (Rome: Ateneo, 1989), pp. 13–30. See notes 11 and 43. On the Greeks' insistence on their own innovativeness see G. E. R. Lloyd, *The Revolutions of Wisdom* (Berkeley: Univ. California Press, 1987), pp. 50–108.

[3] E.g., Hippocrates, *On Diseases* 3.17 (7.158.1); *Epidemics* 7.6 (5.378.1); *On Internal Affections* 26, 51 (7.236.5, 294.20); *Diseases of Women 1* 34, 37, 63, 74, 75, 78 (8.80.9–11, 82.2–7, 90.12, 130.5, 158.19, 160.1–6, 168.18–19, 190.15, 202.15); *Diseases of Women 2* 185 (8.366.19–20). Here and elsewhere, references to Hippocratic texts include, in parentheses, volume, page, and line number in Émile Littré, ed., *Oeuvres complètes d'Hippocrate,* 10 vols. (Paris, 1834–1861).

[4] E.g., Socrates' story of the Egyptians Theuth and Thamus in Plato's *Phaedrus* 274c5–275b2 (but cf. Phaedrus's skeptical reaction, 275b3–4); Herodotus, 2.109; Aristotle, *Metaphysics* 1.1.981b23–25; ps.-Plato, *Epinomis* 986e9–988a5 (Syria, Egypt; but "the Greeks work into final perfection whatever they take over from the barbarians"); Diodorus of Sicily, 1.9.6.

[5] Rival Greek traditions about the invention of writing include its attribution to the Phoenicians Agenor or Cadmus, to Syrians or Assyrians, to the Egyptians Theuth (Thoth) or Menon, and to various Greek gods, heroes, or mortals, including Mnemosyne, the Muses, Hermes, Musaeus, Prometheus, Palamedes, the Cretans, etc. See Plato, *Phaedrus* 274c5–275b2; Herodotus, 5.57–59; Aeschylus, *Prometheus* 460; Diodorus, 1.9.2, 1.16.1 (the Egyptian Hermes), 5.67.3, 5.74.1; the Vatican scholia on Dionysius Thrax, *Grammatici Graeci,* Vol. 1, Pt. 3, ed. A. Hilgard (Leipzig: Teubner, 1901), pp. 182–185; Clement of Alexandria, *Stromateis* 1.16.75; scholia vetera on *Iliad* 7.185; Pliny, *Natural History* 7.192–193; and Scamon of Mytilene (a fourth-century B.C. historian), cited under "Phoinikēïa grammata" in the *Suda* and in Photius's *Lexicon;* Critias in Athenaeus, *Deipnosophistae* 1.28b, see Hermann Diels and Walther Kranz, *Die Fragmente der Vorsokratiker,* 7th ed., 3 vols. (Berlin: Weidmann, 1954), Vol. 2, 88B2.10. See also notes 27 and 32.

[6] "Amasis" is a Hellenized form of the common Egyptian name Ahmosis. See John Boardman, *Athenian Black Figure Vases* (London: Thames & Hudson, 1974), pp. 52–54 and figs. 64–71, 77–91, 99; and Jeffrey M. Hurwit, *The Art and Culture of Early Greece, 1100–480 B.C.* (Ithaca, N.Y./Lon-

Furthermore, Homer's superhuman Egyptian physicians are not the only examples of Greek expressions of admiration for Egyptian medicine. The historian Herodotus describes Egypt as "teeming with physicians" and the Egyptians as "the healthiest of all humans." Although some Greeks also record implicit and explicit criticisms of Egyptian medicine, the first-century B.C. Greek historian Diodorus of Sicily still includes in his universal history not only a positive version of Egyptian medicine but also a favorable, though partly derivative, account of early Egyptian culture in general. And in the first century A.D. Pliny the Elder, drawing ultimately on Greek sources, continues to raise the possibility that medicine was invented by the Egyptians.[7]

Indeed, some ancient Greeks record traditions that not only locate the origin of all human culture in Egypt but also try to establish the superiority of Egyptian culture and the derivation of Greek culture from it. Such views are advocated, for example, by the Greek historian Hecataeus of Abdera in his *Aegyptiaca,* probably written at the end of the fourth century B.C. Hecataeus in part recapitulates and elaborates positions developed by some of his Greek predecessors—for example, by Herodotus and by characters in some Platonic dialogues. Moreover, Hecataeus has in common with several Greek writers that he even locates the origin of the entire human species in Egypt.[8]

Reflecting on the relation of Greek to non-Greek, many other Greeks, from classical poets and Sophists to Hellenistic scientists and philosophers, also radically question the legitimacy of the division of all humans into "Greek" and "barbarian." Although this dichotomy entails a largely negative characterization of "barbarian" in the wake of the Persian Wars of the early fifth century B.C., by the end of the century it becomes challenged in a variety of ways. Thus Euripides puts on stage both "noble barbarians" and "barbaric Greeks"—and, of course, "barbaric barbarians" and "noble Greeks." More explicitly, his Athenian contemporary, the Sophist and dream theorist Antiphon, claims in his *On Truth* that "by nature we are *all* constituted similarly in *all* respects, *barbarians and Greeks alike.*" Some Hippocratic writers, at times insisting on a single nature (*physis*) of all human beings, likewise claim that there are no *innate* differences between

don: Cornell Univ. Press, 1985), pp. 125–202, 250–252. "Amasis" also appears in Herodotus, 1.30, 2.43, 3.39–47; Plato, *Timaeus* 21e; Diodorus, 1.60.1, etc. Especially in the visual art of the Bronze and Iron Ages, the interactions are such that "Greek art" is often a problematic label. See note 14.

[7] Herodotus, 2.77.3, 2.84, 3.1; Diodorus, 1.82 (perhaps drawing on Hecataeus of Abdera; see note 8); and Pliny, *Natural History* 7.50.196. But see also the story of the Greek physician Democedes of Croton, who cured a royal ailment that had left the Egyptian doctors of the Persian king Darius baffled (Herodotus, 3.129–138); Galen, *De simplicium medicamentorum temperamentis ac facultatibus* 7, prooemium, in *Claudii Galeni Opera omnia,* ed. Carl Gottlieb Kühn, 20 vols. in 22 (Leipzig, 1821–1833), Vol. 11, pp. 792–793.

[8] For Hecataeus of Abdera see Felix Jacoby, *Fragmente der griechischen Historiker,* IIIA (Leiden: Brill, 1940), esp. 264F1–6 and F25; and Diodorus, 1.10–98. But see also Plato, *Timaeus* 20d–26e: according to a wise ancient "Egyptian" tale, "retold" third- or fourth-hand by the Platonic character Critias, the original, lost Greek civilization was much older than Egyptian culture, and the same goddess, Athena, founded both Athens and the venerable ancient Egyptian city Saïs, Athens being a thousand years older. See Plato, *Critias* 113a; Aristotle, *Meteorologica* 1.14.352b20–22; and notes 4 and 5. On Diodorus's version (which relies heavily on Hecataeus) see Thomas Cole, *Democritus and the Sources of Greek Anthropology* (American Philological Association Monographs, 25) (Cleveland: Western Reserve Univ. Press for the American Philological Association, 1967), pp. 176–191; and Eduard Schwartz, *Realencyclopädie der Altertumswissenschaft,* Vol. 5 (Stuttgart: Metzler, 1905), cols. 670–672.

Greeks and barbarians (though human beings can be affected differently by different environments).[9] And an ancient source reports as follows on the views of Eratosthenes, the great third-century B.C. scientist and polymath:

> Eratosthenes denied praise to those who divide the total number of human beings into two groups—Greeks and barbarians—and also to those who advised Alexander [the Great] to treat the Greeks as friends but the barbarians as enemies. And, he says, it is better to make such divisions according to virtue [*aretē*] and vice [*kakia*]; for, he says, many of the Greeks are bad [*kakoi*] while many of the barbarians are civilized [*asteioi*, "urbane"], such as Indians and Arians, and likewise Romans and Carthaginians, all of whom govern themselves so admirably.[10]

Inscriptional and archaeological evidence, like the oft-heterogeneous Greek views of non-Greeks, indicates, however, that the historical realities of cultural, political, and economic interactions between Greeks and non-Greeks were far more complex than suggested either by the Greeks' positive verbal and visual representations of non-Greeks or by their negative portrayals.[11]

More significantly, many Greeks were aware of the problematic nature of rival claims, both Greek and foreign, concerning "origins" and "discoveries." Diodorus of Sicily, for example, observes that "with respect to the antiquity of the human race not only are the Greeks in disagreement among themselves but so are many of the barbarians, *all claiming that . . . they themselves were the first* of all humans to become *inventors-discoverers* of the things that are useful in life."[12] It is significant that Diodorus, like some other Greeks, professes agnosticism as to which of these claims to precedence might be valid.

The ancient Greeks are not alone in having reflected on the permeability of their cultural boundaries, on their rich indebtedness to other cultures, and on certain fundamental similarities between Greeks and non-Greeks. For more than a century modern archaeologists, linguists, and anthropologists, like historians of astronomy, art, myth, and religion, have contributed much to an understanding of the numerous cultural interactions between ancient Greece and the Near East, from the earliest to the latest periods of antiquity.[13] Indeed, in certain subfields,

[9] Diels and Kranz, *Fragmente der Vorsokratiker* (cit. n. 5), Vol. 2, 87B44B, col. 2 (quotation), from Oxyrhynchus papyrus 1364 (here and elsewhere, all translations are my own unless otherwise indicated). See also *ibid.*, 86B6, on the claim attributed to the Sophist Hippias of Elis that he used "both Greek and barbarian treatises" when composing his own work, selecting from all of these "the greatest things and those which are kindred"; and Plato, *Protagoras* 337c6–d1. Among Hippocratic writings emphasizing the single nature of human beings see, e.g., *Airs Waters Places* 16. See also Hall, *Inventing the Barbarian* (cit. n. 2), pp. 211–222; and H. C. Baldry, *The Unity of Mankind in Greek Thought* (Cambridge: Cambridge Univ. Press, 1965), pp. 35–37.

[10] Strabo, *Geography* 1.4.9, C66 (= Eratosthenes, fragment II.C24, in H. Berger, *Die geographischen Fragmente des Eratosthenes* [Leipzig, 1880; rpt. Amsterdam: Meridian, 1964]). See Baldry, *Unity of Mankind*, pp. 167–171. It is not unlikely that Eratosthenes was following the lead of some Stoics, on whose influential views see *ibid.*, pp. 151–166, 177–194.

[11] On historical interactions see, e.g., David M. Lewis, *Sparta and Persia* (Leiden: Brill, 1977); John Boardman, *The Greeks Overseas*, 2nd ed. (London/New York: Thames & Hudson, 1980); J. G. F. Hind, "Greek and Barbarian Peoples on the Shores of the Black Sea," *Archaeological Reports*, 1983–1984, *30*:71–97; Marie-Françoise Baslez, *L'étranger dans la Grèce antique* (Paris: Les Belles Lettres, 1984); and Barry W. Cunliffe, *Greeks, Romans, and Barbarians: Spheres of Interaction* (London/New York: Methuen, 1988). See notes 13, 14, and 43.

[12] Diodorus, 1.9.3.

[13] Historians of science, too, have long emphasized these interactions. See, e.g., Benjamin Far-

notably in Greek art and Greek religion, "the orientalizing period" has become an established term to refer especially to the seventh century B.C.[14] A vast body of scholarly work has long been offering rich instruction on these cultural interrelations.

Yet the charge of Hellenocentrism continues to be made, not only against generalists who neglect early Near Eastern, Asian, and New World science while claiming to speak of "ancient science," but also against those whose scholarly focus is purposely and explicitly restricted to the distinctive features of Greek science. To some extent the charge is legitimate, but the term "Hellenocentrism" and its cognates often obscure more fundamental and recalcitrant issues that extend beyond the interpretation of Greek science. Indeed, "Hellenocentrism" is but one manifestation of these problems. They are notably problems that arise from the oft-tacit interpretative deployment of notions of "affinity," "continuity," and "origin"—notions conditioned in part by the history of the modern reception of classical antiquity.

II

It perhaps goes without saying that different epochs tend to have their own dominant, in part culturally conditioned, versions and visions of ancient Greece, and that one tends to be most blind to the "shape" or "style" of one's own time.[15] Yet for all the interepochal and intergenerational squabbles of the last few centuries, there are strong similarities between dominant constructions of Greece from the Second Sophistic to the present, through the Renaissance, Neoclassicism, and

---

rington, *Greek Science* (London: Penguin, 1953), p. 13: "That Greek science, like Greek civilization as a whole, was deeply indebted to the older civilizations of the Near East is certain"; Asger Aaboe, "On the Babylonian Origin of Some Hipparchian Parameters," *Centaurus,* 1955, *4*:122–125; P. Schnabel, "Kidenas, Hipparch und die Entdeckung der Präzession," *Zeitschrift für Assyriologie,* 1927, *37*:1–60; B. L. van der Waerden, "Aegyptische Planetenrechnung," *Centaurus,* 1972, *16*:65–95; J. B. de C. M. Saunders, *The Transitions from Ancient Egypt to Greek Medicine* (Lawrence: Kansas Univ. Press, 1963); M. L. West, *Greek Philosophy and the Orient* (Oxford: Oxford Univ. Press, 1971); G. E. R. Lloyd, *Methods and Problems in Greek Science* (Cambridge: Cambridge Univ. Press, 1991), pp. 278–298; Alexander Jones, "The Development and Transmission of 248-Day Schemes for Lunar Motion in Ancient Astronomy," *Archive for History of Exact Sciences,* 1983, *29*:1–36; Jones, "The Adaptation of Babylonian Methods in Greek Numerical Astronomy," *Isis,* 1991, *82*:441–453; Bernard R. Goldstein and Alan C. Bowen, "A New View of Early Greek Astronomy," *ibid.,* 1983, *74*:330–340, on pp. 339–340; Lloyd, *Magic, Reason, and Experience* (Cambridge: Cambridge Univ. Press, 1979), pp. 229–234; F. Rochberg-Halton, "Elements of the Babylonian Contributions to Hellenistic Astrology," *Journal of the American Oriental Society,* 1988, *108*:51–62 (which rightly emphasizes the transformative role of Greek philosophy); and Edwin Rosner, "Terminologische Hinweise auf die Herkunft der frühen griechischen Medizin," in *Medizingeschichte in unserer Zeit: Festgabe für Edith Heischkel Artelt und Walter Artelt,* ed. Hans-Heinz Euler *et al.* (Stuttgart: Ferdinand Enke Verlag, 1971), pp. 1–22. See notes 14 and 28.

[14] See Walter Burkert, *Die orientalisierende Epoche in der griechischen Religion und Literatur* (Heidelberg: C. Winter, 1984); *Élements orientaux dans la religion grecque ancienne: Colloque de Strasbourg, 22–24 mai 1958* (Paris: Presses Univ. France, 1960); Sarah P. Morris, *The Black and White Style: Athens and Aigina in the Orientalizing Period* (New Haven, Conn.: Yale Univ. Press, 1984); Morris, *Daidalos and the Origins of Greek Art* (Princeton, N.J.: Princeton Univ. Press, 1992); and Tom Rasmussen, "Corinth and the Orientalising Phenomenon," in *Looking at Greek Vases,* ed. Rasmussen and N. Spivey (Cambridge: Cambridge Univ. Press, 1991), pp. 57–78. See note 6.

[15] "Shape" as used by George Kubler, *The Shape of Time: Remarks on the History of Things* (New Haven, Conn.: Yale Univ. Press, 1962). Hugh Kenner—I do not recall in which work—makes similar use of the "style of our time." My statement about "blindness" of course accommodates a self-referential moment.

Romanticism. Even the revolutionary so-called fathers of modernity—Marx, Nietzsche, Freud—in their versions of the ancient world succumbed to the fundamental metaphors and ideological commitments of Romantic Philhellenism. And the recuperative strains of eighteenth- and nineteenth-century Philhellenism continue to reverberate in the restorative aspirations of twentieth-century philosophers (e.g., Martin Heidegger) and historians of Greek culture (e.g., Werner Jaeger's "Third Humanism").[16] Though all too obvious, this is an issue rarely articulated, let alone faced squarely, in the Archimedean *pou stō* of the historian of ancient science. What is the "antiquity" we have received, institutionally and culturally? What is its impact on our acts of historical interpretation? And which visible and invisible contemporary or inherited ideological commitments mediate both our choice of ancient texts or artefacts and our readings of them, be they Chinese, Indian, Near Eastern, Mayan, or Greek?

A central feature of dominant constructions of antiquity has been the notion, at times explicit, more often implicit, of *affinity*. It is not only assumed, to some extent justifiably, that there are central affinities between antiquity and modernity. More insidious is the kindred, unspoken assumption that such affinities are especially, or even uniquely, worthy of our historiographic efforts. Even on the part of those who reject the discredited notion of "anticipation" as a model for writing history of science, this results in the continuing, albeit increasingly sophisticated, retelling of ancient "science" in our own scientific image. It is reflected in the oft-implicit insistence that, broadly speaking, a self-consciously articulated "rationality" and "empiricism" are the most "important" or "interesting" features of the science of any ancient culture. It also is amply evident in what one might call the "teleological fallacy" implicit in related uses of "the growth of science."[17] And it is echoed in the subsidiary insistence that *Greek* "science," unlike the sciences of other ancient cultures, was the product of a unique democracy with which we readily resonate (more on this in Section III). In short, the ancients, and especially the Greeks, are constructed as having been exceptionally capable of scientific "highlights," above all because they were like us.[18]

[16] On Romantic Philhellenism see E. M. Butler, *The Tyranny of Greece over Germany* (Cambridge: Cambridge Univ. Press, 1935); Heinrich von Staden, "Greek Art and Literature in Marx's Aesthetics," *Arethusa,* 1975, *8*:119–144; and von Staden, "Nietzsche and Marx on Greek Art and Literature: Case Studies in Reception," *Daedalus,* 1976, *105*:79–96. For twentieth-century examples see Beat Näf, "Werner Jaeger's *Paideia:* Entstehung, kulturpolitische Absichten und Rezeption," in *Werner Jaeger Reconsidered* (*Illinois Classical Studies,* Suppl. 3), ed. William M. Calder III (Atlanta: Scholars Press, 1992), pp. 125–146; and Donald O. White, "Werner Jaeger's 'Third Humanism' and the Crisis of Conservative Cultural Politics in Weimar Germany," *ibid.,* pp. 267–288.

[17] Cf., e.g., Farrington, *Greek Science* (cit. n. 13), p. 11: "The science of the earliest Greek period resembles ours"; and John Scarborough's account of the growth of Hippocratic humoral theory, "Medicine," in *Civilization of the Ancient Mediterranean: Greece and Rome,* ed. Michael Grant and Rachel Kitzinger, 3 vols. (New York: Scribner's, 1988), Vol. 2, pp. 1229–1232. See note 21.

[18] On the relation of democracy to Greek science see Jean-Pierre Vernant, *The Origins of Greek Thought* (London/Ithaca, N.Y.: Cornell Univ. Press, 1982), esp. pp. 130 ff.; Vernant, *Myth and Thought among the Greeks* (London/Boston: Routledge & Kegan Paul, 1983), pp. 176–210, 343–374; Pierre Vidal-Naquet, *The Black Hunter: Forms of Thought and Forms of Society in the Greek World,* trans. A. Szegedy-Maszak (Baltimore: Johns Hopkins Univ. Press, 1986), pp. 249–261; Lloyd, *Revolutions of Wisdom* (cit. n. 2), pp. 78–83; Lloyd, *Magic, Reason, and Experience* (cit. n. 13), pp. 240–267; and Lloyd, *Demystifying Mentalities* (Cambridge: Cambridge Univ. Press, 1990), pp. 58–67,

The issue is, of course, not "presentism," nor whether or not there are significant conceptual, methodological, and other continuities between ancient and modern science. There *are,* and scholars have pointed these out in meticulous, nuanced detail.[19] Nor is the immense influence of, for example, Plato's *Timaeus,* Aristotle, Euclid, and Galen on subsequent epochs of Western science at issue. Rather, the issue is the distorting consequences entailed by the historiographic habit of brightly foregrounding elements of legitimate continuity, with which we tend to be culturally more at home, while relegating discontinuities to a shadowy, obscure background at best.

Selectivity and elision cannot be avoided in the history of ancient science. As in the scientific text itself, and as in its translation, so too in the modern historiographic text that interprets ancient science, it is through simplification and amplification, through elision and supplementation, that certain meaningful historical relationships are revealed.[20] But there has been inadequate reflection on the cultural conditions that have shaped modern historians' selections and elisions. These cultural conditions include, centrally, two mutually reinforcing collective experiences: the modern reception of ancient Greece as the fountainhead of our culture and, second, modern Western scientific culture as our lodestar.

These long-lived, collective constructions of fountainhead and lodestar have led to concrete, entrenched consequences in the modern history of science. Thus the Hippocratic treatise *On Sacred Disease,* with its criticism of magic and with its overt questioning of etiologies that resort to the divine, is known to practically all historians of ancient science, having been translated often (and anthologized even more often), whereas there is no English translation of the Hippocratic gynecological treatises, which are replete with the "otherness" of Greek science—and which constitute a far larger part of the Hippocratic corpus. Privileging, historiographically, Hippocratic methodological debates (*On Ancient Medicine*), criticisms of magic (*On Sacred Disease*), impressively detailed observational accounts (*Epidemics*), and attempts at measurement and quantification over the "irrational" or "folkloristic" or "magical" within Hippocratic writings; privileging the Aristotelian hypothetico-deductive version of what science is over rival ancient versions; privileging Ptolemy's "astronomical" *Syntaxis* over his "astrological" *Tetrabiblos*[21]—all such acts of privileging might make eminent sense from the point of view of a modern version of "science." If

122–126. On "highlight" history of ancient science see J. T. Vallance, "Marshall Clagett's *Greek Science in Antiquity:* Thirty-five Years Later," *Isis,* 1990, *81*:713–721.

[19] See esp. Lloyd, *Revolutions of Wisdom;* and Lloyd, *Demystifying Mentalities.*

[20] On "idealisations and elisions" in ancient scientific texts see the superb analysis in Lloyd, *Revolutions of Wisdom,* pp. 285–329.

[21] E.g., Marshall Clagett, *Greek Science in Antiquity* (London: Abelard-Schuman, 1957), seems to mention magic only in the context of Egyptian science (pp. 7, 20) and relegates astrology and alchemy to a brief discussion in a chapter entitled "Science and Spiritual Forces in Late Antiquity." Morris R. Cohen and I. E. Drabkin, *A Source Book in Greek Science* (Cambridge, Mass.: Harvard Univ. Press, 1958), offers an even more sanitized version of Greek science. A laudable exception is G. E. R. Lloyd, *Science, Folklore, and Ideology* (Cambridge: Cambridge Univ. Press, 1983). Cf. John Scarborough, "The Pharmacology of Sacred Plants, Herbs, and Roots," in *Magika Hiera: Ancient Greek Magic and Religion,* ed. Christopher A. Faraone and Dirk Obbink (New York/Oxford: Oxford Univ. Press, 1991), pp. 138–174; and Scarborough, ed., *Folklore and Folk Medicines* (Madison, Wis.: American Institute of the History of Pharmacy, 1987). See note 23.

a primary historiographic motivation is to establish continuities between ancient and modern science, or to search for the "origins" of the forms of discourse, of the terminological categories, and of the methodological distinctions with which it has become customary to identify modern Western "science," then such "affinity"-driven acts of privileging could be useful. But such exercises, while (often unwittingly) recapitulating central features of the modern reception of classical antiquity, elide so much of ancient science, also of *Greek* science, that they lure the inexpert, and sometimes the expert, into misrepresentations of the multiple, contrary, and contradictory discourses and voices that sound those disharmonies that constitute "ancient science."

It might be objected that these concerns have been adequately addressed by three useful emphases in recent scholarship: the heterogeneity of Greek science itself, the absorption into Greek science of popular beliefs, and comparative cultural perspectives. Yet, significant though they are, none of these emphases responds fully to the issues raised above.

Several useful recent studies of polemics in Greek scientific texts have underscored the fundamentally agonal, heterogeneous character of Greek science, much of which develops through vigorous debate between opposing points of view. Such debates, it has been recognized, led to the clarification of central categories of discourse and to rhetorical, conceptual, epistemological, and methodological acts of fencing off scientific discourse from other forms of discourse, such as myth. Even these modern discussions of heterogeneity tend, however, to privilege one or two strands within Greek science (e.g., methodology and epistemology) or to favor a few assertive voices (notably those of hard-core "rationalists") over all others. They thus paradoxically locate "heterogeneity"—an indubitable feature of Greek science—within a single general form of "worldmaking" used by contentious rival participants in roughly the same discourse.[22] Debates between philosophico-scientific rivals (e.g., Presocratics or Peripatetics) tend to reflect only the heterogeneity of *viewpoints* within one major type of discourse (philosophico-scientific), rather than the heterogeneity of *discourses* found within ancient "science" (e.g., philosophy, religion, magic, alchemy, folklore, mysticism, allegorism).

The concerns raised above also are not necessarily addressed adequately by observing that, in ancient Greece as elsewhere, the rhetoric of scientific innovation or of rational coherence at times partly conceals the absorption into science of popular beliefs or of a prevailing ideology.[23] Often implicit even in such helpful

[22] "Worldmaking" in the sense developed by Nelson Goodman, *Ways of Worldmaking* (Indianapolis: Hackett, 1978). For studies of polemics in Greek scientific texts see J. Ducatillon, *Polémiques dans la Collection hippocratique* (Lille: Université de Lille, 1977); Lloyd, *Revolutions of Wisdom* (cit. n. 2), pp. 87–102; and Lloyd, *Demystifying Mentalities* (cit. n. 18), pp. 31–35, 57–60. For the demarcation of conceptual and other categories accomplished by such debates see Lloyd, *Revolutions of Wisdom*, Chs. 1, 4; and Marcel Detienne, *The Creation of Mythology*, trans. Margaret Cook (Chicago: Univ. Chicago Press, 1986), Ch. 3.

[23] See Lloyd, *Science, Folklore, and Ideology* (cit. n. 21), pp. 201 ff.; Lloyd, *Magic, Reason, and Experience* (cit. n. 13), pp. 29 ff.; Lloyd, *Revolutions of Wisdom*, Ch. 2; Scarborough, "Pharmacology of Sacred Plants" (cit. n. 21); Heinrich von Staden, "Matière et signification: Rituel, sexe et pharmacologie dans le Corpus hippocratique," *L'Antiquité Classique*, 1991, *60*:42–61; Giuliana Lanata, *Medicina magica e religione popolare in Grecia* (Rome: Ateneo, 1967); and Fridolf Kudlien, "Early Greek Primitive Medicine," *Clio Medica*, 1968, *3*:305–336.

modern analyses is a distinction between what can "legitimately" be regarded as ancient "scientific" (="rational") inquiries (i.e., ones similar, in principle at least, to our own) and that which becomes categorized as the "unscientific" or "pseudo-scientific" or "alien" popular beliefs of ancient societies. In such analyses, "science" thus again becomes subtly distilled from a more complex ancient intellectual and pragmatic activity, and "popular belief" is created as the "assimilated" opposite of a "science" thus isolated and distilled. (And, it should be noted, such interpretative strategies are of course adopted not only by historians of Greek science.[24]) Rather than the texture of the activity, as a whole, being interpreted as a form of "science," the fabric's threads of "nonaffinity," though acknowledged, become relegated to "mere" or "residual" or "problematic" popular belief and hence need no closer analysis, while the threads of "affinity" are accorded "scientific" and originary status and hence become the subject of detailed investigation. Though at times useful, the polarity "science"/"popular belief" thus often is valorized in such a way as to entail a lodestar-driven extraction of affinities and an interpretative deprivileging of discontinuities between ancient and modern science.

Nor are the concerns that arise from the "affinity" or "continuity" models of interpretation necessarily addressed adequately simply by introducing a comparative perspective. Even some excellent comparative studies remain vulnerable on at least two grounds. First, comparative analyses of what is distinctive or unique about Greek science again tend to crystallize precisely those features of Greek "science" that are continuous with our modern notion of "science," at the relative expense of the numerous elements that are alien to modern versions of "science."[25] They too tend, for example, to harp on the ancient Greeks' introduction of mathematical proof, on their pursuit of quantitative precision, on their application of mathematics to problems in natural science, on their development of crucial linguistic and conceptual categories, on their self-conscious epistemological and methodological advances in investing science with an authority that could compete successfully with other forms of "worldmaking," and so forth—all to the neglect of the "unscientific" elements in ancient science. Second, when deployed to find the "origin" of features that are continuous with modern science, the notion of a comparatively established "distinctiveness" tends to slide over into at least implicit claims of superiority.

Analyses of "distinctiveness" are, of course, not necessarily misleading. Nor are discontinuities between ancient and modern science necessarily any more significant than continuities. Rather, I am pleading for the oft-uncomfortable embrace of a partly alien complexity, and for embedding, in each local case (or, for the more audacious, in each epoch or in each culture), this complexity in the cultural or intercultural contexts that seem to have informed and produced it (see Section III). This would entail *inter alia* giving scientists' uses of magic, astrol-

[24] See, e.g., Gary M. Beckman's expert article on Hittite medicine, "Medizin B. bei den Hethitern," in *Reallexikon der Assyriologie,* Vol. 7 (Berlin: De Gruyter, 1987–1990): "The modern student of Hittite healing differentiates between therapies based upon purely physiological techniques . . . and those of magical character. . . . While only the former will be discussed here, the Hittites made no such distinction" (p. 630); and "The contexts in which physicians are active in Hittite documents reflect the *failure* to distinguish the magical from the medical in healing" (p. 631, italics added).

[25] The most thoughtful recent comparative study is Lloyd's *Demystifying Mentalities* (cit. n. 18), which offers an excellent analysis of ancient Greek and Chinese science.

ogy, alchemy, myth, birth prognoses, physiognomy, dream symbolism, omens, Hermeticism, inexactitude within the "exact" sciences, the "irrational," the "nonempirical," and what C. R. Hallpike and other anthropologists again are calling the "primitive"[26] as much place in our interpretations of ancient "science" as a rigorous, philologico-historical conscience demands in each context, rather than using cultural comparison only to reaffirm, with ever-increasing sophistication and precision, the roots or "anticipations" of key features of our scientific selves in any given ancient culture, be it an Aegean, African, Asian, or early New World culture.

III

The issues that lurk in "affinities" and in Hellenocentrism often intersect with those raised by the quest for the "origins" of science, as suggested above. Already in antiquity this search generated a genre of suspect literature,[27] and modern quests display disquieting similarities to some ancient constructions of "origins." For example, like their ancient counterparts, such as those mentioned by Diodorus of Sicily, the rival modern quests often target different ancient cultures or different geographic areas as their privileged destinations. And, as in antiquity, the rival modern choices of originary destination often are mediated by current ideological or ethnic interests. (The many myths of scientific autochthony propagated by the former Union of Soviet Socialist Republics represent an egregious but structurally common modern example.) As in antiquity, current quests for origins produce rival acts of retrospectivist "genealogical" privileging.[28]

Since historians of science, in part legitimately, have invested Greek science with many crucial elements of originality, the quest for *its* origins sooner or later beckons most specialists (and many a nonspecialist). While several useful explanatory perspectives have been developed, they prompt the unease reflected in my remaining observations.

Like some ancient Greeks, I believe that the permeability of their cultural boundaries permitted a richly productive two-way flow. Moreover, I believe that, during much of the history of Greek science, the "non-Greek" and the distinc-

[26] C. R. Hallpike, *The Foundations of Primitive Thought* (Oxford: Oxford Univ. Press, 1979); so too J.-F. Lyotard, "À propos de Cl. Lévi-Strauss: Les indiens ne cueillent pas les fleurs," *Annales Économies Sociétés Civilisations*, 1965, *20*:62–83, e.g., on pp. 63–64, 82. But see R. Schweder, "On Savages and Other Children," *American Anthropologist*, 1982, *84*:354–366.

[27] A. Kleingünther, *Prōtos Heuretēs* (*Philologus*, Suppl. 26.1, 1933); Cole, *Democritus* (cit. n. 8), pp. 48 ff.; and M. Kremmer, *De catalogis heurematum*, Inaugural-Dissertation, philosophische Fakultät, Univ. Leipzig, 1890.

[28] Contrast, e.g., Cheikh Anta Diop, *Civilization or Barbarism: An Authentic Anthropology*, trans. Yaa-Lengi Meema Ngemi (New York: Lawrence Hill, 1991); Diop, *The African Origin of Civilization: Myth or Reality*, trans. Mercer Cook (New York: Lawrence Hill, 1974); or Martin Bernal, *Black Athena: The Afroasiatic Roots of Classical Civilization* (New Brunswick, N.J.: Rutgers Univ. Press, 1987–) with the efforts by Vernant, Vidal-Naquet, and Lloyd (cited in note 18) to find the origins of central features of Western science in Greek political and judicial arrangements. For a critical response by a distinguished Egyptologist to the works of Diop and Bernal see John Baines, "Was Civilization Made in Africa?" *New York Times Book Review*, 11 Aug. 1991, pp. 12–13. For recent responses by classicists see Emily Vermeule, "The World Turned Upside Down," *New York Review of Books*, 26 Mar. 1992, *39*(6):40–43; Mary Lefkowitz, "Not Out of Africa: The Origins of Greece and the Illusions of Afrocentrists," *New Republic*, 10 Feb. 1992, pp. 29–36; and the wide-ranging collection of essays in *The Challenge of Black Athena, Arethusa* (special issue, fall 1989).

tively Greek coexist in inextricably intertwined, creative, but often opaque relations,[29] as do the cross-cultural and the local, the traditional and the original, the borrowed and the indigenous, influence and resistance to influence. The nature of Greek scientific culture therefore cannot be accounted for adequately either by searching out only the indubitably non-Greek sources—Mesopotamian, Iranian, Egyptian, Indian, and so forth—of certain ideas, procedures, data, inventions, and ingredients deployed by Greek science, or by analyzing only those features that distinguish it from the science of other ancient cultures. To do either, or to do neither, is to engage in major elisions. Furthermore, merely identifying foreign borrowings—whether in Greek science or in another ancient science (e.g., Greek elements in Indian astronomy)—can be misleading, since the original semantic, structural, theoretical, or functional context of the borrowed often is usurped and transformed by new cultural contexts, in science as in art and myth.

A further source of my unease is that the quest for the "origins" of science often is tacitly accompanied by a search for ancient motivations that resemble modern scientific ones. These might include motivations such as seeking an enduring, and possibly useful, scientific truth; doing so by obtaining a reliable data base; ferreting out unexamined assumptions in the quest for generalizable theories; aiming at quantitative accuracy; and corroborating hypotheses by putting them to a test. In itself, a search for such motivations in ancient science can be illuminating and unobjectionable, but it often is conducted at the expense of motivations now labeled "unscientific." Such "unscientific" motivations might include, for example, the need to interpret "divine" celestial "omens" for rulers (a well-known motivation that inspired many early Babylonian astronomical observations and, indeed, early mathematical astronomy). Similarly, complex sociocultural motivations inspired the widespread "birth prognoses" (predictions especially of fertility or infertility and of the gender of an unborn child) endorsed by some Hippocratic writers and practiced by several Near Eastern cultures. Or again, at times a scientist might have been motivated by the expectations of an often-skeptical public to accommodate religious and magical beliefs in his texts. There likewise were political and economic motivations at work in several ancient cultures: to please or counsel an autocratic ruler on whose benevolence a scientist's livelihood or life crucially depended.[30] Or, again, the aspiration to predict human predispositions and destinies motivated the fusion of Greek philosophy (including Aristotelian physics), Hellenistic astronomy, and Babylonian celestial divination into the comprehensive mechanistic theory of cause and effect known as Greek astrology.

These and similar "unscientific" sociocultural, political, religious, and psychological motivations should be accommodated in accounts of the "origins" of scientific activities and discoveries. Yet they have been relatively neglected, as other approaches have recently dominated the search for origins. I am referring

[29] For specific examples see note 13; see also Heinrich von Staden, *Herophilus: The Art of Medicine in Early Alexandria* (Cambridge: Cambridge Univ. Press, 1989), pp. 10, 14, 15–22, 149.

[30] On birth prognoses see Gary M. Beckman, *Hittite Birth Rituals,* 2nd ed. (Wiesbaden: Harrassowitz, 1983), e.g., p. 13; E. Iversen, "Papyrus Carlsberg No. VIII, with Some Remarks on the Egyptian Origin of Some Popular Birth Prognoses," *Historisk-filologiske Meddelelser Kgl. Danske Videnskabernes Selskab,* 1939, *36*(5):1–31; Hippocrates, *On Barren Women* 214 (8.414); and Hippocrates, *Aphorisms* 5.59 (4.554). See also Hippocrates, *Regimen,* book 4 (dream symbolism, prayers); and Herodotus, 3.129–138 (for a scientist's dependence on an autocratic ruler).

not only to the fertile multicultural perspectives that have reanimated the—paradoxically Eurocentric—expropriation and appropriation of African (i.e., Egyptian) origins for Greek science and hence for Western science. In the quest for origins two other issues have dominated: systemic political factors, and literacy in its motley ancient guises.

The role of literacy in the development of ancient science has been the subject of lively and, at times, nimble speculation. I offer one example. In a series of increasingly nuanced contributions, Jack Goody has revived the argument that the development and growth of critical, innovative attitudes, methods of investigation, and styles of reasoning are centrally attributable to the development of writing.[31] Long before the "literate revolution" in Greece, significant pre-Greek Mesopotamian and Egyptian scientific achievements already depended crucially on the existence of writing as a technology of communication, as Goody concedes. In Greece, however, he finds a distinctive, much more clearly articulated, and more self-conscious development of a "scientific attitude" than in Mesopotamia or Egypt. This he attributes to a feature of literacy that is unique to Greece among ancient societies. Not just writing per se but the spread of literacy among the Greeks, he claims, is a necessary condition for the development of the critical attitudes and methods that shape, underpin, and distinguish Greek science and philosophy. This "democratization of literacy" in Greece in turn is attributable in large part, Goody argues, to the Greek consonantal-vocalic system of alphabetic writing inspired by Phoenician writing.[32]

Unlike ancient nonalphabetic scripts, the Greek alphabet, with a mere twenty-four letters, is simple to master and hence, in principle, readily accessible to large numbers of people. Furthermore, Greek alphabetic writing was free to function also as a secular, nonsacred, and unofficial tool of communication. Unlike writing in some other ancient cultures, it was neither the "property" nor the "secret" of a small scribal or priestly class. These special social and mechanical features of Greek literacy, Goody argues, centrally account for the distinctiveness, within the ancient world, of Greek science and philosophy as enterprises that consistently, self-consciously present themselves as nonmagical, as nonreligious, as critical, and, at times, as empirical. Written texts, once free of a sacred or official or other "obliging" dimension, and especially in the context of the "democratization" of literacy, invite leisurely scrutiny and hence generate challenge, disagreement, revision, complication, and innovation to an extent that the transience of an oral text produced from memory cannot. The "characteristic" Greek

[31] See esp. Jack Goody and Ian Watt, "The Consequences of Literacy," in *Literacy in Traditional Societies*, ed. Goody (Cambridge: Cambridge Univ. Press, 1968), pp. 27–68; Goody, *The Domestication of the Savage Mind* (Cambridge: Cambridge Univ. Press, 1977); Goody, *The Logic of Writing and the Organization of Society* (Cambridge: Cambridge Univ. Press, 1986); and Goody, *The Interface between the Written and the Oral* (Cambridge: Cambridge Univ. Press, 1987), pp. 1–109. See also Oswyn Murray, "The Word Is Mightier than the Pen," *Times Literary Supplement*, 16–22 June 1989, p. 655.

[32] On the Phoenician origins of the Greek alphabet see note 5 and Alfred Heubeck, *Schrift* (Archaeologica Homerica 30[10]) (Göttingen: Vandenhoeck & Ruprecht, 1979), pp. 73–109; William V. Harris, *Ancient Literacy* (Cambridge, Mass.: Harvard Univ. Press, 1989), pp. 45–46, 278–279; Barry B. Powell, *Homer and the Origin of the Greek Alphabet* (Cambridge: Cambridge Univ. Press, 1990); and Sir Hugh Lloyd-Jones, "Becoming Homer," *N.Y. Rev. Books*, 5 Mar. 1992, *39*(5):52–56, on p. 56.

interest in notions such as observation, examination, scrutiny, testing, experience, inference, evidence, and natural causation hence, it is said, is essentially attributable to the democratization of literacy.

These theses, like those of Eric Havelock and others,[33] are, however, rich in controversial implications, and, not surprisingly, they have not gone unchallenged. Some recent studies persuasively express significant reservations. Goody's critics have argued, for example, that not all cultures exposed to literacy produced original science. More constructively, they argue that while Goody's interpretation of the impact of literacy might help account for certain features of Greek science, political factors are more responsible than literacy for the distinctive characteristics of early Greek science. Following the lead of Jean-Pierre Vernant and Pierre Vidal-Naquet, G. E. R. Lloyd, for example, has argued that "the very considerable experience that many Greek citizens acquired in the evaluation of evidence and arguments in the contexts of politics and law are reflected in the distinctive development of Greek science."[34] There are indeed many parallels between the deployment of notions of evidence, "witnessing," testing, scrutiny, accountability, and interpretation in the largely oral context of Athenian law and politics, on the one hand, and science on the other. Furthermore, as Lloyd shows convincingly, the contestability of tradition, so richly attested in many parts of Greek culture (though resisted in others), the overt and emphatic innovativeness in some Greek scientific texts, and the frequent authorial intrusions in many scientific treatises all are features of Greek science that do indeed have their counterparts in the Athenian political and judicial domain.

To conclude from such valid observations, however, that these features, well attested only for classical *Athenian* democracy (which, of course, strongly resonates with the modern Western experience), somehow "explain" distinctive features of *Greek* science (such as mathematical proof, the Aristotelian theory of scientific method, systematic dissection and experimental vivisection in Alexandria), or, even more problematically, that they account for the "origins" of Western science, is a step vulnerable on several counts (as Lloyd fully recognizes).[35]

Most ancient Greek science was neither manifestly born out of Athenian democracy nor borne by it. No major Presocratic philosopher-scientist, with the exception of Pericles' close associate Anaxagoras (an Ionian from Clazomenae, who eventually was expelled by the Athenians for impiety[36]), spent substantial time in "democratic" Athens. Most Presocratics did not share in the experience of the Athenian Agora or the Pnyx. Indeed, most Presocratics lived under oligarchies, and some under monarchies and tyrannies, about whose judicial practices

[33] E.g., Eric A. Havelock, *Preface to Plato* (Cambridge, Mass.: Harvard Univ. Press, 1963); Havelock, *The Literate Revolution in Greece and Its Cultural Consequences* (Princeton, N.J.: Princeton Univ. Press, 1982); Havelock, *The Muse Learns to Write* (New Haven, Conn.: Yale Univ. Press, 1986); and Kevin Robb, ed., *Language and Thought in Early Greek Philosophy* (LaSalle, Ill.: Hegeler Institute, 1983). For the Athenian evidence see Rosalind Thomas, *Oral Tradition and Written Record in Classical Athens* (Cambridge/New York: Cambridge Univ. Press, 1989).

[34] Notable among studies expressing reservations are Lloyd, *Magic, Reason, and Experience* (cit. n. 13), pp. 239–240; and Lloyd, *Revolutions of Wisdom* (cit. n. 2), pp. 70–78, 79 (quotation).

[35] E.g., Lloyd, *Demystifying Mentalities* (cit. n. 18), p. 59; Lloyd, *Revolutions of Wisdom*, pp. 78–81; and Lloyd, *Magic, Reason, and Experience*, pp. 260–264. See notes 39, 41.

[36] Plutarch, *Life of Pericles* 4–6, 8, 32; Plutarch, *Life of Nicias* 23; and Diogenes Laertius, 2.12–14.

next to nothing is known. I am, of course, not suggesting that Athens was the only fifth-century Greek polis with a democratic constitution. The Euboean city Eretria, for example, introduced a democratic constitution as early as the late sixth century B.C.; Ephesus and Cos made a transition from pro-Persian tyrannies to some form of democracy in 492 and 490 B.C., respectively; Argos had an agrarian democracy by the early 480s; and, not uncharacteristically, the Athenians violently imposed "democracy" on Samos in 440 B.C. with the help of a local faction. But it is far from clear just how widespread democracy was. And where "democracy" did exist, the evidence, though thin on detail, suggests that the motivations, origins, institutions, scope, and ideological commitments of *dēmokratia* in different Greek city-states varied greatly. Aristotle's famous characterization of *dēmokratia* as existing "whenever the free and the poor, being a majority, are in control of the ruling power" (*Politics* 4.4.1290b17–19) does not apply to all "democratic" Greek states and, indeed, not even to Athens without significant qualification. Furthermore, about the *judicial* institutions and *legal* practices of most "democratic" non-Athenian cities—institutions and practices crucial to the "democratic" account of Greek science—we are very poorly informed. To use "democracy" in its only well-documented classical Greek form—the fifth-century Athenian version—to explain the rise of Presocratic science in the sixth and early fifth centuries B.C. in Ionia, Sicily, and southern Italy therefore is problematic in several respects.

It is also noteworthy that democracy, in whatever form, was not universally applauded by Greek scientists. Not only are many violent, revolutionary Greek uprisings (prompted both by external pressures or interventions and by internal class or clan struggles) directed against democracy in the classical period, but prominent Greek scientists and other intellectuals, among them notably Aristotle and Plato, display ambivalence or hostility to democracy. For example, although Aristotle offers one of the more thoughtful ancient analyses of democratic principles, he expresses regret about the development of democracy in Athens after the Persian Wars, he views democracy as inevitably and deplorably unstable, and he firmly relegates it to the ranks of the deviant constitutions. His references to the fitful history of *dēmokratia* outside Athens—for example, in Thebes, Megara, Mantinea, Rhodes, Syracuse, Cyrene, Cyme, Cos, and Heraclea—are replete with descriptions of its failings.[37] Although a scientist's rejection of democracy of course does not in and of itself preclude the possibility that democracy, ulti-

[37] For Aristotle's allusions to democracy outside Athens see, e.g., *Politics* 3.15.1286b20–22, 4.4.1291b7–13, 4.11.1296a22–23, 5.1.1301b39–40, 5.3.1302b25–33, 5.4.1304a27–29, 5.5.1304b25–1305a1, 5.7.1307b23–24, 5.12.1316a32–33, 6.4.1318b6–1319b32. See also Hans-Joachim Gehrke, *Jenseits von Athen und Sparta: Das dritte Griechenland und seine Staatenwelt* (Munich: Beck, 1986). For divergent explanations of the frequent uprisings, including the antidemocratic revolutions, see Eberhard Ruschenbusch, *Untersuchungen zu Staat und Politik in Griechenland vom 7.-4. Jh. v. Chr.* (Bamberg: Aku Fotodruck und Verlag, 1978); G. E. M. de Ste Croix, *The Class Struggle in the Ancient Greek World from the Archaic Age to the Arab Conquests* (London: Duckworth, 1981); Andrew Lintott, *Violence, Civil Strife, and Revolution in the Classical City, 750–330 B.C.* (Baltimore: Johns Hopkins Univ. Press, 1982); M. I. Finley, *Politics in the Ancient World* (Cambridge: Cambridge Univ. Press, 1983); and Hans-Joachim Gehrke, *Stasis: Untersuchungen zu den inneren Kriegen in den griechischen Staaten des 5. und 4. Jahrhunderts v. Chr.* (Munich: Beck, 1985). It is perhaps worth noting that, except in later Hellenistic works assimilated to the Hippocratic corpus, Athens and Athenians are almost never mentioned by Hippocratics; see *Epidemics* 5.9–10 (5.208.18, 210.1).

mately or proximately, made his science possible, such rejections, too, caution against precipitate claims of causal correlations between "democracy" and "science."

Furthermore, in the century after Aristotle, Archimedes probably passed his life in Syracuse and Alexandria, both ruled by monarchs. Euclid's world likewise was not democratic, at least not in the sense associated with Periclean Athens. Herophilus became the first systematic dissector and experimental vivisector of the human body in part through the active intervention of an autocratic Greek-speaking Macedonian monarch in Alexandria. Galen's most productive years were spent in Rome under the Antonine emperors. Ptolemy, Galen's contemporary, lived in Alexandria, likewise under the Imperial principate. In fact, the very paradigm of ancient democracy, Periclean Athens, itself produced remarkably few major *native* scientists: the astronomer Meton comes to mind, and, among lesser figures, so do Euctemon, Archelaus, and Antiphon.[38]

At times the significance of a powerful ruler's patronage for scientific discovery is more readily demonstrable in ancient Greece than is the significance of classical Athenian forensic experience. The "political and judicial domain" very often was not "democracy" but oligarchy, monarchy, tyranny, and the Imperial principate, all in varied modulations. None of these ancient political forms had a monopoly on debate, on the weighing of evidence or "testimony," on challenging tradition, on innovation, or on critical questioning. And none precluded such activities. Conversely, like other ancient forms of government, Athenian "democracy" prosecuted scientists for their theories, banished intellectuals, and burned their books.[39] Socrates' tragic ordeal was not the only Athenian trial of an innovative intellectual.

Furthermore, *early* Greek science antedates the institution of democracy, and *most* Greek science postdates democracy. It has been argued—in order to rescue the "democratic" determinants of early Greek science—that the "key moves in the development" of Greek democracy are Solon's Athenian reforms of 594 B.C. and that, because these antedate Athenian democracy by almost a century, archaic Greek science too (though it was *not* Athenian) grew under the spell of Athenian democracy. One is left uneasy, however, when an Aristotelian attribution of the earliest beginnings of a few features of Athenian democracy to Solon is deployed to argue that *early* Greek science, that is, science practiced under far-flung oligarchies and tyrannies and monarchies, from Sicily and southern Italy

[38] See von Staden, *Herophilus* (cit. n. 29), p. 187 (text 63a); and Alan C. Bowen and Bernard R. Goldstein, "Meton of Athens and Astronomy in the Late Fifth Century B.C.," in *A Scientific Humanist: Studies in Memory of Abraham Sachs,* ed. E. Leichty *et al.* (Occasional Publications of the Samuel Noah Kramer Fund, 9) (Philadelphia: University Museum, Univ. Pennsylvania, 1988), pp. 39–81. Only in the fourth century B.C., after the demise of Periclean Athens, do Plato and his Academy begin to make Athens a major center of *scientific* research (despite the lively presence in fifth-century Athens of the—mostly *non*-Athenian—Sophists). On the ideology and workings of the Athenian democracy see Mogens Herman Hansen, *The Athenian Democracy in the Age of Demosthenes* (Oxford: Blackwell, 1991); Kurt Raaflaub, "Des freien Bürgers Recht der freien Rede," in *Studien zur antiken Sozialgeschichte: Festschrift Friedrich Vittinghoff,* ed. Werner Eck, Hartmut Galsterer, and Hartmut Wolff (Cologne: Böhlau, 1980), pp. 7–57; and W. R. Connor, M. H. Hansen, K. A. Raaflaub, and B. S. Strauss, *Aspects of Athenian Democracy* (Copenhagen: Museum Tusculanum Press, 1990). See also notes 41, 45.

[39] See Plutarch, *Life of Pericles* 32; Plutarch, *Life of Nicias* 23; and Diogenes Laertius, 2.12–14, 9.52. At least one proponent of the political hypothesis, Lloyd, is keenly aware of these difficulties; see the works cited in note 35.

to Asia Minor, was centrally influenced by Athenian "ideas associated with this [democratic] ideology" and by the subsequent Athenian propaganda of openness. Even if it is granted that the ideology of democracy was more important to Athenian intellectual life than "the often grim [presumably "nondemocratic"?] realities of actual political life," it remains to be demonstrated that this ideology is responsible for all or even most of the scientific and philosophical innovations that arose in nondemocratic, non-Athenian Greek communities, especially before democracy flourished and after it withered.[40] On this political hypothesis it also remains to be explained, as suggested above, why "democratic" Athens itself in its "Golden Age" produced remarkably few original *native* scientists but several brilliant poets and visual artists. (At issue is, of course, not the originality, beauty, and efficacy of certain features of Athenian democracy, restricted though its scope was, nor its partial similarity to modern liberal democracy,[41] but rather causal deployments of "democracy" in the interpretation of the origins of ancient science.)

Politics, then, like literacy or a non-Greek fountainhead, is hard put to account adequately either for the distinctiveness of Greek science or for its—or "our"—"origins." Explanatory hypotheses that single out one paramount causal factor rarely function satisfactorily when dealing with such complex questions (as Goody and his critics concede). Instead, a constellation of numerous *interactive* factors has to be accommodated on our horizon of understanding. These include well-known elements such as the spread of trade;[42] the comparative perspectives stimulated by exposure to other belief systems (as reflected, for example, in Presocratic criticisms of ethnocentrism and in the early, often extensive ethnographic observations made by Herodotus, by the Hippocratics, and by others);[43] the technological advances and cultural transformations stimulated by various forms of urbanization;[44] the exceptional geographic factors in Greece that facilitated local autonomy and hence the proliferation of local Greek constitutions; the

[40] Aristotle (?), *Constitution of Athens* 41.2 (but see also Ch. 9). The authorship of this work is disputed; see P. J. Rhodes, *A Commentary on the Aristotelian Athenaion Politeia* (Oxford: Oxford Univ. Press, 1981), pp. 58–63. See also Lloyd, *Demystifying Mentalities* (cit. n. 18), pp. 60–62, 62, 63 (quotations).

[41] On the ideological and, to a lesser degree, institutional similarities between Athenian democracy and modern liberal democracy see Mogens Herman Hansen, "The Tradition of the Athenian Democracy A.D. 1750–1990," *Greece and Rome*, 1992, *39*:14–30; and Hansen, *Was Athens a Democracy? Popular Rule, Liberty, and Equality in Ancient and Modern Political Thought* (Det Danske Videnskabernes Selskab., Historisk-Filosofiske Meddelelser, 1989, *59*). Hansen aptly cautions that "similarity" does not necessarily entail "influence."

[42] Exclusively economic explanatory hypotheses are no more satisfactory than exclusively political ones. See Lloyd, *Magic, Reason, and Experience* (cit. n. 13), p. 236, for an effective comparative argument against an economic interpretation.

[43] See, e.g., Xenophanes, in Diels and Kranz, *Fragmente der Vorsokratiker* (cit. n. 5), Vol. 1, 21B5–16; Herodotus, 3.38; and Hippocrates, *Airs Waters Places* 12–24. Cf. Lloyd, *Magic, Reason, and Experience*, pp. 236–239; *Hérodote et les peuples non grecs* (Entretiens sur l'Antiquité Classique, 35) (Geneva: Fondation Hardt, 1990). But intercultural contact should not be misread as always inevitably producing profound influence; see, e.g., Arnaldo Momigliano, *Alien Wisdom: The Limits of Hellenization* (Cambridge: Cambridge Univ. Press, 1975); and Naphtali Lewis, *Greeks in Ptolemaic Egypt* (New York: Oxford Univ. Press, 1986). See notes 2 and 11.

[44] A theme championed with reference especially to the ancient Near East by V. Gordon Childe in *Man Makes Himself* (London: Watts, 1936), *What Happened in History* (New York: Penguin, 1946), and *The Prehistory of European Society* (Harmondsworth: Penguin, 1958).

polis-structure itself (regardless of whether the polis was oligarchic, monarchic, tyrannical, or democratic); the correlative lack of an overriding, central pan-Hellenic authority with codifying aspirations;[45] the political and cultural ambitions of certain rulers; and cultural habits and traditions that reach back at least to the earliest archaic period, when rulers were "kings," more than a century before Solon's reforms and more than two centuries before the institution of Athenian "democracy."

Such archaic Greek habits include the toleration, or rather the active creation, of rival views and competing explanations, initially expressed mainly through rival mythographic traditions. At least since Homer and Hesiod, and probably much earlier, the Greeks did not, for example, "know *the* myth of Heracles"; rather, they told overtly competing versions of episodes associated with Heracles (and some might also have known versions of the myth of his Sumerian-Akkadian analogue Ninurta).[46] Long before "history," "philosophy," and various *technai,* including medicine, tried to fence themselves off from myth, magic, and poetry in the fifth century B.C.,[47] singers of Greek myth themselves already overtly engage in the criticism of rival claims. "It never was [true] that there was only one kind of Strife," Hesiod asserts polemically in his epic poem *Works and Days* (late eighth century B.C.?). Other non-Athenian archaic poets, too, such as Archilochus and Stesichorus, were not loath to criticize traditions, rivals, and precursors overtly, well before the putative impact of Athenian democratic ideology could have played a role. The fundamental contestability and revisability of tradition therefore are very much part of the agonal culture of Greece. They reflect habits in evidence from its early written traces in the archaic, "prescientific," "predemocratic" epoch too, though their origins remain shrouded in obscurity.

The emphatic overtness of an innovative authorial presence, which is a salient and distinctive feature of certain Greek scientific texts, similarly has been linked causally to the openness and agonal contentiousness of Athenian democracy.[48] Yet already Hesiod not only conspicuously introduces his own name into his poem *Theogony,* but also brings into play his construction of the autobiographical foundations of his claims to authority and truth. And as early as about 700 B.C. Greek artists mark their "authorial" claims by signing their own names to their works.[49]

It is, of course, not being suggested that, from a scientific point of view, there is no difference between Hesiod and, for example, the contentiously egocentric Galen, but rather that certain features that are said to be at once centrally dis-

[45] See the useful observations on the relation between polis-structure and ideology in Kurt Raaflaub, "Die Anfänge des politischen Denkens bei den Griechen," *Historische Zeitschrift,* 1989, *248*:1–32; and Raaflaub, "Die Anfänge des politischen Denkens bei den Griechen" and "Politisches Denken bei den Griechen," in *Pipers Handbuch der politischen Ideen,* Vol. 1 (Munich: Piper, 1985), pp. 189–368.

[46] But see Walter Burkert, "Oriental and Greek Mythology: The Meeting of Parallels," in *Interpretations of Greek Mythology,* ed. Jan Bremmer (London: Croom Helm, 1987), pp. 10–40.

[47] See Lloyd, *Revolutions of Wisdom* (cit. n. 2), Chs. 1, 4; Detiene (cit. n. 22), Ch. 3.

[48] Lloyd, *Revolutions of Wisdom* (cit. n. 2), Ch. 2.

[49] Hesiod, *Theogony* 22–35; see also Hesiod, *Works and Days,* lines 1–12 (esp. the emphatic *egō,* "I," in line 10), 654–662. For "authorial" claims by artists see R. M. Cook, *Greek Painted Pottery* (Chicago: Quadrangle Books, 1960), pp. 256–268; Boardman, *Athenian Black Figure Vases* (cit. n. 6), pp. 11–12; and Heubeck, *Schrift* (cit. n. 32), pp. 122–126.

tinctive of Greek science *and* attributable to the ideology or experience of Athenian democracy are in fact continuous with prominent features of archaic Greek culture, that is, of the prescientific, predemocratic Greece ruled by "kings" (*basileis*).

These early Greek habits—habits of contentiousness, of authorial overtness, of receptivity to other cultures, and of a boisterous, strained political and cultural pluralism—all interacted with the other "predemocratic" geographic, socioeconomic, cultural, political, and ideological factors enumerated above. This interactive, multileveled process probably facilitated *both* the development of distinctive features of Greek science in several different parts of the Greek world *and* the emergence of richly varied local political arrangements, including Athenian democracy. In other words, rather than construing Athenian democracy as the cradle of all significant Western science, one might better understand the varied forms of Greek democracy and oligarchy as local products of a constellation of dynamically interactive forces that, over a much larger spatial and temporal span, also yielded that distinctively pluralistic amalgam of the cross-cultural and the local, of the freely borrowed and the startlingly original, of the traditional and the revolutionary, that we retrospectively label "Greek science."

Animadversions on the Origins of Western Science

By Martin Bernal

I SPENT THE FIRST FIFTY YEARS of my life trying to escape from the shadow of my father, John Desmond Bernal, and hence, among other things, from science and the history of science. Therefore, the trepidation that is proper for anyone who is neither a scientist nor a historian of science writing for *Isis* is multiplied manyfold in my case. Nevertheless, I am grateful for the invitation to put forward my views on the origins of Western science.

Any approach to this question immediately stumbles over the definition of "science." As no ancient society possessed the modern concept of "science" or a word for it, its application to Mesopotamia, Egypt, China, India, or Greece is bound to be an arbitrary imposition. This lack of clarity is exacerbated by the clash between historians, like David Pingree, who are concerned with "sciences" as "functioning systems of thought" within a particular society and those who apply transhistorical standards and see "science" as "the orderly and systematic comprehension, description and/or explanation of natural phenomena . . . [and] the tools necessary for the undertaking including, especially, mathematics and logic."[1] I should add the words "real or imagined" after "natural phenomena."

Pingree denounces the claims of what he calls "Hellenophilia" that "science" is an exclusively Greek invention owing little or nothing to earlier civilizations and that it was passed on without interference to the Western European makers of the "scientific revolution." Puzzlingly, the work of Otto Neugebauer—and his school, including Pingree himself—on the extent and sophistication of Mesopotamian astronomy and mathematics and Greek indebtedness to it, as well as M. L. West's demonstration of the Near Eastern influences on the Presocratic cosmologies, appears to have left this kind of thinking unscathed.[2]

I could not have begun, let alone completed, this paper without many years of patient help and encouragement from Jamil Ragep, who, it should be pointed out, is far from accepting all my conclusions.

[1] For Pingree see "Hellenophilia versus the History of Science," a lecture originally presented at the Department of History of Science, Harvard University, 14 November 1990, and now published in this special section. For the passage cited see G. E. R. Lloyd, *Early Greek Science: Thales to Aristotle* (New York/London: Norton, 1970), p. 1, quoting Marshall Clagett, *Greek Science in Antiquity: How Human Reason and Ingenuity First Ordered and Mastered the Experience of Natural Phenomena,* new corrected ed. (New York: Collier, Macmillan, 1962), p. 15.

[2] See Otto Neugebauer, *The Exact Sciences in Antiquity* (New York: Dover, 1969); and M. L. West, *Early Greek Philosophy and the Orient* (Oxford: Clarendon, 1971)—*pace* John Vallance, "On Marshall Clagett's *Greek Science in Antiquity,*" *Isis,* 1990, *81*:713–721, on p. 715. See also G. S. Kirk, "Popper on Science and the Presocratics," *Mind,* 1960, *69*:318–339, esp. pp. 327–328; and Kirk,

There are still defenders of the claim that "Thales [seen as a Greek] was the first philosopher scientist"—the word "scientist" being used here in the positivist sense. According to G. E. R. Lloyd, the Greeks were the first to "discover nature," "practice debate," and introduce such specifics as the study of irrational numbers (notably $\sqrt{2}$) and geometrical modeling for astronomy. Lloyd sees the discovery of nature as "the appreciation of the distinction between 'natural' and 'supernatural,' that is the recognition that natural phenomena are not the products of random or arbitrary influences but regular and governed by determinable sequences of cause and effect."[3] However, it is clear that at least by the second millennium B.C. Mesopotamian astronomy and Egyptian medicine, to take two examples, were concerned with regular and, if possible, predictable phenomena with relatively little supernatural involvement.[4]

It is true that Egyptian medicine contained some religion and magic. At one point even the "scientific" Edwin Smith Papyrus on surgery turns to magical charms. However, E. R. Dodds and others have shown how isolated the natural philosophers' criticism was against the widespread Greek belief in the efficacy of magic.[5] Even Hippocratic medicine, which is generally regarded as highly rational, was institutionally centered on the religious cult of Asclepius and his serpents, which laid great emphasis on the religious practice of incubation. Both the cult and the practice, incidentally, had clear Egyptian roots.[6]

On the question of the alleged uniqueness of Greek "scientific" debate, as we can see from those in Gilgamesh, "debates" are at least as old as literature. Some, such as the "Dispute between a man and his *Ba*," which dates back to Middle Kingdom Egypt, contain quite profound philosophy. It is also clear that different Mesopotamian, Syrian, and Egyptian cities had not merely different gods but distinct cosmogonies, most of which involved abstract elements or forces without cults, of which the priesthoods of the others were aware. There were also attempted and actual syncretizations, suggesting that there had been debates.[7] This situation resembles that plausibly reconstructed for the cosmological disputes of the Presocratics.

Later Greek philosophical and scientific debates clearly owed a great deal to the Sophists, who came from the Greek tradition of "persuasion," with its close

"Common-Sense in the Development of Greek Philosophy," *Journal of Hellenic Studies,* 1961, *81:* 105–117, pp. 105–106.

[3] Lloyd, *Early Greek Science* (cit. n. 1), p. 8.

[4] Neugebauer, *Exact Sciences in Antiquity* (cit. n. 2), pp. 29–52; Paul Ghalioungui, *The House of Life: Per Ankh Magic and Medical Science in Ancient Egypt,* 2nd ed. (Amsterdam: B. M. Israël, 1973); and Ghalioungui, *The Physicians of Ancient Egypt* (Cairo: Al-Ahram Center for Scientific Translation, 1983).

[5] E. R. Dodds, *The Greeks and the Irrational* (Berkeley/Los Angeles: Univ. California Press, 1951); G. E. R. Lloyd, *Magic, Reason, and Experience* (Cambridge: Cambridge Univ. Press, 1979), pp. 10–58, 263–264; and Garth Fowden, *The Egyptian Hermes: A Historical Approach to the Late Pagan Mind* (Cambridge: Cambridge Univ. Press, 1986), pp. 81–82.

[6] J. B. de C. M. Saunders, *The Transition from Ancient Egyptian to Greek Medicine* (Lawrence: Univ. Kansas Press, 1963), p. 12.

[7] See E. A. Wallis Budge, *The Gods of the Egyptians: Studies in Egyptian Mythology,* 2 vols. (London: Methuen, 1904), Vol. 1, pp. 282–287; and Marshall Clagett, *Ancient Egyptian Science,* Vol. 1: *Knowledge and Order,* 2 pts. (Philadelphia: American Philosophical Society, 1989), pp. 263–372. For an annotated translation of the "Dispute between a man and his *Ba*" see Miriam Lichtheim, *Ancient Egyptian Literature,* 3 vols., Vol. 1: *The Old and Middle Kingdoms* (Berkeley/Los Angeles: Univ. California Press, 1975), pp. 163–169.

association with legal disputes. Oratory, persuasion, and justice are highly valued in nearly all cultures, but, interestingly, they received particular emphasis in Egypt. The central scene in Egyptian iconography is the judicial weighing of the soul of the dead person, and the legal battle between Horus and Seth is a central episode in its mythology. One of the most popular Egyptian texts was that of *The Eloquent Peasant,* which its most recent translator into English, Miriam Lichtheim, describes as "both a serious disquisition on the need for justice and a parable on the utility of fine speech."[8]

I have written elsewhere on the centrality of the image of Egyptian justice to both Mycenaean and Iron Age Greece, and there is no doubt that Greeks of the Classical and Hellenistic periods saw Egyptian law as the ultimate basis of their own. As Aristotle wrote at the end of the *Politics:* "The history of Egypt attests the antiquity of all political institutions. The Egyptians are generally accounted the oldest people on earth; and they have always had a body of law and a system of politics. We ought to take over and use what has already been adequately expressed before us and confine ourselves to attempting to discover what has already been omitted."[9]

While the first attestation of written law in Egypt comes from the tomb of Rekhmire in the fifteenth century B.C., there is no reason to doubt that it existed much earlier.[10] In any event, the Egyptian New Kingdom is sufficiently old by Greek standards. It is clear that what Aristotle was recommending had not hitherto been carried out. Nevertheless, it would seem likely that Aristotle was conventional in his belief that, even though Egyptian and Greek law were very different in his own day, the true foundation of Greek law and justice lay in Egypt.

The emphasis on law is important both because of its promotion of argument and dialectic and because of the projections of social law into nature and the establishment of regularities.[11] There is no doubt that the Egyptian Mȝʿt (*Maat:* "truth," "accuracy," "justice") was central to both social and natural spheres in the same way as the Greek *Moira,* which derived from it. Similarly, it is clear that the Egyptians applied the "justice" of scales to social and legal life at least as early as the Middle Kingdom.[12]

To return to some of the specific claims made for the originality of Greek science, there is now no doubt that Babylonian scholars were concerned with $\sqrt{2}$ and Pythagorean triples as well as having a good approximation of π. The Egyptian estimate for π was even more accurate. The standard use in land mea-

[8] Lichtheim, *Ancient Egyptian Literature,* Vol. 1, p. 169.

[9] Aristotle, *Politics,* 7.10, trans. Ernest Barker in *The Politics of Aristotle* (Oxford: Oxford Univ. Press, 1958), p. 304. See Martin Bernal, "Phoenician Politics and Egyptian Justice in Ancient Greece," in *Der Anfang des politischen Denkens bei den Griechen,* ed. Kurt Raaflaub (Munich: Historisches Kolleg, forthcoming).

[10] A. Théodoridès, "The Concept of Law in Ancient Egypt," in *The Legacy of Egypt,* ed. J. R. Harris (Oxford: Clarendon Press, 1971), pp. 291–313; and Anne Burton, *Diodorus Siculus Book I: A Commentary* (Leiden: Brill, 1972), pp. 219–225.

[11] See Joseph Needham, *Science and Civilization in China,* Vol. 2 (Cambridge: Cambridge Univ. Press, 1956), pp. 518–583.

[12] Bernal, "Phoenician Politics and Egyptian Justice" (cit. n. 9). Clagett subtitles Volume 1 of *Ancient Egyptian Science* "Knowledge and Order" (cit. n. 7), pp. xi–xii. These words are translations of the Egyptian *rḫt* and *Mȝ*ʿt, which he sees as the Egyptian "rudimentary" science. See the constant references to balances and plumblines as symbols of justice in *The Eloquent Peasant,* trans. in Lichtheim, *Ancient Egyptian Literature* (cit. n. 7), Vol. 1, pp. 170–182.

surement of the diagonal of a square of one cubit was the so-called double *remen*, that is to say $\sqrt{2}$ times the cubit.[13] Thus the irrational number par excellence was employed in Egypt from the beginning of the second millennium B.C. at the latest; whether or not its irrationality was proved in Euclidean fashion, its use provides circumstantial evidence that Egyptian scribes were aware of the incommensurability of the side and the diagonal.

Modern scholars have poured scorn on the widespread ancient tradition holding that Egyptians had known of the "Pythagorean" triangle. However, the very cautious Gay Robins and Charles Shute have accepted Beatrice Lumpkin's argument that knowledge of it is shown by the use in Late Old Kingdom pyramids of a *seked* of $5\frac{1}{4}$ palms, which imposed "a half-base width to height of 3:4 and so could have been modelled on a 3:4:5 right angled triangle."[14]

I shall discuss the strong possibility that geometry, thought to be typically Greek, came from Egypt. However, at this point it would seem difficult to argue that before the second half of the fourth century B.C. any aspect of Greek "science"—with the possible exception of axiomatic mathematics—was more advanced than that of Mesopotamia or Egypt.

WAS NEUGEBAUER RIGHT TO DISMISS ANCIENT TRADITIONS OF EGYPTIAN SCIENCE?

In this section I should like to take it as given that R. O. Steuer, J. B. de C. M. Saunders, and Paul Ghalioungui have established not merely that Egyptian medicine contained considerable "scientific" elements long before the emergence of Greek medicine, but that Egyptian medicine played a central role in the development of Greek medicine.[15] Similarly, the work of Neugebauer and his school has made it impossible to deny that some Mesopotamian mathematicians and astronomers were "scientific" in the positivist sense and that Mesopotamian "science" in these areas was crucial to the creation of Greek mathematics and astronomy. However, I should like to challenge these scholars' dismissal of claims that there was an Egyptian mathematics that could have had a significant influence on Greek thinkers.

[13] For the triples see the bibliography at the end of Olaf Schmidt, "On Plimton 322: Pythagorean Numbers in Babylonian Mathematics," *Centaurus*, 1980, *24*:4–13, on p. 13. On the Egyptian estimate for π see Richard Gillings, *Mathematics in the Time of the Pharaohs* (New York: Dover, 1972), pp. 142–143; and Gay Robins and Charles Shute, *The Rhind Mathematical Papyrus: An Ancient Egyptian Text* (London: British Museum Publications, 1987), pp. 44–46. On the double *remen* see Gillings, *Mathematics in the Time of the Pharaohs*, p. 208.

[14] Gay Robins and Charles Shute, "Mathematical Bases of Ancient Egyptian Architecture and Graphic Art," *Historia Mathematica*, 1985, *12*:107–122, on p. 112; and Beatrice Lumpkin, "The Egyptian and Pythagorean Triples," *ibid.*, 1980, *7*:186–187. See also Gillings, *Mathematics in the Time of the Pharaohs*, app. 5.

[15] See Ghalioungui, *House of Life* (cit. n. 4); Ghalioungui, *Physicians of Ancient Egypt* (cit. n. 4); R. O. Steuer and J. B. de C. M. Saunders, *Ancient Egyptian and Cnidian Medicine: The Relationship of Their Aetiological Concepts of Disease* (Berkeley/Los Angeles: Univ. California Press, 1959); and Saunders, *Transition from Ancient Egyptian to Greek Medicine* (cit. n. 6). *Pace* J. A. Wilson, "Medicine in Ancient Egypt," *Bulletin of Historical Medicine*, 1962, *36*(2):114–123; Wilson, "Ancient Egyptian Medicine," editorial in *Journal of the International College of Physicians* (Sect. 1), June 1964, *41*(6):665–673; and G. E. R. Lloyd, introduction to *The Hippocratic Writings*, ed. G. E. R. Lloyd (London: Penguin, 1983), p. 13n. Even the skeptical Heinrich von Staden, who is very reluctant to concede Egyptian influences on Hellenistic medicine, admits that the study of pulses and their timing by water clocks, for which his subject Herophilus of Alexandria was famous, probably came from the Egyptian tradition: von Staden, *Herophilus: The Art of Medicine in Early Alexandria* (Cambridge: Cambridge Univ. Press, 1989), p. 10.

Despite his early passion for ancient Egypt and his considerable work on Egyptian astronomy, throughout his long life Neugebauer insisted that the Egyptians had no original or abstract ideas and that mathematically and scientifically they were not on the same level as the Mesopotamians. He claimed that the accurate astronomical alignments of the pyramids and temples in Egypt and the use of π and ϕ could all be explained as the results of practical knacks rather than of profound thought. An example of this approach is the following: "It has even been claimed that the area of a hemisphere was correctly found in an example of the Moscow papyrus, but the text admits also of a much more primitive interpretation *which is preferable*."[16]

In his *Exact Sciences in Antiquity* Neugebauer did not argue with the pyramidological school; he simply denounced it, recommending that those interested in what he admitted to be "the very complex historical and archaeological problems connected with the pyramids" read the books by I. E. S. Edwards and J. F. Lauer on the subject.[17]

While Edwards does not involve himself with the pyramidologists and their calculations, the surveyor and archaeologist Lauer did, in the face of opposition from Egyptologists, who were "astonished that we should give so much importance to the discussion of theories which have never had any credit in the Egyptological world." Lauer's work had a certain contradictory quality. He admitted that the measurements exhibited by the pyramids do have some remarkable properties; that one can find such relations as π, ϕ, and Pythagoras's triangle from them; and that these facts generally bear out the claims Herodotus and other ancient writers made for them. On the other hand, he denounced the "fantasies" of pyramidologists and claimed that the formulas according to which the pyramids were aligned and the extraordinary degree of sidereal accuracy they exhibited were purely the result of "intuitive and utilitarian empiricism."[18]

A conflict between the acceptance of the extraordinary mathematical precision of the Great Pyramid and a "certainty" that the Greeks were the first "true" mathematicians runs throughout Lauer's many writings on the subject. The strain is made still harder to bear by Lauer's awareness that some Greeks had been told about many of this pyramid's extraordinary features and that they believed the Egyptians to have been the first mathematicians and astronomers. Moreover, there was the problem that so many of the Greek mathematicians and astronomers had studied in Egypt. Lauer's honest attempt to deal with these difficulties was the following:

> Even though up to now, no esoteric Egyptian mathematical document has been discovered, we know, if we can believe the Greeks, that the Egyptian priests were very jealous of the secrets of their science and that they occupied themselves, Aristotle tells us, in mathematics. It seems then reasonably probable that they had been in possession of an esoteric science erected, little by little, in the secrecy of the temples during the long centuries that separate the construction of the pyramids, towards the

[16] Neugebauer, *Exact Sciences in Antiquity* (cit. n. 2), p. 78 (italics added). I shall challenge this interpretation below.

[17] *Ibid.*, p. 96.

[18] J. F. Lauer, *Observations sur les pyramides* (Cairo: Institut Français d'Archéologie Orientale, 1960), pp. 11, 10, 4–24 (here and elsewhere, translations are my own unless otherwise indicated).

> year 2800 [I should put it two hundred years earlier], to the eve of Greek mathematical thought in the sixth century B.C. As far as geometry is concerned, the analysis of buildings as famous as the Great Pyramid would take a notable place in the researches of these priests; and it is perfectly conceivable that they could have succeeded in discovering in it, perhaps long after their erection, chance qualities that had remained totally unsuspected to the constructors.[19]

The question of when Egyptians developed this sophisticated mathematical knowledge is not directly relevant to the topic of this article. However, apart from the precision and intricacy of many of the architectural constructions of the Old Kingdom, there is another argument for the existence of relatively "advanced" mathematics in the first half of the third millennium B.C. This is that although the two great mathematical texts that have survived, the Moscow and the Rhind papyri, come from the Middle Kingdom in the twentieth and nineteenth centuries B.C., some of the problems set in them use measures that belong to the Old Kingdom, which had been discarded by this time.[20]

Lauer's solution still allowed some later Egyptians to have been capable of relatively advanced thought. He continued:

> For the whole length of the three thousand years of her history, Egypt thus, little by little, prepared the way for the Greek scholars who like Thales, Pythagoras, and Plato came to study, then even to teach, like Euclid at the school in Alexandria. But it was in their philosophic spirit, which knew how to draw from the treasure amassed by the technical positivism of the Egyptians, that geometry came to the stage of a genuine science.[21]

Even this degree of recognition was too much for Neugebauer. As he put it at one point: "Ancient science was the product of a very few men and these few happened not to be Egyptians." In 1981 he published his note "On the Orientation of Pyramids," in which he showed how accurate alignments could be made without sophisticated astronomy, simply by measuring and turning the shadow of a model pyramid or the capstone over a period of some weeks. There is no evidence, one way or the other, whether this was the method used, but it would seem plausible, if only because pyramids appear to have had solar rather than stellar cultic associations. Nevertheless, the requirement of what Neugebauer concedes to be "remarkable accuracy of . . . orientation of the Great Pyramid," a structure of extraordinary sophistication, indicates very serious religious and theoretical concerns.[22] Thus, Neugebauer's choice of the word "primitive" to describe the alignment seems inappropriate; the word is—as we shall see—indicative of his general opinion of the ancient Egyptians.

There is little doubt that this modern view of the Egyptians' lack of mathematics and science has been influenced by a distaste for the theology and metaphysics in which much of Egyptian—and Platonic—knowledge was embedded and by progressivist views that no one who lived so early could have been so sophisti-

[19] *Ibid.*, pp. 1–3. For a more skeptical view of this see Robins and Shute, "Mathematical Bases" (cit. n. 14), p. 109.

[20] Robins and Shute, *Rhind Mathematical Papyrus* (cit. n. 13), p. 58.

[21] Lauer, *Observations sur les pyramides* (cit. n. 18), p. 10.

[22] Neugebauer, *Exact Sciences in Antiquity* (cit. n. 2), p. 91; and Neugebauer, "On the Orientation of Pyramids," *Centaurus,* 1980, *24*:1–3, on p. 2.

cated. It may also have been reinforced by assumptions, almost universal in the nineteenth and early twentieth centuries, that no Africans of any sort could have been capable of such great intellectual achievements.

An indication that such attitudes may have had an impact even on such a magnificent champion of liberalism and foe to racism as Neugebauer comes in one of his bibliographical notes, where the first book he recommended "for a deeper understanding of the background that determined the character of Egyptian arithmetic" was Lucien Lévy-Bruhl's *Fonctions mentales dans les sociétés inférieures*. Lévy-Bruhl was far from the worst of his generation. Nevertheless he belonged to it, and it was appropriate that his work was translated into English as *How Natives Think*.[23]

Having said this, there is no doubt that Neugebauer had some substantial arguments to back his case. The strongest of these were his claims that none of the surviving mathematical papyri from pharaonic Egypt contained what he believed to be sophisticated calculations and that the Egyptians' systems of numbers and fractions were too crude for profound mathematical and astronomical thought of the kind that had been attributed to them. There are seven major arguments against this position.

1. The strong possibility that—*pace* Neugebauer—the surviving Egyptian papyri do contain "advanced" mathematics.
2. Parallels from Mesopotamia and Ptolemaic Egypt showing that one cannot rely on the papyrological record to gauge the full range of pharaonic Egyptian "science."
3. The general agreement that Egyptian geometry was equal to or better than that of Mesopotamia, in conjunction with the conventional wisdom that one of the chief contributions of the Greeks to Mesopotamian "arithmetic" was geometric modeling, which suggests that the geometrical input may well have come from Egypt.
4. The coordination of sophisticated geometry and computation in Egypt with extraordinary practical achievements.
5. The Greek insistence that they learned mathematics—and medicine—not from Mesopotamia but from Egypt.
6. The Greek adoption of an Egyptian rather than a Mesopotamian calendar.
7. The facts that much of Hellenistic and Roman science took place in Egypt, not Greece, and that although they wrote in Greek some of its practitioners, including the astronomer Ptolemy, were Egyptian.

The first argument is buttressed by the fact that, as we have seen, Neugebauer preferred "more primitive interpretations" and therefore could have overlooked evidence of more sophisticated work. Thus we must allow for the possibility that the surviving texts contain or refer to elements that are more sophisticated than he and some other twentieth-century historians of science have supposed. There is little doubt about the employment of irrational numbers, mentioned above, and the use of arithmetical and geometrical progressions in the Rhind Papyrus problems 40 and 79.[24]

The Soviet scholar V. V. Struve, who was the first to study the Moscow Mathematical Papyrus, was much more respectful than Neugebauer. He wrote, for instance, that "we must admit that in mechanics the Egyptians had more knowl-

[23] Neugebauer, *Exact Sciences in Antiquity*, p. 92; and Lucien Lévy-Bruhl, *How Natives Think* (New York: Knopf, 1926).

[24] Robins and Shute, *Rhind Mathematical Papyrus* (cit. n. 13), pp. 42–43, 56.

edge than we wanted to believe." He was convinced that this papyrus and the Rhind Mathematical Papyrus demonstrated a theoretical knowledge of the volume of a truncated pyramid, and he has been followed in this interpretation by later scholars. Given the many pyramids successfully constructed during the Old and Middle Kingdoms, this would not in itself seem unlikely. Archimedes, however, maintained in the third century B.C. that the volumes of pyramids were first measured by Eudoxos of Knidos a hundred years earlier.[25]

Here, as in some other instances, Archimedes was knowingly or unknowingly mistaken. Even so, it is possible that Eudoxos was the first to transmit the formulas to Greece. Eudoxos spent many years in Egypt and was reported to have learned Egyptian and to have made translations, some of which may well have come from the *Book of the Dead,* into Greek. As Giorgio de Santillana pointed out, it is unlikely that Eudoxos translated these texts merely for their entertainment value; it is much more probable that he believed that they contained esoteric astronomical information.[26] This raises the important suggestion that Egyptian religious and mystical writings and drawings may well contain esoteric mathematical and astronomical wisdom.

To return to earth with the particular case of the measurement of the surface area of either a semicylinder or a hemisphere in the Moscow Papyrus: Richard Gillings, who believes the measurement refers to the latter, describes the Egyptian operations and writes:

> If this interpretation . . . is the correct one, then the scribe who derived the formula anticipated Archimedes by 1,500 years! Let us, however, be perfectly clear [that] in neither case has any proof that either $A_{cylinder} = \frac{1}{2}\pi dh$ or $A_{hemisphere} = 2\pi r^2$ been established by the Egyptian scribe that is at all comparable with the clarity of the demonstrations of the Greeks Dinostratos and Archimedes. All we can say is that, in the specific case in hand, the mechanical operations performed are consistent with these operations which would be made by someone applying these formulas even though the order and notation might be different.[27]

In general, it is clear that the specifically mathematical papyri give considerable indications of sophisticated operations. As Struve put it in the conclusion of his study of the Moscow Papyrus:

> These new facts through which the Edwin Smith and Moscow papyri have enriched our knowledge, oblige us to make a radical revision of the evaluation made up to now of Egyptian "science" [*Wissenschaft*]. Problems such as the research into the functions of the brain or the surface area of a sphere do not belong to the range of practical "scientific" questions of a primitive culture. They are purely theoretical problems.

[25] V. V. Struve, "Mathematischer Papyrus des staatlischen Museums der schönen künste in Moskau," *Quellen und Studien zur Geschichte der Mathematik* (Pt. A), 1930, *1*:184; and Paul Ver Eecke, *Les oeuvres complètes d'Archimède* (Paris: Blanchard, 1960), p. xxxi. Among later scholars who have supported Struve's view (on pp. 174–176) see Gillings, *Mathematics in the Time of the Pharaohs* (cit. n. 13), pp. 187–194; and Robins and Shute, *Rhind Mathematical Papyrus,* p. 48.

[26] Giorgio de Santillana, "On Forgotten Sources in the History of Science," in *Scientific Change: Historical Studies in the Intellectual, Social, and Technical Conditions for Scientific Discovery and Technical Invention, from Antiquity to the Present,* ed. A. C. Crombie (London: Heinemann, 1962), pp. 813–828, on p. 814.

[27] Gillings, *Mathematics in the Time of the Pharaohs* (cit. n. 13), p. 200.

Or earlier:

> The Moscow Papyrus . . . confirms in a striking way the mathematical knowledge of the Egyptian scholars and we no longer have any reason to reject the claims of the Greek writers that the Egyptians were the teachers of the Greeks in geometry.[28]

Objections by Neugebauer and others to Struve's specific interpretation of the surface area of a hemisphere have now been answered.[29] Similarly, as mentioned above, claims for the use of "Pythagorean" triangles and the sophistication of the measurement of the volume of the truncated pyramid have both survived earlier skepticism. If these bases of Struve's general case still stand, should one accept Neugebauer's dismissal of it?

Even if one were to concede Neugebauer's argument that the mathematics contained in these papyri is merely practical and primitive, there is the second argument: the strong likelihood that more sophisticated work was recorded on others that have not been preserved. Lauer raised the point that all reports indicate that the Egyptian priests were secretive about their writings; therefore there would have been few copies and the chances of their survival would have been slim. It should be emphasized that relatively few papyri of any kind have survived. This is very different from Mesopotamia, where the baked clay tablets are remarkably durable and hundreds of thousands of them have been discovered. The problem with Mesopotamian texts is not a lack of them but the difficulty of finding enough Assyriologists to read and publish them. Even here, however, there are gaps in what exists. Neugebauer points out that the "great majority" of mathematical tablets come from one of two periods, the Old Babylonian period—of two hundred years—in the first half of the second millennium B.C. and the Seleucid period. Continuities between the two sets of texts make it clear that sophisticated mathematics was carried out in the twelve or more centuries that intervened. However, there is no record of this.[30]

The situation is far worse in Egypt, and there is no doubt that most of the papyri written and all of those that have survived were texts used for teaching scribes techniques that were useful for practical accounting rather than "state of the art" advanced mathematics.[31] An instructive parallel can be seen in the Ptolemaic period. Many more mathematical papyri have been found from these few centuries than from the whole pharaonic period, yet none of these go beyond book 1 of Euclid or give any indications of the extraordinary sophistication of the work we know from textual transmission to have been taking place in Hellenistic Egypt. Thus the argument from silence, which should always be applied sparingly, should be used with particular caution in evaluating the absence of textual *proof* of advanced Egyptian mathematics.

Against this, it is argued that the few texts that do exist show a consistency of techniques and notation that makes it impossible for the Egyptians to have produced sophisticated mathematics. This brings us to the third argument against skepticism: Egyptian numerical notation may not have been as flexible and help-

[28] Struve, "Mathematischer Papyrus" (cit. n. 25), pp. 183, 185.
[29] Gillings, *Mathematics in the Time of the Pharaohs* (cit. n. 13), pp. 194–201.
[30] Neugebauer, *Exact Sciences in Antiquity* (cit. n. 2), p. 29.
[31] Robins and Shute, *Rhind Mathematical Papyrus* (cit. n. 13), p. 58.

ful as that of the Mesopotamians, but it was, if anything, better than that in which the Greeks wrote their sophisticated formulas. There is no doubt that Egyptian mathematics was based on very simple principles; on the other hand, the existing papyri show that extraordinarily elaborate mathematical structures were erected upon them.

Neugebauer admits that while the Egyptians were not as good in their arithmetic as the Babylonians, their geometry was equally good; and if we are to believe other scholars' interpretations of the Moscow Papyrus, Egyptians were able to carry out geometrical operations that were beyond those of the Mesopotamians. The notion that the Egyptians were the better geometers fits both with their unparalleled architectural achievements and with their reputation among the Greeks as the founders of geometry and their teachers in it.[32]

Given this concern with geometry, it is not surprising that there are many direct and indirect proofs that Egyptians relied on plans for their architectural constructions. Struve may have been exaggerating when he wrote, "The Egyptian plans are as correct as those of modern engineers."[33] Nevertheless, there is no reason to suppose that they were inferior to those of the Greeks and Romans.

According to the Egyptians, the tradition of making plans went back to Imhotep, at the beginning of the third dynasty, circa 3000 B.C., but most modern scholars have understood this claim merely as a mythical projection onto the deified prototype of all architects. However, it is now proven that architectural plans were used during the Old Kingdom and that Imhotep did design the Step Pyramid and the elaborate complex of buildings around it. Furthermore, an ostracon found at the Step Pyramid does contain measurements for a vault.[34]

This coordination of geometry and computation with architecture constitutes the fourth argument against modern denials that the Egyptians possessed a superior mathematics. While the textual evidence for such knowledge can be construed as ambiguous, the case for it is greatly strengthened by the architectural evidence. In addition to the pyramids there were temples, granaries, and irrigation networks on huge scales that required extraordinary planning and the ability to visualize these structures in advance on writing or drawing surfaces.

The fifth reason for supposing that the Egyptians had sophisticated mathematics is that the Greeks said so. Writers on the subject were unanimous that Egyptian mathematics and astronomy were superior to their own and that while only two Greek mathematicians were supposed to have studied in Mesopotamia, the majority of Greek scientists, astronomers, and mathematicians had studied or spent time in Egypt.

These reports are treated with skepticism by modern historians of science, who *know* that there was no Egyptian science or mathematics worth studying. How-

[32] See Herodotos, 2.109; Diodoros Sikeliotes, 1.69.5, 81.3, 94.3; Aristotle, *Metaphysics* 1.1. (981b); Hero, *Geometria* 2; Strabo, 16.2, 24 and 17.1, 3; and Clement of Alexandria, *Stromateis* 1.74.2. See also Cheikh Anta Diop, *Civilization or Barbarism: An Authentic Anthropology*, trans. Yaa-Lengi Meema Ngemi (New York: Lawrence Hill, 1991), pp. 257–258.

[33] Struve, "Mathematischer Papyrus" (cit. n. 25), pp. 163–165, on p. 165.

[34] Sergio Donadoni, "Plan," in *Lexikon der Ägyptologie*, ed. Wolfgang Helck and Eberhard Otto, 5 vols. (Wiesbaden: Harrassowitz, 1977–1984), Vol. 4, cols. 1058–1060. For my dating see Martin Bernal, *Black Athena: The Afroasiatic Roots of Classical Civilization*, Vol. 2: *The Archaeological and Documentary Evidence* (London: Free Association Books; New Brunswick, N.J.: Rutgers Univ. Press, 1991), pp. 206–216.

ever, as de Santillana wrote about Eudoxos, who undoubtedly studied in Egypt: "We are asked to admit, then, that the greatest mathematician of Greece learned Egyptian and tried to work on astronomy in Egypt without realizing that he was wasting his time."[35]

There is little doubt that after the Persian conquests the mathematics and astronomy of Egypt drew from both Egyptian and Mesopotamian sources. However, the Greek belief that it was an Egyptian tradition strengthens the case that the native component was significant.

The sixth argument against the skeptics is the fact that the Greeks adopted an Egyptian rather than a Mesopotamian calendar. Apart from the greater convenience of the Egyptian calendar, this adoption is indicative of what seems to have been a wider Greek tendency to draw from nearby Egypt rather than more distant Mesopotamia.

The final argument is that in Hellenistic times, while Athens remained the center of Greek philosophical studies, nearly all "Greek" science took place in Egypt. This was partly the result of Ptolemaic patronage, but if we are to believe Greek and Roman sources, they also drew and built on Egyptian wisdom. It is striking that Euclid worked in Egypt at the very *beginning* of the Ptolemaic period, that is to say a mere fifty years after Eudoxos had felt the need to learn Egyptian in order to study mathematics and astronomy. Thus, it would seem more accurate to view Euclid's work as a synthesis of Greek and Egyptian geometry than as an imposition of the Greek rational mind on muddled oriental thinking.

While it is true that Babylonian mathematics and astronomy flourished under the Seleucids, as already noted, most of the great "Greek" scientists wrote in Greek but lived in Egypt, and some indeed may have been Egyptian. For example, the astronomer Ptolemy was known in early Arabic writings as ᵓAṣ Ṣaᶜid, "the Upper Egyptian."[36]

It seems to be generally accepted that the great Greek contribution to mathematics and astronomy was the introduction of geometric modeling, in particular the transposition of Mesopotamian arithmetical astronomical cycles into rotating spheres.[37] However, the Greeks themselves believed that geometry developed in Egypt, a view supported by Egyptian architectural sophistication and the mathematical papyri. Furthermore, those most responsible for the Greek view of the heavens as spinning spheres, Plato and Eudoxos, were reported to have spent time in Egypt and were known for their deep admiration of Egyptian wisdom.[38] We have seen how particularly close Eudoxos's association was with Egyptian priests, and it was precisely Eudoxos who established the new astronomy of complex concentric spheres.

I believe that these seven arguments present a very strong case indeed that there were rich mathematical—particularly geometrical—and astronomical traditions in Egypt by the time Greek scholars came in contact with Egyptian learned priests. After the first Persian conquest of Egypt, in the sixth century B.C., Egyp-

[35] De Santillana, "Forgotten Sources" (cit. n. 26), p. 814.

[36] J. F. Weidler, *Historia astronomiae* (Wittenberg: Gottlieb, 1741), p. 177.

[37] Pingree, "Hellenophilia versus the History of Science" (cit. n. 1).

[38] See the bibliography in Whitney Davis, "Plato on Egyptian Art," *Journal of Egyptian Archaeology*, 1979, *66*:121–127, on p. 122, n. 3.

tian mathematics and astronomy were substantially influenced by Mesopotamian "scientific" thought, a process which continued in Ptolemaic and Roman Egypt.[39] The Egyptian medical tradition appears to have been less affected by Mesopotamia. In general, the "scientific" triumphs of Hellenistic Egypt would seem to be the result of propitious social, economic, and political conditions and the meeting of three "scientific" traditions, those of Egypt, Mesopotamia, and Greece. However, the two former were much older than the third, reaching back to the third millennium or beyond, and more substantial. It should also be noted that the point at which the Greeks "plugged into" Near Eastern "science" was Egypt; this was the reason that the Greeks always emphasized the depth and extent of Egyptian wisdom.

The arbitrariness of the application of the word "science" to ancient civilizations was noted at the beginning of this essay. I suppose, like Humpty-Dumpty, we can use words more or less as we please. However, the only way to claim that the Greeks were the first Western scientists is to define "science" as "Greek science." If less circular definitions are used, it is impossible to exclude the practice and theory of some much earlier Mesopotamians *and Egyptians*.

[39] A clear example of this can be seen in the fragment discussed by Neugebauer in his exquisite swan song: Neugebauer, "A Babylonian Lunar Ephemeris from Roman Egypt," in *A Scientific Humanist: Studies in Memory of Abraham Sachs,* ed. E. Leichty *et al.* (Occasional Publications of the Samuel Noah Kramer Fund, 9) (Philadelphia: University Museum, Univ. Pennsylvania, 1988), pp. 301–304.

A New View of Early Greek Astronomy

By Bernard R. Goldstein and Alan C. Bowen

THE STANDARD HISTORY OF GREEK ASTRONOMY emphasizes the role of planetary theory in its earlier stages, by supposing that the early astronomers aimed primarily to explain planetary phenomena. This view derives from a passage in Simplicius (sixth century A.D.), based on the lost *History of Astronomy* by Eudemus (fourth century B.C.), in which Plato is said to have set astronomers the task of saving planetary phenomena by means of uniform circular motions.[1] Many scholars doubt Simplicius's authority and question Plato's putative role in the development of astronomical theory.[2] But their alternative accounts, while none too clear, likewise assume that the early period of Greek astronomy was concerned mainly with planetary motion. Some assign Plato's role to the Pythagoreans who preceded him, following Geminus (first century A.D.), who reports:

> It is presupposed in all astronomy that the sun, the moon, and the five planets move in circular orbits with uniform speed in a direction opposite to that of the universe. For the Pythagoreans, who were the first to apply themselves to investigations of this kind, supposed the motions of the sun, the moon and the five planets to be circular and uniform. . . . For which reason, they put forward the question as to how the phenomena might be explained by means of circular and uniform motions.[3]

In our view, however, planetary theory was not central to early Greek astronomy as Simplicius and Geminus suggest; consequently, interpretations that stress it minimize the true significance of what is universally acknowledged to be a major scientific achievement of the period, Eudoxus' homocentric model for planetary motion.

We contend that Greek astronomy falls into distinct phases, that the earliest

[1] Simplicius, *In libros Aristotelis de caelo Commentaria* 488.18–24, 492.28–493.5, in J. L. Heiberg, ed., *Commentaria in Aristotelem Graeca,* Vol. VII (Berlin: Academia Litterarum Regia Borussica, 1894). Trans. in M. R. Cohen and I. E. Drabkin, *A Source Book in Greek Science* (Cambridge, Mass.: Harvard Univ. Press, 1948), p. 97.

[2] See, e.g., O. Neugebauer, *The Exact Sciences in Antiquity* (2nd ed., 1957; New York: Dover, 1969), p. 152.

[3] Geminus, *Elementa astronomiae* 10.2–21, ed. C. Manitius, (Leipzig: Teubner, 1898) (our trans.). Geminus is followed by B. L. van der Waerden, "The Earliest Form of the Epicycle Theory," *Journal for the History of Astronomy,* 1974, *5*:175–185: cf. van der Waerden, "The Motion of Venus, Mercury and the Sun in Early Greek Astronomy," *Archive for History of Exact Sciences,* 1982, *26*:99–113. On Geminus's date, see Otto Neugebauer, *A History of Ancient Mathematical Astronomy* (hereafter *HAMA*) (New York/Heidelberg/Berlin: Springer-Verlag, 1975), pp. 579–581.

phase extends from the eighth century B.C. to the time of Plato, and that Eudoxus is the starting-point of the second phase. This division is based on philological considerations and on the observation that, in the earlier period, the motives and problems of Greek astronomy proper differed from those of early cosmology and that any scientific concern with planetary motion was characteristic of the later period.

In studying the history of Greek astronomy, one should recognize that the late evidence on which one must rely may be contaminated by later developments. The most important witness to Greek astronomy is Ptolemy's *Almagest,* written in the middle of the second century A.D. Its successful reception occasioned the loss of most of the earlier texts on Greek astronomy, including all those of Hipparchus (second century B.C.) save his commentary on the astronomical poem of Aratus. The earliest surviving works are those of Autolycus and Euclid (both ca. 300 B.C.). For the period prior to Autolycus, and, to a lesser extent, for that between Autolycus and Ptolemy, historians must depend on fragments preserved in diverse ancient and medieval sources. This state of affairs has made writing the history of Greek astronomy extremely difficult and has introduced into it a considerable amount of reconstruction, much of it highly speculative. Moreover, it has encouraged attempts to locate the ideas, models, and parameters of Ptolemy (or, for some, Copernicus) in the theories of his predecessors, as if their goals and those of Ptolemy (or Copernicus) were identical.

The proper way, however, to recover the history of Greek astronomy is to treat the subject in each period on its own terms. Greek astronomy, as the very name implies, began as the organization of the fixed stars into constellations. The purpose was to construct a calendar by correlating dates and weather phenomena with the risings and settings of the fixed stars or constellations.[4] This tradition, which is the framework of Hesiod's *Works and Days,* persists throughout Greek history and is well represented in Ptolemy's *Phaseis.* The astronomical calendar of risings and settings, weather phenomena, and seasons took the form of a *parapêgma,* quite possibly invented by Meton.[5] Ptolemy cites twelve parapegmatists, beginning with Meton, Euctemon, and Democritus in the fifth century and Eudoxus, Callippus, and Philippus in the fourth.[6] In some of these calendars the dates are simply listed in zodiacal months. (The sun stays in each of the twelve zodiacal signs for one month: these months should not be confused with lunar months, which are defined with respect to the first visibility of the new moon.) The stars, moreover, are not assigned coordinates, and there is no allusion to a celestial sphere: no precise measurement was involved in

[4] See P. Tannery, *Recherches sur l'histoire de l'astronomie ancienne* (Paris: Gauthier-Villars, 1893; New York: Arno Press, 1976), pp. 1–25; and H. Diels and W. Kranz, *Die Fragmente der Vorsokratiker* (hereafter DK) (6th ed., Berlin/Zurich: Weidmann, 1951), Ch. 4 (Hesiod), Ch. 6 (Cleostratus of Tenedos).

[5] See Neugebauer, *HAMA,* p. 622; T. L. Heath, *Aristarchus of Samos: The Ancient Copernicus* (Oxford: Clarendon Press, 1913), p. 295, n. 1. On the tradition, see Hesiod, *Opera et dies,* e.g., lines 383–387, 614–623; Plato, *Respublica* 527d1–4, *Symposium* 188a1–b6, *Gorgias* 451c5–9; Hippocrates, *De aeribus aquis locis* 2.14–26; Ptolemy, *Phaseis,* in *Claudii Ptolemaei Opera quae exstant omnia,* ed. J. L. Heiberg (Leipzig: Teubner, 1897–1898), Vol. II: *Opera astronomica minora;* see also Neugebauer, *HAMA,* pp. 926 ff.

[6] In the *Phaseis:* see Neugebauer, *HAMA,* pp. 588, 929.

determining their risings and settings. Finally, there is no mention of planetary motion or eclipses in the *parapêgmata*.

Since the accounts of celestial motions found in the tradition of cosmological speculation, a tradition which was independent of Greek astronomy in its earliest phase, are, as we shall explain, tangential to the strictly scientific treatment of the same phenomena, it appears that all aspects of astronomy prior to Eudoxus concern the calendar and calendaric cycles.[7] For periods less than a day, although it is difficult to ascertain the methods used, a few passages suggest that the water clock, or *klepsydra,* served to measure relatively brief spans of time. Anaxagoras and Empedocles (both fifth century) are credited with a physical explanation of how this device works.[8] There is also in Herodotus (late fifth century) a report that the Greeks learned of the *polos,* the gnomon, and the twelve-fold division of the day from the Babylonians. Since the word *polos* certainly signifies a hemispherical bowl, it follows that the Greeks had a sundial in the fifth century. Whether these items were indeed borrowed from the Babylonians is a question that need not concern us here.[9]

The person largely responsible for turning astronomy into a mathematical science, that is, a deductive mathematical explanation of physical phenomena, was Eudoxus of Cnidus (ca. 390–337 B.C.).[10] His inspiration was drawn from two sources. The first was the science of music, a theory that analyzed an aesthetic phenomenon, melodious sound, by means of whole-number ratios. The invention of this science is usually associated with Pythagoras but given the state of our sources, this is unverifiable. Nevertheless, a Pythagorean science of music was well established in the fifth century, for Archytas of Tarentum, a Pythagorean contemporary of Plato, presents his own harmonic theory in terms of an older and authoritative tradition.[11] As for Eudoxus' knowledge of this science, there are fragments attesting his familiarity with Archytas's work in this domain.[12]

Second, we propose that Eudoxus was influenced by cosmological speculation, particularly that of the Pythagoreans and Plato. For, in their view, the circular motions of the heavenly bodies manifested a *moral* order that was ul-

[7] See Neugebauer, *HAMA,* pp. 588, 619–624; Heath, *Aristarchus,* pp. 130–133.

[8] On Anaxagoras, see Aristotle, *De caelo* 309a19–21; *Physica* 213a22–27 (= DK 68 A68): [Aristotle], *Problemata* 2.16, 914b9–915a25 (= DK 68 A69). On Empedocles, see Aristotle, *De respiratione* 473a15–474a6 (= DK 31 B100).

[9] Herodotus, *Historiae* 2.109; see also Neugebauer, *Exact Sciences,* pp. 81–91. On the *polos,* see W. W. How and J. Wells, *A Commentary on Herodotus* (Oxford: Clarendon Press, 1928), Vol. I, pp. 221–222; and Ernst Maass, *Aratea,* Vol. XII of *Philologische Untersuchungen,* ed. A. Kiessling and U. v. Wilamowitz-Moellendorff (Berlin: Weidmann, 1892), pp. 123–140. On the question of Babylonian origins, see, e.g., B. L. van der Waerden, *Science Awakening, II: The Birth of Astronomy* (New York: Oxford Univ. Press, 1974), pp. 70–71, 285; together with B. R. Goldstein's review in the *Journal of Near Eastern Studies,* 1978, *37*:275–277.

[10] See George de Santillana, "Eudoxus and Plato: A Study in Chronology," *Isis,* 1940, *32*:248–262.

[11] See Heath, *Aristarchus,* pp. 319–320: for correction and discussion of the text Heath translates, see A. C. Bowen, "The Foundations of Early Pythagorean Harmonic Science: Archytas, Fragment 1," *Ancient Philosophy,* 1982, *2*:79–104.

[12] See fragments D63, D64 (= DK 47 A19a) in F. Lasserre, *Die Fragmente des Eudoxos von Knidos* (Berlin: De Gruyter, 1966) (hereafter cited as *Eudoxus* followed by a fragment number); compare Porphyry, *In Ptolemaei harmonica Commentaria* 93.5–17 (= DK 47 B2). Diogenes Laertius, *Vitae philosophorum* 8.8.86, states that Eudoxus learned geometry from Archytas (= *Eudoxus* T7).

timately analyzable by means of the same whole-number ratios as melodious sound. As Aristotle claims:

> Contemporaneously with these thinkers and before them [i.e., the atomists], those called Pythagoreans, who were the first to engage in the sciences, advanced them; and, having been trained in the sciences, they thought that scientific principles were the principles of all things. Since, of these principles, numbers are by nature first, and since they expected to see in numbers many likenesses to things which are and which come to be, more than in fire, earth and water—because such a property of numbers as this is justice and such a property as that is soul and intelligence, and another is opportunity and likewise so to speak for each of the others; and further, since they saw the properties and ratios of *harmonia* in numbers—since, then, all other things appear in their whole nature to resemble numbers and numbers are first in the whole of nature, they supposed the elements of numbers to be the elements of all things and the entire heaven to be a *harmonia* and number.[13]

The Pythagoreans regarded the explanation of the heavenly motions in terms of these ratios as knowledge of the speeds, risings, and settings of the celestial bodies; and Plato called it astronomy. But, though such speculation did relate celestial movement and number, it would be wrong to see in this any attempt at precise measurement of what is observed. The *explanandum* in these theories is not so much a physical phenomenon as the ethical and aesthetic order it supposedly exhibits. Thus, for the Pythagoreans, the numerical analysis belongs to the reduction of all things to number and of number to the *tetraktys* of the decad; whereas, for Plato, it is part of his theory of the virtuous soul and its *harmonia*.[14]

Eudoxus' innovations in astronomy include the use of homocentric spheres to explain planetary motion (for which he is best known today) and an improved sundial, both of which are discussed below. But we maintain that his fundamental contribution was the two-sphere model: an arrangement of two concentric spheres in which the inner sphere represents the earth, and the outer, the orb of the fixed stars. (By "model," we mean the mathematical analogy of the celestial and terrestrial spheres, an analogy which was exploited on the principle that the properties of one must correspond to the properties of the other. We are not here concerned with the ontological and epistemological status of this model.)[15] It is unlikely that any one of the components of this model was novel: the curvature of the celestial vault or its depiction as a concave hemispherical bowl, the sphericity of the cosmos as well as of the earth, and the belief that the fixed stars rotate about the earth in circular orbits are all attested prior to Eudoxus. Moreover, the myth of Er in Book X of Plato's *Republic* and the construction of the world-soul in his *Timaeus* suggest a multisphere model of

[13] Aristotle, *Metaphysica* 985b23–986a3; our trans. from W. D. Ross, *Aristotle's Metaphysics* (rev. ed., Oxford: Clarendon Press, 1953), Vol. II. On Aristotle's reconstruction of Pythagorean thought, see A. C. Bowen, *The Preface to Plato's Dialectic: The Science of Music,* Ch. 5 (forthcoming).

[14] For defense of this account of Pythagorean astronomy and explanation of Plato's assessment of it in the *Republic,* see Bowen, "Archytas"; A. C. Bowen, "Plato on Science and the Sciences," *Ancient Philosophy,* 1983, *3*(2), forthcoming; and Bowen, *Plato's Dialectic,* Ch. 6.

[15] For a different assessment of Eudoxus' contribution to astronomy, see E. Maula, "The Constants of Nature," *Philosophia* (Athens), 1974, *4*:211–246; and Maula, "The Spider in the Sphere: Eudoxus' Arachne," *Philosophia* (Athens), 1975/76, *5/6*:225–257. For an instance of the theoretical revision of the celestial sphere as a consequence of change in the conception of the terrestrial sphere, see B. R. Goldstein, "The Obliquity of the Ecliptic in Ancient Greek Astronomy," *Archives Internationales d'Histoire des Sciences* (forthcoming).

the heavenly motions. But these passages, like those in which the other elements of Eudoxus' scheme are found, are not texts in astronomy: they belong to a tradition of cosmological-moral theory. Eudoxus' originality lay in using the two-sphere model, with its fixed spherical earth and its rotating sky, to account for the risings and settings of stars, to provide a framework for geographical studies, and to justify a more mathematically sophisticated sundial. At the same time, he laid the foundations for the application of geometrical argument to the study of celestial phenomena, an application manifested a generation later in the extant texts of Autolycus and Euclid.

The fullest definition of the two-sphere model (see Figs. 1 and 2) appears in Strabo's *Geography* (first century B.C.):

> Thus we must take as an hypothesis that the heavens have five zones, and that the earth also has five zones, and that the terrestrial zones have the same names as the celestial zones. . . . The limits of the zones can be defined by circles drawn on both sides of the equator [equinoctial circle] and parallel to it, namely, by two circles which enclose the torrid zone, and by two others, following upon these, which form the two temperate zones next to the torrid zone and the two frigid zones next to the temperate zones. Beneath each of the celestial circles falls the corresponding terrestrial circle which bears the same name: and, in like manner, beneath the celestial zone, the terrestrial zone.

The same idea is found in Aristotle's *Meteorology:*

> There are two inhabitable sections of the earth: one near our upper, or northern pole, the other near the other or southern pole; and their shape is like that of a tambourine. If you draw lines from the center of the earth they cut out a drum-shaped figure. The lines form two cones; the base of one is the tropic, of the other the ever-visible circle, their vertex is at the center of the earth. Two other cones towards the south pole give corresponding segments of the earth.[16]

The following passage makes it possible to date the invention of this two-sphere model within rather narrow limits. In the *Meteorology* one reads:

> Again in the archonship of Nicomachus a comet appeared for a few days about the equinoctial circle (this one had not risen in the west), and simultaneously with it there happened the storm in Corinth. That there are few comets and that they appear rarely and outside the tropic circles more than within them is due to the motion of the sun and the stars. For this motion does not only cause the hot principle to be secreted but also dissolves it when it is gathering.[17]

The passage mentions an event in the year 341 B.C.; the equinoctial circle referred to here is the celestial equator. Aristotle speaks of two other comets, but both are prior to this date (427/426 and 373/372 B.C.) and neither is defined with respect to circles in the heavens;[18] we doubt, therefore, that the two-sphere model

[16] Strabo, *Geographica* 2.5.3, from H. L. Jones, ed. and trans., *The Geography of Strabo* (Loeb Classical Library) (Cambridge, Mass.: Harvard Univ. Press, 1917), Vol. I, pp. 425–427 (see Germaine Aujac, "L'image du globe terrestre dans la Grèce ancienne," *Revue d'Histoire des Sciences,* 1974, *27:*193–210); Aristotle, *Meteorologica* 362a32–b5, trans. from E. W. Webster, *Aristotle: Meteorologica,* in W. D. Ross, ed., *The Works of Aristotle,* Vol. III (Oxford: Clarendon Press, 1931).

[17] *Ibid.,* 345a1–8.

[18] *Ibid.,* 343b1–7.

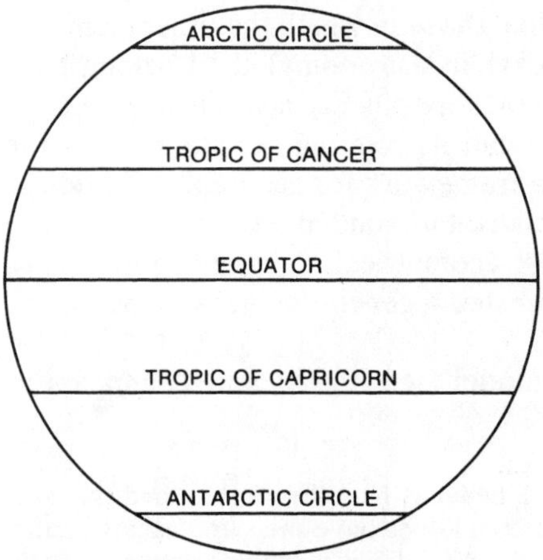

***Figure 1.** The zones of the earth in cross section.*

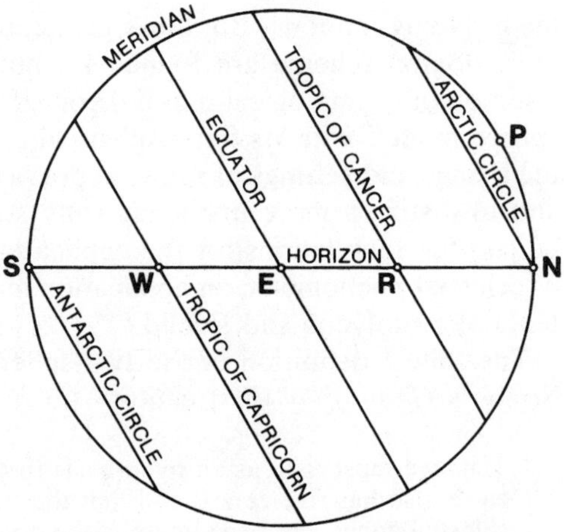

***Figure 2.** The zones of the sky for a latitude in the northern hemisphere in cross section, where* P *is the pole of the equator,* S *the south point and* N *the north point on the horizon.*

was developed until shortly before 341 B.C. It is highly implausible that the reference to the equinoctial circle was added to the observational report at a later date. Moreover, the model can reasonably be dated after 372 B.C., and probably after the composition of Plato's *Timaeus,* which was written sometime between 367 and 347 B.C.[19]

That Eudoxus used the two-sphere model is clear from passages in which he locates on the celestial sphere the tropics, the equator or equinoctial circle, the arctic and antarctic circles—or, as he terms them, the circles of the ever-visible and the ever-invisible stars—according to the fixed stars that lie on them, and defines locations on earth by reference to the tropical circles. Moreover, in three different sources, including Galen's *History of Philosophy* (second century A.D.), we are told: "Eudoxus says that the [Egyptian] priests claim that the flood-waters [of the Nile] occur according to the change of the seasons. When we who live beneath the summer tropical circle have summer, those who live on the opposite side [of the equator] beneath the winter tropical circle then have winter, from which regions the flood-waters break forth."[20]

That Eudoxus actually invented the two-sphere model is supported by Vitruvius's discussion of sundials. According to Vitruvius, Eudoxus or Apollonius

[19] See F. M. Cornford, *Plato's Cosmology: The "Timaeus" of Plato* (London: Routledge & Kegan Paul, 1937), p. 1; H. F. Cherniss, "The Relation of the *Timaeus* to Plato's Later Dialogues," *American Journal of Philology,* 1957, *78*:225–266 (rpt. in *Harold Cherniss: Selected Papers,* ed. L. Tarán [Leiden: Brill, 1977], pp. 298–339).

[20] Galen, *De philosopha historia* 23, in C. G. Kühn, ed., *Claudii Galeni Opera omnia,* Vol. XIX (Leipzig: Teubner, 1830; rpt. Hildesheim: Olms Verlag, 1965) (= H. Diels, *Doxographi Graeci* [henceforth, *DG*], 4th. ed. [Berlin: De Gruyter, 1965], p. 386) (= *Eudoxus* F288). See *Eudoxus* F1–2 (celestial sphere: cf. T14, F3b), F11 and 76 (north pole), F64a–b (ever-visible or arctic circle), F64–65 (summer tropic or Tropic of Cancer), F69 and 71 (equinoctial circle or equator), F72–73 (winter tropic or Tropic of Capricorn), F74 (ever invisible or antarctic circle), F76–78 (solstitial and equinoctial colures). Eudoxus also used this two-sphere system to define geographical latitude (*clima*): see *Eudoxus* F350, F67–68.

invented the *arachnê,* a device which Vitruvius seems to think is a sundial. But we take the *arachnê* (literally, the spider web) to be a network of intersecting lines or curves and, for sundials, this is the projection on the dial surface of the celestial equator and tropics, which are intersected by the lines for the twelve seasonal hours (of which six lie to the east of the meridian and six to the west).[21] Since such lines are found on extant dials of the third century B.C., Vitruvius's alternative suggestion that Apollonius (second century B.C.) may have devised the *arachnê* is not to be given serious consideration.[22] Such a network of lines allows one to tell time with moderate accuracy. But it should not be confused with a coordinate system; indeed, there is no evidence of such a system in early Greek astronomy.

On what sort of sundial did Eudoxus put his *arachnê?* We think a hemispherical bowl most likely. It is "natural" to transfer the equator and the tropics from the celestial vault to a bowl-shaped sundial, which is simply the vault of the sky turned upside down. The day arc of the sun in the sky thus becomes an arc of a circle in the bowl, and the hour lines for sunrise and sunset are well defined.[23] The horizontal plane-dial is, in fact, much more complicated and requires considerable knowledge of conic sections before the appropriate curves can be drawn on it.

According to Vitruvius, Aristarchus (third century B.C.) invented the hemispherical bowl-dial, and there is independent evidence that Eratosthenes (third century B.C.) used one whose rim lay in the horizontal plane.[24] In his list of inventors of sundials, Vitruvius assigns the first sundial, a hemispherical bowl-dial cut for the latitude of its location, to Berosus the Chaldean who, he tells us elsewhere, transmitted Babylonian astrology to the Greeks.[25] Berosus (ca. 300 B.C.) wrote a history of Babylonia in Greek, preserved only in fragments, that concerns the period from creation to Alexander the Great; he is also supposed to have founded a school of astrology on the island of Cos. His scientific activity is reported by Pliny, Seneca, and Vitruvius; and none of their remarks suggests that he had the mathematical ability required to design a cut hemispherical dial. For this reason, we agree with Albert Rehm in mistrusting Vitruvius's claim that Berosus invented this sundial. Moreover, since the *polos* referred to in the passage from Herodotus cited above was a hemispherical bowl and not a plane-dial, we also doubt that the former was invented by Aristarchus

[21] Vitruvius, *De architectura* 9.8.1 (= *Eudoxus* D17), trans. in F. Granger, *Vitruvius: On Architecture* (Loeb Classical Library) (Cambridge, Mass.: Harvard Univ. Press, 1931–1934), Vol. II, p. 255; *arachnê* may mean either spider or spider web; Granger wrongly prefers the former. A seasonal hour is 1/12 of the period of daylight or night (i.e., from sunrise to sunset or from sunset to sunrise); it varies in length from day to day throughout the year (on seasonal hours, see Neugebauer, *Exact Sciences,* pp. 81–91).

[22] See Sharon L. Gibbs, *Greek and Roman Sundials* (New York/London: Yale Univ. Press, 1976), pp. 5, 10: for examples of spherical sundials dating from the third century B.C., see pp. 123, 158.

[23] *Ibid.,* pp. 12–14.

[24] Vitruvius, *De architectura* 9.8.1. On Eratosthenes, see Cleomedes, *De motu corporum caelestium* 94.23–100.23, trans. in I. Thomas, *Greek Mathematical Works* (Loeb Classical Library) (Cambridge, Mass.: Harvard Univ. Press, 1941), Vol. II, pp. 267–273. On the design of this sundial, see Gibbs, *Sundials,* pp. 14–18, 66–71, 122–123.

[25] Vitruvius, *De architectura* 9.6.2 (Granger, *Vitruvius,* Vol. II, p. 247). Here Vitruvius uses the term "astrology" unambiguously to signify a system of divination. The modern discovery of Babylonian texts has led to skepticism concerning the reliability of Greek and Latin reports on the relationship of Hellenistic astrology to its Babylonian predecessors: see Neugebauer, *Exact Sciences,* pp. 187–189; Neugebauer, *HAMA,* pp. 474, 609–610.

and that the cut hemispherical sundial was invented first, and follow Rehm in saying that by "inventor" Vitruvius simply meant the earliest person known by him to have described or used a certain kind of dial.[26]

As to whether some other astronomer who was a contemporary or predecessor of Eudoxus could have devised the two-sphere model and used it in this way, the available evidence suggests not. Both Meton and Democritus might be thought to have the requisite credentials, but there are fatal difficulties in either case. Meton is the earliest authority cited in the *parapêgma* tradition; and he is credited with an observation made on 27 June 432 B.C. of the summer solstice, which served as the starting point for the Greek astronomical calendar. Furthermore, he is considered the discoverer of the nineteen-year calendaric cycle, known as the Metonic cycle, in which nineteen years correspond to 235 synodic months.[27] But, in none of this is there any hint of a model. Even if it is suspected that Meton was influenced by the Babylonians in setting up the nineteen-year cycle and in deciding the date of the summer solstice, a suspicion that has arisen since the recovery of Babylonian astronomical texts in this century, it still remains true that no model is presupposed. Indeed, the Babylonians used an arithmetical scheme to determine the summer solstices in a nineteen-year cycle. This scheme was based, apparently, on the date in their calendar for the heliacal rising of the star Sirius: summer solstice was thought to occur 21 days before this rising.[28] Meton was also understood to be concerned with the division of the year into seasons. In the *Ars Eudoxi,* a papyrus of the second century B.C. that exhibits the state of astronomy in the third century, it is reported that Euctemon, whose accomplishments are not well distinguished from those of Meton in our sources, divided the year into four unequal seasons of 90, 90, 92 and 93 days, beginning with summer.[29] In Simplicius's commentary on Aristotle's *De caelo,* the inequality of the seasons is ascribed to both Meton and Euctemon. (Rehm is probably correct in arguing that Meton and Euctemon simply posited 7 consecutive zodiacal months of 30 days each, and 5 consecutive zodiacal months of 31 days each; that is, that he had no intention of describing the length of the seasons.)[30] The *Ars Eudoxi* also states that Eudoxus and Democritus pre-

[26] See A. Rehm, "*Horologium,*" col. 2419, in A. Pauly, G. Wissowa, W. Kroll, eds., *Real-Encyclopädie der klassischen Altertumswissenschaft* (Stuttgart: Metzler, 1913), Vol. VIII. On Berosus' date, see Neugebauer, *HAMA,* p. 607. For his work, some of which seems more pertinent to astrology than astronomy, see F. Jacoby, *Die Fragmente der Griechischen Historiker* (hereafter *FGrH*) (Leiden: Brill, 1959), Part III.C, Vol. 1, pp. 367–393, No. 680, esp. frags. T5b, T5c, T6, F22a, F16b, F21, F18, F19, F17. The fragments of Berosus' writings are also collected in P. Schnabel, *Berossus und die babylonisch-hellenistische Literatur* (Leipzig: Teubner, 1923).

[27] See Heath, *Aristarchus,* pp. 293–295; Neugebauer, *HAMA,* p. 622; B. R. Goldstein, "A Note on the Metonic Cycle," *Isis,* 1966, *57*:115–116. Theophrastus, in *De signis pluvium, ventorum, tempestatis et serenitas* 4 (= DK 6 A1), states that Meton, an Athenian, learned of this cycle from Phaeinos, a resident alien in Athens who observed the solstices from Mount Lycabettus. Nothing else is known of Phaeinos, but could Babylonian data have been transmitted to Meton through him? As for the observation itself, Meton is supposed to have used a *heliotropium* near the wall of the Pnyx: see Maass, *Aratea,* pp. 13–14; Heath, *Aristarchus,* p. 294, n. 3.

[28] See Neugebauer, *HAMA,* pp. 357–365 (esp. p. 363).

[29] The *Ars Eudoxi* or *Didascalia caelestis* by Leptines is translated into French in Tannery, *Recherches,* pp. 283–294: see p. 294; cf. Neugebauer, *HAMA,* pp. 627–628, 686–687. The same source also reports that Callippus, a successor of Eudoxus, divided the year unequally into seasons. But this division does not conform to that of Euctemon and was presumably based on different considerations.

[30] Simplicius, *In de caelo* 497.17–24 (trans. in Heath, *Aristarchus,* p. 213); Rehm's argument is summarized in Neugebauer, *HAMA,* pp. 628, 1352 and Fig. 3.

ferred to keep the seasons equal, so Meton's proposal for the length of the zodiacal months did not find acceptance among his immediate successors. But, once again, there is no reason here to connect Meton with the two-sphere model.

Democritus flourished at the close of the fifth century and is often associated with Eudoxus in the extant sources. His mathematical achievement, his work in the *parapêgma* tradition, his speculation about the nature and causes of celestial as well as meteorological phenomena, are all preserved in fragmentary form by later writers. But however remarkable his credentials in mathematics, astronomy, and even geography, he did not conceive of the two-sphere model: he believed that the earth was disclike, basically flat but hollowed out slightly in the center.[31]

Eudoxus is, therefore, the most likely figure to have conceived of using a two-sphere model to account for the stellar risings and settings associated with the calendar. It is also possible to reconstruct how Eudoxus might have based his well-known analysis of planetary motion using homocentric spheres on the two-sphere model. His interest in planetary motion was unusual: Plato states that the periods of the planets had been understood by only a very few; and these few, we suggest, were not those astronomers concerned with *parapêgmata* but Pythagorean cosmologists or philosophers.[32]

To extend the two-sphere model so as to account for the sun's motion seen as the set of circular paths of its shadow on the hemispherical sundial, it suffices to observe that this path shifts from day to day and that it is limited by the circles representing the tropics on the dial. Two spheres for the sun can account for this motion: one sphere which rotates once each day, and another whose axis is inclined to it, which completes one revolution in the opposite direction each year. Both spheres rotate uniformly but at different speeds and in contrary directions. This discovery could occasion a general investigation of the behavior of two such pairs of spheres rotating uniformly with axes inclined to one another, where the speeds are either equal or unequal, and the directions, either the same or opposite. Inclined spheres rotating at the same speed but in opposed directions about different poles lead to the hippopede, a path shaped roughly like the figure 8. Eudoxus' next step would be to realize that an additional rotation would allow the hippopede to account for the retrograde arcs that are characteristic of planetary motion.[33]

According to Aristotle, Eudoxus used three spheres for the sun and moon, and four for each of the five planets. In all seven cases, the first sphere was that of the fixed stars, its function being to supply the daily rotation, and the second sphere accounted for motion along the ecliptic. The third sphere for the sun and moon was to take care of their motion in latitude: as Aristotle says,

[31] See Aristotle, *De caelo* 294b13–21; Simplicius, *In de caelo* 511.22–25; Diels, *DG*, p. 377. Democritus is said to have written a geographical survey: see Agathermus, *Hypotyposes geographiae* 1.1 (= *Eudoxus* F273a). Compare Democritus's account of the seasonal flooding of the Nile in Galen, *De philosopha historia* 23 (= Diels, *DG*, p. 385) with Eudoxus' explanation cited above: Democritus makes no reference to a two-sphere model.

[32] Plato, *Timaeus* 39c5–d2; see A. E. Taylor, *A Commentary on Plato's "Timaeus"* (Oxford: Clarendon Press, 1928), ad loc. Plato surely cannot mean the parapegmatists (see *Respublica* 527d1–528a5, 528e6–529c6), and there is no evidence that the Pythagoreans as such ever involved themselves in the construction of *parapêgmata:* see notes 6, 11, and 14 above.

[33] See A. Aaboe, "Scientific Astronomy in Antiquity," *Philosophical Transactions of the Royal Society of London,* 1974, *A.276:*21–42; Neugebauer, *HAMA*, pp. 677–685, 1357–1358.

"the circle in which the moon moves is inclined to the zodiac at a greater angle than that in which the sun moves." That Eudoxus allowed the sun a small variation in latitude is confirmed by Hipparchus: "And that this occurs even Eudoxus admits. Indeed, in his *Enoptron* he declares that the sun appears to vary in its positions at the solstices as well, a variation which is rather obscure and extremely slight."[34] One may suppose that Eudoxus has here accepted a traditional claim about solar motion, and that it did not affect his basic analysis of the apparent motion of the shadow cast by the gnomon on the surface of a hemispherical sundial. Four spheres for the planets were needed to account for their phenomena, including stationary points and retrogradations.

In the period after Eudoxus, the following astronomers and mathematicians studied Eudoxus' model for planetary motions: Menaechmus, who is better known for his work on conic sections; Callippus, who altered Eudoxus' model by adding more spheres; Autolycus, who criticized it for failing to account for the apparent variation in the sizes (i.e., brightnesses) of Mars and Venus, a physical consideration that Eudoxus may have been prepared to ignore; and Polemarchus, a student of Eudoxus.[35] Aristotle had a theory of nested homocentric spheres as well. But of these, Autolycus seems to have been the most influential: there is no evidence of any adherence to homocentric models by ancient astronomers after him. Yet the two-sphere model was never abandoned in antiquity. During the third century, astronomers shifted attention from planetary models to the study of the sizes and distances from the earth of the moon and the sun, of the phases of the moon, and of topics that relate to the general investigation of eclipse phenomena.[36] Interest in stellar risings and settings continued, but it no longer defined the activity of astronomers as before.

By the time of Hipparchus (second century B.C.), Babylonian astronomical data reached the Greek world, although the mode of transmission is unknown. Babylonian astronomy was centuries old and, when it reached the Greeks, it was highly successful at predicting planetary phenomena and lunar eclipses: though they lacked geometrical models, the Babylonians had achieved precise knowledge of planetary and eclipse cycles, and their science was supported with copious quantitative data. The result was a radical transformation of Greek astronomy: Hipparchus not only had access to this data, but also acquired from the Babylonians the notion of quantitative prediction. For he took the Greek models of planetary motion, models whose explanatory function had hitherto been qualitative, and, by using Babylonian data, specified their parameters so as to adapt them for quantitative prediction.[37] This influx of Babylonian data

[34] Aristotle, *Metaphysica* 1073b17–32. *Eudoxus* F63b: cf. F62, F63a. See Neugebauer, *HAMA,* pp. 629–631.

[35] Simplicius, *In de caelo* 504.17–506.3 (trans. except for 505.11–17 in Heath, *Aristarchus,* pp. 221–223). See also G. Aujac et al., *Autolycus de Pitane: La sphère en mouvement, levers et couchers héliaques, testimonia* (Paris: Société Guillaume Budé, 1979), pp. 178–182.

[36] For references to the two-sphere model, in addition to the passages cited above, see those from Posidonius (ca. 100 B.C.) and Ptolemy in Cohen and Drabkin, *Source Book,* pp. 156–181. For later work, see, e.g., Aristarchus's treatise on the sizes and distances of the sun and moon in Heath, *Aristarchus,* pp. 351–414; Alan E. Shapiro, "Archimedes's Measurement of the Sun's Apparent Diameter," *J. Hist. Astron.,* 1975, *6*:75–83; the report on the accounts of lunar phases given by Berosus and Aristarchus in Vitruvius, *De architectura* 9.2.1–4; and, on the treatment of eclipses in the *Ars Eudoxi,* Neugebauer, *HAMA,* p. 688.

[37] See, e.g., Asger Aaboe, "On the Babylonian Origin of Some Hipparchian Parameters," *Centaurus,* 1955, *4*:122–125.

and the resultant transformation of Greek astronomical science effectively obscured the earlier history of this science from later writers. As a consequence, the historian must now exercise care in relying on late testimony in his effort to understand the development of Greek astronomy from the time of Hesiod and Homer to the time when Babylonian science intruded.

The preceding account of the development of early Greek astronomy shows that it may be divided into two phases. The first is defined by the construction of calendars in the form of *parapêgmata*. It is contemporaneous with a tradition of cosmological theorizing in which were treated questions about the nature of the heavenly bodies and their motions. The second begins with Eudoxus' synthesis of these two sorts of activity, and his introduction of homocentric spheres as models to explain the motions of the celestial bodies. Not long thereafter, such models for planetary motion were rejected, and astronomers turned their attention to the study of eclipses, the determination of the sizes and distances of the sun and moon, and the explanation of lunar phases.

Figure 1. *A Babylonian tablet (B.M. 37236) listing undated phases of Mars according to the System A scheme. By permission of the Trustees of the British Museum.*

The Adaptation of Babylonian Methods in Greek Numerical Astronomy

By Alexander Jones

THE DISTINCTION CUSTOMARILY MADE between the two chief astronomical traditions of antiquity is that Greek astronomy was geometrical, whereas Babylonian astronomy was arithmetical. That is to say, the Babylonian astronomers of the last five centuries B.C. devised elaborate combinations of arithmetical sequences to predict the apparent motions of the heavenly bodies, while the Greeks persistently tried to explain the same phenomena by hypothesizing kinematic models compounded out of circular motions. This description is substantially correct so far as it goes, but it conceals a great difference on the Greek side between the methods of, say, Eudoxus in the fourth century B.C. and those of Ptolemy in the second century of our era. Both tried to account for the observed behavior of the stars, sun, moon, and planets by means of combinations of circular motions. But Eudoxus seems to have studied the properties of his models purely through the resources of geometry. The only numerical parameters associated with his concentric spheres in our ancient sources are crude periods of synodic and longitudinal revolution, that is to say, data imposed on the models rather than deduced from them.[1] By contrast, Ptolemy's approach in the *Almagest* is thoroughly numerical.[2] In its theoretical aspect, Ptolemy's astronomy set out to describe the apparent motions of the sun, moon, and planets according to models composed of parts, the dimensions and speeds of which were all calculated from quantitative observations. Its practical aspect expressed the behavior of the models in the form of numerical tables by which the apparent positions and other observable phenomena of the heavenly bodies could be predicted for a specified time.

The research for this article was carried out under a Canada Research Fellowship (Social Sciences and Humanities Research Council of Canada).

[1] Simplicius, *Commentary on Aristotle's De Caelo*, in *Commentaria in Aristotelem Graeca*, Vol. VII: *Simplicii in Aristotelis De caelo Commentaria*, ed. J. L. Heiberg (Berlin, 1894), pp. 493–507. I choose Eudoxus for this comparison because we are relatively well informed by Simplicius about the arrangement of his models, the astronomical phenomena they were intended to explain, and the grounds on which they were first criticized. The same conditions, as I intend to argue elsewhere, probably still applied to the astronomy of Apollonius (ca. 200 B.C.).

[2] G. J. Toomer, *Ptolemy's "Almagest"* (London: Duckworth; New York: Springer, 1984).

The earliest Greek astronomer known to have extensively applied numerical methods to geometrical models is Hipparchus (active ca. 150–125 B.C.).[3] In the *Almagest* Ptolemy passes on much information about how Hipparchus determined the parameters of his solar and lunar models, both when he agrees with Hipparchus's results (as in the measurement of the solar eccentricity and apsidal line) and when he finds fault with them (as in finding the dimensions and periods of the lunar model).[4] Not only do these passages prove the numerical character of Hipparchus's work; but the fact that Ptolemy nowhere discusses comparable measurements by anyone other than Hipparchus is a weighty argument that Hipparchus was the founder of this methodology. This conclusion also finds support in the lack of evidence that any Greek before Hipparchus's time possessed three of the foundations of Greek numerical astronomy: sexagesimal arithmetic, degree measure, and a table of chords for trigonometric computations.[5]

The Mesopotamian origin of the first two of these three resources points to another respect in which Hipparchus differed from his predecessors: his extensive use of elements of Babylonian astronomy. This is not to deny that Babylonian astronomical concepts and conventions, for example the ecliptic and its division into twelve equal zodiacal signs, had entered Greek astronomy from the time of Eudoxus on. But Hipparchus had access to Babylonian data, both observational and theoretical, on a much larger scale and of a much more technical nature than anything that we find in earlier Greek sources. In 1900 F. X. Kugler showed from the recently deciphered astronomical cuneiform texts that the fundamental period relations of Hipparchus's lunar theory, namely his value for the mean synodic month and its ratio to the anomalistic and draconitic months, were parameters taken from the heart of the complex Babylonian System B lunar theory.[6] Significant as it was, this discovery was only the beginning: Babylonian elements have since been recognized in numerous aspects of Hipparchus's astronomical work.[7] At the same time, the study of astrological texts, Greco-Egyptian papyri of the Roman period, and Indian treatises descended from Greek originals has revealed extensive traces of Babylonian predictive schemes, to the degree that it is now evident that many of the most advanced methods of Babylonian

[3] A good general survey of Hipparchus's work is G. J. Toomer, "Hipparchus," in *Dictionary of Scientific Biography,* ed. C. C. Gillispie, 16 vols. (New York: Scribners, 1970–1980) (hereafter *DSB*) Vol. XV, pp. 207–224. For details of the topics discussed here see also section 4 of A. Jones, "Models and Tables of Ancient Astronomy, 200 B.C. to A.D. 300," Part 37.4 of *Aufstieg und Niedergang der Römischen Welt,* ser. II, *Principat* (Berlin/New York: de Gruyter, in press).

[4] Ptolemy, *Almagest* 3.4 (solar model), 4.2, 7, and 9 (lunar periods), and 4.11 (lunar eccentricity).

[5] Degrees and sexagesimal arithmetic first appear in Greek in the *Anaphorikos* of Hypsicles, who wrote a generation or so later than Apollonius, i.e., roughly contemporary with Hipparchus if not a decade or two earlier. See V. De Falco, M. Krause, and O. Neugebauer, *Hypsikles: Die Aufgangszeiten der Gestirne* (Abhandlungen der Akademie der Wissenschaften in Göttingen, Philol.-hist. Kl., 3rd Ser., 62) (Göttingen, 1966); and for the date, G. L. Huxley, "Studies in Greek Astronomers," *Greek, Roman, and Byzantine Studies,* 1963, *4*:83–105. The chord table was probably Hipparchus's invention, as argued by G. J. Toomer, "The Chord Table of Hipparchus and the Early History of Greek Trigonometry," *Centaurus,* 1973, *18*:6–28.

[6] F. X. Kugler, *Die babylonische Mondrechnung* (Freiburg, 1900), pp. 20–40.

[7] Most of the evidence is collected and discussed in G. J. Toomer, "Hipparchus and Babylonian Astronomy," in *A Scientific Humanist: Studies in Memory of Abraham Sachs,* ed. E. Leichty, M. deJ. Ellis, and P. Gerardi (Occasional Publications of the Samuel Noah Kramer Fund, 9) (Philadelphia, 1988), pp. 353–362.

planetary and lunar theory were substantially transmitted into Greek and practiced widely in the Hellenistic world.[8]

How the Greeks came by this knowledge remains controversial, and indeed no single channel of transmission seems adequate to explain the whole range of Babylonian astronomical lore attested in Greek sources. Hipparchus (and through him, Ptolemy) had access to reports of lunar eclipses from Babylon ranging from 747 B.C. to the early fourth century B.C., if not later, as well as Babylonian planetary observations from the second half of the third century B.C. The source of these reports was almost certainly the astronomical archive of Babylon. Significant numbers of observational texts from this archive are extant and are in the course of publication.[9] We can now see that the reports in Ptolemy's *Almagest* represent only a selection from a mass of information so enormous that a translation into Greek of the whole, or even of a large fraction of it, is inconceivable. The only plausible alternative seems to be that some Greek, having considerable technical competence and a specific idea of what sort of observations he wanted, extracted reports from the archive with the collaboration of the astronomers of Babylon. Perhaps this happened more than once, for the Greeks had a long-standing interest in eclipse observations, and the conventions according to which the Babylonian reports in the *Almagest* were translated and dated vary. G. J. Toomer plausibly suggests that Hipparchus himself was responsible for fetching some of the Babylonian observations.[10] That he organized the reports in the published form in which Ptolemy found them is practically certain.[11]

Direct contact with Babylonian astronomers could also explain the Babylonian theoretical parameters in Hipparchus's works. It does not suffice, however, to account for the fact, coming to be fully appreciated only in recent years, that the Babylonian predictive schemes were themselves practiced in Ptolemy's day in substantially unaltered form. The most striking proof of this is a second-century papyrus published by Otto Neugebauer in 1988 that, for all that it uses Greek numerals, is effectively a fragment of a Babylonian System B lunar "ephemeris" in which visibility or eclipse phenomena were computed for a succession of syzygies.[12] This document was a surprise to historians of ancient astronomy, although the shock was perhaps greater than it need have been; for there was already evidence that the rather less complicated Babylonian schemes for predicting planetary phases were used in Roman Egypt.[13]

By what means, then, did the Babylonian predictive schemes pass into the hands of the astrologers of Ptolemy's day? It is at the very least superfluous to

[8] Jones, "Models and Tables" (cit. n. 3), sect. 3.

[9] A. J. Sachs and H. Hunger, *Astronomical Diaries and Related Texts from Babylonia*, 2 vols. to date (Österreichische Akademie der Wissenschaften, Phil.-hist. Kl., 195 and 210) (Vienna, 1988–1989). For the contents of the observational texts see also A. J. Sachs, "A Classification of the Babylonian Astronomical Tablets of the Seleucid Period," *Journal of Cuneiform Studies*, 1948, *2*:271–290; and Sachs, "Babylonian Observational Astronomy," *Philosophical Transactions of the Royal Society of London*, 1974, Series A, *276*:43–50.

[10] Toomer, "Hipparchus and Babylonian Astronomy" (cit. n. 7), pp. 357–360.

[11] Ptolemy, *Almagest* 9.2, refers to an "edition" by Hipparchus of older planetary observations.

[12] O. Neugebauer, "A Babylonian Lunar Ephemeris from Roman Egypt," in Leichty *et al.*, *Scientific Humanist* (cit. n. 7), pp. 301–304. The papyrus is in a private collection.

[13] The evidence is most compelling for Mars and Venus. See B. L. van der Waerden, "Aegyptische Planetenrechnung," *Centaurus*, 1972, *16*:65–91; and A. Jones, "A Second-Century Greek Ephemeris for Venus," *Archives Internationales d'Histoire des Sciences*, in press.

insist that they had to be published in treatises by Greek astronomers before reentering the domain of practical use. One need only assume that scribes trained in the Babylonian astronomical centers occasionally carried their skills elsewhere, and that the methods of astronomical computation that we find in Roman Egypt and still later in India belong to a continuous tradition in which a change of language was an event of little practical significance.[14] Hipparchus and later Greek astronomers could have obtained details of Babylonian theory either in Babylon or from "Chaldeans" residing elsewhere.

Thus two parallel streams of mathematical astronomy emerged in the Hellenistic world during the second century B.C.: an arithmetical astronomy originating in Babylonia and carried on by astrologers, and a numerical-geometrical astronomy founded by Hipparchus and continued by "scientific" astronomers such as Ptolemy. Until quite recently one could have maintained that the two streams maintained separate courses until at last, in late antiquity, the arithmetical approach dried up. Hipparchus, it is true, had assigned to the revolutions in his models certain period relations taken from the Babylonian schemes, but then periodicity was almost the only theoretical assumption shared by the Babylonian schemes and the Greek kinematic models.[15] On a more general level, the very inspiration for Hipparchus's quantification of the models might be traced to his knowledge of what the Babylonians had accomplished.[16] But surely the two traditions, based as they were on fundamentally divergent principles, had little else of substance to offer each other!

However, the assumption that the Greek arithmetical and geometrical traditions developed independently has less foundation in actual documents than in a tacit assumption that the two astronomies had the same scope and the same methodological homogeneity in Hipparchus's time as the original Babylonian systems on the one hand and Ptolemy's system on the other. The harmonious pairing of a theoretical process from observations to quantitative models and a practical process back from the models to predictive schemes or tables is such a prominent feature in the *Almagest* and its modern successors that one is tempted to take it as a pattern for the astronomies of earlier times. It does apply to the period when the Babylonian schemes evolved, probably the fifth and fourth centuries B.C., through the discovery of mathematical models that reproduced the observed distribution of astronomical phenomena.[17] Here theory was the servant of prediction, and we have little reason to believe that the Babylonian astronomers made any important theoretical advances after their computational schemes had attained the more or less canonical forms exhibited in the cuneiform texts of the last three centuries B.C. Once the predictions had attained an accuracy at least as good as that of the observations, there was little motive for deeper investigation. The astronomy that the Hellenistic Greeks received from the hands of the Babylonians was by then more a skill than a science: the quality of the predictions was

[14] In Egyptian astronomical papyri, Greek and Demotic Egyptian are used indiscriminately.

[15] See O. Neugebauer, "Problems and Methods in Babylonian Mathematical Astronomy, Henry Norris Russell Lecture, 1967," *Astronomical Journal*, 1967, *72*:964–972, on p. 972.

[16] Toomer, "Hipparchus and Babylonian Astronomy" (cit. n. 7), p. 361.

[17] For the principles involved see A. Aaboe, "Observation and Theory in Babylonian Astronomy," *Centaurus*, 1980, *24*:14–35.

proverbial, but in all likelihood the practitioners knew little or nothing of the origins of their schemes in theory and observations.

Meanwhile, Greek geometrical astronomy had developed as a "philosophical" science devoted to reconciling eternally valid theoretical models with continually verified observational facts. The date of a particular appearance of Venus as morning star, the magnitude of a particular eclipse, the longitude of Jupiter on a particular date were events the prediction of which possessed no scientific interest. This state of affairs changed with Hipparchus, but less than one might expect.

Hipparchus's use of small numbers of carefully selected observations to determine the parameters of his models had no precedent, so far as we know, in either Greek or Babylonian astronomy. But the resemblance of these procedures to those of Ptolemy in the *Almagest* should not seduce us into thinking that Hipparchus and Ptolemy had the same goals. Ptolemy carefully designed the *Almagest* to give the deduction of his models an apparently ineluctable logical sequence. The solar theory is established first, and purely from observations of solstices and equinoxes; he next works out the lunar theory using the solar theory, then the star catalogue and theory of precession using the solar and lunar theories, and finally the planetary theory using all the preceding work. The plan of each part is itself determined by a chain of cumulative dependence. Ptolemy's treatment of the solar and preliminary lunar models in *Almagest* 3–4 is characteristic. He must first get at least an approximation of the periods of uniform circular motion in his model from a selection of observations over long intervals. Second, he has to measure the ratio of radii of the circular components by applying these mean motions to a selection of observations over short intervals. Third, he determines the instantaneous configuration of the model for an epoch date, from which all subsequent configurations can be extrapolated using the mean motions. Last, he computes tables by which the apparent position of the body can be predicted for a given date. Each step depends on the steps that preceded. If any subsequent corrections are made to the parameters (as happens with the lunar mean motions), Ptolemy takes care to show that his initial assumption of less accurate data introduced no significant errors.

The order of deduction in the *Almagest* is logical but not necessarily chronological, and Ptolemy makes no claim that he actually worked out the stages of his system in exactly the order that he presents. In fact, the notorious solstices and equinoxes that he adduces to confirm the length of the tropical year and the long-term constancy of the lengths of the seasons in *Almagest* 3.1 were allegedly observed in A.D. 139/140, so that the observations on which his solar theory (and therefore the whole system) rests are among the very latest he recorded.[18] What is essential is that the subordinate theories could not be *published* until Ptolemy had empirically justified all the data assumed in their derivation.

Although Hipparchus solved many of the individual problems of establishing parameters of the solar and lunar models in essentially the same way as Ptolemy,

[18] Ptolemy records his own observations from A.D. 127 to 141. For my argument the genuineness of Ptolemy's solstice and equinox observations is irrelevant; this vexed topic is expertly investigated by J. P. Britton, "On the Quality of Solar and Lunar Observations and Parameters in Ptolemy's *Almagest*" (Ph.D. diss., Yale Univ., 1967).

Table 1. Hipparchus's writings on solar and lunar theory, and their counterparts in the *Almagest*

	Almagest	Hipparchus
Solar theory		
Length of the year	3.1	"On the Shifting of the Solsticial and Equinoctial Points," "On the Length of the Year"
Eccentricity and apogee	3.4	Determination of eccentricity and apogee
Epoch and tables	3.2, 6–8	No known work
Lunar theory		
Periods of mean motion	4.2–3	Confirmation of Babylonian System B period relations
Size of epicycle	4.6	Measurement using eccentric model Measurement using epicyclic model
Epoch and tables	4.4, 8, 10	No known work

his researches manifestly had no such broad plan. In the first place, Hipparchus did not write a single comprehensive treatise on astronomy; his contributions to the theory of the sun's and moon's motions were made in several monographs, each devoted to a specific topic. Table 1 correlates the relevant chapters of Books 3 and 4 of the *Almagest* with Hipparchus's writings, identified either by title (where this is known) or by content. The chronology of these monographs is not certain, but it is possible to deduce an approximate sequence from their relation to each other and to the Hipparchian observations reported by Ptolemy.[19] The work on the solar model, corresponding to *Almagest* 3.4, was published probably not long after the vernal equinox of 146 B.C. but before the first measurement of the lunar anomaly, and the second measurement of the lunar anomaly was made before 128 B.C. The works on the length of the year and precession, however, were written later than the vernal equinox of 128 B.C. This means that Hipparchus did not establish the primary element of the solar theory, the period of the sun's mean motion, until near the end of his career, and hence his published researches on the lunar theory cannot have relied on a completely worked out solar model.

Another glaring difference between the Hipparchian and Ptolemaic programs is the apparent absence of predictive tables in Hipparchus's work. Only one author, the astrologer Vettius Valens (late second century), speaks of Hipparchian tables for the sun, and lunar tables published by Hipparchus are nowhere mentioned.[20] If the solar tables that Vettius Valens knew were truly Hipparchus's, and not a later fabrication based perhaps on Hipparchus's solar theory, the point still remains that Hipparchus could scarcely have composed them when he wrote his treatises on the lunar theory; for at this stage he had no confirmed value for any of the fundamental periods (the sidereal, tropical, and anomalistic years) of the solar model.

[19] For the details see Jones, "Models and Tables," sect. 4; and Toomer, "Hipparchus" (both cit. n. 3).

[20] Vettius Valens, *Vettii Valentis Antiocheni Anthologiarum Libri Novem*, ed. D. Pingree (Leipzig: Teubner, 1986), p. 339. In the context it seems unlikely that Valens would have omitted a reference to Hipparchian lunar tables if he had known of any.

But if Hipparchus did not make tables, how did he find the computed positions of the sun and moon that he needed for his theoretical investigations? Considering how generous Ptolemy is with information about other aspects of Hipparchus's theoretical work, he is oddly taciturn whenever reference has to be made to Hipparchus's methods of calculating these positions. I suspect, indeed, that Hipparchus often merely stated the results without discussing how he got them. Only through recent analyses of certain passages of the *Almagest* has the surprising fact emerged that Hipparchus used Babylonian arithmetical schemes to compute both solar and lunar longitudes.[21]

The evidence for Hipparchus's method of computing solar positions comes from Ptolemy's account of Hipparchus's measurements of the ratio of radii between the circles in his lunar model, in which the initial data were the intervals of time and longitude separating three observed lunar eclipses. The intervals of time could be deduced from the observation reports, but the longitudes had to be calculated because parallax interferes with the observed positions. When Ptolemy adopts Hipparchus's procedure in *Almagest* 4.6, he uses his already established solar theory and tables to compute the sun's longitude for the three times of mid-eclipse; the moon's longitude is of course exactly 180° from the sun's. Ptolemy elsewhere (*Almagest* 4.11) quotes the intervals of longitude that Hipparchus assumed in his two measurements of the lunar radial ratio, and it is easy to confirm that these numbers cannot have resulted from a correct trigonometric calculation based on Hipparchus's solar model, with or without tables. The Babylonian System A lunar scheme (with some slight modifications to be discussed presently) furnished him with a much easier approach. In this scheme the synodic arc, that is, the progress in longitude from one opposition (or conjunction) to the next, is determined solely by the sun's longitude at the former date: the ecliptic is divided into two zones, and a constant synodic arc is associated with each zone. Given an initial solar longitude for the first eclipse (possibly estimated from proximity to a solstice or equinox), Hipparchus was able to find the longitudes for the other two eclipses by simple arithmetic.

For Hipparchus's lunar computations, we must turn to the measurements of positions of fixed stars that he is known to have made at various stages of his career. Two conditions determined how he had to find stellar positions: he could easily observe a star's location relative to the moon when the moon passes nearby, but unfortunately the lunar theory was still too insecure to permit an accurate direct computation of the moon's own position at the time of the observation. In exceptional instances, Hipparchus was able to sight the relative positions of the moon and stars during lunar eclipses. For such observations the solar longitude, and hence also the true lunar longitude, could be estimated from the interval of time between the eclipse and the nearest equinox or solstice; the lunar position would then be corrected for parallax. But Ptolemy describes in *Almagest* 7.2 a more general procedure that could be used on any night. During the day before or after the stellar observation, one observes the apparent elongation be-

[21] A. Jones, "Hipparchus's Computations of Solar Longitudes," *Journal for the History of Astronomy,* 1991, *22*:101–125; and Jones, "The Development and Transmission of 248-Day Schemes for Lunar Motion in Ancient Astronomy," *Archive for History of Exact Sciences,* 1983, *29*:1–36, on pp. 23–27.

tween the sun and moon. The solar longitude itself is either calculated or measured using an instrument such as Ptolemy's armillary sphere. The apparent lunar longitude resulting from adding the observed elongation to the sun's longitude must then be corrected for parallax to obtain the moon's true longitude at the time of the diurnal sighting. To this one adds (or subtracts) the product of the moon's current daily motion in longitude by the fraction of a day separating the diurnal and nocturnal sightings to get the true longitude at the time of nocturnal sighting, which must be corrected again for parallax to obtain the apparent position. Since only a few degrees in this final result are contributed by the lunar theory, the error due to the imperfections of this theory can be counted on to be small.

Although Ptolemy does not tell us so, it is very probable that this procedure too goes back to Hipparchus. Ptolemy actually quotes (*Almagest* 4.3 and 5) three diurnal observations of the elongation between the sun and moon that Hipparchus made in 128/127 B.C. Because Ptolemy uses these observations to measure the second anomaly of the moon, modern historians have suggested that Hipparchus too suspected the existence of a second anomaly and was attempting to investigate it.[22] However, it is doubtful whether Hipparchus had actually refined his lunar model to the point where he could have detected the second anomaly. What confirms that these three sightings were associated with *stellar* observations is the fact that Ptolemy has preserved (unnecessarily, for his own purposes) Hipparchus's citation of the moon's daily motion for one of the sightings.[23] The way that this is expressed is critical. Hipparchus writes that the moon's "course" (*δρόμος*) was the 241st, an expression that long eluded explanation until Toomer recognized that it referred to the index number of the day in question in a table of daily motion over a 248-day lunar anomalistic period.[24] This table originally formed part of a Babylonian arithmetical scheme for predicting lunar longitudes for a sequence of consecutive days, associated with the lunar System B.[25] The exact pattern of the Babylonian daily motion table (a linear zigzag function) is described by Geminus in the first century, while adaptations of it are well attested in the Roman period. Without further information it seems safest to assume that Hipparchus used the unmodified Babylonian table.

The debt of Hipparchus's theoretical researches to Babylonian astronomy thus extends beyond his use of Babylonian observations and period relations, justifiable from Ptolemy's point of view, to a reliance on predictive schemes that were not derived from his assumed theoretical models. For Ptolemy such practices would have been utterly irreconcilable with the apodictic and didactic purpose of the *Almagest*. Hipparchus apparently was satisfied to have recourse to the already existing schemes on the grounds that they yielded sufficiently accurate

[22] E.g., O. Neugebauer, *A History of Ancient Mathematical Astronomy* (Berlin: Springer, 1975), p. 309; and Toomer, *Ptolemy's "Almagest"* (cit. n. 2), p. 217 n. 2.

[23] The only dated Hipparchian stellar observation in the *Almagest* is one of Regulus in 128 B.C. (*Almagest* 7.2), and this is also the year that Ptolemy uses for the epoch of Hipparchus's stellar observations in general. G. Grasshoff, *The History of Ptolemy's Star Catalogue* (New York: Springer, 1990), p. 154, concludes on statistical grounds that the bulk of Hipparchus's stellar observations were made about ten years earlier, in the early 130s.

[24] Toomer, *Ptolemy's "Almagest,"* p. 224 n. 14.

[25] Jones, "248-Day Schemes" (cit. n. 21), pp. 2–11.

predictions (or so he believed). For his immediate needs, correct numbers were all that mattered.

Hipparchus put the Babylonian schemes to use in still another way. In the absence of direct observations, he could resort to the schemes to argue for the existence or nonexistence of various astronomical phenomena. It would not be surprising if the prominence of the zodiacal anomaly in the Babylonian planetary schemes provided Hipparchus with ammunition in his polemic against the Greek planetary models hypothesized in his time, since by Ptolemy's account (*Almagest* 9.2) these models generated a constant synodic anomaly. The eclipse schemes were still more useful because they analyzed the solar and lunar motions into factors that could be interpreted in the light of geometrical models. We have one elegant instance of this principle in Hipparchus's extant *Commentary on Aratus,* where he uses it to refute Eudoxus's hypothesis of a solar motion in latitude.[26] Hipparchus argues that if the sun exhibited a noticeable latitudinal deviation from the ecliptic, then the earth's shadow would also deviate from the ecliptic. Yet, he goes on, the eclipse predictions of the "astrologers" (ἀστρολόγοι) assume no such deviation, and lead to calculated eclipse magnitudes in error by never more than two digits, which is insignificant. At the date he was writing, the only methods of prediction that could fit this description were the Babylonian schemes.

If the arithmetical methods offered themselves as ready-made substitutes for missing components in the new Hipparchian astronomy, the geometrical theory might in turn be the source of improvements in schemes that had been cut off from their empirical origins. Probably no Greek astronomer had the faintest notion of why the particular structures that occurred in the Babylonian schemes had been chosen, but this was no obstacle to tinkering superficially with the numerical parameters associated with these structures. The phenomenon can already be seen in Hipparchus's calculations of the solar longitudes for the eclipses used in his two measurements of the lunar radial ratio.[27] When making his first measurement he was not content to take over the rules of the System A lunar scheme as he found it. Without altering the constant synodic arcs associated with the two zones of the ecliptic, he changed the definition of the zones so that they were exactly equal and centered on the apsidal line that he had determined theoretically for his solar model. Making the zones equal was a mistake, because it introduced a systematic error in the predicted longitudes that was large enough to throw his first measurement of the lunar model seriously off. For his second measurement, Hipparchus also changed the synodic arcs, probably to conform with the maximum and minimum apparent velocities resulting from the eccentricity in his solar model. This correction brought about a considerable improvement in the accuracy of the ensuing calculations. If Hipparchus did eventually publish solar tables for the prediction of the sun's longitude on given dates, they could have used a similar zone structure with the constant increments scaled to correspond to days instead of synodic months.[28]

[26] Hipparchus, *Hipparchi in Arati et Eudoxi Phenomena Commentarium,* ed. K. Manitius (Leipzig: Teubner, 1894), pp. 88–90.

[27] Jones, "Hipparchus's Computations" (cit. n. 21).

[28] A pre-Ptolemaic solar table has now turned up in an unpublished papyrus, *P. Oxy. Inv.* 32 4B.1/E(1-2) a. A zone scheme much more refined than Hipparchus's two-zone scheme apparently underlies the table, but the associated speeds are derived from Hipparchus's model.

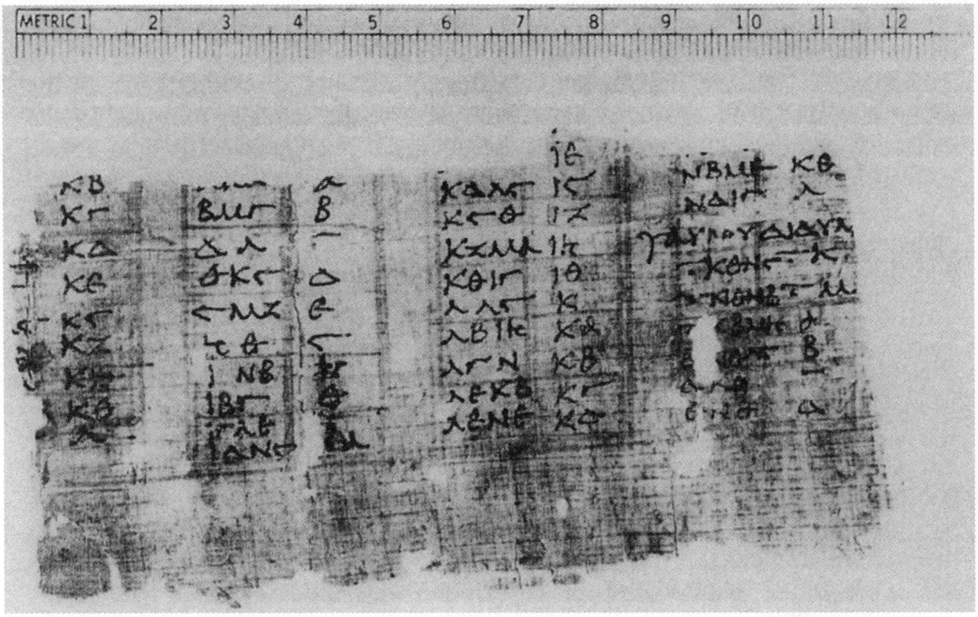

Figure 2. Part of an auxiliary table (P. Mich. 151) belonging to a Greek adaptation of the System A Mars scheme. By permission of the Department of Rare Books and Special Collections, University of Michigan Library.

A papyrus dating from the third century of our era, *P. Heid. Inv.* 4144 + *P. Mich.* 151 (cf. Figure 2), testifies to the existence of a scheme for Mars related to the Babylonian System A scheme in much the same way as Hipparchus's solar schemes were related to their Babylonian ancestor.[29] Like other schemes from Greco-Egyptian papyri that will be discussed below, this scheme cannot be dated precisely or attributed to any particular astronomer, although we are obviously dealing with developments after Hipparchus but uninfluenced by Ptolemy. Mars's pronounced orbital eccentricity results in a very uneven distribution of occurrences of the planet's phases over the ecliptic. The Babylonian scheme reproduced this distribution very successfully by dividing the ecliptic into six equal zones, with associated synodic arcs varying from a minimum of 30° to an ideal maximum of 90°. Through B. L. van der Waerden's analysis of the data for Mars in Egyptian astronomical "Almanacs," we know that the original Babylonian scheme, employing these synodic arcs, was known in Roman Egypt.[30] In this papyrus, however, a new set of values for the synodic arcs are associated with the six Babylonian zones, exhibiting a much smaller range of variation, from $40\frac{3}{4}°$ to 57°. The modified scheme generates such poor predictions of the longitudes of Mars's phases that it clearly cannot have been based on, or even confronted with, systematic observations of the phases. On the other hand, the numbers agree closely with the synodic arcs that one would get from a geometrical

[29] A. Jones, "Babylonian and Greek Astronomy in a Papyrus Concerning Mars," *Centaurus*, 1991, *34* (in press). Earlier discussions: O. Neugebauer, "A New Greek Astronomical Table (P. Heid. Inv. 4144 + P. Mich. 151)," *Historisk-filosofiske Meddelelser: Danske Videnskabernes Selskab*, 1960, *39*(1); and Neugebauer, *History of Ancient Mathematical Astronomy* (cit. n. 22), pp. 946–948.

[30] Van der Waerden, "Aegyptische Planetenrechnung" (cit. n. 13), pp. 77–87.

model in which Mars travels on an epicycle revolving around an eccentric deferent, with eccentricity of $\frac{1}{10}$ or less. The astronomer who devised the papyrus scheme, I conclude, had settled on this excessively small eccentricity for Mars's orbit, used it to calculate synodic arcs for six initial longitudes at 60° intervals, and imposed these values on the System A zones.

The Babylonian schemes were directed primarily at the prediction of planetary phases, eclipses, and other lunar phenomena connected with syzygies, but even before their transmission into Greek they had been extended to address the problems of predicting day-to-day positions and the dates when a planet entered a new zodiacal sign. The growth of Hellenistic horoscopic astrology undoubtedly encouraged the development of more handy, if not always more accurate, methods. Daily motion in Babylonian astronomy was characteristically handled by interpolation between phases. The simplest hypothesis, linear motion, gave rise to "velocity schemes" that broke up a planet's synodic period into a sequence of intervals of constant speed; the patterns varied through the ecliptic in order to account for the effects of zodiacal anomaly. There is little evidence for velocity schemes in the extant Greek documents, but they do turn up in early Indian texts that depend on Greek sources.[31] In most cases the Indian velocity schemes are intended to apply throughout the ecliptic; thus all attempt to describe the zodiacal anomaly is abandoned in the interest of simplicity. In other respects these schemes seem to have been modeled on Babylonian patterns without interference from Greek kinematic theory; a possible exception is an elaborate (and fragmentarily preserved) velocity scheme for Venus that, as van der Waerden has shown, gives a good approximation of epicyclic motion.[32]

A more refined treatment of daily motion attested in cuneiform texts is to interpolate between phases using higher-order difference sequences, that is, to assume that the daily motion, or even the daily *change* in motion, increases or decreases by constant differences.[33] Bridging the interval in time and longitude between two given phases with a second- or third-order sequence of terminating sexagesimal numerals is not a trivial problem, and the two surviving examples, for Jupiter and Mercury, may be scribal exercises rather than specimens of normal Babylonian astronomical practice. Nevertheless we have two second-century papyri, *P. Dem. Carlsberg* 32 and *P.S.I.* 1492, that show similar methods applied to the daily motion of Mercury and Saturn.[34] Algebraic difficulties are eliminated in these "templates" by tabulating a single standard mean synodic cycle composed of stretches of second-order and linear motion. The template was used in conjunction with a list of epoch longitudes corresponding to the beginning of each synodic period; one obtained the longitude for the day in question by reading off from the template the progress in longitude corresponding to the number of days

[31] For summaries and references see D. Pingree, "History of Mathematical Astronomy in India," in *DSB*, Vol. XV, pp. 533–633, on pp. 540–542.

[32] B. L. van der Waerden, "The Motion of Venus in Greek, Egyptian and Indian Texts," *Centaurus*, 1988, *31*:105–113.

[33] P. Huber, "Zur täglichen Bewegung des Jupiter nach babylonischen Texten," *Zeitschrift für Assyriologie*, 1957, N.S., *18*:265–303.

[34] O. Neugebauer and R. Parker, *Egyptian Astronomical Texts*, 3 vols. (Providence, R.I.: Brown Univ. Press, 1969), Vol. III, pp. 240–241 and pl. 79B; and A. Jones, "A Greek Saturn Table," *Centaurus*, 1984, *27*:311–317.

since epoch, and adding it to the epoch longitude. There may have been a separate correction for zodiacal anomaly, but we have no evidence for it. The numerical patterns of the two preserved templates seem to be uninfluenced by geometrical models.

New theoretical considerations did find their way into a similar template scheme for lunar motion.[35] This lunar scheme seems to have been very popular during the first three centuries of our era, and enough relevant texts survive to permit an almost complete reconstruction. The template is a descendant of the Babylonian table of lunar daily motion over a 248-day anomalistic period, which we have already met in connection with Hipparchus. In both the Greek scheme and its Babylonian ancestor the lunar daily motion is described as a linear zigzag function, alternately increasing and decreasing by constant differences between fixed maxima and minima. The assumed periods of longitude and anomaly are more accurate in the template scheme than in the Babylonian version, and the difference between maximum and minimum daily motion has been reduced to about $\frac{5}{7}$ of the Babylonian value. The last change results in a pattern of longitudinal motion that closely fits a simple epicyclic model with a radial ratio of about $5\frac{1}{3}$:60, which is very near Ptolemy's ratio of the epicycle radius, $5\frac{1}{4}$:60. This suggests that the Babylonian function was adapted to bring it into accord with a post-Hipparchian model correctly describing the moon's longitudes at syzygies (when the second anomaly vanishes), that is, a model based on eclipse observations.

Entirely without precedent in the Babylonian sources is the treatment of lunar latitude in the Greek template scheme. The epoch list and the template tabulate not only the moon's progress in longitude but also its argument of latitude, that is, its elongation from the northern limit of its orbit. The coordinated variation in the argument of latitude and the daily motion shows that the inventors of the template scheme were fully aware of the interdependence of lunar anomaly and latitude, an effect that has no counterpart in the Babylonian lunar schemes but follows directly from consideration of a kinematic model. From an obscure passage in Pliny's *Natural History* (2.68–76) it appears that planetary template schemes existed in the first century that described latitudinal motion in a similar way; the extant papyrus templates concern only longitudinal motion.[36]

The pre-Ptolemaic astronomy that emerges from the study of the fragmentary contemporary documents is a strange symbiosis of arithmetical and geometrical methods that will be sure to disconcert anyone who expects to find a record of steady scientific progress from the Babylonians through Hipparchus to Ptolemy. Theoretical work operated with geometrical hypotheses but did not disdain to employ initial data derived from arithmetical schemes; the predictive schemes, for their part, readily absorbed new parameters from geometrical theory without casting off their arithmetical structures. For Hipparchus the availability of the Babylonian predictive methods was, on the whole, a boon. It must be remembered that, unlike Ptolemy, Hipparchus could not use his Greek predecessors' work as a first approximation in determining the parameters of his models; without the arithmetical schemes he would have had to invent new ways of dealing

[35] Jones, "248-Day Schemes" (cit. n. 21), pp. 14–23.
[36] A. Jones, "Pliny on the Planetary Cycles," *Phoenix*, 1991, *45* (in press).

with subsidiary problems that were only incidental to the topics of his researches. On the other side, the efforts of Hellenistic astronomers to accommodate Babylonian schemes to their models at first gave valuable experience in astronomical table making; the lunar template scheme is an especially successful fusion of Babylonian and Greek concepts that well deserved its wide reception in antiquity. The attempts to tamper with the planetary schemes were less happy and mark a step backward, both in accuracy of prediction and in theoretical sophistication, from the Babylonian originals.

During this same period, methods of astronomical prediction were developed that actually described, rather than merely being fitted to, geometrical models, using mean motions and equation tables calculated by the correct trigonometrical formulas. These methods are known to us primarily through the Indian astronomical traditions, which are practically silent about the empirical basis of the parameters of their models. To date, comparable tables have not turned up in any papyrus; but an astronomical inscription discovered at Keskinto on Rhodes, probably dating from about 100 B.C., evinces the use of huge compound periods of planetary anomaly characteristic of the Indian systems.[37] Perhaps the "aeon-tables" mentioned in various ancient sources belonged to this tradition.[38] The Indian tables already approach Ptolemy's ideal of "exhibiting the uniform circular motion,"[39] but Ptolemy seems to have been the first to understand the importance of working out the entire system of the heavenly bodies according to a logical deductive plan and a consistent methodology. The *Almagest* was so successful in this respect that it is only with much difficulty that we are coming to appreciate its originality.

[37] Neugebauer, *History of Ancient Mathematical Astronomy* (cit. n. 22), pp. 698–705.

[38] B. L. van der Waerden, "Ewige Tafeln," in *Arithmos-Arrythmos: Skizzen aus der Wissenschaftsgeschichte: Festschrift für Joachim Fleckenstein zum 65. Geburtstag,* ed. K. Figala and E. H. Berninger (Munich: Minerva, 1979), pp. 285–293; and Toomer, *Ptolemy's "Almagest"* (cit. n. 2), p. 422 n. 12.

[39] The phrase is quoted from *Almagest* 9.2; Ptolemy describes the concept as one of his goals in tabular presentation near the end of 3.1.

Greek Medicine as Science and Craft[§]

By Owsei Temkin

IF asked to give a sketch of the development of Greek medicine that could count on general acceptance, I should offer the following outline: Little is known of Greek medicine before the fifth century B.C.; the earliest medical writings that have come down to us are represented by the so-called works of Hippocrates, although this collection is not the work of one man. Its individual books originated approximately between 450 and 350 B.C. They deal with internal diseases, surgery, and gynecology; they offer rational physiological as well as pathological theories often at discord with one another; clinical observations, with emphasis on the prognostic significance of signs, are found side by side with a schematic cataloguing of diseases. Surgical therapy of dislocations and fractures is often quite modern, the materia medica well developed, but where internal diseases are concerned, the dietetic treatment stands in the foreground. The medical literature of the subsequent 400 years is lost; events during this period have to be reconstructed from fragmentary remains and the accounts of later authors. Yet during this period, some decisive changes took place in Alexandria and Rome. The great scientist physicians of Alexandria in the early third century B.C., notably Herophilus and Erasistratus, consolidated the dogmatic school with its stress on research and theory of disease. But the dogmatic school was rivalled by the new empirical sect that banished the search for hidden causes and referred the physician to experience (his own and that of others), supplemented by analogical reasoning in unprecedented cases. After Greek medicine had established a foothold in Rome, it again gave rise to a new sect fitting the new environment. The methodists opposed the dogmatists' scientific theories as well as the empiricists' trust in experience. Instead, they stressed what is common in diseases, laxity and tenseness of the body as a whole or its parts and periods in their course, and advocated a methodical treatment by astringents and laxatives, and a metasyncritic cycle in chronic conditions.

The split into sects is brought to a certain standstill towards the turn of the first to the second Christian centuries. The available literature is more abundant: the herbal of Dioscorides, some works of Rufus of Ephesus and of the methodist Soranus; the book of Aretaeus the Cappadocian may also hail from this time. Then, of course, with the second century A.D., the wealth of the Galenic writings allows us an insight into the medical conditions of this last scientifically creative century. Galen himself appears as a careful anatomist, although monkeys and other animals are the sole material for his dissections. His pretended physiology of man is a synthesis of animal anatomy and experiment, and speculation, and it is constructed from a rigidly teleological point of view. As a pathologist he is the self-confessed heir of Hippocrates whom he interprets as the author of the system of the four humors, blood, phlegm, yellow and black bile, embodying pairs of the contrasting qualities warm-cold and moist-dry. Galen's therapy although rich in dietetic and pharmacological detail is yet dogmatically systematized to correspond to his humoral pathology. With Galen ancient medicine has reached its turning point. What follows is on the one hand, encyclopedias and shorter treatises serving practical purposes, and on the other, com-

§ Paper (modified) read at the joint session of the American Historical Association and the History of Science Society at Washington, D. C., 28 December, 1952.

mentaries explaining the classical authors, which means Hippocrates and Galen, for the use of the schools. In this encyclopedic and scholastic form, Greek medicine is bequeathed to the peoples of the East.

As it stands, the sketch might have the approval of many scholars not only of today, but even of fifty years ago, since it has carefully avoided any controversial points. Its chief weakness is its almost exclusive attention to Greek medicine as it found literary expression. True, the books are by far the main sources on which we have to rely, but did Greek medicine as a whole really follow the development mirrored by the books? Is it not conceivable that as a craft Greek medicine remained unchanged during the thousand years our sketch encompassed? In short, what is the relationship between science and craft in Greek medicine?

A tendency to distinguish two types of medical practitioners is discernible in Hippocratic times, for we find designations for medical men which hint at some difference of status.[1] Aristotle flatly claims that "'physician' means both the ordinary practitioner, and the master of the craft, and thirdly, the man who has studied medicine as part of his general education."[2] This statement is not surprising if taken as an evaluation. We need only think of the interest in medical matters evinced from pre-Socratic philosophers down to laymen like Celsus to recognize Aristotle's third type of physician. Likewise, the physician Diocles of Carystus who, as Werner Jaeger[3] has shown, was associated with Aristotle and indebted to his philosophy, obviously appeared as a master of his craft who could explain what he did, if compared with the ordinary practitioners.

But Aristotle's terminology does not necessarily indicate a distinction between two social classes of professional physicians. Here we may refer to a passage in Plato's *Laws* which has received attention.[4] Plato contrasts the free-born doctor who deals chiefly with "the ailments of free men" with those practitioners, often slaves themselves, who find their patients among the slaves. Slave doctors are, of course, a known institution throughout antiquity. Since highly educated persons could become slaves, there is no reason to assume that physicians belonging to this social class were *eo ipso* of inferior skill or knowledge. Varro states that farmers near a city or near large estates "prefer to have in their neighborhood men whose services they can call upon under a yearly contract — physicians, fullers, and smiths — rather than to have such men of their own on the farm; for sometimes the death of one artisan wipes out the profit of a farm."[5] The "medicus" on the farm obviously would be a slave; the one hired from the neighborhood might be slave or free. In either event the social difference would hardly be decisive for any difference in quality.[6] Nor was it altogether decisive for Plato, according to whom those who deal with the ailments of free men do so "by investigating them from the commencement and according to the course of

[1] See chs. 1 and 7 of *On Ancient Medicine* and the edition by A.-J. Festugière, *Hippocrate: L'ancienne médecine*, Paris, C. Klincksieck, p. 27 f., notes 6 and 7.

[2] Aristotle, *Politics* III, 6; 1282 a (translation by H. Rackham, Loeb Classical Library, p. 227).

[3] Werner Jaeger, *Diokles von Karystos*, Die griechische Medizin und die Schule des Aristoteles, Berlin, De Gruyter, 1938.

[4] Plato, *Laws*, IV; 720 (translation by R. G. Bury, Loeb Classical Library, vol. 1, pp. 307–309), also IX; 857, C–D; Werner Jaeger, *Paideia*, 3 vols., New York, 1939–44; see vol. 3, p. 215; Ludwig Edelstein, The Rôle of Eryximachus in Plato's Symposium, *Transactions of the American Philological Association*, 1945, vol. 76, p. 99; O. Temkin, in *Social Medicine, Its Derivations and Objectives*, ed. Iago Galdston, New York, Commonwealth Fund, 1949, p. 4 f.; and T. A. Sinclair, Class Distinction in Medical Practice: a Piece of Ancient Evidence, *Bull. Hist. Med.*, 1951, vol. 25, p. 386 ff.

[5] Varro, *On Agriculture*, I, 16, 4 (translation by W. D. Hooper, revised by H. B. Ash, Loeb Classical Library, p. 221; adapted).

[6] Joseph Schumacher, *Die naturphilosophischen Grundlagen der Medizin in der griechischen Antike*, Berlin, De Gruyter, 1940, p. 238.

nature," while the others "whether they be free-born or slaves, acquire their art under the direction of their masters, by observation and practice and not by the study of nature — which is the way in which the free-born doctors have learnt the art themselves and in which they instruct their own disciples." [7] Whatever else Plato may have in mind here, one of the distinctions he establishes is the study from nature by one type of physician, and learning by rote by another. It is possible that illiteracy was also found among the latter type. On the whole, however, the great number of medical books written in antiquity hints at an equally great number of physician authors as well as a large reading public.[8] When Galen says that "the majority of those nowadays pursuing medicine or philosophy cannot even read well," [9] the association of physicians with philosophers is consoling and indicates that the statement must not be taken too literally. It is somewhat reminiscent of the remark supposedly made by the head of the medical school of Harvard when President Eliot proposed written examinations for the M.D. degree: "I had to tell him that he knew nothing about the quality of Harvard medical students. More than half of them can barely write." [10]

Let us then say that there were two types of medical men in antiquity and, for convenience's sake, let us designate them as physician and leech. After all, our term physician is derived from physicus, the student of nature, and the name seems appropriate for those who study diseases according to nature, as Plato says. We shall, however, beware of thinking of physician and leech as two exclusive classes. We must admit the possibility of a large intermediary group, just as we must admit the possibility of the existence of good leeches and bad physicians. Our distinction is made with an eye on the nature of the physician's science and the leech's craft.

Now it has to be said that our knowledge of the content of Greek medical science has not changed considerably during the last thirty years. That is to say, the comparison between our knowledge and that of the Greeks in anatomy, physiology, pathology, and materia medica was made rather thoroughly by the older medical historians, and we have added details only. Nor can there be any doubt that many modern scientific methods, taken as such, were known to Greek physicians. Heidel has shown this very well for the Hippocratic collection.[11] Cohen's and Drabkin's source book [12] demonstrates it for the whole period. I would even go so far as to say that some ancient works contain valuable and still unexploited information from the modern point of view. Thus I doubt whether we possess a description of the muscular system of the Barbary ape which, in fullness of detail, can compare with that in Galen's "Anatomical Administrations."

Nevertheless, we are no longer able to speak of Greek medical science with the assurance of previous generations. This uncertainty is not due so much to considerations of content as of form and purpose. Obviously, Greek medicine was somehow different from modern medicine. We can name the disciplines and their methods which make modern medicine scientific, but can we do the same for antiquity? The answer to this question depends very much on our judgment of the Hippocratic writings. Even at the end of Greek civilization they were still classical textbooks. To deny their scientific nature means to deny that Greek medicine had a scientific orientation throughout its history.

[7] See above, footnote 4.

[8] I. E. Drabkin, On Medical Education in Greece and Rome, *Bull. Hist. Med.*; 1944, vol. 15, p. 350.

[9] Galen, De libris propriis, *Scripta minora*, vol. 2, ed. I. Mueller, Lipsiae, Teubner, 1891, p. 91.

[10] Quoted from Esther Lucile Brown, *Physicians and Medical Care*, New York, Russell Sage Foundation, 1937, p. 17.

[11] William Arthur Heidel, *Hippocratic Medicine, Its Spirit and Method*, New York, Columbia University Press, 1941.

[12] M. R. Cohen and I. E. Drabkin, *A Source Book in Greek Science*, New York, McGraw-Hill, 1948.

The hostility of many Hippocratic works to magic has always been considered a factor in favor of their scientific spirit. The rationalism of the treatise *On the Sacred Disease* with its condemnation of supernatural explanations is shared by many of the other works. And since the appearance of Edelstein's *Asclepius*[13] we have also obtained a better understanding of the relationship of this rationalism to religion. There is neither competition nor enmity between the god and the physician. On the contrary, the Asclepieion takes care of many cases with which the physician cannot cope. Although it is probable that Asclepius, before becoming a healing god, was the patron saint of the medical men, this relationship did not affect the lay character of Greek medicine.

Rationalism is a necessary but insufficient condition of science. In the quest for an additional feature, a passage in Plato's *Phaedrus* defining the method of Hippocrates himself has long offered a ready answer. Socrates asks: "Now do you think that one can acquire any appreciable knowledge of the nature of the soul without knowing the nature of the whole?" And Phaedrus replies: "If Hippocrates the Asclepiad is to be trusted, one cannot know the nature of the body, either, except by this method."[14] "This method" refers to "knowing the nature of the whole." And if by "the whole" is meant the universe, then Hippocrates thought that the understanding of the body presupposed an understanding of surrounding nature. All that remained was to identify this method in those Hippocratic writings in which it was used in order to find both science and the genuine works of Hippocrates.[15] It is an example of very much depending on very little, viz. the interpretation of the words "the whole." If the words "the whole" are understood differently, we arrive at a meaning, embodied in Jowett's translation, where Phaedrus' answer reads: "Hippocrates the Asclepiad says that the nature even of the body can only be understood as a whole."[16] The interpretation of this passage is a matter of dispute,[17] but the tendency to regard genuine Hippocratic science as interested in a study of the surrounding universe was, and still is, predominant. According to Nestle,[18] the physician's practical preoccupation with prognostic signs impressed upon him the idea of the regularity of nature and led to an investigation of causal relationships. Hippocrates, in particular, assumed that man must be considered as under the influence of surrounding nature, and elaborated this view theoretically and as applied to individual and mass diseases. Having rejected all speculative elements, medicine became an independent science and conceived of itself "as a superpersonal, spiritual unit, a work of many individuals and generations in a process of continuous change, correction, and perfection."[19] Medicine

[13] Emma J. Edelstein and Ludwig Edelstein, *Asclepius*, a Collection and Interpretation of the Testimonies, 2 vols., Baltimore, The Johns Hopkins Press, 1945. See especially vol. 2, pp. 53 ff., 139, 169.

[14] Plato, *Phaedrus*, 270 C (translation by H. N. Fowler, Loeb Classical Library, p. 549, adapted).

[15] Attempts in this direction go back at least as far as Galen; see W. H. S. Jones, *Philosophy and Medicine in Ancient Greece*, with an Edition of Peri archaies iatrikes, Baltimore, The Johns Hopkins Press, 1946, p. 16 ff. (*Supplements to the Bulletin of the History of Medicine*, no. 8).

[16] *The Works of Plato*, selected and edited by Irwin Edman, Modern Library, p. 317. The interpretation of "the whole" as referring to the body has been given in detail by Ludwig Edelstein, *Peri aeron und die Sammlung der hippokratischen Schriften*, Berlin, Weidmann, 1931 (*Problemata*, Heft 4), pp. 118–135.

[17] Karl Deichgräber, *Die Epidemien und das Corpus Hippocraticum*, Berlin, Akademie der Wissenschaften, 1933, p. 151, in argumentation against Edelstein has tried to uphold "the whole" as meaning the universe. For a very detailed discussion of the Phaedrus passage see P. Kucharski, La 'Méthode d'Hippocrate' dans le Phèdre, *Revue des Études Grecques*, 1939, t. 52, pp. 301–357. Edelstein has maintained his interpretation against Deichgräber and others in the article "Hippocrates," Pauly-Wissowa, *Real-Encyclopädie*, etc., Supplementbd. VI, 1935, col. 1318 ff.

[18] Wilhelm Nestle, Hippocratica, *Hermes*, 1938, vol. 73, pp. 1–38.

[19] *Ibid.*, p. 38.

shared the method of strict etiological research with history, and the spiritual father of this method was the philosopher Democritus.[20]

In this view, etiological research appears as the essentially scientific element of Hippocratic medicine. The counterpart to this form of scientific medicine is represented by the treatise *On Ancient Medicine*. The book has interested historians of Greek culture, because its author starts out with a sophistic theory of the origin of civilization.[21] Man raised himself from a brutish condition when, by trial and error, he learned to prepare food more suitable to himself. Likewise by trial and error he learned that sick people needed a diet different from that of the healthy. Thus medicine arose and with it the only promising method for further medical research. All attempts to base medicine on empty hypotheses as used in talk about heaven and the regions below are futile.

Some forty years ago, Diels [22] called the author a positivist and pointed out that he paid attention to individual cases rather than general laws. This is not quite correct. In the first part of the treatise, its author argues from a therapeutic point of view and sees the perfection of medicine in an ever-increasing approximation of treatment to the manifold reactions of the human body.[23] In the last part, however, he interprets these reactions as founded in the quality of humors and the structure of organs.[24] He seems to think that by reasoning backwards from ascertained therapeutic facts, a science of nature will be possible. For this reason Gomperz,[25] as Littré before him, believed to have found here the truly Hippocratic method alluded to by Plato.[26]

The attack upon mere hypotheses and the high value set upon trial and error in the first part, have made the treatise appear a kind of "Introduction to experimental medicine." And since today the experimental approach in science enjoys aristocratic privileges, so to speak, it is not surprising that attention has focussed upon *Ancient Medicine* which in the course of a few years has been the subject of three monographs.[27]

These two examples show the difficulty of identifying Greek medicine with any of the forms of modern scientific medicine. They will also have prepared us for Edelstein's attempt at separating medicine and science in ancient Greece. Edelstein proceeded from an analysis of the role of prognosis in Hippocratic medicine. In his opinion this role has to be understood from the social position of the physician as a craftsman not backed by a degree or license.[28] Apart from its psychological significance,[29] correct prognosis was the means of acquiring a reputation and of warding off suspicion in

[20] *Ibid.*, p. 31. The role of Democritus (next to Empedocles) for the etiological research of Greek physicians has been stressed by Hans Diller, *Wanderarzt und Aitiologe*, Leipzig, 1934 (*Philologus*, Supplementband 26, Heft 3); while Max Pohlenz, *Hippokrates und die Begründung der wissenschaftlichen Medizin*, Berlin, De Gruyter, 1938, pp. 26 and 62, tries to make Democritus dependent upon the physicians. See also Diller's review of Pohlenz in *Gnomon*, 1942, vol. 18, pp. 65–88.

[21] For details see Felix Heinimann, *Nomos und Physis*, Basel, Reinhardt, 1945 (*Schweizerische Beiträge zur Altertumswissenschaft*, H. 1).

[22] H. Diels, Über einen neuen Versuch, die Echtheit einiger Hippokratischen Schriften nachzuweisen, *Sitzungsberichte d. königl. Preussischen Akademie d. Wissensch.* Philos.-hist. Classe, 1910, vol. 53, p. 1141, footnote 1.

[23] See ch. 9 and Walter Müri, in *Hermes*, 1936, vol. 71, p. 468, and Jones, *op. cit.*, p. 73. Edelstein, Empirie und Skepsis in der Lehre der griechischen Empirikerschule, *Quellen und Studien zur Geschichte der Naturwissenschaften und der Medizin*, 1933, Bd. 3, Heft 4, p. 49, refers this to the individual reactions of each case.

[24] See ch. 19 and, especially, chs. 22–24.

[25] Theodor Gomperz, Die hippokratische Frage und der Ausgangspunkt ihrer Lösung, reprinted in *Hellenika*, 2. Bd., Leipzig, 1912, pp. 324–354; see p. 325.

[26] *Ibid.*, p. 330, where Gomperz asserts his belief that Plato, in the *Phaedrus*, when speaking of Hippocrates, had in mind the author of *Ancient Medicine*.

[27] Festugière, *op. cit.* (see above note 1), W. H. S. Jones, *Philosophy and Medicine in Ancient Greece* (see above, note 15), and Hermann Wanner, *Studien zu Peri archaies iatrikes*, Diss. Zürich, Zürich, 1939.

[28] Edelstein, *Peri aeron* etc., chs. 2 and 3.

[29] *Ibid.*, p. 65 f.

fatal cases.[30] The interest and knowledge of the physician were practical, not scientific.[31] If he talked [32] and wrote about the nature of man, of health and disease, he borrowed his concepts from contemporary, i.e. pre-Socratic, philosophy, and this relationship between medicine and philosophy remained basically unchanged.[33] Medicine was the practical art, acquired by apprenticeship and practiced "in accordance with traditional views and usages." [34] The sects were philosophical in character. The dogmatists were adherents of Aristotle and the Stoics, the empiricists followed the Academy in their denial that nature was understandable, the methodists accepted the scepticism of Pyrrho.[35] The dogmatists were connected with Aristotelian thought and thus it was in developing Aristotle's arguments that the Alexandrian physicians turned to the study of human anatomy.[36] But at no time before Galen did Greek medicine incorporate dissection and experimentation as indispensable parts; even Galen did not consider all his anatomical and physiological research essential for the physician *qua* professional healer.[37]

Edelstein's views as recently summarized [38] have a profound bearing upon the whole relationship between Greek medicine and culture. We have now two almost diametrically opposed views. On the one hand, Greek physicians are considered as the great scientific teachers, to whom even the philosophers are greatly indebted and whom they quote from the time of Plato. This claim is denied by Edelstein who thinks that scientifically the doctors were the recipients and that the philosophers quoted them mainly as an example. People were willing to subject their bodies to the orders of the doctor, unpleasant as these orders often were. They should be willing to accept the commands of the philosophers who treated their souls.

I believe that Edelstein's work has provided a new basis for our understanding of Greek medicine and that its fruitfulness is demonstrated not least by the questions it permits the historian of medicine to ask. One of them is concerned with a closer scrutiny of the practitioner's role in this development. We shall attempt to draw a rough picture of ancient leechcraft and then ask ourselves for the possible factors that promised to make the leech a better practitioner by turning him philosopher.

Immediately we find ourselves embarrassed by the scanty information about ancient leechcraft. Jones [39] proposed to think of it in the likeness of a medieval guild, which refers to the master-apprentice relationship rather than to any social organization about which we know next to nothing. The question then before us is: what kind of medicine did the master teach and the apprentice learn? Or to simplify even further: what did the leech treat and how did he treat it?

---

[30] *Ibid.*, pp. 81 and 88.

[31] *Ibid.*, p. 93.

[32] *Ibid.*, p. 104 f.

[33] L. Edelstein, The Relation of Ancient Philosophy to Medicine, *Bull. Hist. Med.*, 1952, vol. 26, pp. 299–316, especially p. 304, where it is said (with reference to Aristotle, *De respiratione* 480 b 26 ff. and *De sensu*, 436 a 19 ff.): "physicians concerned with biology and physiology . . . took their departure from philosophy, just as truly systematic natural philosophy ended in a consideration of the principles of life, of health and disease, and of the constituents of the body."

[34] *Ibid.*, p. 301; see also Edelstein, *Peri aeron*, p. 92 f.

[35] Edelstein, Empirie und Skepsis, *op. cit.*, see also his article "Methodiker" in Pauly-Wissowa, *Real-Encyclopädie*, Supplementbd. VI, 1935, col. 367 ff.

[36] Edelstein, Die Geschichte der Sektion in der Antike, *Quellen u. Studien zur Geschichte d. Naturwissenschaften und der Medizin*, Bd. 3, Heft 2, Berlin, 1932, see p. 95 ff.; see also The Development of Greek Anatomy, *Bull. Hist. Med.*, 1935, vol. 3, pp. 235–248.

[37] Edelstein, The Relation of Ancient Philosophy to Medicine, *loc. cit.*, p. 304. G. Lebouque, Une anatomie antique du coeur humain, *Revue des Études Grecques*, 1944, t. 77, p. 7–40, believes that Philistion dissected human bodies and refers to the Hippocratic *On the Heart* which he thinks renders a lecture by Philistion.

[38] In Edelstein's last quoted article, *Bull. Hist. Med.*, 1952, vol. 26, pp. 299–316.

[39] W. H. S. Jones, *The Medical Writings of Anonymus Londinensis*, Cambridge, University Press, 1947, p. 161.

In the *Republic*, Plato says: "A carpenter, when he is sick expects his physician to give him a drug which will operate as an emetic on the disease, or to get rid of it by purging or the use of cautery or the knife." [40] Plato quotes this carpenter as an example of a patient who cannot afford complicated treatment. His physician may well be a leech and since drugs are mentioned, we may look for our answer to an herbal, perhaps the *Materia medica* of Dioscorides of the first century A.D. It is a large herbal, but then we want to get a picture of a good leech, not necessarily an ignorant one.[41] Opening the book at random and leaving out some botanical and pharmaceutical details we read that the caper-plant "disturbes the belly, is good for the stomach and causes thirst, but is easier on the stomach when eaten cooked than raw. Its fruit makes the spleen shrink . . . it also moves the urine and causes bloody discharge, helps in sciatica and paralysis . . . and against ruptures and convulsions, likewise it moves the menses and purges away phlegm, and stops tooth ache. . . . The dry bark of the root is suitable for the aforesaid (conditions), and cleanses all chronic, dirty, and indurated sores. With barley meal it is used as a poultice in those suffering from the spleen, if chewed it helps the tooth ache, finely ground and mixed with vinegar it wipes off leprous spots and vitiligo." [42]

Most of this is immediately understandable. The drug is useful in a number of complaints, most of them symptoms or syndromes with which we too are acquainted. Of others, especially the "leprous spots and vitiligo" we are not quite sure, but we can assume that the Greek leech had learned to diagnose them. The reference to shrinking of the spleen indicates that he knew where the spleen is and how to palpate it. In view of the prevalence of malarial fevers this is not surprising. Most of the symptoms mentioned like sores, paralysis, tooth ache, can easily be described without reference to complicated or theoretical concepts. We meet with more serious difficulties when reading about "sufferers from the spleen," which may refer to a disease entity, and "purging away phlegm," where we are confronted with a concept of humoral pathology. Disregarding the latter example for a moment, and generalizing somewhat broadly, we may say that the leech was acquainted with some disease entities and a large number of symptoms. The diseases are indicated by names like peripneumonia, pleurisy, phthisis, phrenitis, epilepsy, melancholy, many of which seem of popular origin.[43] We meet them in the Hippocratic collection as well as in later works. The leech can recognize diseases and symptoms and has remedies for them, partly drugs, partly simple dietetic measures; possibly he has also learned to take into consideration the strength, age, and sex of his patient. He treats internal as well as surgical complaints, relying on some anatomical knowledge, and an accumulated store of manual skills. This knowledge is passed on from teacher to pupil; it can be increased by contact with other physicians or by accidental observations.

Nothing prevents us from imagining Greek leechcraft being thus practiced at all times of Greek history with the addition of little else. But we must also realize that there are medical limitations to this type of practice. The limitations are those of any technique without theory. The leech may know very much, but his knowledge will always consist of a certain number of skills. Beyond this number he cannot go, that is to say he cannot deal with what is unfamiliar to him and he cannot individualize

[40] *Republic*, book III, ch. 15; 406 D (P. Shorey's translation, Loeb Classical Library, vol. 1, p. 275).

[41] I have purposely chosen a late and sophisticated work, just because I want to show that even here one can still imagine the existence of a core of practical knowledge relatively little affected by the course of time except for the broadened coverage of drugs.

[42] Dioscorides, *Materia medica*, II, 173; ed. M. Wellmann, vol. 1, Berlin, Weidmann, 1907, p. 241.

[43] Henry E. Sigerist, *Antike Heilkunde*, München, Heimeran, 1927, p. 20.

his treatment. And while he may go in quest of new recipes and manipulations, he has no principle of research. Now we have had an indication that the leech was not quite without the help of some theoretical concepts; for we heard Dioscorides recommend the caper-plant as purging phlegm. Similar concepts, related to theory and yet not dependent on special theories, existed much earlier as we learn from the case histories in the first book of the so-called Hippocratic *Epidemics*. In the description of symptoms in individual patients, terms occur which are used like sense data although they transcend the realm of mere description. For instance, discharges are described as "thin, bilious, biting, scanty, black," [44] etc., or as "uncocted as it were." [45] The reader is supposed to know what marks them as "bilious" or "uncocted." But the former term also refers to a humor and the latter involves the analogy between the process of digestion and cooking. If it is said that a woman having a sore throat showed a persistent "acrid, smarting and salt flux," [46] this means a matter sensibly defined by these attributes, but it also refers to this matter as flowing. Besides, the reader is expected to have some anatomical knowledge, and be able to understand the note that a patient "at first suffered in the cardia and the right hypochondrium." [47] Similar concepts are used in other Hippocratic writings in the manner of a terminology common among all medical men at that time.[48] It is quite possible that one or the other term is of relatively late origin; [49] nevertheless it seems likely to me that some of them at least belong to a previous archaic stage. Just as some names of diseases, they may even reflect popular usage. This assumption would explain how the author of *Epidemics I* can describe the "flux" in the woman with the sore throat, how the author of *Ancient Medicine* can speak of fluxes "that turn to the throat," [50] and how Herodotus can relate the measures the Lybians take to protect their children from a phlegm "flowing down" from the head [51] — and how all three can count on immediate understanding. Such an assumption would relieve us of unproductive attempts at tracing medical comments in Greek writers to the few medical books that happened to come down to us.

With this broad picture in mind of a leechcraft as the core of the healing art providing a certain uniformity of practice, we ask why some leeches became physicians. In antiquity, medicine was a profession, philosophy was not — unless the philosopher became a teacher. This means that the philosopher might belong to any profession,

---

[44] Hippocrates, *Epidemics I*, case 14 (translation, adapted, by W. H. S. Jones, Loeb Classical Library, vol. 1, p. 211).

[45] *Ibid*, case 2 (p. 191).

[46] *Ibid*., case 5 (p. 197).

[47] *Ibid*., case 4 (p. 193). L. von Brunn, Hippokrates und die meteorologische Medizin, *Gesnerus*, 1946, vol. 3, pp. 151–173, and 1947, vol. 4, pp. 1–18 and 65–85 has shown how supposedly empirical data of the *Epidemics* are imbued with theoretical notions, and how the firm belief in a definite order leads the Hippocratic author to doubtful generalizations. In the first book of the *Epidemics* to which I have confined myself (for reasons stated below, n. 73) the evidence is not as striking as in the third book. Schumacher, *op. cit*., has given a very detailed interpretation of Hippocratic and pre-Hippocratic medicine in concepts of natural philosophy.

[48] Deichgräber, *Die Epidemien* etc., pp. 40–51, sketches the physiological and etiological system of the author of the *Epidemics*. While it is quite possible that the author consciously held these tenets, he did not deem it necessary to explain his system and obviously counted on the understanding of his readers. The close relationship with *Prognostics* has also been noted by Deichgräber; *Prognostics* speaks a similar language and also takes understanding for granted.

[49] Plato, in the *Republic*, III, 14; 405 D refers to a medical neologism. He blames people who because of slothful living are full of winds and fluxes and thus "compel the clever Asclepiads to give to diseases the names of gases (physai) and catarrhs." To which is added the comment that these are indeed "new and absurd disease names." From this passage it evinces that "flux" was in common usage, whereas to label diseases "gases" (physai) or "catarrh" was considered new.

[50] *On Ancient Medicine*, ch. 19.

[51] Herodotus, IV, 187; the description as "flowing down" must not be identified with Plato's criticism of catarrh as a disease name.

and the intellectual adventure of pre-Socratic philosophy may have attracted leeches among others. The Hippocratic author who argued that all diseases, including the sacred disease, were natural and that magic had no place in their cause or treatment may well have been a leech stimulated by Democritus as Diller thinks.[52] The competitive character of Greek medicine certainly induced many leeches to borrow from the philosophers in order to construct medical theories and systems which could form the content of speeches to impress patients and whole audiences.[53] After Plato and Aristotle had declared a mere craftsman inferior to the master who could give reasons for what he did,[54] social ambition was strong enough to make a leech appear in the philosopher's dress. All these possibilities have to be conceded. But it does not follow that philosophy and science were mere epiphenomena to true leechcraft. I believe that there were medical reasons combined with social incentives that induced the leech to become a physician and that made it imperative, later on, for every physician to belong to a medical sect. These reasons I find in the rise of dietetic medicine which Edelstein has so well described,[55] and the individualization of Greek life.[56]

The rise of the dietetic physician who regulates the whole life of his patient seems to fall within the fifth century. The ancients themselves thought so; [57] Plato ascribes his emergence to Herodicus of Selymbria,[58] who from a gymnastic trainer turned into a physician. Without limiting ourselves to a single name, we can yet see a connection between the gymnast who builds up the athlete and the physician who protects the health of man and cures him by similarly regulating food and drink, work, rest, and sleep.[59] The dietetic physician who believes he knows "how to cause in men by regimen moist or dry, hot or cold," believes himself able to cure even the sacred disease [60] once he has learned from its course and causes the right moment at which to interfere with the propitious and inimical factors. Dietetic medicine is mainly concerned with the treatment of internal diseases, those which affect the whole man. There are physicians who identify medicine with dietetic medicine. It is a curious thing that the author of *Ancient Medicine* who tries to explain the development of medicine, makes no references to pharmacology and surgery at all. Asclepiades who, in the first century B.C., laid down a doctrine which was eventually to lead to the methodist sect "dispensed with the use of [medicaments] for the most part," as we are informed.[61] On the other hand, many dogmatists and the empiricists set great store by drugs. These examples suffice to show that dietetics was the great dynamic element within Greek medicine. In the dietetic treatment of diseases radical differences are sometimes noticeable; the methodists, in particular, vary widely from the other two sects. And among the dogmatists and empiricists, individual differences appear which cannot merely be laid down to difference in training and experience, but in which theory plays a guiding

---

[52] Hans Diller, *Wanderarzt und Aitiologe*, p. 111 ff.

[53] See above, n. 32.

[54] Plato, *Gorgias*, 465 ff., Aristotle, *Metaphysics* I, 2.

[55] Ludwig Edelstein, Antike Diätetik, *Die Antike*, 1931, vol. 7, pp. 255–270; and his description of the dietetic physician in *Peri aeron* etc. and, *ibid.*, p. 114, the comments on individualization.

[56] Max Neuburger, *History of Medicine*, translated by Ernest Playfair, vol. 1, London, 1910, p. 133, says of the Hippocratic physician that he found his goal in the sick individual "as Socrates individualised abstractions, as Thucydides placed personality in the foreground."

[57] Menon begins his account of dietetic etiologies of disease with Euryphon (Jones, *The Medical Writings of Anonymus Londinensis*, p. 32).

[58] Plato, *Republic*, III, 14; 406 E.

[59] On the competition between physician and gymnast see Ludwig Englert, *Untersuchungen zu Galens Schrift Thrasybulos*, Leipzig, Barth, 1929 (*Studien z. Geschichte d. Medizin*, Heft 18).

[60] Hippocrates, *On the Sacred Disease*, last chapter (translation by Jones, Loeb Classical Library, vol. 2, p. 183).

[61] Celsus, *De medicina*, V, preface (translation by W. G. Spencer, Loeb Classical Library, vol. 2, p. 3).

role. Greek medicine became split into sects not over preferences for one or the other drug or one or the other surgical technique, but over different explanations of diseases [62] or the very need for such explanations.[63]

As we remarked before, theory is a guide to the unknown [64] and allows individualization in treatment. Now there is no more individualized form of therapy than the minute regulation of a man's life in all its details.[65] And in the fifth century B.C. Greek life became increasingly individualistic,[66] especially with the advent of the sophistic movement when the flow of medical literature too set in. Plato's free doctor treats his patients by "learning himself from the sufferers and imparting instruction to them . . . and gives no prescription until he has gained the patient's consent" [67] etc. Plato's remark is corroborated by the following beginning of a Hippocratic work: "He who wishes to interrogate properly concerning treatment and answer him who interrogates and also object properly, must ponder the following things": [68] whereupon follows a long list of theoretical questions.

Once the expansion of medicine beyond the limits of a craft had become a necessity, this expansion could follow different directions. Systems could be built in which anatomy, crude as it was, formed a cornerstone, or where the older semi-popular concepts of humors were elaborated, or qualities emphasized. In all these attempts, the contact with philosophy was close; they were to lead to the science of the dogmatist school, of Galen, and, subsequently, to the modern medical sciences of anatomy, physiology, and pathology. This line of development is understandable to us, above all where anatomical reasoning in surgical cases occurs.[69] But we can even understand how an individual case might be diagnosed in terms of deranged humors and then treated correspondingly,[70] though we no longer believe in the theory of humors. Peculiarly enough, we experience greater difficulties with some of the clinical writings of the Hippocratic collection where we would expect an immediate contact. When we read the individual case histories of the first book of *Epidemics* we are at some loss to determine the purpose of their composition. Instinctively we try to "diagnose" these cases, i.e. to identify them with one or the other disease entity we are familiar with. But even when we succeed, we realize that a diagnosis of this kind was not the author's aim. It has been suggested with some degree of probability, that the book has a prognostic character.[71] From these illustrative case histories, the reader is to

---

[62] This accounts largely for the many sects among the dogmatists. As to the different explanation and treatment of disease by dogmatists see Caelius Aurelianus' (i.e. Soranus) criticism in his works on acute and chronic diseases and the so-called Anonymus Parisinus, edited by R. Fuchs, *Rheinisches Museum für Philologie*, 1894, N.F., vol. 49, pp. 532–558 and 1895, vol. 50, pp. 576–599.

[63] Even before the advent of the empiricists, Diocles claimed that one should not listen to people who believe that they must state a cause for everything. This he thinks is often quite unnecessary for practice and one should rather put one's trust in things well known from long experience (Werner Jaeger, *Diokles von Karystos*, p. 37 f.). But this criticism does not prevent Diocles from being a dogmatist.

[64] It is significant that to the empiricists the reference to mere experience became problematic in cases where the physician was confronted with a new kind of disease.

[65] I find it very difficult to imagine the complicated hygienic prescriptions for healthy persons as emanating from leechcraft. According to Adolf Palm, *Studien zur Hippokratischen Schrift Peri Diaites*, Diss. Tübingen, 1933, p. 38 f., the first zoological system in Greece developed among the physicians of the late fifth century B.C. when they endeavored to study the dietetic significance of different animals.

[66] On the individualization of Greek life and the concomitant release of intellectual energy, see E. R. Dodds, *The Greeks and the Irrational*, Berkeley, 1951.

[67] See above, footnote 4.

[68] *On Diseases, I*, ch. 1; ed. Littré (*Oeuvres complètes d'Hippocrate*), vol. 6, p. 140.

[69] The surgical works of the Hippocratic collection have thus remained the most modern ones; see Owsei Temkin, The Role of Surgery in the Rise of Modern Medical Thought, *Bull. Hist. Med.*, 1951, vol. 25, p. 248.

[70] That is by diminishing the quantity of the humor through bleeding, phlegmagogues etc. or acting upon the quality involved according to the principle of *contraria contrariis curantur*.

[71] See Edelstein, *Peri aeron* etc. p. 79–81;

learn how to evaluate the appearance and disappearance of symptoms in their significance to the outcome of the disease and moreover how to recognize the right "moment" for therapeutic interference.[72] But whatever the underlying motive may have been,[73] I believe that in their form these case histories offer an explanation of Plato's saying that the free doctor treated diseases "by investigating them from the commencement and according to the course of nature." The origin, course, and outcome of sickness in the individual are under observation and the symptoms are evaluated in relation to one another. That means that the physician can learn from the comparison of many case histories, but must judge for himself in each case and cannot rely on routine as his helper may do.

The individual approach to disease as documented in these case histories has so often been stressed that it needs no further elaboration. But it is worth pointing out that here, more clearly than ever before, disease is conceived as a process in time.[74] Thereby, disease becomes more than a symptom or a syndrome, designated by a popular or scientific name. The time-implied character of disease is also evident in that (first) part of *Epidemics I* which has led Deichgräber to coin the term "meteorological medicine."[75] Here the author described climatic events and the course of such diseases as showed seasonal variations of prevalence and character as they occurred during certain periods on the island of Thasos. Sporadic diseases are excluded and this part is less individualized. But even where the author deals with disease entities he does not view them as mere static combinations of symptoms but ascribes to them beginning and end, severity, complications, etc., in short imparts to the diseases a "history." So obvious is all this to us that we easily forget how essential the temporal concept of disease is for clinical medicine.[76] This concept could survive the particular prognostic motives from which it had originated as well as the physiological notions of the author,[77] and remain as a pattern of clinical study independent as it were from the philosophical "sciences." Small wonder then that the *Epidemics* became of interest to the empirical sect which denounced science. This interest, however, shows that the empirical sect does not represent a return to mere leechcraft. There is studied method in the empiricist's gathering of experience; it is merely a different form of research. Least of all does the methodist fall back upon the tradition of the leech, for he more than all others upsets the accustomed ways. The great

Deichgräber, *Die Epidemien* etc. p. 10 f. Von Brunn (*Gesnerus*, 1947), p. 14, doubts the exclusively prognostic character.

[72] Cf. Edelstein, *ibid.* Neuburger, *op. cit.*, p. 133, finds the essence of Hippocratic art "in the fact that it knew how, in individual cases, to decide which impressions collectively permitted of conclusions upon prognosis and indicated the occasion for therapeutic measures and the form they should take."

[73] One of the great difficulties presenting itself is the disregard of therapy. This has long been noticed and "Hippocrates" has been criticized for neglecting his patients. However, the patients received nourishment or they did not; in either case their "diet" — even if not prescribed by the physician — was right or wrong and somehow influenced the course of the disease. This consequence seems to follow from the premises of ancient dietetic medicine. Since the attitude to therapy is not quite the same in *Epidemics* I and III, I have limited myself to the former in spite of the generally accepted connection of the two books. The Hippocratic case histories have recently been discussed in detail by Pedro Lain Entralgo, *La Historia Clinica*, Madrid, Consejo Superior de Investigaciones Cientificas, 1950.

[74] Sigerist, *Antike Heilkunde*, p. 17, with regard to *Epidemics* writes: "Die ägyptische, die babylonische Literatur geben uns nur Augenblicksbilder. Schilderungen vom Verlauf einer Krankheit würden wir dort vergeblich suchen."

[75] Deichgräber, *op. cit.*, p. 12 f.

[76] See Otto Guttentag, On the Clinical Entity, *Annals of Internal Medicine*, 1949, vol. 31, p. 490, and Two Diagrams on the Clinical Entity, *The Journal of Pediatrics*, 1950, vol. 37, pp. 530–534.

[77] Thus the empiricists reinterpreted Hippocrates; see Karl Deichgräber, *Die griechische Empirikerschule*, Berlin, Weidmann, 1930, p. 319. The *Epidemics* are, of course, not the only Hippocratic work where disease is seen as a process; see e.g. the writing *On the Sacred Disease* which gives a complete pathogenesis culminating in the idea of the *kairos in treatment*. This shows how closely related etiological and prognostic writings could be (see p. 221).

methodist, Soranus, demands even of the future midwife that she should be literate "so that she may be able to grasp the art by theory too." [78]

Accustomed as we have become to scientific precision in modern medicine and its statistically proved therapeutic results, we have difficulties in conceiving of the clashes of the ancient sects as more than quibbles. But within the framework of ancient civilization the practitioner who wishes to overcome the narrowness of the craft in medical matters has no choice but to join a sect. This alone gives him the chance not only to enhance his prestige but to become a better, because broadly thinking, physician.[79] A final argument for this thesis seems to lie in the fact that even a dogmatist like Galen who loudly proclaimed "that the best physician must also be a philosopher" appears sincere in his medical demands and limitations upon philosophy. In the booklet bearing that title,[80] Galen claims that the majority of physicians, though would-be admirers of Hippocrates, do nothing to emulate him. They study neither astronomy nor its basic science, geometry, and even despise those who do so. They are ignorant of anatomy and think it useless to divide diseases into species and genera. They are so lacking in prognostic skill that "if anybody foretells a hemorrhage or a sweat, they call him a sorcerer and a narrator of marvels." [81] They are hardly able to ordain the sick man's regimen with a view to the future climax of the disease. Finally, they cannot even explain things properly.

It is not without interest that Galen thinks that the majority admire Hippocrates and, as he says, "declare him to be the first of all." *Hoi polloi* are not philosophical opponents to science but practical men who cannot be bothered. For this reason Galen does not simply admonish the physician to devote himself to scientific studies. The prerequisite for such study is a contempt for money. The true physician must have a balanced mind and follow truth. He must practice the logical method to know how many diseases there are as to species and genera and how to find the therapeutic indication for each. The same method will teach him the composition of the body of elements, tissues, and organic parts, and will be needed to prove things concerning the use and function of each part. "What then remains that the physician worthy of Hippocrates in his practice of the art be a philosopher? If, in order to discover the nature of the body, the difference of diseases, and the indications of treatment he must be trained in the theory of logic, if in order to practice these things diligently and steadfastly he must disdain money and must exercise self control, then indeed he has all the parts of philosophy: logic, natural science, and ethics." [82] At the end of the booklet, Galen says it would be foolish to argue over words. But as a matter of fact, his ideal physician is a limited philosopher, a man who is interested in ethics and science, and in logic because it is a tool for the latter.

Certain agnostic and eclectic trends in Galen's thought have long been obvious [83] and have received further confirmation from Galen's Synopsis of Plato's *Timaeus*, preserved in Arabic translation and quite recently published by Kraus and Walzer. Galen knows that Plato and Aristotle ascribe divinity to the stars, but he does not mention it. Likewise, he does not confirm, on his part, that the world has a soul.[84]

---

[78] Soranus, *Gynaecia* I, 3, ed. I. Ilberg, 1927, p. 4, 19 (*Corpus Medicorum Graecorum*, IV).

[79] This holds good in principle; individually a good leech might have been a better practitioner than a theorizing physician.

[80] I used the edition by Ernst Wenkebach, Der hippokratische Arzt als das Ideal Galens, *Quellen u. Studien zur Geschichte d. Naturwissenschaften und der Medizin*, 1933, vol. 3, pp. 363–383. For a French translation see Ch. Daremberg, *Oeuvres anatomiques, physiologiques et médicales de Galien*, vol. 1, Paris, 1854, pp. 1–7.

[81] *Ibid.*, p. 379, 8 f.

[82] *Ibid.*, p. 382, 8–13.

[83] Eduard Zeller, *Die Philosophie der Griechen in ihrer geschichtlichen Entwicklung dargestellt*, 3. Theil, 1. Abtheilung. Dritte Auflage, Leipzig, 1880, p. 830 f.

[84] *Galeni Compendium Timaei Platonis* etc.

This goes well together with his consistent refusal to pronounce on the essence of the human soul and the question of its immortality. "To know this," Galen states elsewhere, "is not necessary for the healing of diseases or the preservation of health; nor for ethical, practical and political philosophy. Indeed every one may call the latter as he wishes, separating it from theoretical philosophy." [85] Kraus and Walzer conclude that in cosmology and psychology, Galen tended towards the view of the sceptics.[86] I do not wish to argue the formal correctness of this statement. But I would think that Galen's indifference to contemplative philosophy is tinged by his being a physician. The physician must be a philosopher in so far as philosophy represents science and ethics; science will give him an understanding of nature and disease. Ethics will give him the attitude he needs to be a student of nature, and a guide for men whom he is to treat by diet, i.e. by regulating their way of life. And ethics, in Galen's mind, demands recognition of a divine providence.[87] Thus Galen can even admit that the immediate practical value of anatomy for the physician is limited, and can yet cultivate anatomy and physiology as studies which reveal nature and thus glorify God.[88]

We can now sum up by returning to our original problem. Having given a brief sketch of its history, we asked whether Greek medicine as a whole really followed the development mirrored in the books of its physician authors, or whether as a craft it might not have remained unchanged? Our answer was that Greek leechcraft, unaided by science or broad generalizing concepts may always have existed. But we suggested that, however great his skill, the mere leech was inferior to the physician whose philosophical indebtedness expressed itself by his membership in one of the sects. We may imagine a broad stratum of practical men between the leaders whose ideas the literature reflects, and the practitioner bound by mere tradition. But even the latter will not have remained entirely unaffected by the thought trickling down to him; in this sense I believe that the literature from Hippocrates to Galen, incomplete as it is, contains the essential history of Greek medicine. And to this we may add an epilogue. In the Greek East, the date of Galen's death coincides with the end of the empiric and methodist sects. By the middle of the fourth century Galen has emerged as the great medical authority next only to Hippocrates, and his true heir. From now on there is only one medical science, based on just those parts of philosophy which Galen accepted. And since, moreover, this science at an increasing rate is acquired from books, the break is prepared between the mere craftsman, the mechanic, and the learned doctor. In antiquity this break was never complete within the medical profession. Even the most learned physician was still expected to have acquired the necessary skill and practical insight. But together with the systematization of Galen a tradition of medical science was formed that was to influence deeply the Latin West, even beyond the Middle Ages. The doctor is expected to have studied logic, he is to know astronomy, which often means astrology, and the study of human anatomy and physiology are integral parts of medicine, as are pathology, hygiene, and therapeutics. The transmittal of that tradition to the West transformed the physician of 400 B.C. into a physicus in the literal sense, while the craftsman found his own way of life in the guild.

edd. Paulus Kraus et Richardus Walzer, Londinii, 1951 (*Plato Arabus*, vol. 1), p. 13.

[85] Kühn edition, vol. 4, p. 764; the passage is quoted by Kraus and Walzer, *op. cit.*, p. 15, foonote 4. See also Zeller, *loc. cit.*

[86] *Op. cit.*, p. 15.

[87] *De placitis Hippocratis et Platonis*, IX, chs. 7 and 8 (ed. I. Mueller, Lipsiae, 1874, especially pp. 799 ff.).

[88] *On the Use of Parts*, XVII, 1. Galen claims that the search for the usefulness of parts constitutes the principle of a perfect theology which is of greater importance than all medicine. I think that this statement has to be understood in connection with Galen's pragmatic evaluation of "theology" as indicated in *De placitis Hippocratis et Platonis*, *loc. cit.*

Science and the Early Christian Church

By David C. Linberg

> The pagan party . . . asserted that knowledge is to be obtained only by the laborious exercise of human observation and human reason. The Christian party asserted that all knowledge is to be found in the Scriptures and in the traditions of the church; that, in the written revelation, God had not only given a criterion of truth, but had furnished us with all that he intended us to know. . . .
>
> The Church thus set herself forth as the depository and arbiter of knowledge; she was ever ready to resort to the civil power to compel obedience to her decisions. She thus took a course which determined her whole future career; she became a stumbling block in the intellectual advancement of Europe for more than a thousand years.[1]

THUS WROTE JOHN WILLIAM DRAPER (1811–1882) in a polemic against the excesses of organized Christianity, especially Catholicism. In so doing, he gave shape to what has become a very widespread (probably the dominant) interpretation of the relationship between science and the early church: that the church, if it did not entirely stamp out science, surely retarded its progress. This

I wish to thank the following for helpful comments: William B. Ashworth, William J. Courtenay, Gary B. Deason, Edward Grant, Ronald L. Numbers, Robert Siegfried, Nicholas H. Steneck, Robert S. Westman and Keith E. Yandell. A forthcoming history of the relations of science and Christianity, edited by David C. Lindberg and Ronald L. Numbers, will include a revised version of this paper.

[1] John William Draper, *The Conflict between Religion and Science,* 7th ed. (London: Henry S. King, 1876), pp. 51–52. On Draper and his career, see Donald Fleming, *John William Draper and the Religion of Science* (Philadelphia: Univ. Pennsylvania Press, 1950).

theme was echoed near the end of the nineteenth century by Andrew Dickson White (1832–1918). After quoting Augustine on the necessity of yielding to Scriptural authority, White commented: "Following this precept of St. Augustine there were developed, in every field, theological views of science which have never led to a single truth—which, without exception, have forced mankind away from the truth, and have caused Christendom to stumble for centuries into abysses of error and sorrow."[2] Draper and White believed that Christianity waged war on science in two ways. First, the early church fathers denigrated the investigation of nature for its own sake: with the kingdom of heaven just around the corner, there was no time or energy to waste on irrelevancies. Second, whatever truth was discovered through patient observation and reasoning was forced to yield to the puerile opinions extracted by dogmatic churchmen from sacred writings. The result was a tyranny of ignorance and superstition that "perverted" and "crushed" true science.[3]

The thesis of Draper and White has given way to a spectrum of scholarly opinion in the twentieth century. Some scholars continue to affirm, although (in most cases) somewhat less militantly, the Draper-White view.[4] Others have gone to the opposite extreme, arguing that Christianity was good for science—indeed, that modern science would not have come into existence without it.[5] And some have sought middle ground.[6] But this is scholarship; in popular opinion the Draper-White view still prevails. It frequently appears in books aimed at the general reader; moreover, Draper's and White's own works continue to be reprinted, purchased, and presumably read and believed.[7]

This essay will reassess the evidence, while shunning as far as possible the

[2] Andrew Dickson White, *A History of the Warfare of Science with Theology in Christendom,* 2 vols. (New York: Appleton, 1896), Vol. I, p. 325. On White, see the biography by Glenn C. Altschuler, *Andrew D. White—Educator, Historian, Diplomat* (Ithaca: Cornell Univ. Press, 1979); also the excellent historiographical essay by James R. Moore, *The Post-Darwinian Controversies* (Cambridge: Cambridge Univ. Press, 1979), Ch. 1.

[3] See White, *Warfare,* Vol. I, pp. 97, 376.

[4] E.g., George Sarton, *Introduction to the History of Science,* 3 vols. (Baltimore: William & Wilkins, 1927–1948), Vol. I, pp. 17, 21; Sir William Dampier, *A History of Science and its Relations with Philosophy and Religion* (Cambridge: Cambridge Univ. Press, 1929), pp. 66–73; Charles Singer, *Religion and Science Considered in their Historical Relations* (New York: Robert McBride, [1932]), pp. 52–56; J. D. Bernal, *Science in History,* 4 vols. (London: C. A. Watts, 1954), Vol. I, pp. 258–262; Ernst Mayr, *The Growth of Biological Thought: Diversity, Evolution, and Inheritance* (Cambridge, Mass.: Belknap Press of Harvard Univ. Press, 1982), pp. 307–309.

[5] R. Hooykaas, *Religion and the Rise of Modern Science* (Grand Rapids: Eerdmans, 1982); Eugene M. Klaaren, *Religious Origins of Modern Science: Belief in Creation in Seventeenth-Century Thought* (Grand Rapids: Eerdmans, 1977); Stanley L. Jaki, *The Road of Science and the Ways to God* (Chicago: Univ. Chicago Press, 1978). None of these authors addresses himself in any serious way to the patristic period, but since their larger view is that Christianity was a necessary condition for the development of science, it follows that the advent of Christianity was a necessary first step.

[6] Marshall Clagett, *Greek Science in Antiquity* (London: Abelard-Schuman, 1957), pp. 130–145; E. J. Dijksterhuis, *The Mechanization of the World Picture,* trans. C. Dikshoorn (Oxford: Clarendon Press, 1961), pp. 89–95; G. E. R. Lloyd, *Greek Science after Aristotle* (London: Chatto & Windus, 1973), pp. 167–171. Lloyd, however, does regard Christianity as having had an overall detrimental affect on science.

[7] See Will Durant, *The Age of Faith* (Story of Civilization, Vol. IV) (New York: Simon & Schuster, 1950), pp. 78–79; Thomas Goldstein, *Dawn of Modern Science: From the Arabs to Leonardo da Vinci* (Boston: Houghton Mifflin, 1980), p. 43. Between them, Draper's and White's books have gone through more than eighty editions. Draper's book has been translated into Spanish, French, German, Polish, and Russian; White's into French, Italian, Swedish, and Japanese. Draper's book was last published in 1970; White's is still in print.

polemical and ideological goals that motivated Draper and White and continue to motivate many discussions of early Christianity and science. Instead of using the historical problem as an occasion for attacking or defending Christianity for its detrimental or beneficial affects on science, we must endeavor to understand the complexity and subtlety of their interaction.

Discussions of our subject have frequently suffered from the assumption that in antiquity there was an intellectual discipline having more or less the same methods and the same lines of demarcation as modern science, to which the term "science" can be properly and unambiguously applied. Thus it was modern science, or its immediate antecedent, that Draper and White and their followers hold Christianity to have retarded. But the truth is far more complicated. Several of the subdivisions of modern science did exist as recognizable disciplines in antiquity, for example, medicine (with some associated biological knowledge) and mathematics (including astronomy and other branches of mathematical science). But there was nothing in antiquity corresponding to modern science as a whole or to such branches of modern science as physics, chemistry, geology, zoology, and psychology. The subject matters of these modern disciplines all belonged to natural philosophy and thus to the larger philosophical enterprise. Even such distinctions as existed (for example, that between natural philosophy and mathematics) could be easily overlooked, since the disciplines thus distinguished did not represent clearly defined social roles or professions.[8] The natural philosopher and the mathematician were often the same person, and professionally he would probably have identified himself neither as natural philosopher nor as mathematician, but simply as a teacher; and his teaching would likely have extended beyond mathematics and natural philosophy to all manner of other philosophical issues. It is true that there were always professional physicians, but even here the lines of demarcation were fuzzy, since many a physician was also a philosopher. The world of the intellect had a unity in antiquity that it does not have today.

The methodology of this scholarly enterprise has often been misrepresented. Members of the Draper-White School (as the opening quotation instances) have portrayed ancient philosophy as an early version of modern scholarship—the embodiment of the ideals of rationality, objectivity, and whatever other traits they find praiseworthy in themselves. There were indeed ancient philosophers and philosophical schools for which such a characterization is more or less apt, and the rationalism of Greek thought was surely one of the great achievements of antiquity.[9] But in late antiquity philosophy was changing. The philosophical classics of the past continued to be available and influential, but within

[8] On the classification of theoretical knowledge see James A. Weisheipl, O.P., "The Nature, Scope, and Classification of the Sciences," in *Science in the Middle Ages,* ed. David C. Lindberg (Chicago: Univ. Chicago Press, 1978), Ch. 14. On the organization of science in antiquity, see Ludwig Edelstein, *Ancient Medicine: Selected Papers* (Baltimore: Johns Hopkins Univ. Press, 1967), pp. 429–439 (the final section of an article "Recent Trends in the Interpretation of Ancient Science," rpt. from *Journal of the History of Ideas,* 1952, *13*:573–604); G. E. R. Lloyd, *Early Greek Science: Thales to Aristotle* (London: Chatto & Windus, 1970), pp. 125–130; and Thomas W. Africa, *Science and the State in Greece and Rome* (New York: Wiley, 1968).

[9] On Greek rationalism, see Bruno Snell, *The Discovery of the Mind: The Greek Origins of European Thought* (Cambridge, Mass.: Harvard Univ. Press, 1953); E. R. Dodds, *The Greeks and the Irrational* (Berkeley: Univ. California Press, 1951); G. E. R. Lloyd, *Magic, Reason and Experi-*

the living, contemporary tradition the focus of attention was shifting toward ethics, metaphysics, and theology; and, in some of its manifestations, philosophy was becoming progressively more like a religion, based on inspired authorities, with mystical illumination and personal salvation among its principal goals. Thus Neoplatonic authors such as Porphyry, Iamblichus, Proclus, and Damascius (third through sixth centuries) accepted the *Chaldean Oracles* (esoteric religious writings, devoted to theurgy, demonology, and other forms of magic) as an authoritative source of revealed truth, beyond the reach of rational discussion and debate.[10] Those who would characterize the early Christian tradition as "superstitious" must apply the same term to aspects of contemporary pagan philosophy.

There is also a tendency, within the Draper-White tradition, to see pagan philosophy as tolerant, committed to a free market of ideas—in contrast to the intolerance of Christians. Indeed, those who regret the triumph of Christianity frequently view the struggle between Christianity and pagan philosophy as a battle for freedom of thought. But this too is a misconception. One need only recall that Plato demanded solitary confinement (and, in extreme cases, execution) for those who denied the existence of the gods and their involvement in human affairs.[11] Intolerance was (and is) a widely cultivated trait, shared about equally by pagans and Christians. Moreover, each party was capable of employing coercive measures when it possessed the political power to do so; Christians, in fact, appear to have yielded to the temptation less often than did pagans.[12]

This is the pagan philosophical culture that the early church confronted. It comprised both a contemporary philosophical tradition and a collection of philosophical classics; it dealt with an enormous range of philosophical issues, covering the spectrum from epistemology to politics; and it furnished the technical tools for reasoned discourse. It did not look very much like modern science. Although we will eventually focus our attention on that portion of its content that pertained to nature, we must (if we wish to understand the church's response) begin by considering it as a whole. We must ask: How did the early church regard pagan intellectual culture? How and to what extent did Christians make use of it? Was Christian theology ever a rational enterprise, employing the tools of Greek philosophy, or did it always involve a retreat from philosophy into the claims of revelation? How were Christian thought and the pagan philosophical tradition affected by their encounter? We must, in short, begin by examining the problem of reason and revelation—the question raised by Tertullian when he inquired: "What has Athens to do with Jerusalem?"

---

ence: Studies in the Origins and Development of Greek Science (Cambridge: Cambridge Univ. Press, 1979).

[10] Hans Lewy, *Chaldean Oracles and Theurgy: Mysticism, Magic, and Platonism in the Later Roman Empire* (Cairo: Institut Français d'Archéologie Orientale, 1956), pp. 67–76; John J. O'Meara, *Porphyry's Philosophy from Oracles in Augustine* (Paris: Etudes Augustiniennes, 1959); Richard Walzer, *Galen on Jews and Christians* (London: Oxford Univ. Press, 1949), pp. 55–56; Henry Chadwick, *The Early Church* (Harmondsworth: Penguin, 1967), p. 112. E. R. Dodds, *Pagan and Christian in an Age of Anxiety* (Cambridge: Cambridge Univ. Press, 1965), p. 122, points out that while Christians were substituting reason for authority, pagans were substituting authority for reason.

[11] Plato, *Laws* 10.

[12] This can be gathered from Chadwick, *Early Church*, pp. 171–173.

ATHENS AND JERUSALEM

Christianity spread rapidly outward from Jerusalem and surrounding Judaea during the first century of the Christian era, particularly to the north and west, into Syria, Asia Minor, and as far as Rome. Beginning as a Jewish sect appealing largely to Hellenized Jews, it first broadened its reach into the Gentile world through the efforts of the Apostle Paul. It was too insignificant to attract serious attention from the Roman authorities until Christians were blamed by the Emperor Nero for the great fire that destroyed Rome in A.D. 64. Considered antisocial for their unwillingness to participate in the traditional religious practices, Christians thereafter attracted sporadic persecution through the third century—persecution that seems to have contributed to the growth rather than the decline of Christianity. The conversion of the Emperor Constantine early in the fourth century marked the beginning of a radical change in the political fortunes of Christianity; by the end of the fourth century it had become the state religion.[13]

From the beginning Christianity attracted converts from a wide social and intellectual spectrum. The Apostle Paul is representative of highly educated Hellenized Judaism; one finds within his writings ample evidence of familiarity with Greek philosophical systems. As people committed to sacred writings, Christians were in need of literacy and therefore at least elementary education, but it was not until the second century that Christianity, under the influence of doctrinal dissent within and attack from without, developed a significant intellectual tradition. The first major Christian apologist was Justin Martyr, of Greek descent, born in Samaria near the beginning of the second century and martyred in Rome between 162 and 168. Justin studied Stoic, Aristotelian, Pythagorean, and Platonic philosophy, finding satisfaction only in the latter. Later he converted to Christianity and became convinced of the fundamental compatibility between Christian doctrine, Platonic metaphysics, and Stoic ethics. He firmly rejected pagan polytheistic religion, but welcomed such pagan philosophy as was consistent with Biblical teaching. Justin explained the impressive parallels between Christianity and pagan philosophy (particularly Platonism, which not only could be construed as monotheistic, but also taught the immortality of the soul and the formation of the world at a point in time) by proposing that the Greek philosophers had studied the Old Testament, and also that all of humankind, insofar as it thinks rightly, does so through participation in the universal rational power, the divine *logos,* Christ. Justin thus expressed deep confidence in the power of reason—a divine gift.[14]

The apologetic work begun by Justin was continued by Athenagoras, Theophilus of Antioch, Clement of Alexandria, Origen, and others. There were varieties of opinion among these apologists, of course, but all were familiar with

[13] See Chadwick, *Early Church,* an excellent short history. Longer histories are innumerable; I have made greatest use of Louis Duchesne, *The Early History of the Church,* 3 vols. (New York: Longmans, Green & Co., 1914–1924).

[14] On Justin, see Henry Chadwick, *Early Christian Thought and the Classical Tradition: Studies in Justin, Clement, and Origen* (New York: Oxford Univ. Press, 1966), Ch. 1; Chadwick, "The Beginning of Christian Philosophy: Justin: The Gnostics," in *The Cambridge History of Later Greek and Early Medieval Philosophy* (henceforth *Cambridge History*), ed. A. H. Armstrong (Cambridge: Cambridge Univ. Press, 1970), Ch. 9; Frederick Copleston, S.J., *A. History of Philosophy,* 9 vols. (Westminster, Md.: Newman Press, 1946–1975), Vol. II, pp. 16–18.

Greek philosophy, esteemed portions of it, particularly Platonic philosophy (for its many affinities with Christian theology), and put it to apologetic use whenever possible. Clement (d. between 211 and 215), a teacher in the catechetical school in Alexandria, regarded Greek philosophy as absolutely essential for the defense of the faith against heresy and skepticism and for the development of Christian doctrine. Central to Clement's thought was the doctrine that truth is one—that ultimately all truth, wherever it may be encountered, is God's truth.[15] In his *Stromateis* he claimed that "barbarian and Greek philosophy have torn off a piece of the eternal truth . . . from the theology of the Logos who eternally is. And he who brings together again the divided parts and makes them one, mark well, shall without danger of error look upon the perfect Logos, the truth.[16] Like Justin, he argued that Greek philosophy partook of the truth because it was plagiarized from the Old Testament, and also because pagan philosophers employed their God-given rational capacities to obtain a portion of divine truth. Thus did Clement endeavor to rescue Christianity from the charge of intellectual obscurantism (already leveled against it in antiquity, as discussed below) by affirming the value of Greek philosophy.

The attitude of Origen (ca. 185–ca. 254) toward Greek philosophy, particularly Platonic, was even more liberal than Clement's. Origen, also an Alexandrian teacher, possessed a thorough knowledge of Greek philosophy—Aristotelian, Platonic, Stoic, and Epicurean. He adopted the basic elements of Plato's theology, cosmology, and psychology, while borrowing his terminology and definitions from Aristotle. His student Gregory Thaumaturgus, commenting on Origen's teaching methodology, reveals Origen's remarkable openness to pagan philosophical sources:

> He required us to study philosophy by reading all the existing writings of the ancients, both philosophers and religious poets, taking every care not to put aside or reject any . . . , apart from the writings of the atheists. . . . He selected everything that was useful and true in each philosopher and set it before us, but condemned what was false. . . . For us there was nothing forbidden, nothing hidden, nothing inaccessible. We were allowed to learn every doctrine, non-Greek and Greek, both spiritual and secular, both divine and human; with the utmost freedom we went into everything and examined it thoroughly, taking our fill of and enjoying the pleasures of the soul.[17]

But not all Christians shared the opinion of Justin, Clement, and Origen. Tatian, a Syrian Christian who apparently studied under Justin, was skeptical of the value of Greek philosophy and launched an attack on the teaching of the

[15] E. F. Osborn, *The Philosophy of Clement of Alexandria* (Cambridge: Cambridge Univ. Press, 1957), pp. 117–126. On Clement see also Chadwick, *Early Christian Thought,* Ch. 2; Charles Bigg, *The Christian Platonists of Alexandria* (Oxford: Clarendon Press, 1886), Lect. 2; H. B. Timothy, *The Early Greek Apologists and Greek Philosophy, Exemplified by Irenaeus, Tertullian, and Clement of Alexandria* (Assen: Van Gorcum, 1973), pp. 59–80, 88–98.

[16] Quoted by Osborn, *Philosophy of Clement,* p. 124.

[17] Gregory Thaumaturgus, *In Origenem Oratio,* quoted by M. L. Clarke, *Higher Education in the Ancient World* (London: Routledge & Kegan Paul, 1971), pp. 126–127 (with minor changes in punctuation). On Origen's borrowings see Dodds, *Pagan and Christian,* pp. 127–129; Charles N. Cochrane, *Christianity and Classical Culture: A Study of Thought and Action from Augustus to Augustine* (Oxford: Clarendon Press, 1940), p. 226. See also Chadwick, *Early Christian Thought,* Ch. 3; Bigg, *Christian Platonists,* Lects. 4–6.

pagan schools.[18] In the third century the author of the *Didascalia apostolorum* warned Christians against the dangers of pagan literature: "Shun all heathen books. Of what concern to you are strange ideas or laws or pseudo-prophets, which often lead inexperienced men into error? What is lacking to you in God's word, that you should turn to that heathen nonsense?" And in the fourth century John Chrysostom, who was willing to concede that pagan schools were not without value, nevertheless pointed out that "the study of eloquence requires good morals, but good morals do not require eloquence."[19] Whether or not such views were typical of Christians, they were portrayed as such by pagan opponents. Celsus (fl. ca. 177–180) accused Christians of enjoining: "Let no one educated, no one wise, no one sensible draw near. For these abilities are thought by us to be evils. But as for anyone ignorant, anyone stupid, anyone uneducated, . . . let him come boldly."[20] Eusebius, probably reporting Porphyry's view, refers to those who "have supposed that Christianity has no reason to support itself but that those who desire the name confirm their opinion by an unreasoning faith and an assent without examination." And the Emperor Julian, who promoted a restoration of paganism in the fourth century, argued as follows against the Christians (at the same time giving paganism somewhat more than its due): "Ours are the reasoned arguments and the pagan tradition which comprehend at the same time due worship of the gods; yours are want of reason and rusticity, and all your wisdom can be summed up in the imperative 'Believe.' "[21]

The church father who is generally taken to epitomize the anti-intellectualism of the early church is Tertullian (ca. 155–ca. 230), whose views we must therefore consider with some care. Tertullian was a native of Carthage in Roman Africa, of pagan parentage, well educated in philosophy, medicine, and law, and able to write in either Greek or Latin. Etienne Gilson has portrayed him as an implacable foe of pagan philosophy, the archetype of those Christian theologians who wished to substitute faith for reason.[22] In a celebrated denunciation of philosophy, Tertullian exclaims:

> What indeed has Athens to do with Jerusalem? What concord is there between the Academy and the Church? What between heretics and Christians? Our instruction comes from "the porch of Solomon," who had himself taught that "the Lord should be sought in simplicity of heart." Away with all attempts to produce a mottled Christianity of Stoic, Platonic, and dialectic composition! We want no curious disputation after possessing Christ Jesus, no inquisition after enjoying the gospel! With our faith, we desire no further belief. For once we believe this, there is nothing else that we ought to believe.

[18] Copleston, *History of Philosophy,* Vol. II, p. 18; Etienne Gilson, *Reason and Revelation in the Middle Ages* (New York: Scribners, 1938), pp. 11–12.

[19] Quoting *Didascalia et constitutiones apostolorum* 10, ed. Franciscus X. Funk, Vol. I (Paderborn: Ferdinand Schoeningh, 1905), pp. 12–14; and Chrysostom from M. L. W. Laistner, *Christianity and Pagan Culture in the Later Roman Empire* (Ithaca: Cornell Univ. Press, 1951), p. 53 (for another trans. of the *Didascalia* see *ibid.,* p. 50).

[20] Origen, *Contra Celsum,* trans. Henry Chadwick (Cambridge: Cambridge Univ. Press, 1953), p. 158.

[21] Both quoted by Walzer, *Galen on Jews and Christians,* p. 54.

[22] Gilson, *Reason and Revelation,* pp. 5–11; cf. Cochrane, *Christianity and Classical Culture,* pp. 227–230.

Elsewhere Tertullian attacks vain curiosity about nature:

> Now, pray tell me, what wisdom is there in this hankering after conjectural speculations? What proof is afforded to us, notwithstanding the strong confidence of its assertions, by the useless affectation of a scrupulous curiosity, which is tricked out with an artful show of language? It therefore served Thales of Miletus quite right, when, star-gazing as he walked with all the eyes he had, he had the mortification of falling into a well. . . . His fall, therefore, is a figurative picture of the philosophers; of those, I mean, who persist in applying their studies to a vain purpose, since they indulge a stupid curiosity on natural objects, which they ought rather [intelligently to direct] to their Creator and Governor.[23]

There can be no doubt that Tertullian was not an enthusiast for secular learning; but neither was he the uncompromising opponent of reasoned discourse that these passages, if allowed to stand alone, might seem to imply. It was apparent to him that philosophy led easily to heresy, especially the Gnostic heresy (a dualistic system, which radically separated the transcendent deity from a dark and evil material world). What he therefore opposed was not philosophy generally, but heresy or the philosophy that gave rise to it.[24] When not engaged in polemic against heresy, he could express quite a favorable view of mankind's rational capacities. On one occasion he argued that "reason . . . is a thing of God, inasmuch as there is nothing which God the Maker of all has not provided, disposed, ordained by reason—nothing which He has not willed should be handled and understood by reason." He even defended the possibility of rational knowledge of divine things: "One may no doubt be wise in the things of God, even from one's natural powers. . . . For some things are known even by nature: the immortality of the soul, for instance, is held by many; the knowledge of our God is possessed by all."[25] Tertullian is frequently quoted as having said, of the resurrection of Christ, "I believe it because it is absurd." However, scholars have adequately established, first, that this is a misquotation; but more importantly that Tertullian was simply making use of a standard Aristotelian argumentative form, maintaining that the more improbable an event, the less likely is anybody to believe that it has occurred without specific supporting evidence; therefore, the very improbability of an alleged event, such as Christ's resurrection, is evidence in its favor.[26] Thus, far from seeking the abolition of reason, Tertullian must be seen as appropriating Aristotelian rational techniques and

[23] Tertullian, *On Prescription against Heretics* 7 (my own translation of the final sentence) and *Ad Nationes* 2.4, both trans. Peter Holmes, in *The Ante-Nicene Fathers,* ed. Alexander Roberts and James Donaldson, 10 vols. (New York: Scribners, 1896–1903), Vol. III, pp. 246, 133.

[24] This theme is nicely developed by Robert H. Ayers, *Language, Logic, and Reason in the Church Fathers* (Hildesheim: Georg Olms, 1979), pp. 25–34. For another useful discussion of Tertullian, see Timothy, *Early Greek Apologists,* pp. 40–58.

[25] Tertullian, *On Repentance* 1, trans. S. Thelwall, and *On the Resurrection of the Flesh* 3, trans. Peter Holmes, both in *The Ante-Nicene Fathers,* Vol. III, pp. 657, 547. I was led to a number of the quotations from Tertullian by Ayers.

[26] That is, resurrection of the dead is so improbable an event that the apostles would not have believed in the resurrection of Christ if they had not been faced with incontrovertible evidence that indeed, on this occasion, the improbable had occurred. This truth makes the resurrection of Christ more probable than some other event, the occurrence of which might have been accepted merely on the basis of general plausibility. See Ayers, *Language, Logic, and Reason,* pp. 21–24; Robert Grant, *Miracle and Natural Law in Graeco-Roman and Early Christian Thought* (Amsterdam: North-Holland, 1952), pp. 194, 209; and Aristotle, *Rhetoric* 2.23.22.

putting them to apologetic use. Philosophy, despite its dangers, had a place in the armory of the Christian. But Tertullian never forgot to put first things first.

In the long run, the most influential statement of the Christian attitude toward philosophy and reason was that of Augustine (354–430). Augustine obtained a rich classical education, heavily literary in its orientation, in the North African schools of Thagaste, Madaura, and finally Carthage. In his early manhood Augustine yielded to the appeal of Manichaean religion, with its dualism between good and evil and its unwavering rationalism; later he discovered the works of Plotinus and became deeply embued with Neoplatonic philosophy. Meanwhile he converted to Christianity and began the effort, which occupied him through much of his life, of accommodating Neoplatonism and Christianity.[27]

Augustine's position has frequently been misunderstood as an attempt to substitute faith for reason. But this was surely never his purpose: philosophy and the philosophical life were not to be replaced or repudiated, but to be Christianized.[28] Esteem for human rational capacities pervades Augustine's writings. Reason is a divine gift, which distinguishes humankind from brutes, and its exercise is to be assiduously cultivated. In a letter to Consentius, Augustine spoke of his wish "to arouse your faith to a love of understanding, to which true reason conducts the mind and for which faith prepares it." He went on to point out that heretical reasoning about the Trinity "is to be shunned and detested, not because it is reasoning, but because it is false reasoning; for if it were true reasoning, it would surely not err. Therefore, just as you would be ill advised to avoid all speaking because some speaking is false, so you must not avoid all reasoning because some reasoning is false." Reasoning is indispensable if the faith is to be defended and its content understood; it is also required if one is to grasp that portion of truth that has not been revealed. Augustine's letter to Consentius is a remarkable expression of these themes:

> You say that truth is to be grasped more by faith than by reason. . . . Therefore, according to your rule you ought in this matter . . . to follow only the authority of the saints and not seek understanding by asking me for reasons. For when I begin . . . to lead you to an understanding of such a great mystery, . . . I will simply be giving you reasons, insofar as I am able. But if it is not unreasonable for you to beg me or some other teacher to help you to understand what you believe, then you ought to correct your rule—not to the point of overturning faith, but of permitting you to discern in the light of reason what you already firmly hold by faith.
>
> Heaven forbid that God should hate in us that by which he made us superior to the other animals! Heaven forbid that we should believe in such a way as not to accept or seek reasons, since we could not even believe if we did not possess rational souls. Therefore, in certain matters pertaining to the doctrine of salvation that we cannot yet grasp by reason—though one day we shall be able to do so—faith must precede reason and purify the heart and make it fit to receive and endure the great light of reason; and this is surely something reasonable. Thus it is reasonable for the Prophet [Isa. 7:9] to have said: "Unless you believe, you will not understand." Here he was doubtless distinguishing between these two things and advising us first to

[27] See Peter Brown, *Augustine of Hippo: A Biography* (Berkeley: Univ. California Press, 1969); on Augustine's education, see Ch. 3. On Augustine's Neoplatonism, see esp. A. H. Armstrong, *St. Augustine and Christian Platonism* (Villanova, Pa.: Villanova Univ. Press, 1967); reprinted in Armstrong, *Plotinian and Christian Studies* (London: Variorum Reprints, 1979), Ch. 11. Late in his career Augustine moved away from Neoplatonic assumptions.

[28] See Cochrane, *Christianity and Classical Culture,* which has deeply influenced me.

> believe, so that afterwards we might understand what we believe. It is thus reasonable to require that faith precede reason. . . . If, therefore, it is reasonable for faith to precede reason in certain matters of great moment that cannot yet be grasped, surely the very small portion of reason that persuades us of this must precede faith.[29]

Despite his reference in the final lines of this passage to reason that precedes faith, Augustine usually stresses the movement from faith to reason or understanding. Without faith, there will be no understanding; once faith is achieved, the quest for understanding is obligatory.

Can we then say that Augustine subordinates reason to faith? Yes, in the sense that ultimate authority rests with revelation. Augustine had no hesitation in proclaiming his resolve "never to deviate in the least from the authority of Christ."[30] But to concentrate exclusively on the question of authority is to adopt a modern perspective and to overlook Augustine's. Augustine viewed faith not as a taskmaster to which reason must submit, but as the condition that makes genuine rational activity possible. Christian faith provides the foundation, the blueprint, and the materials, without which no sound philosophical structures can be built. Philosophy thus finds its fulfillment within the framework of faith. For Augustine the highest goal is understanding: "Our Lord . . . says to believers, 'Seek and ye shall find' [Matt. 7:7]. But one cannot speak of that being found which is believed without knowledge." The relationship between faith and reason is therefore that between the precondition and the ultimate objective, the means and the end; in that sense, we find in Augustine the subordination not of reason to faith, but of faith to reason.[31]

What, then, has Athens to do with Jerusalem? No uniform answer will suffice for all of the fathers of the Christian church. There was a spectrum of attitudes toward pagan culture, from deep mistrust to high enthusiasm. But few would have rejected pagan philosophical culture totally. For the great majority, whether or not to philosophize was no issue; the question was rather how and about what to philosophize. The classic answer given by Augustine and destined for enormous influence set philosophy to work on revelation and the content of Christian belief. But this required philosophical education. Christians were thus committed to education—and in practical terms this meant secular education, since no Christian educational system was developed until much later. Gregory of Nazianzus (329–389) revealed how liberal the Christian position could be, when he wrote:

> I take it as admitted by men of sense, that the first of our advantages is education; and not only this our more noble form of it, which disregards rhetorical ornaments and glory, and holds to salvation, and beauty in the objects of our contemplation: but even that external [i.e., pagan] culture which many Christians ill-judgingly abhor, as treacherous and dangerous, and keeping us afar from God. For as we ought not

[29] Augustine, Letter 120, ed. A. Goldbacher, in *Corpus scriptorum ecclesiasticorum latinorum*, Vol. XXXIV (Vienna: F. Tempsky, 1895), quoting pp. 708, 709, 705–707.

[30] Augustine, *Against the Academics* 3.20.43; quoted from *Fathers of the Church*, Vol. I (Washington: Catholic University of America Press, 1948), by Vernon J. Bourke, *The Essential Augustine* (New York: New American Library, 1964), p. 25.

[31] Quoting Augustine, *On Free Choice* 2.2.6, trans. Carroll Mason Sparrow, as quoted by Burke, *The Essential Augustine*, p. 25. On faith as a condition for understanding, see Cochrane, *Christianity and Classical Culture*, Ch. 10–11; Brown, *Augustine of Hippo*, Ch. 10.

> to neglect the heavens, and earth, and air, and all such things, because some have wrongly seized upon them, and honour God's works instead of God: but to reap what advantage we can from them for our life and enjoyment, while we avoid their dangers; not raising creation, as foolish men do, in revolt against the Creator, but from the works of nature apprehending the Worker, and, as the divine apostle says, bringing into captivity every thought to Christ: and again, as we know that neither fire, nor food, nor iron, nor any other of the elements, is of itself most useful, or most harmful, except according to the will of those who use it; and as we have compounded healthful drugs from certain of the reptiles; so from secular literature we have received principles of enquiry and speculation, while we have rejected their idolatry, terror, and pit of destruction. . . . We must not then dishonour education, because some men are pleased to do so, but rather suppose such men to be boorish and uneducated, desiring all men to be as they themselves are, in order to hide themselves in the general, and escape the detection of their want of culture.[32]

There are no data to permit a judgment regarding the average educational levels of pagans and Christians, but it is clear that the Christian intelligentsia were at least as well educated as, and the intellectual equals of, their pagan opponents.[33]

THE CHURCH AND NATURAL SCIENCE

Although science was neither an autonomous discipline nor a profession during the patristic period, we can nonetheless investigate the relationship between Christianity and those aspects of the philosophical enterprise that were concerned with nature. Did science (in this limited sense) benefit or suffer from the appearance and triumph of Christianity? Did Christianity, with its other-worldliness and its emphasis on Biblical authority, stifle interest in nature, as the old stereotype proclaims? Or was there a more ambiguous and subtle relationship?

We must begin our inquiry by briefly surveying the state of science in late antiquity. Was there, in fact, a decline of science for which Christianity might be held responsible? The answer is not simple. Surely there are instances of important scientific work in the early centuries of the Christian era. Ptolemy's work in astronomy and Galen's in medicine (both in the second century A.D.) and Diophantus's mathematical efforts (in the third century) are outstanding examples. And, as we shall see, John Philoponus presented a major and important reassessment of Aristotelian physics and cosmology as late as the sixth century. Nevertheless, it is agreed by most historians of ancient science that creative Greek science was on the wane, perhaps as early as 200 B.C., certainly by A.D. 200. Science had never been pursued by very many people; it now attracted even fewer. And its character shifted away from original thought toward commentary and abridgement. Creative natural science was particularly scarce in the Roman world, where scholarly interests leaned in the direction of ethics and

[32] Gregory of Nazianzus, *The Panegyric on St. Basil,* trans. Charles G. Browne and James E. Swallow, in *A Select Library of Nicene and Post-Nicene Fathers of the Christian Church,* Ser. 2, 14 vols., ed. Philip Schaff and Henry Wace (New York: Christian Literature Company, 1890–1900), Vol. VII, pp. 398–399.

[33] A. H. Armstrong, "Reason and Faith in the First Millenium A.D.," in *Scholasticism in the Modern World* (Proceedings of the American Catholic Philosophical Association, 40) (Washington, D.C.: American Catholic Philosophical Association, 1966), p. 107; rpt. in Armstrong, *Plotinian and Christian Studies,* Vol. XII.

metaphysics; such natural science as Rome possessed was largely confined to fragments preserved in handbooks and encyclopedias.[34]

Can Christianity be held responsible in any way for this decline? Let us first consider Christian other-worldliness. In antiquity there was a broad spectrum of attitudes toward the material world. At one end of the spectrum was pagan cosmic religion, constructed from a mixture of Pythagorean, Platonic, Aristotelian, and Stoic doctrines. This religion saw the material cosmos, or at least its upper, heavenly part, as a perfect expression of divine creativity and providence, "the supreme manifestation of divinity," and indeed itself a divine being. Moreover, study and contemplation of the cosmos were judged the only ways to God; natural philosophy and theology had been merged. At the other end of the spectrum was the Gnostic attempt to equate the material world with evil. The cosmos was viewed by "pessimistic" Gnostics as a disastrous mistake, the scene of disorder and sin, the product of evil forces, the antithesis of the divine, and a prison from which the soul must escape in order to make its way to its true home in the spiritual realm.[35] Finally, between these extremes there was Platonic philosophy (or, in the hands of certain Neoplatonists, Platonic religion), which distinguished clearly between the transcendent world of eternal forms and their imperfect replication in the material cosmos. Neoplatonists by no means considered the world to be evil; it was the product of divine intelligence and, as A. H. Armstrong puts it, the "best possible universe that could be produced under difficult circumstances."[36] Contemplation of it was even held to play a positive, albeit small, role in leading the soul upward to the eternal forms. Nevertheless, Neoplatonism was fundamentally other-worldly; the material world for all its beauty, remained the scene of imperfection and disorder; and it had to be escaped before humanity could achieve its highest good, the contemplation of eternal truths.

There was, of course, no unitary Christian view of the material world. But orthodox Christianity, as it developed, emphatically rejected the extremes; nature was neither to be worshipped nor to be repudiated. Christianity was deeply influenced by Neoplatonic philosophy, and most Christian thinkers adopted some form of the Neoplatonic attitude. Gregory of Nyssa (ca. 331–ca. 396) believed deeply in the unreality and deceitfulness of the material world and yet recognized that it could provide signs and symbols that would lead mankind upward to God.[37] Augustine insisted that sin is situated not in the body, but in the will. This was a point of extraordinary importance, because it helped to

[34] The best histories of ancient science are by G. E. R. Lloyd: *Early Greek Science: Thales to Aristotle* (cit. n. 8) and *Greek Science after Aristotle* (cit. n. 6). On science in late antiquity, see Lloyd, *Greek Science after Aristotle,* Ch. 10; Clagett, *Greek Science in Antiquity,* Chs. 8–11. (Benjamin Farrington, *Greek Science* [Harmondsworth: Penguin, 1961], would put the decline of Greek science even earlier, in the 4th century B.C.) On Roman science, see William Stahl, *Roman Science: Origins, Development, and Influence to the Later Middle Ages* (Madison: Univ. Wisconsin Press, 1962).

[35] For cosmic religion I have closely followed A. H. Armstrong, "The Material Universe," Ch. 4 of Armstrong and R. A. Markus, *Christian Faith and Greek Philosophy* (London: Darton, Longman & Todd, 1960); see esp. pp. 31–33, 39–40; quoting here p. 31. On Gnosticism, see Hans Jonas, *The Gnostic Religion,* 2nd ed. (Boston: Beacon Press, 1963); Robert M. Grant, *Gnosticism and Early Christianity,* 2nd ed. (New York: Columbia Univ. Press, 1966).

[36] Armstrong, "Material Universe," p. 34.

[37] I. P. Sheldon-Williams, "St. Gregory of Nyssa," in *Cambridge History,* ed. Armstrong, pp. 447–456.

liberate western Christendom from the notion that the soul is contaminated by its contact with the body—and therefore that matter and flesh must be inherently evil.[38] Nevertheless, in a well-known passage from the *Enchiridion,* Augustine expressed serious doubt about the value of natural science:

> When it is asked what we ought to believe in matters of religion, the answer is not to be sought in the exploration of the nature of things, after the manner of those whom the Greeks called "physicists." Nor should we be dismayed if Christians are ignorant about the properties and the number of the basic elements of nature, or about the motion, order, and deviations of the stars, the map of the heavens, the kinds and nature of animals, plants, stones, springs, rivers, and mountains; about the divisions of space and time, about the signs of impending storms, and the myriad other things which these "physicists" have come to understand, or think they have. . . . For the Christian, it is enough to believe that the cause of all created things, whether in heaven or on earth, whether visible or invisible, is nothing other than the goodness of the Creator, who is the one and the true God.[39]

Yet, insofar as scientific knowledge is required, it must be taken from the pagan authors who possess it:

> It frequently happens that there is some question about the earth, or the sky, or the other elements of this world, the movement, revolutions, or even the size and distance of the stars, the regular eclipses of the sun and the moon, the course of the years and seasons; the nature of the animals, vegetables, and minerals, and other things of the same kind, respecting which one who is not a Christian has knowledge derived from most certain reasoning or observation. And it is highly deplorable and mischievous and a thing especially to be guarded against that he should hear a Christian speaking of such matters in accordance with Christian writings and uttering such nonsense that, knowing him to be as wide of the mark as, to use the common expression, east is from west, the unbeliever can scarcely restrain himself from laughing.[40]

A view broadly the same as Augustine's was presented by Pope Leo the Great (440–461). Leo argues that the material world is not to be denigrated:

> Man, awake, and recognize the dignity of your own nature. Remember that you were made in the image of God; and though it was spoilt in Adam, it has been remade again in Christ. Use these visible creatures as they ought to be used, as you use earth, sea, sky, air, springs and rivers; and praise and glorify the Creator for everything fair and wonderful in them.

But neither should the material creation be allowed to occupy the center of attention:

> Do not devote yourself to the light in which birds and snakes, beasts and cattle, flies and worms delight. Feel bodily light with your bodily senses and clasp with all the strength of your mind that true light which "lightens every man coming into this

[38] Armstrong, *Augustine and Christian Platonism,* p. 11. Despite Augustine's influence, the tendency to denigrate the flesh persisted in medieval (especially early medieval) Christendom.

[39] Augustine, *Enchiridion* 3.9, trans. Albert C. Outler (The Library of Christian Classics, 7) (Philadelphia: Westminster Press, 1955), pp. 341–342. The qualification contained in the opening line of this passage ("what we ought to believe *in matters of religion*") is often overlooked by those who quote it.

[40] Augustine, *De genesi ad litteram* 1.19; quoted by Meyrick H. Carré, *Realists and Nominalists* (London: Oxford Univ. Press, 1946), p. 19.

> world." . . . For if we are the temple of God, and the Spirit of God dwells in us, what everyone of the faithful has in his own soul is more than what he admires in the sky. We are not, of course, . . . telling you this to persuade you to despise the works of God, or to think that there is anything against your faith in the things which the good God has made good; but so that you may use every kind of creature, and all the furniture of this world, reasonably and temperately. . . . So, since we are born to the things of this present life but reborn to those of the future life, let us not devote ourselves to temporal goods but be set on eternal ones. . . .[41]

The material world is not to be loved, but to be used; it is not an end in itself, but a means to the contemplation of higher things.

What were the implications of this attitude for the scientific enterprise? If we employ as a standard of comparison some sort of ideal world, a scientist's nirvana, in which social values and resources are all marshaled in support of scientific research, Christianity may be judged harshly: the church was certainly not calling for the establishment of scientific research institutions, nor urging able young men to undertake scientific careers. Most of the pejorative pronouncements regarding the early church in relation to science seem to spring from the anachronistic application of precisely such criteria. But what we must realize is that the early church was thus expressing values obtained from the pagan environment. On a spectrum of *pagan* values, from cosmic religion to Gnostic repudiation of the cosmos, the church fathers chose a middle position. There can be no doubt that Biblical teaching about the creation as God's handiwork was decisive in determining where on the spectrum Christians would land, and therefore it is clear that their Christianity was highly relevant to the issue; but it must be recognized that the alternatives from which they chose were of pagan origin.

It seems unlikely, therefore, that the advent of Christianity did anything to diminish the support given to scientific activity or the number of people involved in it. The study of nature held a precarious position in ancient societies; with the exception of medicine and a little astronomy, it served no practical function and was rarely seen as a socially useful activity. As a result, it received little political patronage or social support, but depended on independent means and individual initiative.[42] With the declining economic and political fortunes of the Roman Empire in late antiquity, people of independent means decreased in number, and initiative was directed elsewhere. Moreover, changing educational and philosophical values were diverting attention from the world of nature. Inevitably the pursuit of science suffered. Christianity did little to alter this situation. If anything, it was a little less other-worldly than the major competing ideologies (Gnosticism, Neoplatonism, and the mystery religions) and offered slightly greater incentive for the study of nature.[43] Christianity regarded science

[41] Leo, *In nativitate Domini sermo VII;* quoted by Armstrong, "Material Universe," pp. 36–37.

[42] Edelstein, *Ancient Medicine*, pp. 434–435; Lloyd, *Early Greek Science*, pp. 125–130.

[43] It is true that cosmic religion attached great importance to the visible world as a manifestation of the divine being, but its preoccupation with the heavenly bodies and its tendency to venerate them precluded its becoming a major patron of scientific activity; indeed, as Armstrong has pointed out, cosmic religion had its own "curious kind of materialized other-worldliness" ("Material Universe," p. 33). Armstrong has also pointed out that the tendency of cosmic religion to merge science and theology was probably not beneficial for either enterprise (*ibid.*, pp. 39–40). Cf. Armstrong, *St. Augustine and Christian Platonism*, pp. 9–24; and Armstrong, "Man in the Cosmos: A Study of

as important only insofar as it served the faith; but at least on occasion it served the faith.

PRACTITIONERS OF SCIENCE

We have been proceeding at the theoretical level. What did Christian involvement in science or natural philosophy amount to in practical terms? How much science was known? What did Christians contribute to its preservation and further development? And how did it interact with their theology? These are extraordinarily difficult questions, because the basic research that would make it possible to answer them has, in general, not yet been undertaken. Nevertheless, let us take some preliminary steps toward answers by considering briefly the work of three Christians representing different degrees of involvement in natural philosophy—Basil of Caesarea, Augustine, and John Philoponus.

Basil (ca. 330–379) was from Cappadocia in eastern Asia Minor, where, in the last decade of his life, he became Bishop of Caesarea. In his *Homilies on the Hexaemeron* (the six days of creation), he brought to bear what natural philosophy he could in defense and elucidation of the Biblical account of creation. Basil begins by attacking the materialists (undoubtedly the Ionians and atomists), who have failed to see that the cosmos is a beautiful and purposeful creation, the work of an intelligent Creator. Basil's own philosophical preferences become clear when he identifies the Creator with Plato's Demiurge and accepts a Platonic hierarchy of celestial intelligences. Also, against the materialists and in concert with Christian and Platonic teaching, he defends a temporal cosmos—that is, one that had a beginning and will have an end.[44]

A good bit of Aristotelian cosmology and physics appears in the *Homilies*—much of it reported (without endorsement) as the useless imaginings and empty noises of the philosophers. Basil accepts the doctrine of the four elements and reports the arguments for a fifth celestial element. He refers to the opinion (of Anaximander and Democritus) that there is an infinity of worlds, but counters with Aristotle's denial, on geometrical grounds, of the possibility of more than a single world.[45] He inquires regarding the position of the earth in the cosmos and recounts the Aristotelian doctrine of a fixed, central earth, situated in the place to which all heavy bodies naturally descend; it is implicit in this account, moreover, that the earth is spherical. However, according to his custom, Basil

---

Some Differences between Pagan Neoplatonism and Christianity," in *Romanitas et Christianitas*, ed. Willem den Boer et al. (Amsterdam: North Holland, 1973), pp. 5–14; reprinted in Armstrong's *Plotinian and Christian Studies*, Vol. XXII. My conclusions about Christian attitudes toward nature are remarkably similar to those recently expressed by Darrel W. Amundsen in regard to Christian attitudes toward secular medicine; see his "Medicine and Faith in Early Christianity," *Bulletin of the History of Medicine*, 1982, *56*, 326–350.

[44] Basil, Homily I, in *Select Library*, ed. Schaff and Wace (cit. n. 32), Vol. VIII, pp. 53–55. See also I. P. Sheldon-Williams, "St. Basil of Caesarea," in *Cambridge History*, ed. Armstrong, pp. 432–438; Lynn Thorndike, *A History of Magic and Experimental Science*, 8 vols. (New York: Columbia Univ. Press, 1923–1948), Vol. I, pp. 481–494; and Frank E. Robbins, *The Hexaemeral Literature: A Study of the Greek and Latin Commentaries on Genesis* (Chicago: Univ. Chicago Press, 1912).

[45] Basil, Homily IV and Homily I (elements), Homily III (worlds), in *Select Library*, Vol. VIII, pp. 74–75, 58, 66.

refuses to commit himself to the truth of this scheme—although he apparently does not think it improbable. He declares simply that we should direct our admiration toward the source, rather than the details, of cosmic order:

> If there is anything in this system which might appear probable to you, keep your admiration for the source of such perfect order, for the wisdom of God. Grand phenomena do not strike us the less when we have discovered something of their wonderful mechanism. Is it otherwise here? At all events let us prefer the simplicity of faith to the demonstrations of reason.[46]

One cannot help being impressed by Basil's considerable command of basic Greek cosmology and natural philosophy—most of it obtained, no doubt, from handbooks and compendia, rather than from the original sources. But he is not overcome with admiration for pagan authors. He sometimes labels their arguments "ridiculous" and refers repeatedly to their inability to agree among themselves: "Why torment ourselves to refute the errors of philosophers, when it is sufficient to produce their mutually contradictory books and, as quiet spectators, to watch the war?"[47] Above all, he is adamant in denying these matters any importance for their own sake.

We can learn more about the way in which Biblical doctrine and pagan philosophy interacted in Basil's thought by considering his discussion of the various heavens. The problem is the apparent discrepancy between Aristotle, who had argued that beyond the planetary spheres there is a single heaven (bearing the fixed stars), and the opening verses of Genesis 1, where reference is apparently made to two heavens, one created on the first day and another on the second. The relevant passage reads:

> In the beginning God created the heaven and the earth. . . . And there was evening and there was morning, one day. And God said, "Let there be a firmament in the midst of the waters, and let it separate the waters from the waters." And God made the firmament and separated the waters which were under the firmament from the waters which were above the firmament. And God called the firmament Heaven. And there was evening and there was morning, a second day.[48]

Basil, who refuses to accept an allegorical interpretation of the passage (and, indeed, attacks his Christian predecessors for so doing),[49] feels compelled to acknowledge the existence of two heavens and, moreover, of a body of supercelestial water between the two. In the long run Basil's distinction among three separate heavenly entities was to give rise to the medieval scheme of three heavens: the outermost or empyreum, which served as the abode of angels, then the aqueous or crystalline heaven, composed of crystallized water, and finally the firmament, to which the stars are affixed.[50] We see clearly how Biblical claims could intrude into natural philosophy and shape cosmological theory.

Augustine, who flourished fifty years after Basil, had a much fuller command

[46] Basil, Homily I, *ibid.*, p. 57.

[47] Basil, Homily III, *ibid.*, p. 70; cf p. 67.

[48] Genesis, 1:1, 5–8, Revised Standard Version.

[49] Basil, Homily III, in *Select Library*, Vol. VIII, p. 71.

[50] See Francis S. Benjamin, Jr., and G. J. Toomer, eds., *Campanus of Novara and Medieval Planetary Theory: Theorica planetarum* (Madison: Univ. Wisconsin Press, 1971), pp. 393–394; Edward Grant, "Cosmology," in *Science in the Middle Ages*, ed. Lindberg, pp. 275–278.

of pagan natural philosophy than did his predecessor. His works reveal a man broadly educated in the full range of the liberal arts. In his *Confessions* he recalls his discovery of Aristotle's *Categories* and reports reading in his youth all the books on the liberal arts that he could obtain, including works on rhetoric, logic, geometry, arithmetic, and music.[51] In *De ordine,* written early in his career, he develops an educational program that includes mathematics and mathematical sciences as studies preparatory to philosophy. And in his *Retractions* he reports that he once intended to write manuals on all of the liberal arts, including arithmetic, geometry, music, and the elements of philosophy.[52] References to many pagan sources are scattered throughout his writings.

Despite his studies, Augustine came in the long run to view natural knowledge (for its own sake) with no greater enthusiasm than had Basil. We are, he advised, to set our hearts on things celestial and eternal, rather than earthly and temporal. Nevertheless, the temporal could serve the eternal, and Augustine frequently acknowledged the utility of natural knowledge for the elucidation of Christian doctrine and the exegesis of scripture. Fragments of Greek natural philosophy are thus sprinkled throughout his works. A good example (but only one among many) is his frequent use of Greek ideas on light and vision for the development of his own theology and epistemology. Augustine employs the Neoplatonic doctrine of emanation, explained by analogy with the radiation of light, to reveal the nature of the trinity.[53] He puts the phenomena of illumination to epistemological use, arguing that just as the sun must illuminate corporeal things in order that they may be seen by the corporeal eye, so intelligible things must be illuminated with a divine light if they are to be grasped by the intellect:

> But distinct from these objects [of intellectual vision] is the light by which the soul is illumined, in order that it may see and truly understand everything, either in itself or in the light. For the light is God himself, whereas the soul is a creature. . . . And when it tries to behold the Light, it trembles in its weakness and finds itself unable to do so. Yet from this source comes all the understanding it is able to attain.[54]

This epistemological use of light is a Platonic motif, taken over and Christianized by Augustine. In the course of his many discussions of the psychology and epistemology of perception, Augustine clearly commits himself to the extramission theory of vision, according to which light emerges from the observer's eye to perceive its object. In his *Literal Commentary on Genesis,* for example, he remarks that

> surely the emission of rays from our eyes is an emission of a certain light. And it can be gathered that this [light] is emitted, since when we look into the air adjacent

[51] Augustine, *Confessions* 4.16. It is important to realize that Augustine's education, like that of almost every other educated man of the period, was heavily literary in orientation. There might be some attention to quadrivial studies (arithmetic, geometry, astronomy, and music), but among Christians and non-Christians alike these would be heavily overshadowed by concern with grammar, rhetoric, and literary classics.

[52] See Henri-Irénée Marrou, *Saint Augustin et la fin de la culture antique,* 4th ed. (Paris, De Boccard, 1958), pp. 187–197; Brown, *Augustine of Hippo,* pp. 121–126. Only the manual on music survives; the others, Augustine notes in *Retractions* 1.5, were only begun and then lost.

[53] See Augustine, *De trinitate* 4.20.

[54] Augustine, *De genesi ad litteram* 12.31.59, trans. J. H. Taylor, S.J.; rpt. in Bourke, *The Essential Augustine,* p. 97.

> to our eyes we observe, along the same line, things situated far away. . . . Nevertheless, this light that is in vision is shown to be so scanty that unless it is assisted by an exterior light, we cannot see anything.[55]

Once again Augustine's sources are Platonic.

A second example of Augustine's use of natural knowledge and efforts to deal with natural questions is his doctrine of *rationes seminales,* or "seedlike principles." The problem is to reconcile the Biblical notion that God created everything in the beginning of time with the observational fact that there is a progressive development of natural (particularly biological) forms. To resolve the difficulty, Augustine calls on a Stoic notion, which Plotinus also appropriated, according to which nature contains germs or seedlike principles that direct and determine its subsequent unfolding. According to Augustine, God created all things in the beginning, some actually and some potentially—the latter as seedlike principles, which later developed into mature creatures, much as a seed develops into a mature plant. Augustine thus uses Greek natural philosophy to resolve an exegetical problem—maintaining that God's creative activity is truly completed in the beginning, and yet taking full account of observational and commonsense notions regarding the development of natural things. It is noteworthy that Augustine applies the doctrine of seedlike principles even to the origin of Adam and Eve.[56]

R. A. Markus has pointed out that from Augustine's doctrine of seedlike principles there follows a conception of natural law. This implication of the theory was acknowledged by Augustine himself in his *Literal Commentary on Genesis:*

> All the normal course of nature is subject to its own natural laws. According to these all living creatures have their particular determinate inclinations . . . and also the elements of non-living material things have their determinate qualities and forces, in virtue of which they function as they do and develop as they do. . . . From these primordial principles everything that comes about emerges in its own time and in the due course of events.

Each thing behaves according to its God-given inclination—the law of its own nature. From this it is but a short step to the distinction between God as first cause and a created order of secondary causes:

> It is one thing to build and to govern creatures from within and from the summit of the whole causal nexus—and only God, the Creator, does this; it is another thing to

[55] Quoted by David C. Lindberg, *Theories of Vision from al-Kindi to Kepler* (Chicago: Univ. Chicago Press, 1976), p. 90. On Augustine's theory of divine illumination, see R. A. Markus, "Augustine: Reason and Illumination," in *Cambridge History,* ed. Armstrong, pp. 362–373; Etienne Gilson, *The Christian Philosophy of Saint Augustine,* trans. L. M. Lynch (New York: Random House, 1960), pp. 77–96. On Augustine's optical knowledge and use of light metaphors, see also François-Joseph Thonnard, "La notion de lumière en philosophie augustinenne," *Recherches Augustiniennes,* 1962, *2,* 125–175; and Lindberg, *Roger Bacon's Philosophy of Nature* (Oxford: Clarendon Press, 1983), pp. xxxix–xli.

[56] Eugène Portalié, S.J., *A Guide to the Thought of St. Augustine,* trans. Ralph J. Bastian, S.J. (Chicago: Henry Regnery, 1960), p. 139. On the seedlike principles, see R. A. Markus, "Augustine: God and Nature," in *Cambridge History,* ed. Armstrong, pp. 398–399; Gilson, *Christian Philosophy of Augustine,* pp. 206–208; Jules M. Brady, S.J., "St. Augustine's Theory of Seminal Reasons," *New Scholasticism,* 1964, *38,* 141–158; Christopher J. O'Toole, *The Philosophy of Creation in the Writings of St. Augustine* (Washington, D.C.: Catholic Univ. America Press, 1944).

> apply externally forces and capacities bestowed by him in order to bring forth at such and such a time, or in such and such a shape, what has been created. For all things were created at the beginning, being primordially woven into the texture of the world; but they await the proper opportunity for their appearance.[57]

There is a kind of double causation: on the one hand, things change and develop according to the natures that God has given them; on the other, God governs His creation "from the summit of the whole causal nexus."

This brings us to the question of miracles. Augustine is not perfectly consistent on the subject. On one occasion he argues that an event is miraculous if caused by direct divine intervention, in violation of the natural pattern embodied in the created secondary causes. More frequently Augustine points out that God's decision to violate the usual order is no less natural (and no more miraculous) than his decision to abide by it: "Just . . . as it was possible for God to create any natures He chose to create, so it is no less possible for Him to change any qualities He chooses to change in any natures He chose to create." Ultimately everything is of divine origin, and the concept of miracle, if it has any meaning at all, represents merely the violation of our expectations:

> A portent means, in ordinary parlance, "something contrary to nature," although, in fact, such happenings are not really contrary to nature, for the simple reason that nothing that happens by the will of God can be "contrary to nature." The "nature" of any particular created thing is precisely what the supreme Creator of the thing willed it to be. Hence, a portent is merely contrary to nature as known, not to nature as it is.[58]

A final example of Augustine's relationship to pagan natural philosophy is his opinion of astrology. In the *City of God,* Augustine mounts a vigorous attack on the science of astrology, particularly its fatalistic teachings. He argues over and over that if twins, conceived at the same instant and born at almost the same time, differ dramatically in personality, character, and course of life, it surely cannot be sensibly maintained that the stars determine a person's fate.

> [The astrologers] have never been able to explain why twins are so different in what they do and achieve, in their professions and skills, in the honors they receive, and in other aspects of their lives and deaths. In all such matters, twins are often less like each other than like complete strangers; yet, twins are born with practically no interval of time between their births and are conceived in precisely the same moment of a single sexual semination.

Augustine can admit stellar influence on physical things, but the human will must be left untouched; only thus can its freedom be preserved. There was, we can see, a theological motivation underlying the discussion. A final question: if the stars do not determine human fate, if astrologers are the perpetrators of

[57] Augustine, *De genesi ad litteram* 9.17.32 and *De trinitate* 3.9.16; quoted by Markus, "Augustine: God and Nature," pp. 399, 400. I owe Markus not only the basic point of this paragrph, but also the quotations.

[58] Augustine, *City of God* 21.8, trans. Gerald G. Walsh, S.J., and Daniel J. Honan, in *Fathers of the Church,* ed. R. Deferrari et al., Vols. I–VIII (New York: Fathers of the Church; Washington: Catholic Univ. America Press, 1947–), Vol. VIII, pp. 362, 359. On Augustine's view of miracles, see Markus, "Augustine: God and Nature" pp. 400–402; Brown, *Augustine of Hippo,* pp. 415–418.

fraud, how can the occasional astrological success be explained? Successful astrological predictions, Augustine maintains, have nothing to do with the casting of horoscopes, but depend on the promptings of evil spirits, "whose business it is to persuade men, and keep them persuaded, of the false and dangerous opinion that men's destinies are settled by the stars."[59]

The last figure in our study is John Philoponus, an Alexandrian Christian of the first half of the sixth century, who illustrates the compatibility between Christianity and very intense involvement in natural philosophy. Philoponus was a professional teacher, holder of a chair in philosophy in the school of Alexandria, and one of the last great ancient commentators on Aristotle. He wrote commentaries on several of Aristotle's logical works, as well as on the *Physics, Meteorology, On the Soul,* and *On Generation and Corruption*—works in which he undertook a major attack on Aristotelian natural philosophy. Philoponus's central goal was to deny Aristotle's dichotomy between the celestial and terrestrial regions. To that end he argued that different stars are of different colors, that difference of color implies variations in composition, that composition implies the possibility of decomposition and decay, from which it follows that the heavens are no more exempt from decay than are things in the terrestrial region. He argued that the sun is composed of fire (a terrestrial substance) rather than a fifth celestial substance, the quintessence, and that astronomy (he clearly has Ptolemaic astronomy in mind) destroys the Aristotelian notion that heavenly bodies possess simple motion about the center of the universe. It follows that the heavens are not divine, and this enabled Philoponus to draw a radical distinction between the Creator and all of his creation (heaven as well as earth). A central Aristotelian doctrine thus fell before Christian doctrine. But this does not mean that the attack was philosophically frivolous; on the contrary, Philoponus proceeded intelligently, with considerable rigor, and (historians of science have been quick to point out) with notable benefit for the future course of cosmology.[60]

There was much more to Philoponus's campaign against Aristotelian philosophy. He attacked Aristotle's doctrine of the eternity of the world. He attempted a reassessment of Aristotle's theory of light.[61] And he undertook a major assault on Aristotelian dynamics, denying that in a medium a body falls with a speed proportional to its weight, that speed of descent would be infinite in a void, and that a projectile is maintained in motion through the action of the medium after it loses contact with the projector. It is noteworthy that Philoponus's attack on Aristotelian dynamics rested not on any kind of theological foundation, but to a very considerable extent on arguments from experience.[62] Philoponus's ef-

[59] Augustine, *City of God* 5.1, trans. Demetrius B. Zema, S.J., and Gerald G. Walsh, S.J., in *Fathers of the Church,* Vol. VI, pp. 243, 254. The "twins" argument against astrology was not original with Augustine. On Augustine's view of astrology, see also Theordore O. Wedel, *The Mediaeval Attitude toward Astrology, Particularly in England* (New Haven: Yale Univ. Press, 1920), pp. 20–24; Thorndike, *History of Magic,* Vol. I, pp. 504–522.

[60] On Philoponus see S. Sambursky, "John Philoponus," *Dictionary of Scientific Biography* (New York: Scribners, 1970–1980), Vol. VII, pp. 134–139; Sambursky, *The Physical World of Late Antiquity* (London: Routledge & Kegan Paul, 1962), pp. 154–175 and passim; and I. P. Sheldon-Williams, "The Reaction against Proclus," in *Cambridge History,* ed. Armstrong, pp. 477–483.

[61] Jean Ann Christensen, "Aristotle and Philoponus on Light" (Ph.D. diss., Harvard University, 1979).

[62] For translations of some of the relevant documents, see Morris R. Cohen and I. E. Drabkin,

forts, some of them motivated by Christian belief and some of them not, clearly confute the claim that Christianity and serious natural philosophy were fundamentally and necessarily antagonistic.

CONCLUSIONS

A sober view of the relationship between Christianity and science in the patristic period has proved remarkably difficult to obtain. One reason for this is that studies of the problem have so often been undertaken with polemical or apologetic purposes in view, as expressions of religious preference. Critics of Christianity have seized upon instances of Christian displeasure with pagan learning and inflated them into a systematic rejection by the religious establishment of the scientific enterprise. Defenders of Christianity, playing the same game, have exaggerated Christian contributions to science into representative episodes and symbols of a positive relationship between Christianity and science. But the true relationship was far more complex than either of the extreme positions reveals.

Our attempt to characterize that relationship must begin with a qualification. When we speak of the Christian position, we mean the "center of gravity" of a distribution of Christian opinion, for great variety existed. The church was not monolithic, and there was no universal Christian view of pagan philosophy or natural science. Christian attitudes toward classical culture were perhaps as diverse as the comparable attitudes of pagans. In each community there were people who valued philosophy and others who denigrated it, people who thought natural science useful and those who considered it a waste of time or even a detriment. Such attitudes were determined not merely by the claims of theology, but by other forces as well. In late antiquity there were social and intellectual forces tending to discourage and alter the character of philosophical discourse, particularly to divert attention from the impracticalities of natural philosophy toward the quest for true happiness and other matters of ultimate concern.[63] Christians, of course, responded differently to these forces, just as they responded differently to the claims of Christian theology.

How, then, should Christian involvement in science or natural philosophy be characterized? Few Christians regarded study of the natural world as of more than secondary, perhaps even tertiary, importance. Next to salvation and the development of basic Christian doctrine, it was decidedly insignificant. There is no cause for alarm, Augustine pointed out, if the Christian "should be ignorant of the force and number of the elements. . . . It is enough for the Christian to believe that the only cause of all created things . . . is the goodness of the Creator."[64] It would be crude distortion to maintain that Christianity offered major stimulus to scientific activity.

But it would also be distortion to create the impression that there was no Christian involvement in natural philosophy or that the church retarded or crushed science. Many fathers of the church not only possessed a significant

---

A Source Book in Greek Science (Cambridge, Mass.: Harvard Univ. Press, 1948), pp. 217–223; for analysis, see Clagett, *Greek Science in Antiquity*, pp. 169–176.

[63] Many scholars have pointed out that in antiquity the term "philosophy" came increasingly to denote the quest for happiness or salvation.

[64] Augustine, *Enchiridion* 3.9 (cit. n. 39).

body of natural knowledge, but also considered it useful for scriptual exegesis and defense of the faith. Augustine, in his efforts to formulate a Christian world view, put considerable portions of Greek natural philosophy (particularly Platonic) to work. Thus the church fathers used Greek natural science, and in using it they transmitted it. We must count this transmission as one of the major Christian contributions to science. Until the twelfth century, when a wave of translation brought an abundance of new sources to the Latin-speaking West, patristic writings constituted a major repository of scientific learning.

What the church transmitted, it also altered—and had its own doctrines altered in return. Christian doctrine and Greek natural philosophy must be viewed not as independent, unchangeable bodies of thought, situated side-by-side in the patristic period, with an occasional exchange of fisticuffs, but as interacting and mutually transforming views of the world. Christianity first transformed the philosophical tradition by performing a selective function. Because the church fathers had a strong preference for Platonic philosophy, they helped to determine that Plato's view of the world would prevail for a thousand years, until direct access to Aristotelian philosophy was gained in the twelfth century. But transformation could also occur when revelation impinged directly on natural questions. For example, the Scriptures directly addressed the eternity of the world, and we have seen how Basil's understanding of the opening verses of Genesis led to the multiplication of celestial spheres. The heavens were also dedivinized as a result of the encounter between natural philosophy and Christian theology. In return, Christians learned to read the Bible with Greek, particularly Platonic, eyes; and Christian theology became thoroughly embued with Greek metaphysics and cosmology.[65] The extent of this mutual transformation was probably unrecognized by the participants and unwanted, but unless we take cognizance of it we cannot begin to understand the subsequent course of Western theology, philosophy, and science.

[65] For discussions of the influence of Greek thought on Christian doctrine, see Edwin Hatch, *The Influence of Greek Ideas on Christianity,* ed. Frederick C. Grant (New York: Harper, 1957); Timothy, *Early Christian Apologists,* pp. 81–98.

Gregory of Tours, Monastic Timekeeping, and Early Christian Attitudes to Astronomy

By Stephen C. McCluskey

WESTERN EUROPE PRODUCED FEW scientific achievements during the period from the fifth to the tenth centuries. Although the early Middle Ages maintained a tenuous contact with the science of antiquity, Roman science itself provided a poor foundation for the construction of a new science.[1] Yet the Dark Ages were succeeded by that remarkable endeavor of scholars from the Latin West to seek out Greco-Arabic science and assimilate it into their own intellectual tradition.

This search for scientific knowledge has significant implications, for we seek only what we know of and value. If there was little scientific progress in the early Middle Ages, a rudimentary scientific activity was nonetheless essential to that later quest for learning. Monasteries were a major source of stability and the principal focus of learning during this unsettled period, and we should look to them if we are to characterize more precisely this scientific activity, its context, and its development.

The rudimentary techniques of astronomical computus that arose to determine the date of Easter have been a recurring topic of discussion. Less well studied, however, are the observational practices concerned with that monastic timekeeping employed to regulate the hours of monastic prayer. This astronomy is applied, not pure, science; it is best understood in the context of the rituals it was intended to regulate. The earliest requirement that the time of prayer be governed by watching the stars is found in John Cassian's discussion of the night offices, which is based largely on practices he had observed in the monasteries of Lower Egypt during the closing decades of the fourth century.[2] Cassian's

Drafts of this paper were read by Anthony Aveni, Bruce Eastwood, Gregory Good, David Lindberg, Wesley Stevens, and a number of anonymous referees. An earlier version was presented at the Seventh Medieval Science Colloquium, Collegeville, Minn., 1–5 June 1987. I am grateful for the criticisms and suggestions I have received, and to the library staff of St. Vincent Archabbey and College, Latrobe, Pennsylvania.

[1] William H. Stahl, *Roman Science: Origins, Development, and Influence to the Later Middle Ages* (Madison: Univ. Wisconsin Press, 1962); see esp. pp. 250–251.

[2] Robert Taft, *The Liturgy of the Hours in East and West* (Collegeville, Minn.: Liturgical Press, 1986), pp. 58–62, 76–80, 96–100; Owen Chadwick, *John Cassian* (Cambridge: Cambridge Univ. Press, 1968), pp. 71–73; and John Cassian, *De institutis coenobiorum* 2.1–2.6, ed. and trans. Jean-Claude Guy (Paris: Editions du Cerf, 1965). Egyptian monastic rules mention a call to assemble for predawn prayers and the discipline of latecomers. There is no indication whether stellar timekeeping was involved: Arnold Villeux, *La liturgie dans la cénobitisme Pachômien au quatrième siècle* (Studia

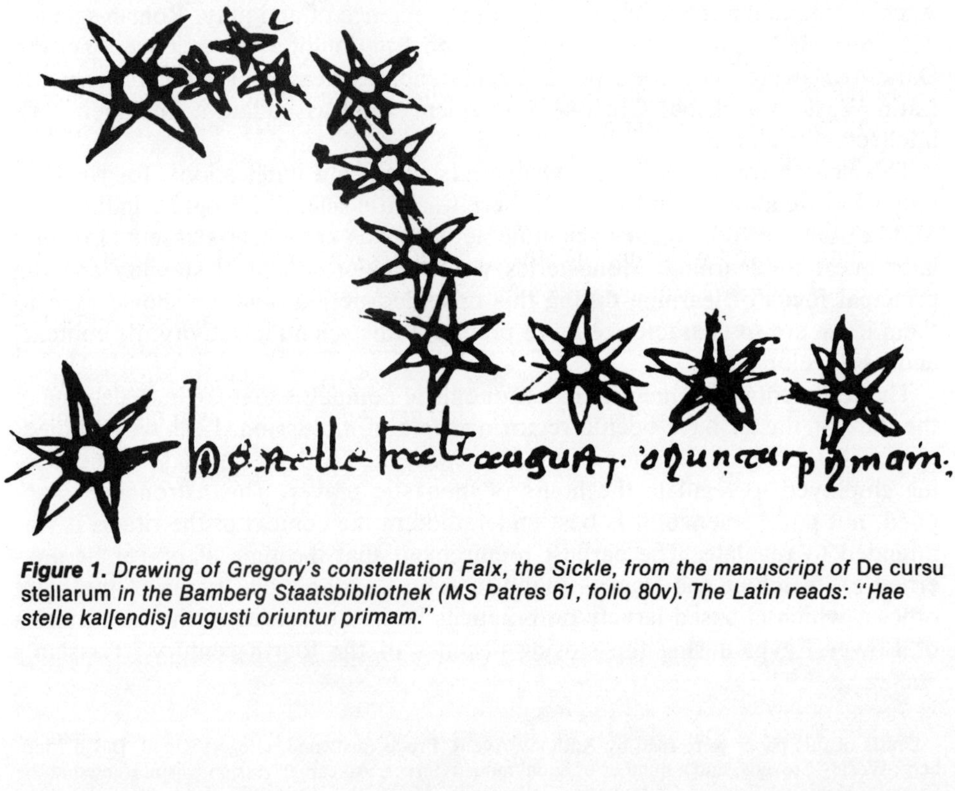

Figure 1. *Drawing of Gregory's constellation Falx, the Sickle, from the manuscript of* De cursu stellarum *in the Bamberg Staatsbibliothek (MS Patres 61, folio 80v). The Latin reads: "Hae stelle kal[endis] augusti oriuntur primam."*

Institutions (*De institutis coenobiorum*) and *Regula*, written in Marseilles in the 420s, disseminated these practices throughout the Latin West. Cassian stipulated that the night offices should be governed, not by a whim or by the sleeplessness of the monk assigned to wake his brothers, but by careful and frequent observation of the course of designated stars that regulated calling the brothers to prayer.[3]

Subsequent monastic rules, inspired in part by Cassian, shared his concern with the regular sequence of monastic prayer. St. Benedict of Nursia in his *Rule* (dated about 550) holds the abbot personally responsible for sounding the signal to wake the community for prayer, although he is allowed to entrust this duty to a careful brother. In the event the monks are wakened too late to complete their prayers, the negligent monk should "make due satisfaction to God in the oratory."[4]

The earlier and more detailed *Rule of the Master* (*Regula Magistri*, dated to the first quarter of the sixth century) designates this duty in a weekly rotation among the deans, or *praepositi*, who stay up in pairs lest one fall asleep, waking the abbot when the time for prayer has arrived. He, in turn, sounds the signal to wake the remaining monks. Isidore of Seville's *Rule for Monks* (*Regula monachorum*, written in the first quarter of the seventh century) assigns the responsibility for nocturnal timekeeping specifically to the sacristan.[5] The mechanics of reckoning the time for prayer, however, are not specified in monastic rules, even Cassian's; that is left to individual arrangements.

Keeping time by watching the stars is a rudimentary kind of astronomy, more akin to the folk astronomies of primitive peoples and of Hesiod's *Works and Days* and Virgil's *Georgics* than to the learned mathematical astronomy of the quadrivium, for which Ptolemy's *Almagest* would provide the paradigm. Yet it does require familiarity with the constellations and some knowledge of their movements with the seasons and of the changing length of day and night through the year. One of the few discussions of these technical details in a monastic context appears in the short treatise *De cursu stellarum* (*On the Course of the Stars*), ascribed since 1853 to Gregory of Tours, an ascription confirmed by subsequent recovery of a fragment of an eighth-century manuscript of the treatise.[6]

Anselmiana, 57) (Rome: Herder, 1968), pp. 292–315. Resolution of Egyptian timekeeping practices would demand study of the largely unexplored Coptic texts.

[3] "Sollicite frequenterque stellarum cursu praestitutum congregationis tempus explorans," Cassian, *De institutis* 2.17; cf. *Regula Cassiani* 16, in H. Ledoyen, "La 'Regula Cassiani' du CLM 28118 et la Règle anonyme de l'Escoriale A.I.13," *Revue Bénédictine*, 1984, *104*:175.

[4] *Regula Benedicti* 45.1, 11.12–13, ed. and trans. Timothy Fry (Collegeville, Minn.: Liturgical Press, 1981), pp. 206–207, 246–247. It may be significant for the spread of stellar timekeeping that Benedict specifically recommends Cassian's *Institutions* as a guide for monastic perfection: *ibid.*, 73.6. On the chronology of and influences among monastic rules see Adalbert de Vogüé, "The Cenobitic Rules of the West," *Cistercian Studies*, 1977, *12*:175–183.

[5] *Regula Magistri* 31–32, ed. Adalbert de Vogüé (Paris: Editions du Cerf, 1964), pp. 168–175; and Isidore of Seville, *Regula monachorum* 20.1, in *Patrologiae cursus completus, series Latina*, ed. J. P. Migne (Paris, 1850), Vol. 83, cols. 889–890.

[6] See Gregory of Tours, *De cursu stellarum*, ed. F. Haase (Wroclaw, 1853), which attributes the text to Gregory because he described having written such a work and it had similarities with his known works; and Vienna, Österreichische Nationalbibliothek, Ser. Nov. 37 (Hill Monastic Microfilm Library, or HMML, 20549), an eighth-century fragment recovered from the bindings of two codices, which attributes the text to Gregorius [Tu]roni. *De cursu stellarum* was also edited by B. Krusch, *Monumenta Germaniae historica, scriptorum rerum Merovingicarum*, Vol. I, Pt. 2 (Hannover: Hahn, 1885), pp. 854–872. The astronomical portion (18–47) is found in two manuscripts:

Gregory's application of astronomical knowledge to the service of God emphasizes the proper completion of the daily round of prayer. Regular communal prayer had been stressed by the canons of the Second Council of Tours, held in 567, only six years before Gregory's consecration as bishop. The canons spelled out the number of psalms to be chanted at each season of the year in the Basilica of St. Martin and stipulated that whoever failed to chant the required number would fast on bread and water until evening prayer.[7] *De cursu stellarum* was written some time after Gregory's consecration as bishop in 573, apparently as part of his episcopal obligation to instruct and guide the clergy and regulate the numerous monasteries in the vicinity of Tours.[8] Gregory's treatment of mundane details compensates for the lack of particulars in either of the monastic rules that may have been used at Tours, whether John Cassian's *Institutions* or the *Rule of the Master*.[9]

In *De cursu stellarum* we see a practical astronomy that has been reduced to its barest essentials. Gregory's stated purpose was to provide only those elements of astronomy that would help a monk determine the proper time to rise in the hours before dawn for nocturnal prayer. But this is not an indigenous Gallic folk astronomy;[10] Gregory was educated in the declining tradition of Roman Gaul, and parallels to his astronomy can be found in a wide range of texts. His selection of particular elements that served his practical religious purpose illustrates how monastic culture, with its faith in a divinely established natural order, shaped the preservation and transformation of astronomical practice.

ASTRONOMY, ASTROLOGY, AND CHRISTIAN CULTURE

As an introduction to astronomy *De cursu stellarum* is little more than an appetizer. No geometric abstractions, no mentions of circles and poles occur in Gregory's treatise. Instead of using abstract celestial coordinates, Gregory defines the positions of the stars in terms that would be directly useful to an observer watching them as they rise.

Bamberg Staatsbibliothek, MS Patres 61 (eighth c.), fols. 75v–82v; and Vatican, Biblioteca Apostolica Vaticana, MS 67 (twelfth c.), fols. 120v–126v. None of the manuscripts used in previous editions name the author.

My thanks to the directors of these libraries and of the Hill Monastic Microfilm Library for providing microfilm copies of manuscripts in their care. I am currently preparing a new edition of the entire treatise based on the eight extant manuscripts.

[7] *Concilia Galliae*, A. 511–A. 695, ed. Carolus de Clercq, *Corpus Christianorum, series Latina* (Turnholt: Brepols, 1963), Vol. 148A, pp. 182–183. The stipulated monthly sequence of psalms is not the same as that of *De cursu stellarum*.

[8] For the exceptional influence of bishops on the foundation and regulation of early monasteries, in contrast to the later independence of abbots under the Rule of St. Benedict, see Elie Griffe, "Saint Martin et la monachisme gaulois," in *Saint Martin et son temps: Mémorial du XVIe centenaire des débuts du monachisme en Gaule, 361–1961* (Studia Anselmiana, 46) (Rome: "Orbis Catholicus"/ Herder, 1961), pp. 17–19. On the monasteries founded near Tours before 590, inspired by the establishment of Marmoutier by the monk Bishop Martin (ca. 316–397), see Friedrich Prinz, *Frühes Mönchtum im Frankenreich: Kultur und Gesellschaft in Gallien, den Rheinland und Bayern am Beispiel der monastischen Entwicklung (4. bis 8. Jahrhundert)* (Munich/Vienna: R. Oldenbourg, 1965), map XIIa.

[9] Prinz, *Frühes Mönchtum*, p. 98, n. 54, and map IVa.

[10] Much remains to be done toward the recovery and interpretation of medieval folk astronomies. See, e.g., Stephen C. McCluskey, "The Mid-Quarter Days and the Historical Survival of British Folk Astronomy," *Archaeoastronomy*, No. 13, supplement to *Journal for the History of Astronomy*, 1989, *20*:S1–S19.

If Gregory's astronomy is not part of the mainstream of Greco-Roman mathematical astronomy, similarities to a number of astronomical traditions suggest direct or indirect influences. An analogue to Gregory's location of constellations in relation to the path of the sun appears in Martianus Capella's widely circulated *Marriage of Philology and Mercury*. Here we find a description of 183 parallel daily circles that the sun travels twice each year. Gregory had read this standard text in the liberal arts and noted that Martianus "taught you . . . to trace the courses of the stars in his astronomy."[11]

Descriptions of the successive risings of the constellations fill the works of Virgil and the other astronomical poets, a genre that Gregory mentions, if only to reject the poets' mythological descriptions of the constellations.[12] The genre includes such calendrical works as Virgil's *Georgics,* Ovid's *Fasti,* Manilius's *Astronomica,* Aratus's *Phaenomena,* and parts of Pliny's *Natural History*. The works are filled with practical astronomical detail, and also with allusions to the power of the stars and their mythological origins that are antithetical to the evolving, more strictly Christian, culture.

It was Christian culture that guided Gregory's relation to the diverse astrologies of his Greek and Roman predecessors and of his Germanic, Celtic, and even some Christian contemporaries with Manichean proclivities.[13] He carefully rejected those constellation names that assigned mythological origins to the constellations or would impute a divine nature to the stars. In *De cursu stellarum* the celestial motions do not represent powers independent of God's dominion over creation; they, like the other natural wonders that Gregory discusses, are under the dominion of God.

These roles of the divine lawgiver, his dominion over nature and nature's ultimate dependence upon him, are found in monastic prayer as well. The predawn office of lauds takes its name from Psalms 148–150, a series of praises of God by his creation that form one of the earliest unvarying elements of monastic prayer. Through the centuries, every day began with praises of God for his dominion:

> Give praise to the Lord in heaven;
>   praise him, all that dwells on high.
> Praise him all you angels of his,
>   praise him, all his armies.
> Praise him, sun and moon;
>   praise him, every star that shines.
> Praise him, you highest heavens,
>   you waters beyond the heavens.
> Let all these praise the Lord;
>   it was his command that created them.
> He has set them there unageing for ever;
>   given them a law that cannot be altered.[14]

[11] Martianus Capella, *De nuptiis Philologiae et Mercurii* 7.872, trans. W. H. Stahl and R. Johnson (New York: Columbia Univ. Press, 1977); and Gregory of Tours, *Historia Francorum* 10.31, trans. O. M. Dalton (Oxford: Clarendon Press, 1927).

[12] *De cursu stellarum* 17.

[13] The semi-Manichaean followers of the Spanish bishop Priscillian of Avila maintained that the stars influenced the fate of the human soul. Priscillian was executed on a charge of sorcery, stemming from his ritual practices and his Manichaean tendencies, in 386, yet he was revered in Spain late into the sixth century. Henry Chadwick, *Priscillian of Avila: The Occult and the Charismatic in the Early Church* (Oxford: Clarendon Press, 1976), pp. 138–148, 195–201.

[14] Psalm 148, lines 1–6; the text is from Ronald Knox's translation of the Vulgate (New York: Sheed & Ward, 1954). Medieval commentators commonly interpreted this passage as referring to an

God's dominion was still questioned among the rustic population of Gregory's time, who continued the practices of their Germanic and Celtic forebears. Eligius, Bishop of Noyon from 641 to 660, reflected Christian perceptions of the pagan threat when he forbade the practice of addressing the sun and moon as "Lords," for they serve the needs of men by divine decree.[15] A syncretistic conflation of Christianity with astral theology is found even in the Bamberg manuscript of *De cursu stellarum*. Accompanying the discussions of the regular change of the length of day through the seasons and of the visibility of the moon through the month are two crude drawings of personifications, or even deifications, of the sun and moon set in circles atop pillars. The image of the sun contains a bearded Christlike figure, crowned with twelve solar rays; the figure of the moon is feminine and crowned with a lunar crescent.[16]

Equally incompatible with the norms of Christianity were the teachings of mathematical astrologers who sought to foretell the future through the stars. Gregory, with many of his contemporaries, saw this as a superstitious art and rejected it, denying especially that the stars were causes of events on the earth.[17] He nonetheless admitted that there could be signs in the heavens. His discussion of the constellations closes with a brief treatment of comets in which he acknowledges them as signs of future events. As examples he describes two recent comets, one of which had preceded a pestilence in the Auvergne while the other preceded the death of King Sigibert in 575. In the same passage he quotes a poem for the Epiphany by Prudentius (348–ca. 410), whose more strident tone reflects his conflict with a still-active Roman paganism. Prudentius interpreted the Christmas star as both foretelling the coming of Christ and symbolizing his dominion over the heavens. To him the Christmas star was like a royal banner of that great ruler who commands the stars, before which the other stars give way.[18] The Christian dispute with astrology was not merely a dispute with pagan superstition; it was also a dispute over a concept of the universe. To the pagan or the Manichaean the world was an arena of conflict among opposing forces; to the Christian all creation was ruled by a single supreme Lord.

To the extent that the Christian universe operated under the dominion of a

---

unchanging natural order. Augustine, *Enarrationes in Psalmos,* Psalm 103, Sermo 4, 2; Psalm 148, 3, ed. Eligius Dekkers and Johannes Fraipont, *Corpus Christianorum* (cit. n. 7), Vol. 40; Cassiodorus, *Expositio Psalmorum* 148.1–6, ed. M. Adriaen, *ibid.,* Vol. 97; an anonymous Irish commentator, in *Glossa in Psalmos: The Hiberno-Latin Gloss on the Psalms of Codex Palatinus Latinus 68,* ed. Martin McNamara (Studi e Testi, 310) (Vatican City: Biblioteca Apostolica Vaticana, 1986), p. 307; and Remigius of Auxerre, *Enarrationes in Psalmos,* Psalm 148, in *Patrologia latina,* ed. Migne (cit. n. 5), Vol. 131, col. 840. See also Robert M. Grant, *Miracle and Natural Law in Graeco-Roman and Early Christian Thought* (Amsterdam: North-Holland, 1952), pp. 19–28.

[15] *Vitae Eligii Episcopi Noviomagensis Liber II,* in *Monumenta Germaniae historica* (cit. n. 6), Vol. IV (Hannover: Hahn, 1902), p. 707.

[16] Bamberg Staatsbibliothek, MS Patres 61, fol. 79r. The twelve rays are an iconographic convention, representing the months or the signs of the zodiac and reflecting the sun's role as governor of the year. On early Christian symbolism of the sun and moon see Hugo Rahner, S.J., *Greek Myths and Christian Mystery,* trans. Brian Battershaw (New York: Harper & Row, 1963), pp. 89–176.

[17] *De cursu stellarum* 16; M. L. W. Laistner, "The Western Church and Astrology During the Early Middle Ages," *Harvard Theological Review,* 1941, *34*:251–275; and R. Bonnaud, "Notes sur l'astrologie latine au VIe siècle," *Revue Belge de Philologie et d'Histoire,* 1931, *10*:557–577.

[18] *De cursu stellarum* 34; and Prudentius, *Cathemerinon* 12.5–40, in *Works of Prudentius,* ed. and trans. H. J. Thomson (Cambridge, Mass.: Harvard Univ. Press, 1949). Cf. Prudentius, *Apotheosis* 611–630, *ibid.,* and Gregory, *Historia Francorum* 4.36(51), 4.24(31) (cit. n. 11). The monks for whom Gregory was writing were members of a society that was at least nominally Christian. On Gregory's evaluation of the significance of signs and prodigies see Giselle de Nie, "Roses in January: A Neglected Dimension in Gregory of Tours' *Historiae,*" *Journal of Medieval History,* 1979, *5*:259–289.

supreme lawgiver, its laws were potentially intelligible. Gregory followed Prudentius in making this careful distinction: he accepted signs in the heavens, but he would not admit that these signs were independent causes that challenged God's dominion over nature and mankind.

THE ASTRONOMY OF *DE CURSU STELLARUM*

De cursu stellarum can be divided into four sections, of which only the last two touch our concern with astronomical practice. The first two sections treat what seems to be a scarcely related topic, the wonders of the world. Here Gregory goes beyond the immediate demands of monastic timekeeping to emphasize God's trustworthy dominion over nature.

The first section deals with the seven wonders, or *miracula,* of the ancient world, and the second with seven natural wonders created by God, which will endure "until the Lord directs the dissolution of the world."[19] Gregory treats these natural wonders largely as spiritual allegories; some, such as the burning springs of Grenoble, are seen as instances where the divine hand has suspended the usual course of nature.[20] Yet the last two wonders, the regular motions of the sun and of the moon, are noted as clear expressions of a divinely established order.[21] This treatment of the sun and moon begins the astronomical discussion and leads into the two following sections, which first describe the motions of a number of constellations and then show how the regular course of the stars can be employed to regulate the regular course of monastic prayer. Despite their differences, all four sections reflect the unifying theme running through the entire treatise: a consideration of how God established the order of creation and how it, in turn, can order the worship of the creator.

The astronomical discussion begins with the regular changes of the length of the day through the seasons, reckoning this *miraculum,* along with the changing phases of the moon, as the sixth and seventh wonders of God's creation. The scheme that Gregory provides for the length of daylight is clearly not indigenous to Frankish Gaul. It reflects a simple method of calculation in which the length of daylight in December is taken as nine hours and posited to increase regularly by one hour per month until June, when it reaches fifteen hours. The same linear process then reverses back to December, when the length of daylight is again nine hours.

Such rudimentary arithmetical schemes were common, especially in the eastern Mediterranean, where they are found in such diverse sources as astrological treatises and the menologia of the Orthodox churches. Many of these schemes, which may represent a survival of early Babylonian linear schemes, employ the same 15:9 ratio of longest to shortest days. This ratio is more appropriate for the Mediterranean region than for the latitude of Tours (47.3°N), where it is approxi-

[19] *De cursu stellarum* 9.

[20] "'Limpharum in gremiis inimicus condidit ignis, communesque ortus imperat alta manus. . . . ' O admirabile potentiae divinae misterium!" *Ibid.*, 14. The opening quotation is from a poem by a certain Hilarius, usually taken to be Bishop Hilarius of Arles (d. 449), but sometimes Bishop Hilarius of Poitiers (fl ca. 315–367). In speaking of the fire hidden in the bosom of the spring, Hilarius and Gregory allude to the conflict between pagan and Christian attitudes toward nature, for Lympha is not just a term for a spring but the name of a goddess of such places.

[21] *De cursu stellarum* 15–16. On Gregory's acceptance of a clear order of nature see de Nie, "Roses in January" (cit. n. 18), pp. 267–279; and Giselle de Nie, "The Spring, the Seed and the Tree: Gregory of Tours on the Wonders of Nature," *J. Medieval Hist.*, 1985, *11*:89–135.

Table 1. Gregory's scheme for determining lunar visibility through the month.

Day	—	1	2	3	4	5	6	7	8	9	10	11	12	13	14	15
Hours visible	0	½	1½	2	3	4	5	5½	6	7½	8	8½	9½	10	11	12
Day	30	29	28	27	26	25	24	23	22	21	20	19	18	17	16	—

mately 15.75:8.25, but because of the simplicity of changing one hour each month, systems employing the 15:9 ratio are widespread.[22]

Gregory uses a similar linear scheme to determine the changing visibility of the moon through the month (see Table 1). Here the moon's visibility increases by about an hour each day until it reaches twelve hours on the fifteenth day of the month; it then decreases with similar uniform steps to invisibility on the thirtieth day of the month. Such schematic thirty-day months are found in the work of Vettius Valens, in Bede's *De temporum ratione,* and in a number of medieval texts used by Bede. The method does not account for seasonal changes, and Gregory makes no attempt at precision, rounding his tabulated values, albeit inconsistently, to the next lower half hour.[23]

De cursu stellarum treats the stars in two separate sections. First is a general astronomical description of fourteen constellations; next is a discussion of Gregory's primary concern: how the course of the stars can be used to regulate the course of nocturnal prayer throughout the year. His liturgical discussion concerns the time to rise for the first of the two night offices: nocturns (called matins in modern monastic practice), which begins a few hours after midnight, and matins (now called lauds), which follows about dawn. Most rules stipulate an interval of rest during winter between nocturns and lauds; from Easter to the autumnal equinox, nocturns leads directly into lauds.[24] The length of the night offices similarly varied to suit the seasons, being shorter in summer and longer during the long nights of winter. This seasonal pattern could be further modified by liturgical requirements. In some months—for example, the penitential periods before Christmas and Easter—Gregory allows both for the regular nocturnal office and for lengthy vigils in which a hundred or more psalms would be chanted.[25]

[22] Otto Neugebauer, *A History of Ancient Mathematical Astronomy* (New York: Springer-Verlag, 1975), pp. 706–711. Pliny (*Natural History* 6.216, 219) describes the longest day appropriate to northern Greece and southern Italy as 15 hours, while he gives the longest day for the Atlantic coast of Gaul as 16 hours, and for Britain, 17 hours. Similar linear functions appear in other early medieval Latin texts. An anonymous martyrology in an eighth-century Salzburg manuscript that also contains Bede's *De temporum ratione* and other computistical works employs a similar linear function stepping one hour twice each month. Although the text is inconsistent for June and December, it appears that the author intended to skip the steps at the solstices and achieve an extreme value of 17:7. John McCullough, "*Martyrologium excarpstum:* A New Text from the Early Middle Ages," in *Saints, Scholars and Heroes: Studies in Medieval Culture in Honour of Charles W. Jones,* ed. Margot H. King and Wesley M. Stevens, Vol. II (Collegeville, Minn.: Saint John's Abbey and Univ., 1979), pp. 179–237.

[23] *De cursu stellarum* 18; Neugebauer, *History of Ancient Mathematical Astronomy,* p. 830; and Bede, *De temporum ratione* 24, in *Bedae opera de temporibus,* ed. Charles W. Jones (Cambridge, Mass.: Medieval Academy of America, 1943), pp. 226–227, 359. Quite different, and inconsistent, values are given in Pliny, *Natural History* 2.11.58; 18.65.323–325. My discussion of Gregory's practice of rounding benefits from a reviewer's comments. Given the irregularities in Gregory's table, there is little justification for connecting it with any particular source.

[24] *Regula Benedicti* 8 (cit. n. 4); and *Regula Magistri* 33 (cit. n. 5).

[25] Lengthy vigils are called for in a number of early monastic rules but are absent from Benedict's *Rule,* which first appeared in southern Gaul about the year 625 and only became significant after 690:

In discussing the constellations, Gregory is careful to avoid any taint of the astral theology or astrology of his pagan Greek and Roman predecessors. He ignores many of the familiar constellation names, and his constellations sometimes overlap the boundaries of several classical constellations. Consequently, even identifying his asterisms requires some attention to minor details.

Gregory's descriptions include as many as three separate elements to assist his reader in identifying the constellations: A sketch of the constellation, an indication of the time of its rising or setting and, in a few cases, a description of the place where it rises that reflects its declination. The drawings of each constellation depict them oriented as if rising on the eastern horizon. As they appear in the Bamberg manuscript the illustrations are artistically crude; the stars are not connected or outlined to indicate the form of the constellation. Yet the configuration of the stars and the brightness of distinctive ones provide useful guides for the monastic observer.[26] The sketches alone clearly identify the larger constellations, although they cannot unambiguously define smaller constellations with fewer stars (see Figure 1).

Further guidance comes from the month of heliacal rising, which Gregory gives for most constellations. For six of them he also tabulates the number of hours that the constellation is visible for each month through the year, mentioning, when appropriate, those months when it is invisible or those when it sets after sunset and rises again before dawn. Some descriptions that do not directly give times of rising note that the constellation rises a given number of hours after another in the series. Some descriptions contain a third element that sets limits on the constellations' declination. In four cases Gregory notes explicitly that the constellation follows the path of the sun for a particular month; at other times there are allusions to it following a northerly or southerly path.

The liturgical section, by discussing five constellations that govern the time of nocturnal prayers, provides further indirect information about the rising of these constellations. For each month Gregory gives several rules of the sort that if you begin when a given star rises you can sing a certain number of psalms before dawn, or if you begin when another star reaches the third hour, you can sing another number of psalms.[27] He gives such instructions over several months for stars in these five constellations, and for most months he uses several stars. This provides enough redundant information to compute the time each constellation rises, which further limits its position in the heavens. Using the times of rising determined by this statistical analysis of the nocturnal offices, and declinations determined by the conditions of the stars' rising, I was able to identify the majority of Gregory's constellations.[28] Most of the clearly identified constellations are

---

Regula Magistri 49; Caesarius of Arles, *Statuta sanctarum virginum* 69, ed. Germanus Morin (Maretioli, 1942); and *Regula Benedicti* (cit. n. 4), pp. 397–398, 407. See also Gérard Moyse, "Monachisme et réglementation monastique en Gaule avant Benoît d'Aniane," in *Sous la Règle de Saint Benoît: Structures monastiques et sociétés en France du moyen âge a l'époque moderne* (Geneva: Librarie Droz, 1982), pp. 3–19; cf. Prinz, *Frühes Mönchtum* (cit. n. 8), pp. 267–273.

[26] These figures differ from the more artistic, but astronomically unintelligible, drawings of the classical constellations in manuscripts of Aratus and Hyginus, e.g., Vienna, Österreichischen Nationalbibliothek 387 (HMML 17564; see n. 6) (ninth c.); Vienna, Österreichischen Nationalbibliothek 51 (HMML 13416) (twelfth c.). This striking difference reflects the concern with actual astronomical practice by readers of *De cursu stellarum*.

[27] *De cursu stellarum* 36–47.

[28] For details of the analysis see Stephen C. McCluskey, "The Constellations of *De cursu stellarum*," unpublished paper.

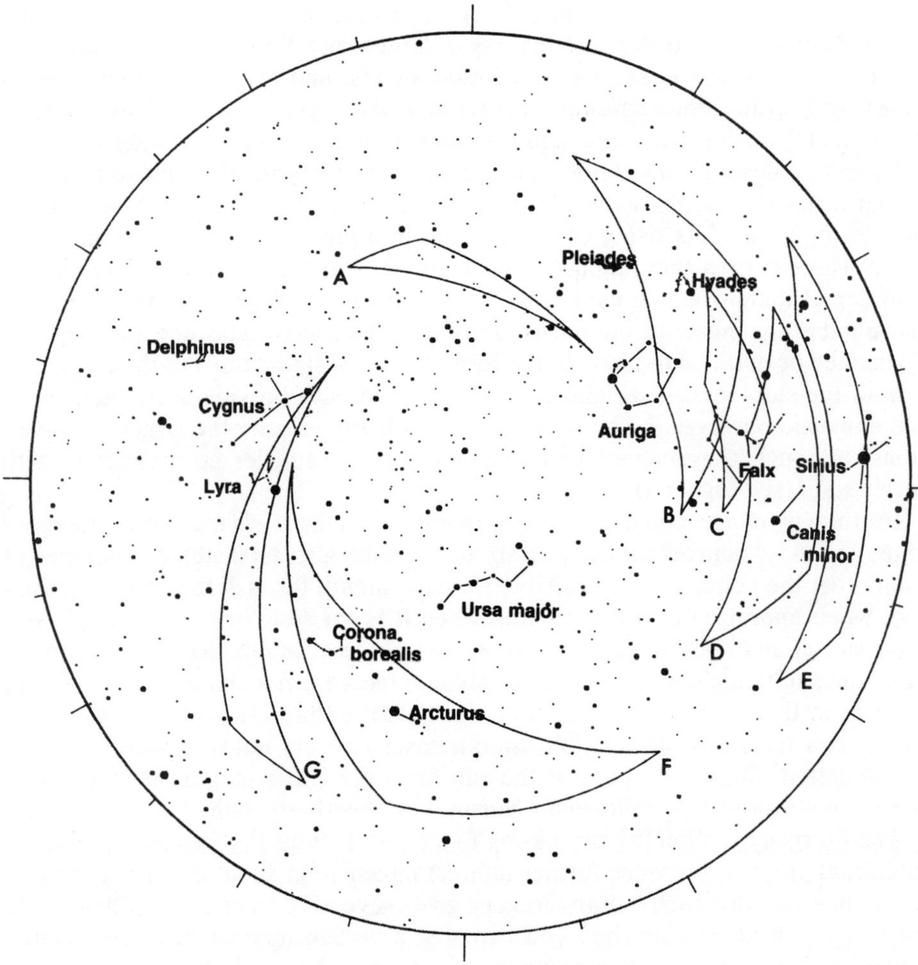

Figure 2. The constellations of De cursu stellarum, *with lettered error envelopes based on the number of psalms sung between the constellations' rising and dawn, and the constellations' declination inferred from Gregory's descriptions. A. Signum Christi. B. Massa (Gregory gives Pliadas and Butrio as synonyms). C. Falx. D. "Duae Falce secuntur." E. Quinio. F. Rubeola. G. Omega.*

traditional, but few of their names are. In a few cases he appears to refer to a new asterism that is not mentioned in classical sources.

Gregory's constellations are here plotted on a star chart (see Figure 2).[29] His Greater Cross is clearly Cygnus, his Alfa or Lesser Cross is Delphinus, and Omega is Lyra; his Rubeola is Arcturus, Quinio is made up of the five stars in Canis Major surrounding Sirius, and his unnamed group of two stars following Falx (the Sickle) are in Canis Minor. His Falx is apparently a group of stars in Gemini and Monoceros. Less well defined are his Signum Christi, which may be the constellation Auriga and the Kids, and the Bier, which may be the Hyades. Drawn from classical astronomy are his Couch (Stefadium), which is our Corona

[29] The constellations of Figure 2 are precessed to the epoch of A.D. 600. Positions are based on a computer tape of the positions, magnitudes, and proper motions of bright stars graciously provided by the U.S. Naval Observatory through the assistance of Dr. LeRoy Doggett.

Borealis (Stephanos), and the Wain, our Ursa Major. Also classical, but less certain, are the stars he calls Massa, Pliadas, or Butrio, which are probably the Pleiades but may be the Hyades instead.[30]

Since the problem of monastic timekeeping, Gregory's discussion of the length of days, and some of his constellation names all reveal oriental influences, we cannot tell a priori whether the techniques in the liturgical section reflect observations actually made in his see in northern Gaul or are merely adaptations of values originating in a center of astronomical or monastic activity to the south. The statistical analyses mentioned above, fitting the time between the computed appearance of the five constellations (as specified in Gregory's instructions for each month) and the computed time of sunrise for that month to the number of psalms Gregory stated could be chanted in that time, resolved this question.

By performing separate analyses for different places and epochs (Table 2), I found the sum of the squared residuals (a measure of error) for these analyses to increase south of the latitude of Tours; for the analysis two centuries earlier, the error also increases. To the north of Tours, stars that Gregory describes as rising become circumpolar, and such analyses are impossible.

In sum, these data place the observations employed in *De cursu stellarum* in northern Gaul near the end of the sixth century. This result strengthens the case for Gregory's authorship, demonstrates that he was not merely repeating received astronomical material, and, most significantly, documents the continued practice of astronomical observations during a time of supposed scientific inactivity.

Observations of the kind that Gregory prescribes to regulate the time of prayer are best carried out when the observer has an unobstructed view of a fairly distant horizon. At Gregory's time monks frequently resided in scattered buildings or caves; Jean Hubert maintains that the quadrangular cloister, walled in by a church, dormitory, refectory, and other buildings, did not become the dominant monastic form in Gaul before the eighth century. St. Martin's foundation at Marmoutier, a few kilometers from Tours, provides a prime example of the earlier building practice. This site on the north bank of the Loire offers the unrestricted view of the horizon that Gregory's astronomy presumes.[31]

This kind of astronomy and this attitude to nature are widely disseminated and generally expressed in the monastic literature of the fifth to the eleventh century. Mere similarity cannot prove direct influence upon or by *De cursu stellarum,* but

[30] The identifications of Omega, Rubeola, Quinio, and the two stars following the Sickle agree with the traditional identifications of the astronomer J. F. Galle, which appear as comments to Haase's edition of *De cursu stellarum* and as brief notes in Krusch's edition (both cit. n. 6). Galle also identified Signum Christi as Capella and the Hyades and Gregory's Massa as the Pleiades; see *De cursu stellarum,* ed. Haase, pp. 42, 44–47. A recent controversy surrounds Galle's identification of Rubeola as Arcturus and of Quinio. See Wolfhard Schlosser and Werner Bergmann, "An Early-Medieval Account on the Red Colour of Sirius and Its Astrophysical Implications," *Nature,* 1985, *318*:45–46; and the discussion between Stephen C. McCluskey, R. H. van Gent, and Schlosser and Bergmann, "The Colour of Sirius in the Sixth Century," *Nature,* 1987, *325*:87–89.

[31] Claire Stancliffe, *St. Martin and His Hagiographer: History and Miracle in Sulpicius Severus* (Oxford: Clarendon Press, 1983), pp. 22–26; and Jean Hubert, "L'érémétisme et l'archaéologie," in *L'eremetismo in Occidente nei secoli XI e XII: Atti della seconda Settimana internazionale de studio, Mendola, 30 agosto–6 septembre 1962* (Milan: Vita e Pensiero, 1965), pp. 473–475, reprinted in Herbert, *Arts et vie sociale de la fin du monde antique au moyen âge* (Geneva: Droz, 1977), [pp. 204–206]. Martin's earlier foundation at Ligugé, near Marseilles, however, was in the ruins of a former Roman villa and employed the enclosed quadrangular plan typical of Roman architecture. See Stancliffe, *St. Martin,* p. 23.

our concern is less with Gregory's sources and influence than with how the concepts found in *De cursu* manifested themselves in the monastic astronomy during this period, even as more reliable means of timekeeping appeared.

Table 2. Analysis of Gregory's figures to test for latitude and epoch.

Place	Latitude	Epoch	Error
Tours	47.3°	590	142.2
Lerins	43.5°	590	163.7
Seville	37.5°	590	175.5
Tours	47.3°	390	149.8

MONASTICISM AND ASTRONOMY AFTER *DE CURSU*

Gregory's instructions suggest that implicit in the earliest monastic rules was an institutional framework for the practice of astronomy. This tradition continued, and the practice of astronomical observation within monasteries provided fertile ground for the reemergence of dormant astronomical theory. In this regard, we must not forget that the many writings of Bede and his successors on astronomy and computus emerged from monastic communities and reflect a similar concern with keeping ritual time, although more in a calendrical than a horological aspect.[32]

The continuing importance of this tradition of monastic timekeeping is reflected in the commentaries on Benedict's *Rule* emerging out of the Carolingian monastic reform. Smagardus of St. Mihiel quoted (after 816) John Cassian's *Regula* on the importance of accuracy in calling the monks to prayer. Unlike Cassian, Smagardus does not propose any method of timekeeping. He mentions the existence of a *horologium,* apparently a sundial, only in a criticism of those who treat fasting as a punishment and "frequently look at the *horologium* and often lift their eyes to heaven, noting the course of the sun."[33]

The commentary of Abbot Hildemar (fl. ca. 850) expresses astronomical concerns more directly, in his discussions of the liturgical hours and matters of monastic discipline. In winter Benedict's *Rule* requires that monks rise at the eighth hour of the night. Hildemar notes that some consider this excessively harsh, since in winter the night is eighteen hours long, which seems to require that monks rise before midnight and pray ten hours until dawn. He resolves this problem by noting that at the equinoxes night and day each have twelve hours, and that Benedict intended this pattern of twelve hours of night to be continued throughout the year.[34]

Hildemar also comments specifically on the practice of timekeeping, making it

[32] See Jones's notes in *Bedae Opera de temporibus* (cit. n. 23), pp. 125–129.

[33] *Smagardi abbatis exposito in Regulam S. Benedicti,* ed. A. Spannagel and P. Englebert (Corpus Consuetudinum Monasticarum, 8) (Sieberg: Schmitt, 1974), pp. 270, 96–97.

[34] *Regula Benedicti* 8.1–3 (cit. n. 4); and *Expositio regulae ab Hildemaro tradita et nunc primus typis mandata,* ed. Rupertus Mittermüller, O.S.B. (Regensburg/New York/Cincinnati: F. Pustet, 1880); pp. 277–279. Cf. the shorter southern Italian commentary in *Pauli Warnefridi diaconi Casinensis in Sanctam Regulam Commentarium* (Monte Cassino: Typis Abbatiae, 1880). On Hildemar's commentary see Wolfgang Hafner, *Der Basiliuscommentar zur Regula S. Benedicti: Ein Beitrag zur Autorenfrage karolingischer Regelkommentare* (Münster: Aschendorff, 1959), pp. 110–111, 143.

clear that observation of the stars was the primary means. In discussing the passage where Benedict imposes the responsibility for announcing the time for prayer upon the abbot or a responsible brother appointed by him, Hildemar notes that the brother should not be chosen arbitrarily but should be responsible and know well the hours of the night. Where Benedict had stipulated that a monk who wakens the community late for prayer should "make due satisfaction to God in the oratory," Hildemar takes the opportunity to discuss how the severity of the offense can vary with circumstances. The least severe penance would be assigned for a case in which clouds obscured the stars and the monk, through his concern about sounding the signal too early, instead sounded it too late. It is striking that Hildemar does not mention any alternative timekeeping device or technique that this conscientious monk could have consulted. Peter Damian (ca. 1067) recommends what was probably the more common practice—that when the stars were not visible, the *significator horarum* should chant psalms to note the passage of time.[35]

The obligation of keeping time by watching the stars is also seen in rules derived from the Cluniac reform, which specifically assigned this duty to the sacristan. As pointed out above, the sacristan as timekeeper can be traced back to Isidore of Seville (see note 5). The sacristan is given similar responsibilities outside the Cluniac houses, as in a tenth-century rule and in Lanfrance of Canterbury's *Monastic Constitutions,* where the sacristan sounds the call to prayer. When clocks became available, the sacristan's special responsibility for timekeeping expanded to include them. Later monastic customaries assign the regulation of the *horologium* to the sacristan, and there are many details concerning clocks in sacristans' records from Norwich Cathedral (1290, 1322), Ely Abbey (1291), St. Albans Abbey (ca. 1327), and Durham Cathedral (1360).[36]

Observation of the stars without instruments, however, seems to have been the principal means of nocturnal timekeeping at most ninth-century monasteries, and to have persisted at least into the thirteenth century. An eleventh-century collection of liturgical cantica from a French monastery, in a pocket-sized volume apparently intended for a monk's private use, includes descriptions of how to observe the changing azimuth of stars over the buildings of the monastic enclosure to determine the time of nocturnal prayers throughout the year. Monasteries that did possess water clocks, such as the prosperous monastery of Fleury (St. Benoît-sur-Loire), still regulated the clepsydra by observing the stars without instruments.[37] A set of instructions written on loose pieces of slate at the Abbey

[35] *Regula Benedicti* 11.11–13, 47.1; Hildemar, *Expositio regulae,* pp. 288–89, 475; and Peter Damiani, *De perfectione monachorum* 17, in *Patrologia latina,* ed. Migne (cit. n. 5), Vol. 145, col. 315).

[36] For the Cluniac rules see Giles Constable, "Horologium stellare monasticum (saec. XI)," in his *Consuetudines Benedictinae variae (Saec. XI – Saec. XIV)* (Corpus Consuetudinum Monasticarum, 6) (Sieburg: Schmitt, 1975), pp. 1–18, on p. 5, n. 8. For the non-Cluniac rules see *Regularis concordia Anglicae Nationis Monachorum Sanctimonialumque* 1.27, 2.29, ed. and trans. Thomas Symons (New York: Oxford Univ. Press, 1953); and Lanfranc of Canterbury, *Monastic Constitutions,* ed. and trans. David Knowles (New York: Oxford Univ. Press, 1951), pp. 82–85. For the sacristan and clocks see P. F. Lefèvre, A. H. Thomas, eds., *Le coutumier de l'Abbaye d'Oigny en Bourgogne au XIIe siècle* (Spicilegium Sacrum Lovaniense, Etudes et Documents, 39) (Louvain, 1976), pp. 50–52; and G. F. C. Beeson, *English Church Clocks* (London: Antiquarian Horological Society, 1971), pp. 15, 16, 18, 20.

[37] See Giles Constable, "Horologium stellare monasticum" (cit. n. 36); Rachael Poole, "A Monastic Star Time Table of the Eleventh Century," *Journal of Theological Studies,* 1915, *16*:98–104; and *Consuetudines saeculi X/XI/XII; Monumenta non-Cluniacensia,* 1. *Consuetudines Floriacenses anti-*

of Villiers about 1267 indicates how many uniform units of time, as measured by a clepsydra, should pass before rising for psalms on different nights of the year. The slates specify how the sacristan was to reset the clepsydra by watching the sun as it appeared at various windows. The hourglass was not used until much later. The earliest evidence of it comes from the beginning of the fourteenth century and in a nautical, rather than monastic, context.[38]

In the matter of astronomical practice, as opposed to astronomical theory, the techniques described here represent a willingness to adapt traditional practices to the new conditions of northern European monasteries. Gregory's earlier Greco-Roman technique of watching risings is more precise with nearly vertical rising over a clear and unobstructed horizon. Now observations were made over nearby structures, employing the more rapid change of azimuth at northern latitudes to compensate for the restrictions on the horizon imposed by the new design of monastic cloisters.

Such openness to change is also reflected in the intellectual curiosity of those monks who did much to recover the astronomy of the Islamic world. Abbot John of Gorze's study of Arabic astronomy during a diplomatic mission to Spain in the 950s is credited with stimulating the earliest dissemination of Arabic science.[39]

The studies of Gerbert of Aurillac (later Pope Sylvester II) connect this interest in Hispano-Arabic science to the tradition of monastic timekeeping. In the 980s, when Gerbert was abbot of Bobbio, he tabulated, somewhat incorrectly, two simple arithmetic schemes, derived from Martianus Capella and more advanced than the linear scheme of *De cursu stellarum,* for computing the changing length of daylight at regions with longest days of eighteen and fifteen hours. His concern with timekeeping is further attested by reports that he constructed a *horologium* for the monastery at Magdeburg, a *horologium arte mechanica compositum* when he was bishop of Rheims, and a *horologium vero aquaticum* (clepsydra) at Ravenna. The first two instruments were most likely astrolabes, although they might have been sundials.[40]

---

quiores (Saec. X), ed. A. Davril and L. Donnat (Corpus Consuetudinum Monasticarum, 7.3) (Sieburg: Schmitt, 1984), p. 42. The subsequent significance of the clepsydra is argued by John D. North, "Monasticism and the First Mechanical Clocks," in *The Study of Time,* ed. Julius T. Fraser and N. Lawrence, Vol. II (New York: Springer Verlag, 1975). Fleury's clepsydra may have been fairly rudimentary, unlike earlier Greek and Roman models. Abbo of Fleury (ca. 945–1004) discusses how to use a clepsydra to time the rising of the signs of the zodiac; the clepsydra he describes has no pointer attached to measure the flow of time, still less an analemmatic dial to account for the varying length of seasonal hours: see G. R. Evans and A. M. Peden, "Natural Science and the Liberal Arts in Abbo of Fleury's Commentary on the Calculus of Victorius of Aquitaine," *Viator,* 1985, *16*:108–127 (A. M. Peden has graciously provided me with a copy of her transcription of this passage from Berlin, MS Phillipps, 1833, fol. 13v.)

[38] Albert d'Haenens, "La clepsydre de Villiers (1267)," in *Klösterliche Sachkultur des Spätmittelalters* (Sitzungsberichte der Österreichischen Akademie der Wissenschaften, Philosophisch-historische Klasse, 367) (Vienna: Österreichische Akademie der Wissenschaften, 1980), pp. 321–342; and, on the hourglass, R. T. Balmer, "The Operation of Sand Clocks and Their Medieval Development," *Technology and Culture,* 1978, *19*:615–632.

[39] James Westfall Thompson, "The Introduction of Arabic Science into Lorraine in the Tenth Century," *Isis,* 1929, *12*:184–193.

[40] Gerbert, *Epistola de horologiis duorum climatum ad fratrem Adam,* in *Gerberti Opera mathematica (972–1003),* ed. Nicolaus Bubnov (Hildesheim: Georg Olms, 1963), pp. 38–41; Thietmar of Merseburg, *Chronicon, ibid.,* p. 382; William of Malmesbury, *De rebus gestis regum Anglorum* 2.168, *ibid.,* p. 388; *Gesta epp. Halbertstadensium, ibid.,* p. 391; and Martianus Capella, *De nuptiis* 7.878 (cit. n. 11). Bubnov interprets the Magdeburg and Rheims accounts as referring to astrolabes; as does

Monastic interest in timekeeping also extended to the planispheric astrolabe. The *De utilitatis astrolabii*, sometimes attributed to Gerbert, is first found in a late tenth- or early eleventh-century collection of astronomical works from the monastery of Santa Maria de Ripoll. The author notes that this new Islamic import can be used by his monastic readers to regulate the celebration of the divine office. Before 1048 another monk, Hermann the Lame, wrote one of the first Latin descriptions of the construction of the planispheric astrolabe, that pocket watch of the Middle Ages.[41] Those concerned with the reckoning of time in monasteries were more than ready to accept a new astronomical timekeeper, just as they would later accept the mechanical clock that succeeded it.[42]

The changes described here in astronomical timekeeping from the fifth to the eleventh centuries reflect an open and evolving tradition. Although such concern with practical astronomy does not exclude the parallel influence of the learned astronomical tradition, monastic resourcefulness in practical matters seems more innovative than does the contemporaneous study of the received corpus of astronomical texts.[43] These connections between astronomy and monasticism might seem fortuitous, since monks or those educated in monasteries constituted much of the educated class in Latin-speaking Europe. Yet we must recall that of all the possible areas of study, educated Latins first chose to focus their scientific interests upon astronomy and the related mathematical disciplines.[44] If Christian monotheism undergirded the faith in an orderly universe, it was monasticism that, through its commitment to reproduce that celestial order in an orderly round of prayer, directed that faith toward the study of the heavens.

Werner Bergmann, *Innovationen im Quadrivium des 10. und 11. Jahrhunderts: Studien zur Einführung von Astrolab und Abakus im lateinischen Mittelalter* (Sudhoffs Archiv, Zeitschrift für Wissenschaftsgeschichte, Beihefte, 26) (Stuttgart: Franz Steiner, 1985), p. 159. Bergmann concludes from its description that the Magdeburg instrument was an astrolabe correctly constructed with a plate for the latitude of Magdeburg after observations of "quadam stella nautarum," apparently the pole star. Although this description could apply equally to the construction of a sundial for a specific latitude, there is no evidence of Gerbert's familiarity with their construction. The phrase *arte mechanica* commonly refers to the technique of dividing the face of a sundial—see Jerome, *Commentariorum in Esaiam* 11.38.4/8 in *Corpus Christianorum Series Latina,* Vol. 73, ed. G. Morin (Turnholt: Brepols, 1963); and Bede, *In Regum Librum XXX Quaestiones* 25, *ibid.*, Vol. 119, ed. D. Hurst (1962)—but it could apply equally well to dividing the face of an astrolabe.

[41] Gerbert, *De utilitatibus astrolabii* 1.2, 5.4, 6.1, in *Opera,* ed. Bubnov, pp. 116, 129–130; Werner Bergmann, "Der Traktat 'De mensura Astrolabii' des Hermann von Reichenau," *Francia,* 1980, *8*:65–103; and Mary Catherine Welborn, "Lotharingia as a Center of Arabic and Scientific Influence in the Eleventh Century," *Isis,* 1931, *16*:188–199. For two views on the attribution of the first item to Gerbert see David C. Lindberg, "The Transmission of Greek and Arabic Learning to the West," in *Science in the Middle Ages* (Chicago: Univ. Chicago Press, 1978), pp. 60–61; and Bergmann, *Innovation im Quadrivium* (cit. n. 40), pp. 148–155, and the sources cited there.

[42] The thesis that monastic timekeeping was the initial driving force behind the late medieval development of clockwork is convincingly presented in David S. Landes, *Revolution in Time: Clocks and the Making of the Modern World* (Cambridge, Mass.: Belknap Press, 1983), pp. 53–70.

[43] The practical focus of early medieval science in a number of different areas has been discussed in A. C. Crombie, *Medieval and Early Modern Science,* Vol. I (Garden City, N.Y.: Doubleday Anchor, 1959), pp. 18–25. Constable, "Horologium stellare monasticum" (cit. n. 39), p. 5, considers that "the practical needs of daily life in the monastery help . . . explain the interest in astronomy shown by many medieval monks."

[44] The centrality of astronomy in this process has been most persuasively advocated by Richard Lemay, *Abu Maᶜshar and Latin Aristotelianism in the Twelfth Century: The Recovery of Aristotle's Natural Philosophy through Arabic Astrology* (American University of Beirut, Publication of the Faculty of Arts and Sciences, Oriental Series, 38) (Beirut, 1962), pp. xiii–xxviii.

Anglo-Saxon Plant Remedies and the Anglo-Saxons

By Linda E. Voigts

IN THE SURVIVING BODY of vernacular writing from Anglo-Saxon England, medical texts, containing mostly remedies to be derived from plants, bulk large. In addition to glossed remedies and short treatises,[1] four long Old English medical works survive, three of them in unique manuscripts: the *Læceboc* (Bald's Leechbook), *Lacnunga,* and *Peri Didaxeon.*[2] The fourth text, the *Herbarium Apulei,* is the Old English translation of the late antique herbal falsely attributed to Apuleius and expanded to 185 plants by the addition of the treatise on betony of Antonius Musa and some of the pseudo-Dioscoridean *Ex herbis femininis* recipes; the plant remedies

This article is a revised version of a paper which was read at the conference on Nature in the Middle Ages, at the State University of New York at Binghamton, October 1976, and at the Missouri Botanical Garden, Saint Louis, Missouri, June 1977. I am grateful for the suggestions offered by Professor Bert Hansen and by the staff of the Missouri Botanical Garden, particularly Dr. Charles Huckins. I also acknowledge with gratitude the support provided for work with relevant manuscripts in British libraries in 1975 by a grant-in-aid of research from the American Council of Learned Societies.

[1] An example of a glossed recipe is the recipe "wið eahwærce" written in the margin of p. 208 of Cambridge, Corpus Christi College 41, a manuscript containing the Old English translation of the *Historia ecclesiastica* of Bede. Shorter medical writings survive not only in glosses but are often sandwiched between other writings as well; e.g., a treatise on the development of the fetus that might properly be called medical is found in a collection of prognostics in British Library Cotton Tiberius A. iii, fols. 40v–41.

[2] All of the longer and many of the shorter Old English medical texts were edited in three volumes by Thomas Oswald Cockayne in *Leechdoms, Wortcunning, and Starcraft of Early England,* Rolls Series, 35 (London, 1864–1866). These were reissued (London: Holland Press, 1961) with a new introduction by Charles Singer, but the reprint, which omits Cockayne's introduction and the key to his *apparatus criticus,* does not replace the Rolls Series volumes. Early in the 20th century some of the discrete texts were re-edited. New editions are currently being prepared for the *Læceboc,* the *Herbarium Apulei,* and *Peri Didaxeon.*

The *Læceboc,* which is made up of three collections of medical recipes that seem to derive from the court of Alfred the Great, is in a MS from the mid-10th century, British Library Royal 12 D. xvii, fols. 1–127v. See N. R. Ker, *A Catalogue of Manuscripts Containing Anglo-Saxon* (Oxford: Clarendon Press, 1957), No. 264, and the facsimile of the codex, introduced by C. E. Wright: *Bald's Leechbook,* Early English Manuscripts in Facsimile, 5 (Copenhagen: Rosenkilde and Bagger, 1955).

Lacnunga, which means "healings" or "cures," differs from the other texts in its high percentage of charms and other magical elements. It occurs in a codex written in the late 10th or early 11th century, British Library Harley 585, fols. 130–151v and 157–193. J. H. G. Grattan and Charles Singer edited the *Lacnunga* in *Anglo-Saxon Magic and Medicine* (London: Oxford University Press, 1952). See the discussion of the MS on pp. 206–209, and Ker, *Catalogue,* No. 231.

Peri Didaxeon, "Concerning Schools of Medicine," is found in a MS which may be as late as 1200 and so is not included in Ker. The codex, British Library Harley 6258b, contains this text on fols. 51v–66v, following the expanded *Herbarium Apulei* on fols. 1–51. The question of whether the language of this text is late Old English or early Middle English is yet unresolved. Max Löweneck's edition in Erlanger Beiträge zur englischen Philologie, 12 (Erlangen: Junge, 1896) emphasized the so-called Salernitan material in the work, but Charles Talbot has since established that the Petrocellus-Gariopontus material is not Salernitan at all and is found in the late-9th-century *Læceboc;* "Some Notes on Anglo-Saxon Medicine," *Medical History,* 1965, *9*:156–169. The arguments concerning the stage of the language of *Peri Didaxeon* are discussed in Peter Bierbaumer, *Der botanische Wortschatz des Altenglischen,* Teil II: *Lācnunga, Herbarium Apuleii, Peri Didaxeon,* Grazer Beiträge zur englischen Philologie, 2 (Bern: Herbert Lang, 1976), p. xii.

are followed by remedies obtained from animals, *De taxone* and the *Medicina de quadrupedibus* of Sextus Placitus.[3] This remedy book survives in four manuscripts, one of which (British Library Cotton Vitellius C. iii) represents the plant and animal sources of the remedies in brightly colored illustrations.[4] Together these manuscripts make up the earliest body of vernacular medical texts in medieval Europe. Closely related to this tradition is the justly famous illustrated Latin manuscript of the Pseudo-Apuleius herbal from Bury Saint Edmunds, at least some of which seems to be a Latin rendering of the Old English text; it also contains Old English glosses.[5]

These Old English texts alone occupy more than five hundred leaves or one thousand manuscript pages, and we cannot estimate the number of codices which failed to survive because of changes in the language or such calamities as the dispersal of monastic libraries in the sixteenth century and the Ashburnham House fire in the eighteenth.[6] What is clear, however, when one considers the number of animal skins necessary for more than a thousand pages of manuscript text, is that these surviving codices represent no small investment of resources and time on the part of Anglo-Saxon monastic houses. That observation alone justifies reopening the question of the practicality of these Anglo-Saxon medical texts. Were they written to be used by those who would heal? Would it have been possible in pre-Conquest England to know and use the plants depicted in them? I think the answer to both questions is yes.

Although that answer may appear obvious to those who do not work in the

[3]The Latin text of the herbal without the pseudo-Dioscoridean recipes is found in *Antonii Musae de herba vettonica, liber Pseudoapulei herbarius, anonymi de taxone liber, Sexti Placiti liber medicinae ex animalibus . . .* , ed. Ernst Howald and Henry Sigerist, Corpus Medicorum Latinorum, 4 (Leipzig: Teubner, 1927). On the attribution to the author of the *Metamorphoses,* see L. E. Voigts, "The Significance of the Name 'Apuleius' to the *Herbarium Apulei,*" *Bulletin of the History of Medicine,* 1978, *52*:214–227. This herbal was quite popular in the early Middle Ages. Howald and Sigerist discuss more than 40 codices and fragments, and Augusto Beccaria lists 25 MSS surviving from before the 12th century alone: *I codici de medicina del periodo Presalernitano* (Rome: Edizione di Storia e Letteratura, 1956). See also Henry Sigerist, "Zum Herbarius Pseudo-Apulei," *Sudhoffs Archiv für Geschichte der Medizin,* 1930, *23*:197–204; *The Herbal of Pseudo-Apuleius from the Ninth Century MS. in the Abbey of Monte Cassino—Codex Casinensis 97—Together with the First Printed Edition of Joh. Phil. de Lignamine—Editio princeps Romae 1481—both in Facsimile,* ed. F. W. T. Hunger (Leyden: Brill, 1935); and the facsimile of Vienna, Nationalbibliothek 93, with commentary by Charles Talbot and Franz Unterkircher, *Medicina antiqua libri quattuor medicinae,* 2 vols. (Graz: Akademische Druck- u. Verlagsanstalt, 1972). On the illustrative tradition of this herbal see Heide Grape-Albers, *Spätantike Bilder aus der Welt des Arztes* (Wiesbaden: Pressler, 1977), a study which lists four additional manuscripts. For a discussion of the textual tradition and an edition of the so-called Sextus Placitus portion of the vernacular text, see Hubert J. de Vriend, *The Old English Medicina de Quadrupedibus,* Dissertation, Groningen 1972 (Tilburg: Giannoten, 1972).

[4]The three codices lacking illustrations are Harley 6258b, fols. 1–51, from the late 12th or early 13th century; Harley 585, fols. 1–129v, from the late 10th or early 11th century; and Bodleian Hatton 76, fols. 68–130 (Ker, *Catalogue,* No. 328), from the mid-11th century, a MS written with spaces that correspond to the illustrations in Cotton Vitellius C. iii. On Cotton Vitellius C. iii, fols. 11–85, a codex which seems to date from after 1050, see Ker, *Catalogue,* No. 219; and L. E. Voigts, "A New Look at a Manuscript Containing the Old English Translation of the *Herbarium Apulei,*" *Manuscripta,* 1976, *20*:40–60, and 1977, *21*:62; and "One Anglo-Saxon View of the Classical Gods," *Studies in Iconography,* 1977, *3*:3–16.

[5]Bodleian Bodley 130, fols. 1–95v (Ker, *Catalogue,* No. 302). For a facsimile and discussion of the MS, see *The Herbal of Apuleius Barbarus from the Early Twelfth-Century Manuscript formerly in the Abbey of Bury St. Edmunds* (*MS. Bodley 130*), described by Robert T. Gunther, Roxburghe Club Publications, 182 (London, 1925). For de Vriend's arguments that the section of Bodley 130 dealing with remedies to be obtained from animals is a Latin rendering of the Old English text, see *The Old English Medicina de Quadrupedibus,* pp. xlv–liii.

[6]The disastrous fire of 1731 which damaged many of the MSS in the Cotton Collection, including Vitellius C. iii, was responsible for the near destruction of Otho B. xi, containing a number of herbal recipes related to the *Læceboc* and some which were independent of it; see Ker, *Catalogue,* No. 180. Fortunately, a transcript of a large portion of the codex, copied by Laurence Nowell in 1562, survives as British Library Add. 43703; the recipes are fols. 261–64v. See Roland Torkar, "Zu den ae. Medizinaltexten in Otho B. XI und Royal 12 D. XVII, mit einer Edition der Unica," *Anglia,* 1976, *94*:319–338.

sometimes arcane field of medieval herbals, it is neither the easy answer nor is it the traditional one. Our understanding of the uses of pre-Salernitan written medicine,[7] mostly drug therapy to be obtained from plants, has been dominated by two views which must be reevaluated. The first is that the surviving codices manifest an uncritical copying of classical texts with no real understanding and no thought to their practical use. The most extreme statements of this view were articulated by Charles Singer, who deplored the inability of "the wilting mind of the Dark Ages" to deal intelligently with classical herbals and who described the entire corpus of Anglo-Saxon medical texts as "the darkest and deliquescent stage of a [*sic*] outdated culture."[8] This view was shared and disseminated by Wilfrid Bonser, who wrote: "Sterile formulae, which could be applied without any exercise of reasoning, alone survived for use during the Dark Ages. It is these, therefore, . . . which appear in Anglo-Saxon medical practice."[9]

The second and related notion has to do with the depictions of plants in Cotton Vitellius C. iii and Bodley 130. Historians of medicine and art historians alike assert that these illustrations can have been of no practical value to the Anglo-Saxon user of these codices for two reasons: first, the illustrations are stylized and botanically inaccurate and are therefore unrecognizable; and second, a number of them represent Mediterranean plants not found today north of the Alps.[10] The issue of botanical inaccuracy and stylization can be dealt with briefly. The illustration accompanying the recipes using sage (*Salvia officinalis* L.) in Bodley 130 has been criticized because the plant is not depicted with opposite leaves.[11] In fact, some of the leaves in the illustration are opposite and others are alternate, but more important they are represented at an angle to the stem, which suggests, along with the coloring, that the artist is portraying the dried plant, from which it is indeed difficult to discern if the leaves are opposite or alternate. (See Fig. 1.) Furthermore, stylistic representations are not by definition useless if one has had any experience with the depicted plant. After all, even today one recognizes the plants on seed packages, knowing full well

[7]A useful though incomplete guide to the body of early medieval medical writing is Walter Puhlmann, "Die lateinische medizinische Literatur des frühen Mittelalters," *Kyklos,* 1930, *3*:395–416; this guide of course omits vernacular texts. Most MSS containing early medieval medical texts are found in Beccaria, *I codici di medicina del periodo Presalernitano.* Although indispensable, Beccaria is not complete. The work omits, e.g., the 9th-century Beneventan MS in Glasgow, Hunter T.4.13.

[8]Charles Singer, "The Herbal in Antiquity," *The Journal of Hellenic Studies,* 1927, *47*:31; Introduction to the reprint of *Leechdoms, Wortcunning and Starcraft,* Vol. I, p. xlvii. See also his "Greek Biology and its Relation to the Rise of Modern Biology," *Studies in the History and Method of Science II,* ed. Charles Singer (Oxford: Clarendon Press, 1921), p. 2.

[9]Wilfrid Bonser, *The Medical Background of Anglo-Saxon England* (London: Wellcome Historical Medical Library, 1963 [prepared for publication 1949]), p. 35. This sort of statement is also pervasive in popular treatments of the subject; e.g., in Ernst and Johanna Lehner, *Folklore and Odysseys of Food and Medicinal Plants* (New York: Tudor, 1962), p. 98, one finds the following statement: "In the Dark Ages many of the herbal manuscripts of early days were destroyed by the martial rulers and their mercenaries who were devoid of any interest in science and culture."

[10]See Thomas Downing Kendrick, *Late Saxon and Viking Art* (London: Methuen, 1949), p. 25, and Kurt Weitzmann, "Das klassische Erbe in der Kunst Konstantinopels," *Alte und neue Kunst,* 1954, *3*:50. Similarly, Erich Bethe argued in *Buch und Bild im Altertum* (Leipzig: Harrassowitz, 1945), a work much praised by Weitzmann (see *Illustrations in Roll and Codex,* Princeton: Princeton University Press, 1947, pp. 9–10), that plant illustration evolved from *Lehrbild* to *Schmuckbild* where the plant representations were valued *qua* pictures, not as guides for acquiring the plants for medical remedies; he insisted (pp. 35–36) that in 12th-century Saxon-Norman herbals the plants could not be recognized from their illustrations. Grape-Albers, whose focus is on the late antique origins of the herbal illustrations *(Spätantike Bilder)* disagrees with Weitzmann regarding the function of the illustrations.

[11]See the commentary on fol. 33v in *The Herbal of Apuleius Barbarus from the Early Twelfth-Century Manuscript.* The leaves of the illustration depicting *salvia* in Cotton Vitellius C. iii (fol. 51r) *are* opposite.

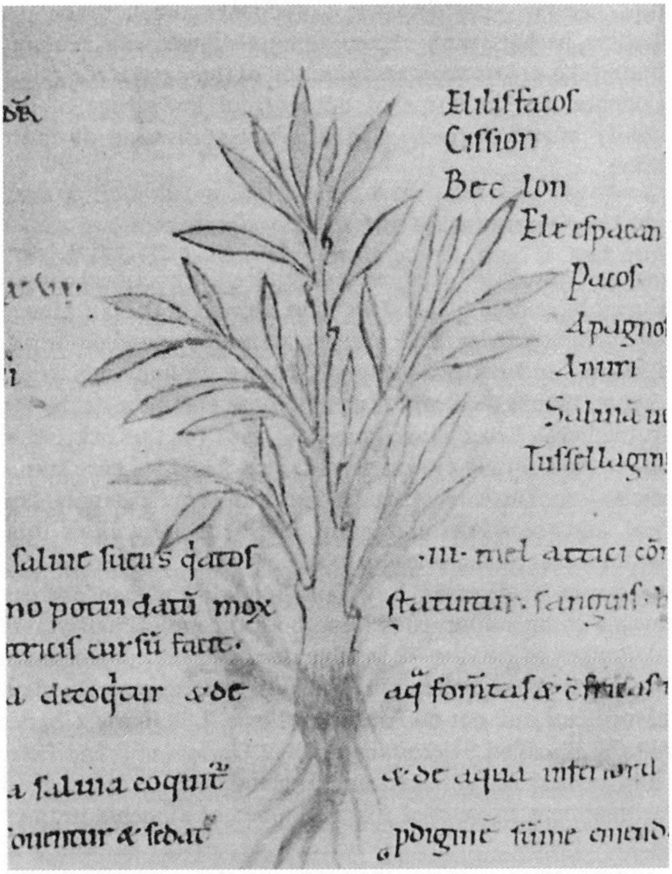

Figure 1. Sage, c. 1100,* Herbarium Apulei, *Bodley 130, fol. 33v. (Reproduced by permission of the Bodleian Library.)

that the plants which one grows will not likely approximate the illustrations.[12]

The question of whether or not Mediterranean plants could have been known in Anglo-Saxon England is a more complex one, to be dealt with at length below. An influential instance of this traditional view is the statement made by Singer more than once that the illustration labeled *hennebelle* in Cotton Vitellius C. iii (fol. 23v) was a meaningless exercise because it represents the Mediterranean plant *Hyoscyamus reticulatus* L., not the *Hyoscyamus niger* L. (henbane) that presently grows in England.[13] I shall return to this example, but the point is clear that the dominant view of Anglo-Saxon medical manuscripts in this century has been that "the whole herbal,

[12]Charles Talbot put it aptly (*Medicine in Medieval England,* London: Oldbourne, 1967, pp. 20–21): "The illustrations to these texts are formal rather than natural, and it has been inferred that those who relied on them could not really distinguish or recognize any of the plants described; one might argue equally well that the Anglo-Saxons could not recognize, from the formalized illustrations of buildings, either a church or a monastery."

[13]"Greek Biology," pp. 76–77. Singer repeats this observation in his introduction to Grattan and Singer, *Anglo-Saxon Magic and Medicine,* p. 28. Similarly, Gunther observes that "plants from which the original pictures were taken belonged to a southern flora and either do not occur in Britain or were not available when the scribe was at work on the manuscript"; *The Herbal of Apuleius Barbarus,* p. xxiv.

text and pictures, became quite useless as a practical manual."[14] Therefore, to argue, as I do, that these herbals were copied for actual use with healing plants, it is necessary to undertake a systematic reevaluation of these received notions in terms of manuscript evidence and in terms of our current knowledge of early medieval medicine, of early medieval trade, and of plant cultivation during the minimal climatic optimum.

MEDICINE IN THE EARLY MIDDLE AGES

The dictum that early medieval medical texts epitomize uncritical copying of antique medical texts must be revised by what we have learned in the last fifteen years about early medieval medicine in general and by a specific examination of progressive improvement in text and codex. Although we are dealing with largely uncharted territory, we know, thanks to recent research, more about the deliberate encouragement in early medieval Benedictine houses of care for the sick (Rule 36) that is therapeutic rather than simply custodial in nature,[15] and we have learned that early medieval medicine—monastic, ecclesiastical, and in some instances, lay—was pragmatic, empirical, and sometimes efficacious. The seminal work of Brian Lawn has made clear that there was a theoretical base to early medieval medicine, albeit not one devoted to the causes of diseases. Lawn's analysis of the standard early medieval medical *summa*—a compendium containing both theoretical and pragmatic texts—surviving in a number of codices, is invaluable. It makes convincing his argument that the contributions of the Salernitan Adelhard of Bath were based on traditional early medieval medicine and not on Arabic medicine.[16] Similarly, Charles Talbot has established that the so-called Salernitan works of Gariopontus and Petrocellus were in fact used and revised in ninth-century England, and recent analyses show the validity of the treatment prescribed for a number of ailments in the Old English remedy books.[17]

That the pragmatic, empirical practice of medicine in the early Middle Ages probably resulted in better care than did the post-Salernitan school medicine devoted to theoretical etiology has been suggested by John Riddle.[18] This recent emphasis on the early medieval focus on care of the sick rather than on cause of disease reminds us of the importance of Soranus and pseudo-Soranic writings to the medieval transmission of medical texts.[19] It may also suggest, without demanding historical conti-

[14]Otto Pächt, "Early Italian Nature Studies and the Early Calendar Landscape," *Journal of the Warburg and Courtauld Institutes,* 1950, *13*:27.

[15]Heinrich Schipperges, *Die Benediktiner in der Medizin des frühen Mittelalters* (Leipzig: St. Benno, 1964); and Erna Patzelt, "Moines-médicins" in *Études de civilisation médiévale (IXe–XIIe siècles): Mélanges offerts à Edmond-René Labande* (Poitiers: Centre d'Études Superiéures de Civilisation Médiévale, 1974), pp. 577–588. For the text of a remarkable defense of medicine written in a Benedictine house c. 750–850, see Karl Sudhoff, "Eine Verteidigung der Heilkunde aus den Zeiten der 'Mönchsmedizin'," (Sudhoffs) *Archiv für Geschichte der Medizin,* 1913, *7*:223–237. Also of value are Hanno Caprez, "Die Klostermedizin," *Ciba Zeitschrift,* 1951, *11*:4638–4666, and two studies of regional monastic medicine: Claudius Franz Mayer, "Das Zeitalter der Mönchs- und Priestermedizin in Ungarn," *Kyklos,* 1930, *3*:354–394; and F. Vercauteren, "Les medecins dans les principautes de la Belgique et du nord de la France, du VIIIe âu XIIIe siècle," *Le Moyen Age,* 1951, *57*:61–92.

[16]Brian Lawn, *The Salernitan Questions* (Oxford: Clarendon Press, 1963), pp. 1–15; 20–21.

[17]Talbot, "Some Notes on Anglo-Saxon Medicine," pp. 156–169, and *Medicine In Medieval England,* pp. 18–20. Stanley Rubin, *Medieval English Medicine* (New York: Harper & Row, 1974); see esp. pp. 128 and 131–137. The study of the efficacy of various treatments used in antiquity bears on the early medieval texts which prescribe similar remedies; see Guido Majno, *The Healing Hand: Man and Wound in the Ancient World* (Cambridge, Mass.: Harvard University Press, 1975). See also Walter Lewis and Memory P. F. Elvin-Lewis, *Medical Botany* (New York: Wiley, 1977).

[18]John Riddle, "Theory and Practice in Medieval Medicine," *Viator,* 1974, *5*:157–184.

[19]Lawn, *The Salernitan Questions,* pp. 8–11.

nuity, that early medieval medicine bears a number of resemblances to the methodist sect of classical medicine, the discipline that disregarded the theories of disease causation and concentrated instead on symptoms and therapeutics.

TEXT AND MANUSCRIPT

In revising the notion that medical texts were academic exercises, these general studies provide a useful background. However, the narrower question of whether or not Anglo-Saxon remedy books were meant to be used is best approached by a consideration of the progressive improvement of texts in the course of their transmission and of manuscripts in the course of their use. Four surviving manuscripts of the Old English translation of the enlarged Pseudo-Apuleius herbal testify to the popularity of this remedy book in Anglo-Saxon England, and an examination of some of the changes made in the transmission of this text of late antique origin proves instructive. In its original form the *Herbarium* seems to have consisted of 130 chapters, each including a depiction of the plant and then a list of remedies provided by the particular simple. In the transmission of this herbal, not only were other plant chapters added, but changes were made in the book to make it a more useful pharmacopeia; lists of synonyms to the plant names were added to each chapter[20] and sometimes information concerning the habitat of the plants. Many chapters, like the one on *scelerata* (*clufþunȝe,* Ch. 9), in the Old English version contain information about the ground where the plant best grows. The vernacular text provides this information before the recipes are given, whereas in the standard Latin text (Ch. 8), it is found at the end of the chapter.[21]

Particularly interesting is the vernacular treatment of the information provided concerning the first plant, betony, source of twenty-nine remedies in the Old English version (forty-seven in the Latin). Betony is simply entered as the first chapter of the Old English herbal, whereas in the Latin it is a separate treatise, "Antonii Musae de herba vettonica liber," almost always found preceding the *Herbarium Apulei.* What is important here is that the Old English text omits the Latin discussions that are not essential for obtaining the plant and using it as a remedy, and it unites at the beginning of the chapter essential information from two discussions. From the introductory Latin material the Old English omits the greetings of Antonius Musa to M. Agrippa and the celebration of betony over other plants; it retains the instructions for gathering betony in August without the use of iron and for drying it. From the concluding material in the Latin text the Old English omits the prayer or incantation to the plant and takes only the statements concerning the use of the plant for soul or body or as a shield against nocturnal visitors and frightening visions and the statement regarding habitat ("nascitur in pratis et in montibus, locis mundis et opacis circa frutices . . ."). This information in the Old English text differs from the Latin in that it does not come after the twenty-nine remedies, however; it is the first information provided on the plant before the instructions on gathering and drying it are given: " Þeos wyrt þe man betonicam nemneð heo biþ cenned on mædum 7 on clænum dunlandum 7 on ȝefriþedum stowum. . . ." (This plant, which one calls betony, is produced in meadows, and on open moor lands, and in shady places. . . .)

[20]The addition of synonyms and tables of contents is discussed in Sigerist, "Zum Herbarius Pseudo-Apulei," pp. 197–198.

[21]For the Old English text, see Cockayne, *Leechdoms,* Vol. I, p. 98, and for the Latin text, see Howald and Sigerist, *Antonii Musae,* pp. 39–40. Bodley 130, containing a text that may be a translation back to Latin from Old English (see above, n. 5), also provides the habitat of the plant at the beginning of the entry.

In short, what we find in the Old English version is what might be called an "improved" text, a version easier to use than the Latin. Inessential information, some of it magical, has been omitted, and the information important to anyone who might want to find, dig, and dry the plant has been combined from two locations in the Latin and placed at the beginning of the chapter. The redactor, whether he was working in the vernacular or composing a Latin intermediary, was concerned with making the text a useful one. That he may well have been successful is indicated by the habitat entry for *Betonica officinalis* L. in Polunin's *Flowers of Europe*: "Meadows, heaths, woods. June–Oct. All Eur."[22]

Another change in the history of *Herbarium Apulei* which made it more practical is the addition of tables of contents to enable the user to find a remedy for an ailment without reading through a codex chapter by chapter. In two of the Latin recensions one finds *tituli morborum,* which provide what strikes the twentieth-century reader as a sophisticated information retrieval system—implying that the information is worth getting to. The *tituli morborum* are listings by ailments in the traditional head-to-foot order; under each entry, like *ad capitis fracturam,* is listed the numbers of those plant chapters containing a remedy for that ailment.[23] In the Anglo-Saxon herbals, however, one finds the simpler contents list which cites the plant chapters and under each plant name gives the list of ailments or symptoms for which the plant supplies remedies. Although this arrangement is less sophisticated than the *tituli morborum,* it still helps the user find a recipe. The user of Cotton Vitellius C. iii could move quickly to an entry in the text by matching the rubricated numeral from the contents list with the rubricated numeral at the top of the text leaf. The contents list in this codex (fols. 12–18v) is written in a color-key system as well; it distinguishes among the recipes listed by ailment under the plant names by varying the colors of the upper-case letters beginning the entries. That a contents list was considered important is testified to by the fact that in two Anglo-Saxon codices of the Pseudo-Apuleius herbal—the Old English Harley 585 (fols. 115–129v) and the Latin Bodley 130 (fols. 73–75)—the contents tables were added after the completion of the text. In Bodley 130 it is clear that the rubricated numerals which identify the chapters were coordinated with the selective list of contents in the manner of Cotton Vitellius C. iii and were added by the later scribes who supplied that table.

Two other observations on the improvement of the Pseudo-Apuleius remedy book should be made. At Saint Gall a ninth-century herbal survives which seems to have originated there (Stiftsbibliothek 217). It contains entries for sixty-two plants. Thirty-six of the chapters derive from the *Herbarium Apulei,* but twenty-six have no known source, and a number of those seem to be the addition of alpine plants. One finds as well entries which vary from the *Herbarium Apulei* in the addition of such details as "nascitur in excelsis montis juxta aquas" (Ch. 13, *ginciana*).[24] It may also be significant that the *Ex herbis femininis* treatise which accompanies the Pseudo-Apuleius herbal in the Old English texts seems to be a sophisticated reworking of Dioscoridean materials by someone in Ostrogothic Italy who knew the plants in the treatise.[25]

Other medical texts surviving from Anglo-Saxon England reward an examination

[22] Cockayne, *Leechdoms,* Vol. I, p. 70; Howald and Sigerist, *Antonii Musae,* pp. 3–4, 11. See Oleg Polunin, *Flowers of Europe* (London: Oxford University Press, 1969), p. 359. The identification for *Betonica officinalis* L. is now *Stachys officinalis* (L.) Trev.

[23] See Sigerist, "Zum Herbarius Pseudo-Apulei," pp. 197–198.

[24] The text of this herbal and a discussion of it are to be found in Erhard Landgraf, "Ein frühmittelalterlicher Botanicus," *Kyklos,* 1928, *1*:114–146; *ginciana* is found on p. 128.

[25] See Riddle, "Theory and Practice," p. 163. Riddle's forthcoming publications which unravel the vexed tradition of *Ex herbis femininis* recipes are of major importance to the study of early medieval herbals.

with conclusions similar to those deriving from the *Herbarium Apulei*. The *Lacnunga* contains entries, to be mentioned again, that prescribe, like the ninth-century Saint Gall antidotary, a plant remedy discovered by the Arabs but unknown to the ancients.[26] Another indication that Latin medical texts were meant to be used by Anglo-Saxons who knew the plants by vernacular names is the various surviving word lists giving vernacular words for Latin plant names. One can find plant names in the better-known general glossaries like the Corpus, Harley, and Antwerp glossaries,[27] but of particular interest are two that are made up of plant lists: the *Durham Herbal Glossary* and the *Laud Herbal Glossary*. Both survive in manuscripts containing Latin medical texts, and it is clear that they have been added to the texts to make the Latin remedies usable for someone who knows the relevant plants by their vernacular names.[28] As has already been mentioned, eight of the plants in the illustrated Latin Bury Saint Edmunds herbal are identified in a contemporary hand by their Old English names, and the Latin herbal in Bodleian Ashmole 1431 contains fifty-seven vernacular glosses.

Perhaps the best instance of an Anglo-Saxon vernacular medical text that gives every indication of being intelligently compiled for practical use is the *Læceboc*, the so-called Leechbook of Bald. Consisting of three books, the first two of which are closely related to the circle of King Alfred, it survives in the unique manuscript which may have been copied at the Old Minster, Winchester, circa 950–960. No single Latin source for any of the three books is known; the books seem to be an intelligent amalgam of various Latin sources, such as Alexander of Tralles, Marcellus of Bordeaux, and the Gariopontus-Petrocellus material already mentioned.[29] The compilation of these books containing herbal remedies and relatively sophisticated surgical procedures may suggest a longstanding tradition of vernacular medicine, for there is some philological evidence that at least part of the compiled material already existed in Old English.[30] A study of the text of the *Læceboc*, like a study of the other Old English medical texts and plant glossaries, makes it difficult to persevere in the

[26] See Bierbaumer, *Der botanische Wortschatz des Altenglischen*, Teil II, s.v. *sideware* (pp. 103–104) and John Riddle, "The Introduction and Use of Eastern Drugs in the Early Middle Ages," *Sudhoffs Arch.*, 1965, *49*:190–192.

[27] On the MS Cambridge, Corpus Christi College 144, a word list from the 8th or 9th century containing more than 2,000 words, see Ker, *Catalogue*, No. 36, and *An Eighth-Century Latin–Anglo-Saxon Glossary*, ed. J. H. Hessels (Cambridge: Cambridge University Press, 1890). Regarding the glossary in British Library Harley 3376, see Ker, No. 240, and *The Harley Latin–Old English Glossary*, ed. Robert T. Oliphant (The Hague: Mouton, 1966); some scattered fragments from this glossary remain unpublished, e.g., Kansas University, Spencer Research Library, Pryce P2A:1. The Latin–Old English glossary in Antwerp, Plantin-Moretus Museum 47, is arranged under subject headings; see Ker, No. 2, and Max Förster, "Die ae. Glossenhandschrift Plantinus 32 (Antwerpen) und Additional 32246," *Anglia*, 1917, *41*:94–161.

[28] On Durham, Cathedral Hunter 100, fols. 82–84v, see Ker, *Catalogue*, No. 110, and "Das Durhamer Pflanzenglossar," ed. Bogislav von Lindheim, *Beiträge zur englischen Philologie*, 1941, *35*:1–81. On the 250 Old English words in the glossary of herb names in Bodleian Laud Misc. 567, fols. 68–73, see Ker, No. 345, and *The Laud Herbal Glossary*, ed. J. Richard Stracke (Amsterdam: Rodopi, 1974).

[29] The extensive analysis of the efficacy of *Læceboc* treatments in Rubin's *Medieval English Medicine* bears mention again as does Talbot's "Some Notes on Anglo-Saxon Medicine." Two useful studies on book production at Winchester in the late 9th and 10th centuries bear on Bald's Leechbook: Audrey Meaney, "King Alfred and his Secretariat," *Parergon*, 1975, *11*:16–24, and M. B. Parkes, "The Palaeography of the Parker Manuscript of the *Chronicle*, Laws and Sedulius, and Historiography at Winchester in the late Ninth and Tenth Centuries," *Anglo-Saxon England*, 1976, *5*:149–171.

[30] Torkar, "Zu den ae. Medizinaltexten," pp. 323, 326–327. A 9th-century fragment containing Old English recipes, Louvain, Bibliothèque de l'Université, Fragmenta H. Omont 3, has recently been noticed; it provides important evidence of early vernacular medical writing. See Angus Cameron and Bella T. Schauman, "A Newly-Found Leaf of Old English from Louvain," *Anglia*, 1977, *95*:289–312; and N. R. Ker, "A Supplement to *Catalogue of Manuscripts Containing Anglo-Saxon*," *Anglo-Saxon England*, 1976, *5*:128 (No. 417).

notion that surviving manuscripts represent mindless copying of sterile formulae.

For further confirmation that early medieval medical texts were not copied as academic exercises, it is necessary to examine the surviving manuscripts for evidence of use and for what might be called "improvement in codex" rather than "improvement in text," that is to say, for additions to manuscripts by users. The addition of contents lists in later hands to two of the Anglo-Saxon herbals (Harley 585 and Bodley 130, where the chapters were also rubricated to agree with the contents list in the manner of Cotton Vitellius C. iii) has already been mentioned, as has the color-key system in the Cotton Vitellius C. iii contents list. That these lists were used and could be personalized can be inferred from the marginal nota sign added to the content listing *wið untrumnesse* on folio 68v of the Pseudo-Apuleius herbal in Hatton 76 and the identical idiosyncratic nota sign penned beside the corresponding recipe on folio 77; surely some user of the remedy book found a recipe valuable and wanted to signal it in both the contents list and the text so he could find it quickly. Similarly, it is possible to turn quickly to the herbal which begins on folio 68 of Hatton 76 without thumbing through the codex; a leather tag has been sewn to the folio.

Marginal notations provide evidence of use, although it is impossible to be certain what meaning the signs had for those consulting the codex; they may have been used to mark a passage for future reference or they may have been made in the process of recopying the codex. Whatever the reason, a number of users have made notations in the *Læceboc,* and these have been largely disregarded in discussions of the manuscript. There are seventeen silver circles painted in the margin of Royal 12 D. xvii at various points and twenty-five notation signs in ink, most of them marks in the shape of an X or a plus sign. There are even more drypoint notation signs, and they are mostly larger than the ink signs; most of them X-marks or cross-marks. In addition, the codex displays nearly thirty elongated nota signs which, Ker suggests (No. 264), indicate that the manuscript was read in the twelfth or thirteenth century. There are other sorts of marginal notations as well; in four instances marginal Latin comments have been scratched out, although one Latin gloss does survive, as do two Old English correction glosses. Marginal neums are also to be found on folios 30v and 89v.

Perhaps the strongest indication that an Anglo-Saxon medical manuscript was considered a living remedy book and not an artifact is the addition of other recipes by later users. Many instances of this practice can be found in the illustrated vernacular herbal in Cotton Vitellius C. iii. Previously unnoticed fragments of later writing around a hole on folio 11 are part of a recipe using *senecion.*[31] On the last leaf of the Sextus Placitus portion of the herbal, folio 82v, begins an accumulation of miscellaneous Latin and Old English recipes in different eleventh- and twelfth-century hands. These recipes end on folio 83v and are followed by a Latin tract on urine in a twelfth-century hand on folios 84–85. Folio 18v, following the contents list and preceding the ornamental title and author portrait, was originally blank, except for three words. The leaf now contains, in the left column and a portion of the right, two Old English recipes with Latin titles in red, "ad vertiginem" and "ad pectoris dolorem." It should also be mentioned that the illustrator or colorist has supplied, in a number of instances, vernacular plant names to label the illustrations.[32] Yet another

[31]It is possible to make out "ad ner. 7dolorecl . . ./cum axungia et alia.senecio. pecto/lib.ru*m* vitica." A relevant recipe is the third remedy in Ch. 76 of the *Herbarium Apulei,* "*Ad* pedum tumorem vel *dolorem* aut idem *nervorum.* Herba *senecion* tunsa *cum axungia* . . ." (italics mine).

[32]See Voigts, "A New Look," pp. 41–42, for a discussion of these labels. One might conclude that those illustrations that bear the artist's vernacular labels (in spelling that sometimes varies from that of the text)

addition is on the leaf facing the beginning of the herbal, folio 10v; this leaf contains in the right column the Latin names of seventy-eight plants in a twelfth-century hand with barely discernible Old English glosses above twenty-one of them. This list consists of those plants treated in the popular Latin poem attributed to Macer (Odo of Meung) and usually titled either "De viribus herbarum" or "De virtutibus herbarum."[33] All of these examples of "improvement in codex" make it clear that not only were Anglo-Saxon medical books intended to be used, they were also subject to alteration and addition to make them better guides to healing.

ECCLESIASTICAL EXCHANGE AND COMMERCIAL TRADE IN THE EARLY MIDDLE AGES

The evidence of the use of these remedy books brings one to the question of the availability of plants discussed and depicted in them. The answer to that question is both another way of countering the first dictum that Anglo-Saxon medical books are academic exercises and of responding to the second, that texts and illustrations were useless where they involve plants not indigenous to transalpine Europe, a reservation that applies to medical texts deriving from antiquity[34] as well as to those early medieval texts which incorporate Arab discoveries like galingale and zedoary.[35] The fact of the matter is that Mediterranean and Eastern plants could be had in Europe throughout the early Middle Ages both in a not-for-profit exchange and by way of commercial trade. First, there is a good deal of evidence of exchange among individual ecclesiastics and between religious houses. Too much should not be made of the (c. 754) letter of Bishop Cynehard of Winchester to Lull in Mainz asking for medical books because those he had dealt with ingredients unknown to him and difficult to obtain.[36] The letter seems less typical to me than one of the several early-eleventh-century letters of Fulbert of Chartres which deal with the dispatch of herbs; Fulbert says in a letter which accompanies a packet of medicines and herbs for Adalbero of Laon that Adalbero can easily find the instructions for the administration of the medicines in his own antidotaria.[37] Other continental examples can be cited. Gregory of Tours refers to the exchange of herbs between monastic houses, and Alcuin writes of the death of one Basil who had formerly conveyed medicines to the recipient of the

---

identify the plants the artist himself knew. Dr. Charles Huckins suggests that one would indeed expect a colorist to be familiar with native plants because much of his pigment was of vegetable origin.

[33] Ker, *Catalogue,* No. 218. The poem was probably written by Odo of Meung in the first half of the 11th century. It consists of 2,269 hexameter lines on the virtues of plants. See *Macer Floridus de viribus herbarum,* ed. Ludwig Choulant (Leipzig: Vossius, 1832); *A Middle English Translation of Macer Floridus de viribus herbarum,* ed. Gösta Frisk (Uppsala: Lindequistska, [1949]); Max Manitius, *Geschichte der lateinischen Literatur des Mittelalters* (Munich: Beck, 1911–1931), Vol. II, pp. 539–547; and Bruce P. Flood, Jr., "The Medieval Herbal Tradition of Macer Floridus," *Pharmacy in History,* 1976, *18*:62–66.

[34] On the use (as opposed to study or transmission) of classical medical texts in the Middle Ages, see Riddle, "The Introduction," pp. 189–198 and a number of articles by Jerry Stannard: "Greco-Roman Materia Medica in Medieval Germany," *Bull. Hist. Med.,* 1972, *46*:455–468; "Medieval Reception of Classical Plant Names," *Revue de Synthèse,* 3rd Ser., 1968, *49–52*:153–162; "Eastern Plants and Plant Products in Medieval Germany," *Proceedings, XIII International Congress of the History of Science, Moscow 1971,* 1974, Sec. 4: 220–225; "Squill in Ancient and Medieval Materia Medica, with Special Reference to its Employment for Dropsy," *Bulletin of the New York Academy of Medicine,* 2nd Ser., 1974, *50* (No. 6): 684–713.

[35] See above, n. 26, and Jerry Stannard, "Benedictus Crispus, an Eighth Century Medical Poet," *Journal of the History of Medicine,* 1966, *21*:41.

[36] Ep. 114 in *Die Briefe des heiligen Bonifatius und Lullus,* ed. Michael Tangl, Monumenta Germaniae Historica, Ep. sel., 1 (Berlin, 1916).

[37] *The Letters and Poems of Fulbert of Chartres,* ed. and trans. Frederick Behrends (Oxford: Clarendon Press, 1976), Letter 47, p. 82; see also Letters 48, 24, 49.

letter.[38] I have found three instances in the correspondence of Boniface where the Roman ecclesiastics specify the herbs and spices they are sending to the Anglo-Saxon archbishop in Germany.[39] A letter survives, in a ninth-century formulary, which may have been written by Walahfrid Strabo, requesting seeds of chive which were not to be bought in France.[40]

While continental practice may suggest an analogous situation in pre-Conquest England, more important for a consideration of Anglo-Saxon medical books is the evidence of ecclesiastical exchange of plants among Anglo-Saxons. Denehard, Lull, and Burchard of the eighth-century Anglo-Saxon mission to Germany wrote of sending pepper and cinnamon along with frankincense to Abbess Cuniburg in England (739–741).[41] Saint Willibald, the widely travelled native of Wessex who became Bishop of Eichstadt around 740, seems to have been an amateur trafficker in controlled substances. In 726 he obtained balsam in Jerusalem and was able to smuggle it through customs at Tyre by passing it off as petroleum in a gourd which did indeed contain petroleum in the outer segment.[42] Another important instance seems to be implied in chapters near the end of the Book II of the *Læceboc* of Bald (Royal 12 D. xvii, fols. 105v–106r) where we are told that the recipes given there had been ordered to be told to King Alfred by Elias III, Patriarch of Jerusalem (c. 879–907). That these recipes, which call for Levantine ingredients—scammony, ammoniacum, tragacanth, galbanum, balsam, and petroleum—were transmitted to Alfred along with the ingredients has been inferred on the basis of Asser's telling of Elias sending gifts as well as letters to Alfred.[43]

As the reference to the customs station deluded by Willibald's strategem implies, commercial trade in medicinal herbs and control thereof certainly occurred in the early Middle Ages alongside the more informal exchanges of churchmen and religious houses. In the late seventh and early eighth centuries the monks of Corbie imported "considerable quantities" of herbs from Marseilles,[44] and a ninth-century shopping list survives from that house as well—a list of twenty-three herbs, mostly exotica, to be purchased "at the market," perhaps in Cambrai.[45] In 793 a Spanish Muslim visiting Mainz marvelled at the quantity of "Indian" spices to be found there.[46] We know the Carolingians obtained herbs from Spain, Africa, and the Levant by trading in Provence, and overland trade through Lombardy during the seventh through the ninth centuries seems to have involved herbs, was closely

[38] Caprez, "Die Klostermedizin," p. 4654. Ep. 45 in *Alcuini epistolae*, ed. E. Duemmler, Monumenta Germaniae Historica, Ep. 4, Carolini Aevi, 2 (Berlin, 1892).

[39] Ep. 62, 84, 85 in *Die Briefe des heiligen Bonifatius und Lullus*.

[40] Wilfrid Blunt, "Walahfrid Strabo," in *The Hortulus of Walahfrid Strabo* (Pittsburgh: Hunt Botanical Library, 1966), p. 12.

[41] Ep. 49 in *Die Briefe des heiligen Bonifatius und Lullus*.

[42] "The Travels of Willibald, A.D. 721–727, Written from his own Recital by a Nun of Heidenheim," in *Early Travels in Palestine*, ed. Thomas Wright (London: Henry G. Bohn, 1848), p. 21. See also Robert S. Lopez, "The Trade of Medieval Europe: The South," in *The Cambridge Economic History of Europe*, Vol. II, ed. M. Postan and E. E. Rich (Cambridge: Cambridge University Press, 1952), p. 263. The identification of balm or balsam is difficult: among the possibilities are *Balanites aegyptiaca* Delile, *Commiphora opobalsamum* Engl., *Populus gileadensis* Rouleau, and *Impatiens balsamina* L.

[43] See Wright's discussion of this passage in the prefatory material to the facsimile, *Bald's Leechbook*, pp. 17–18.

[44] Lopez, "The Trade of Medieval Europe," p. 261. Invaluable for information on the Roman trade system, some of which survived into the Middle Ages, is J. Innes Miller, *The Spice Trade of the Roman Empire: 29 B.C. to A.D. 641* (Oxford: Clarendon Press, 1969); see esp. the table of 84 "Spices of the Roman Empire," pp. 112–118.

[45] For the list, see Riddle, "Introduction," p. 194; see also Lopez, "The Trade," p. 261.

[46] Lopez, "The Trade," p. 273.

regulated, and, by special agreement, favored Anglo-Saxon traders.[47] It is also worth remembering that the depots administered by the government of the Eastern Empire to control foreign trade were called ἀποθῆκαι, the word which came later to mean a storehouse for drugs and in the form we now know it, *apothecary,* to mean a dispensary.[48] To be sure, we cannot assume that the *amarium pigmentarum* of an Anglo-Saxon house contained all the plants for 185 chapters of the *Herbarium Apulei,* but there is certainly evidence to conclude that at least some of the Mediterranean plants would have been available through trade.

PLANT CULTIVATION: THE MINIMAL CLIMATIC OPTIMUM

Even with the explanations of the availability of southern plants to Northern Europe via trade, we are still left with the objection to the usefulness of herbal illustrations which depict southern plants; one has no need for the illustration of a living plant if one can obtain sap, oil, dried leaves, or seeds in trade. However, this traditional judgment, like the first, bears reassessing, in this case in terms of recent research in palaeoclimatology and in terms of a careful study of the monastic herb garden.

Traditional disparagement of Anglo-Saxon herbal illustrations by historians of medicine and art seems to have assumed that the climatic conditions and floristic zones of Europe today prevailed in the Middle Ages as well,[49] and indeed, under present circumstances, manuscript miniatures of Mediterranean plants seem inappropriate for England. Climatic conditions during the Anglo-Saxon period were, however, not the same as they are today, and although most of us have a vague notion that Greenland once was green and that Vinland was not a fantasy in A.D. 1000, that information has had little impact on our understanding of the Anglo-Saxon world. To my knowledge, no one has attempted to accommodate the question of southern plants in Anglo-Saxon herbals to the recent work in palaeoclimatology and its related disciplines, palynology, dendrochronology, and historical phytogeography. It is time to do that.

Speculation on the impact of climate change on historical civilizations is crucial to some recent explanations linking the decay of Mediterranean civilizations in two different instances to periods of Saharalike conditions in that region (accompanied by a corresponding amelioration of weather conditions in the North of Europe). For example, the decline of Byzantine civilization in the seventh through the eleventh centuries has been linked to the famine, plague, and depopulation that were a consequence of the reduced rainfall of the climate change.[50] Of late, similar explanations for historical patterns in Northern Europe have been offered. Behind these explanations lie, for English history at least, the important work in palaeoclimatology of H. H. Lamb and the application of this sort of study to historical interpretation, particularly in regard to continental Europe, by Emmanuel LeRoy Ladurie.[51]

What seems clear now is that the history of Northern Europe was decidedly

[47] *Ibid.,* pp. 271, 279.

[48] *Ibid.,* p. 275. Surely many if not most of the Eastern drugs required in the 9th-century Saint Gall antidotary and discussed in detail by Riddle in "Introduction," pp. 186–189, were obtained through trade.

[49] On the phytogeography of the present-day British Isles, see Harry Godwin, *The History of the British Flora* (2nd ed., Cambridge: Cambridge University Press, 1975), pp. 448–450.

[50] Rhys Carpenter, *Discontinuity in Greek Civilization* (Cambridge: Cambridge University Press, 1966); see esp. pp. 1–17.

[51] See H. H. Lamb, "The Early Medieval Warm Epoch and its Sequel," *Palaeogeography, Palaeoclimatology, Palaeoecology,* 1965, *1*:13–37, and his volume of selected papers, *The Changing Climate* (London: Methuen, 1966); Emmanuel LeRoy Ladurie, "Aspects historiques de la nouvelle climatologie," *Revue*

affected by what is called the "minimal climatic optimum," or the "secondary climatic optimum"—that is to say, a period from around the ninth century to the thirteenth when the climate of Northern Europe was warmer and drier than it has been since, even in the comparatively warm period of the first half of the twentieth century.[52] During the peak of this xerothermic period, circa A.D. 1000–1200, mean annual temperatures in Northern Europe were at least 1° to 2° C higher than they are now,[53] and annual rainfall was 10 percent less than its present amount.[54] This was the era of population buildup and overflow in Scandinavia, the time, in short, of good centuries for the North of Europe and bad for the South.[55] This era was followed by a period from 1200 to 1400 of climatic decline in Northern Europe[56] (and an upswing in Mediterranean civilization[57]) and a period of partial recovery from 1400 to 1550 before the "little Ice Age" of 1550 to 1850.[58]

While these climate changes can be discerned from growth patterns of trees in North America and Japanese annual records of the date of the first cherry blossoms that go back to the ninth century,[59] it is European evidence that pertains here. During the minimal climatic optimum, preglacial trees thrived at heights in the Alps where they have failed to grow since, even in the comparatively warm twentieth century.[60] Of particular importance for the Anglo-Saxon world are the archaeological evidence and the written records of Anglo-Saxon vineyards. The *Domesday Book* recorded in 1085 thirty-eight vineyards in England in addition to those of the king. English wine was considered equal to that of France in terms of quality as well as quantity; there were more than half a dozen vineyards of over five acres, and at least some of them, as in Gloucestershire and at Thorney, were unprotected. A vineyard was maintained at Ely for several centuries.[61] Similarly, remains of medieval tillage, evidenced by surviving ridge and furrow, indicate that plowing went on in Redesdale in Northum-

---

Historique, Jan.–Mar., 1961, *225*:1–20, and *Times of Feast, Times of Famine: A History of Climate since the Year 1000,* trans. Barbara Bray (Garden City: Doubleday, 1971). Of importance for future work in this area will be D. Justin Schove's forthcoming climatic history of Europe from A.D. 800.

[52]Lamb, "The Early Medieval Warm Epoch," *passim,* and three essays in *The Changing Climate:* "On the Nature of Certain Climatic Epochs which Differed from the Modern (1900–39) Normal," pp. 58–112; "Our Changing Climate, Past and Present," pp. 1–20; and "Britain's Climate in the Past," pp. 170–195. See also LeRoy Ladurie, *Times of Feast,* pp. 254–264.

[53]See LeRoy Ladurie, *Times of Feast,* p. 127, and Lamb, "The Early Medieval Warm Epoch," p. 13. It may be noted that in 1959 Lamb asserted that the peak was 800–1000 ("Our Changing Climate, Past and Present," p. 7).

[54]LeRoy Ladurie, *Times of Feast,* pp. 254–257; Lamb, "The Early Medieval Warm Epoch," p. 13. Particularly valuable are Lamb's summer dryness/wetness and winter mildness/severity indices, in *The Changing Climate,* pp. 100, 182, and the tables on pp. 21–27, "The Early Medieval Warm Epoch."

[55]Carpenter, *Discontinuity in Greek Civilization,* p. 15. Scandinavian population expansion to the New World was, of course, possible because of climatic conditions of Greenland; see W. Dansgaard, *et al.,* "One Thousand Centuries of Climatic Record from Camp Century on the Greenland Ice Sheet," *Science,* 1969, *166*:377–381.

[56]Lamb, "Our Changing Climate, Past and Present," pp. 8–10. For records of depopulation and for scientific evidence, including pollen diagrams from bogs, of the consequences for Scandinavia of the cooling period, see Axel Steenberg, "Archaeological Dating of the Climatic Change in North Europe about A.D. 1300," *Nature,* 1951, *168*:672–674.

[57]Carpenter, *Discontinuity in Greek Civilization,* p. 15.

[58]Lamb, "Our Changing Climate, Past and Present," pp. 10–12, and "On the Nature of Certain Climatic Variations," pp. 65–66; LeRoy Ladurie, *Times of Feast,* Ch. 4, "The Problems of the 'Little Ice Age'," pp. 129–226.

[59]LeRoy Ladurie, *Times of Feast,* pp. 23–40, 270, and H. C. Fritts, *Tree Rings and Climate* (New York: Academic Press, 1976); see esp. Fig. 8.9.

[60]LeRoy Ladurie, *Times of Feast,* pp. 244–254.

[61]See Lamb, "Our Changing Climate," in *The Flora of a Changing Britain,* ed. F. Perring (London: The Botanical Society of the British Isles, 1974), p. 13; "The Early Medieval Warm Epoch," pp. 30–31; "Our Changing Climate, Past and Present," pp. 7–8; and "Britain's Climate in the Past," p. 191.

berland up to a height of 1,050 feet for some two centuries until 1250 or 1300. Cultivation has not been attempted there subsequently.[62]

It is difficult to assess precisely the implications of this xerothermic period for floristic zones in Anglo-Saxon England.[63] Many factors other than temperature and rainfall must be considered, factors like biotype variation, soil content,[64] and the Anglo-Saxon plowing which altered floristic zones in England.[65] Furthermore, there are plant distributions, like that of the strawberry tree, *Arbutus unedo* L. (Mediterranean region and three counties in Ireland), that defy explanation.[66] Nonetheless, the period when Anglo-Saxon medical texts calling for Mediterranean plants were copied (10–12th centuries) and when Anglo-Saxon herbals of Mediterranean origin were illustrated (1050 and 1100) was an era of comparatively warm, dry summers, an important characteristic of Mediterranean floristic zones.[67] While I do not suggest that the Mediterranean plants represented in the herbals grew wild in Anglo-Saxon England, I would argue that it would not have been difficult to cultivate them during the minimal climatic optimum, and there is some scientific evidence of the introduction of plants during this era. Palynological evidence indicates that the woad, *Isatis tinctoria* L., prescribed in the Old English *Herbarium Apulei* and mentioned five times in the *Læceboc* and once in the *Lacnunga,* was cultivated in England during the Anglo-Saxon period.[68] Similarly, the Mediterranean peony, *Paeonia mascula* (L.) Mill., a plant celebrated for its medical use in the *Herbarium Apulei,* is found naturalized in England only at Steep Holm, on the grounds of a thirteenth-century Augustinian priory; it was formerly found growing on the site of the abbey near Winchcombe in Gloucestershire, an eighth-century foundation.[69]

PLANT CULTIVATION: THE HERB GARDEN

That the presence in England of two Mediterranean medical plants named in Anglo-Saxon remedy books seems to result from Anglo-Saxon cultivation is a crucial point for this paper and one which raises a final area of concern in the matter of Anglo-Saxon knowledge of Mediterranean plants—that of the monastic herb garden. It is important to emphasize the easily overlooked implications of the minimal climatic optimum that made cultivation of southern plants easier, but the cultivation practices

[62] Lamb, "Our Changing Climate," *The Flora of a Changing Britain,* p. 13.

[63] J. P. Savidge, "Changes in Plant Distribution following Changes in Local Climate," in *The Flora of a Changing Britain,* pp. 25–31.

[64] Ann P. Conolly and Eilif Dahl, "Maximum Summer Temperature in Relation to the Modern and Quaternary Distributions of Certain Arctic-Montane Species in the British Isles," in *Studies in the Vegetational History of the British Isles: Essays in Honour of Harry Godwin,* ed. D. Walker and R. G. Westeds (Cambridge: Cambridge University Press, 1970), pp. 159–167.

[65] Godwin, *History of British Flora,* pp. 477–479. See also Oliver Rackham, *Hayley Wood: Its History and Ecology* (Cambridge: Cambridgeshire and Isle of Ely Naturalists' Trust, 1975), pp. 41–44.

[66] Godwin, *History of British Flora,* p. 294, and Polunin, *Flowers of Europe,* pp. 296–297.

[67] On Mediterranean climate and plant life, see Oleg Polunin and Anthony Huxley, *Flowers of the Mediterranean* (London: Chatto and Windus, 1965), pp. 1–14. See also C. D. Piggott, "The Response of Plants to Climate and Climatic Changes," in *The Flora of a Changing Britain,* pp. 32–44.

[68] Godwin, *History of British Flora,* p. 480. The Old English word is *wad.* For the Old English Pseudo-Apuleius herbal, see Cockayne, *Leechdoms,* Vol. I, p. 174, and for the Latin text, see Howald and Sigerist, *Antonii Musae,* p. 127. See Bierbaumer, *Der botanische Wortschatz des Altenglischen,* Teil I and II, s.v. *wad.* The plant was also used as a source of dye and may have been gathered in this connection, if not cultivated, by the blue-painted Britons encountered by the Romans.

[69] For the Old English text, see Cockayne, *Leechdoms,* Vol. I, pp. 168–170, and for the Latin text, Howald and Sigerist, *Antonii Musae,* pp. 120–121. On its growth on the site of medieval monastic houses, see Nancy Jenkins, "Medieval Monastic Accounts, Medicines and Spices," *The Pharmaceutical Journal,* 1954, 4th Ser., *118* (172): 515–516; Polunin and Huxley, *Flowers of the Mediterranean,* p. 66, and Polunin, *Flowers of Europe,* p. 110.

bear examination in their own right. We are, after all, not so much dealing with native floral populations as we are with pampered plants grown within the cloister walls that radiate heat and provide shelter from the wind—plants that are weeded, mulched, and watered, in many instances annuals, for which the question of winter hardiness is irrelevant.[70] Indeed, the history of cultivated plants is quite a different matter from the history of floristic zones of indigenes.[71]

Medieval herb gardens represent a continuation of an ancient institution. Pliny in the book of the *Historia naturalis* which deals with plants produced "by Mother Earth for medicinal purposes only" tells of having examined nearly all the plants of which he writes in the special garden of the botanical authority Antonius Castor.[72] We know as well that the Romans made some effort to acclimatize delicate plants from other regions; evidence remains of special buildings, with windows of talc and heated by flues, for sheltering plants in winter.[73] While there is no proof of continuity between the Roman and medieval gardens in transalpine Europe, it should be pointed out that gardens were found with Roman villas on the Moselle and the Rhine, and one would assume that villas in Roman Britain carried on the practice as well.[74] The significance of the gardens of antiquity is that the cultivated garden plants of Rome became the cultivated garden plants of Europe, at least until 1560; and as recently as the nineteenth century an analysis of German peasant gardens concluded that the plants contained therein were of Greek or Latin origin, plants mentioned in Dioscorides, Pliny, and other antique writers.[75]

Valuable records of plant cultivation from the early Middle Ages survive in the garden lists from the court of Charlemagne. Two shorter lists are found in the inventories of five royal estates; the estate of Asnapius lists twenty herbs and vegetables and eight trees, and the estate of Treola inventories twenty-seven and ten.[76] Of even greater interest is the *Capitulare de Villis,* circa 800, which reflects the estate reforms of either Charlemagne or his son, Louis the Pious. The *Capitulare* contains a royal command giving the names of seventy-two vegetables and medicinal plants and sixteen fruit and nut trees to be grown in the garden, "Volumus quod in horto omnes herbas habeant, id est lilium, rosas, fenigrecum, costum, . . ."[77] Many of these plants and trees are southern, such as cumin, capers, figs, but there is no reason to think that the emperor was not quite serious in his wish to have the plants grown in his gardens, and indeed, it is not unlikely that many of the plants could have been grown with careful cultivation. It has been suggested that Carolingian gardeners acclimatized southern and eastern plants which Charles acquired as a result of his

[70] E.g., cumin, *Cuminum cyminum* L., listed both in the *Capitulare de villis* of Charlemagne and in the Saint Gall cloister plan, is an annual.

[71] For an invaluable introduction to the history of cultivated plants, see the 1964 Masters Lectures, William T. Stearn, "The Origin and Later Development of Cultivated Plants," *Journal of the Royal Horticultural Society,* 1965, *90*:279–291, 322–340.

[72] *Historia naturalis* XXV 5, 9, ed. W. H. S. Jones, Loeb Classical Library (Cambridge, Mass.: Harvard University Press, 1966), Vol. VII, p. 142. Pliny here uses the word *hortulus* as does Walahfrid some centuries later; the word seems to have a more precise meaning than simply the diminutive form of *hortus.*

[73] Henry Inigo Triggs, *Garden Craft in Europe* (London: Batsford, [1913]), p. 9.

[74] Marie Luise Gothein, *A History of Garden Art,* ed. Walter P. Wright, trans. Mrs. Archer-Hind (New York: Dutton, 1928), Vol. I, pp. 127–130.

[75] Stearn, "Origin," pp. 324–326.

[76] *Capitularia Regum Francorum,* ed. A. Boretius, Monumenta Germaniae Historica, Leg. Sec. 2 (Hanover, 1883). For a brief but useful discussion and bibliography and translation of the inventories, see Henry R. Loyn and John Percival, *The Reign of Charlemagne: Documents on Carolingian Government and Administration* (London: Edward Arnold, 1975), pp. 98, 103, 105. See also Rudolph von Fischer-Benzon, *Altdeutsche Gartenflora* (Keil: Lipsius und Tischer, 1894), pp. 5, 181–182.

[77] *Capitularia Regum Francorum,* Vol. I, pp. 82–91. See also Loyn and Percival, *Reign of Charlemagne,* pp. 64–65, 73; Fischer-Benzon, *Gartenflora,* pp. 5, 183–184; and Stearn, "Origin," pp. 324–325.

dealings with the Caliph Haroun-al-Rashid.[78]

The continental records which are more germane to an Anglo-Saxon focus than royal documents are those describing monastic herb gardens, for literate Anglo-Saxon culture was dominated by monasticism, and the surviving Anglo-Saxon medical texts seem to be the products of monastic scriptoria.[79] Numbered among the continental monastic records is the famous plan for the exemplary Carolingian monastery, the Saint Gall cloister, which depicts elaborate facilities for the care of the sick and a herb garden with sixteen plots designated by the names of medicinal plants. The medical garden, adjacent to the *domus medicorum,* should not be confused with the vegetable garden with designations for eighteen plants, or the orchard-cum-cemetery diagram representing both graves and fruit and nut trees.[80] As in the case of the *Capitulare,* we find cumin and figs named.[81] The most personal account of a monastic herb garden is in the *Hortulus* of Walahfrid Strabo, Abbot of Reichenau, 838–849, a poem which celebrates the labor of the garden and considers twenty-nine plants, twenty-one of which he has cultivated. Recounting the effort involved in hacking out nettles, building up the beds, spreading manure, and carrying water, the poem tells of some plants grown from seed, others from stocks, of training gourds on props, of the superiority of the betony of the garden over that which grows wild, of drying plants to store for the winter. It is clear that the abbot speaks from a genuine involvement with the cultivation of plants when he celebrates in hexameters the medicinal values of the plants he raises.[82] Another poem by a continental monk dealing with remedies to be obtained from seventy-seven plants was also popular in the Middle Ages. In fact, the plant names from the "De virtutibus herbarum" of Macer (Odo of Meung) have already been mentioned; they are found with Old English glosses in Cotton Vitellius C. iii, on the folio preceding the Old English illustrated Pseudo-Apuleius herbal.[83]

Unfortunately, our records of Anglo-Saxon monastic gardens are not so rich as those for the Continent. We do know, however, that there was a garden at Ely from the seventh century on,[84] and that the first abbot was famed for his planting and grafting skills in the garden and orchard.[85] Similarly, another East Anglian house, Thorney, which is not far from Ely and like Ely has already been mentioned for its vineyards, was famous for its tree garden and the luxuriant growth of its plants.[86] The proximity of these two fenland houses with a reputation for horticulture and viniculture may be more than a coincidence. Neither is remote from Bury Saint Edmunds where Bodley 130 was painted, and I have argued elsewhere that the conventions of the monastic dedication page in the vernacular illustrated *Herbarium Apulei* suggest

[78] Gothein, *A History of Garden Art,* Vol. I, p. 190.

[79] Although we do not know the origin of all the surviving Anglo-Saxon remedy books, we do know that Bodley 130 was a product of Bury Saint Edmunds, that the unique manuscript containing the *Læceboc* was copied at the Old Minster, Winchester, and that the monastic dedication page (fol. 11v) in Cotton Vitellius C. iii tells us something about the origin of that codex (for a discussion of the dedication page, see Voigts, "A New Look," pp. 44–60).

[80] See P. Jung, "Das Infirmarium im Bauriss des Klosters von St. Gallen vom Jahre 820," *Gesnerus,* 1949, *6*:1–8, and Figs. 1–2; Walter Horn, "On the Origins of the Medieval Cloister," *Gesta,* 1973, *12*:46–48, and Figs. 1, 48; and Fischer-Benzon, *Gartenflora,* pp. 184–187.

[81] Gothein does not think that the fig or laurel could have been grown at Saint Gall (*A History of Garden Art,* Vol. I, p. 175), but again I believe that judgment overlooks the characteristics of the xerothermic period and the circumstances of monastic horticulture.

[82] *The Hortulus of Walahfrid Strabo* (see above, n. 40), and Fischer-Benzon, *Gartenflora,* pp. 187–188.

[83] See above, n. 33.

[84] Frank Crisp, *Mediaeval Gardens,* ed. Catherine Childs Paterson (New York: Brentanos, [1924]), Vol. 1, p. 95.

[85] Alicia Amherst Cecil, *A History of Gardening in England* (3rd ed., New York: Dutton, 1910), p. 7.

[86] See n. 61 and Gothein, *A History of Garden Art,* Vol. 1, p. 178.

that that codex could have originated at an East Anglian monastery like Peterborough, Croyland, Ramsey, Thorney, or Ely.[87]

While we lack an Anglo-Saxon equivalent of the Saint Gall cloister plan, a post-Conquest diagram of the abbey at Canterbury may well suggest practices from more than a century before. In this drawing half of the rectangular cloister is designated "herbarium," and while it lacks the plant labels of the Saint Gall plan, it does represent the area filled with stylized drawings of plants.[88] The gardener's records from the Norwich chartulary postdate the Conquest by three hundred years and so cannot serve to imply Anglo-Saxon practices, although surely the responsibilities and efforts entailed by the garden are not unique to the later period.[89]

IMPLICATIONS

The time has surely come to put to rest the notion that the Anglo-Saxons who compiled, copied, and illustrated the surviving vernacular and Latin remedy books epitomized the "wilting mind of the Dark Ages." We must discard the first dictum that the surviving texts represent mindless copying of "sterile formulae," and the second that the illustrations were of no worth because they are stylized and depict Mediterranean plants. We must rather look at the sophisticated handling of text by the Anglo-Saxon makers and users of medical books, at the vernacular revision in the *Herbarium Apulei* of the Latin tract on betony, for example, a reorganization that deletes the nonessential, mostly magical, section and unites the important material about habitat and preparation of the plant at the beginning of the chapter. We must also look at the way Anglo-Saxons attempted to make their books as useful as possible, by adding tables of contents and recipes, by marking recipes with marginal notations, by providing Latin-vernacular herbal glossaries to Latin remedy books and glossing Latin plant names with the vernacular labels. We must likewise acknowledge that Anglo-Saxon abbesses and kings acquired exotic drugs through informal exchange, largely through ecclesiastics, and similarly, we must consider that some exotic plants were likely obtained through trade. Most important, we must admit to the probability that Anglo-Saxon monks and nuns cultivated Mediterranean plants with great care and under more auspicious climatic conditions than exist today, remembering that plants like peony, formerly not native to Britain, were naturalized after cultivation during the pre-Conquest era. In short, we must grant that Anglo-Saxons valued healing plants, that they valued books about healing plants, and that they dealt with both intelligently.

Two final examples should make this point. The first is a reconsideration of Chapter 5 in the Old English *Herbarium Apulei,* dealing with hennebelle (Fig. 2). We have been told that this chapter is useless as it stands in Cotton Vitellius C. iii (fol. 23v) because the illustration does not depict the henbane, *Hyoscyamus niger* L., that is now native to England; rather it portrays the Mediterranean *Hyoscyamus reticulatus* L.[90] The fact is that Anglo-Saxons knew full well that the hennebelle depicted was not the dangerous native plant; not only does it not resemble it, but the text carefully

[87] "A New Look," pp. 43–56.

[88] The plan, which seems to have been drawn as a guide to the waterworks, the most prominent element in the sketch, is bound with the Edwine Psalter executed at Saint Augustine's, Canterbury, in the 12th century, now Cambridge, Trinity College R. 16. 2. See Plt. 228 in Crisp, *Mediaeval Gardens,* Vol. II.

[89] These accounts are edited in Cecil, *History of Gardening,* pp. 9–13. Anglo-Saxon *horti* and *hortuli* are listed in the *Domesday Book*: T. H. Turner, "Observations on the State of Horticulture in England in Early Times," *Archaeological Journal,* 1848, *5*:295–297.

[90] See above, n. 13. The Old English recipe for the hennebelle of fol. 23v is found in Cockayne, *Leechdoms,* Vol. I, pp. 94–96, and the Latin (Ch. 4) in Howald and Sigerist, *Antonii Musae,* pp. 32–34.

Figure 2. Hennebelle, c. 1050, **Herbarium Apulei** *(Old English), Cotton Vitellius C. iii, fol. 23v. (Reproduced by permission of the British Library.)*

distinguishes between the two plants: "Þeos wyrt þe man symphoniacam nemneð 7 oðrum naman belone 7 eac sume men hennebelle hatað wihrt on beȝanum landum 7 on sandiȝum landum 7 on wyrttunum. þonne ys oðer þisse ylcan wyrte sweart on hiwe. 7 stiðran leafum 7 eac ætriȝum. ðonne ys seo ærre hwitre 7 heo hæfð þas mæȝnu." (This plant which one calls symphoniaca and others name belone and also some men call hennebelle grows on cultivated lands and on sandy lands and in gardens. Then there is another [of] this sort of plant [which is] dark in color and [has] stiffer leaves and [is] also poisonous. The first one then is the whiter and it has the powers.) Again, the vernacular text has placed at the beginning of the chapter the material found at the end of the Latin text, and this is given pride of place to make certain that the user does not utilize the poisonous native relative of the healing plant by mistake. The compiler and illustrator did not make a stupid mistake; rather they made every effort to direct the user to the correct plant. And a correct plant it is, for after looking through scores of dried varieties of solanaceae one is struck by the verisimilitude of the illustration to the Mediterranean and Turkish varieties of *Hyoscyamus: H. aureus* L., *H. pusillus* L., *H. reticulatus* L.[91]

Finally, I think we must take Walahfrid Strabo, the monastic cultivator of plants, at his word in describing the situation that must have obtained in Anglo-Saxon England as well as at Reichenau. He speaks of paying a great deal of money for Indian pepper, and he commends a recipe using pennyroyal for an ailing stomach, saying he knows for a truth that it works, but he will also grant that wearing a sprig of pennyroyal behind one's ear is simply a matter of custom and habit.[92] Most significant, however, he writes that he knows what he does about plants and gardening because he has learned from common opinion, and from reading ancient books, and from hard labor. That must surely also be the case with Anglo-Saxon users of plant remedies, and we must grant them that.

> Haec non sola mihi patefecit opinio famae
> Vulgaris, quaesita libris nec lectio priscis;
> Sed labor et studium, quibus otia longa dierum
> Postposui, expertum rebus docuere probatis.[93]

[91] For a number of helpful illustrations, see A. I. Poyarkova, "Contributions to the Systematics of the Species Henbane Related to Hyoscyamus Reticulatus L.," *Journal Botanique de l'URSS*, 1942, *27*:117–130. While Pliny discusses four varieties of *Hyoscyamus* (XXV 17, 35–37), I do not think the Anglo-Saxon herbal tradition draws on this discussion. If anything, the Anglo-Saxon treatment resembles in some points Pliny's distinction between black and white hellebore (XXV 21, 48–52).

[92] *Hortulus*, pp. 52–54.

[93] *Ibid.*, p. 24. The poetic translation by Raef Payne, p. 25, reads:

> This I have learnt not only from common opinion
> And searching about in old books, but from experience—
> Experience of hard work and sacrifice of many days
> When I might have rested, but chose instead to labor.

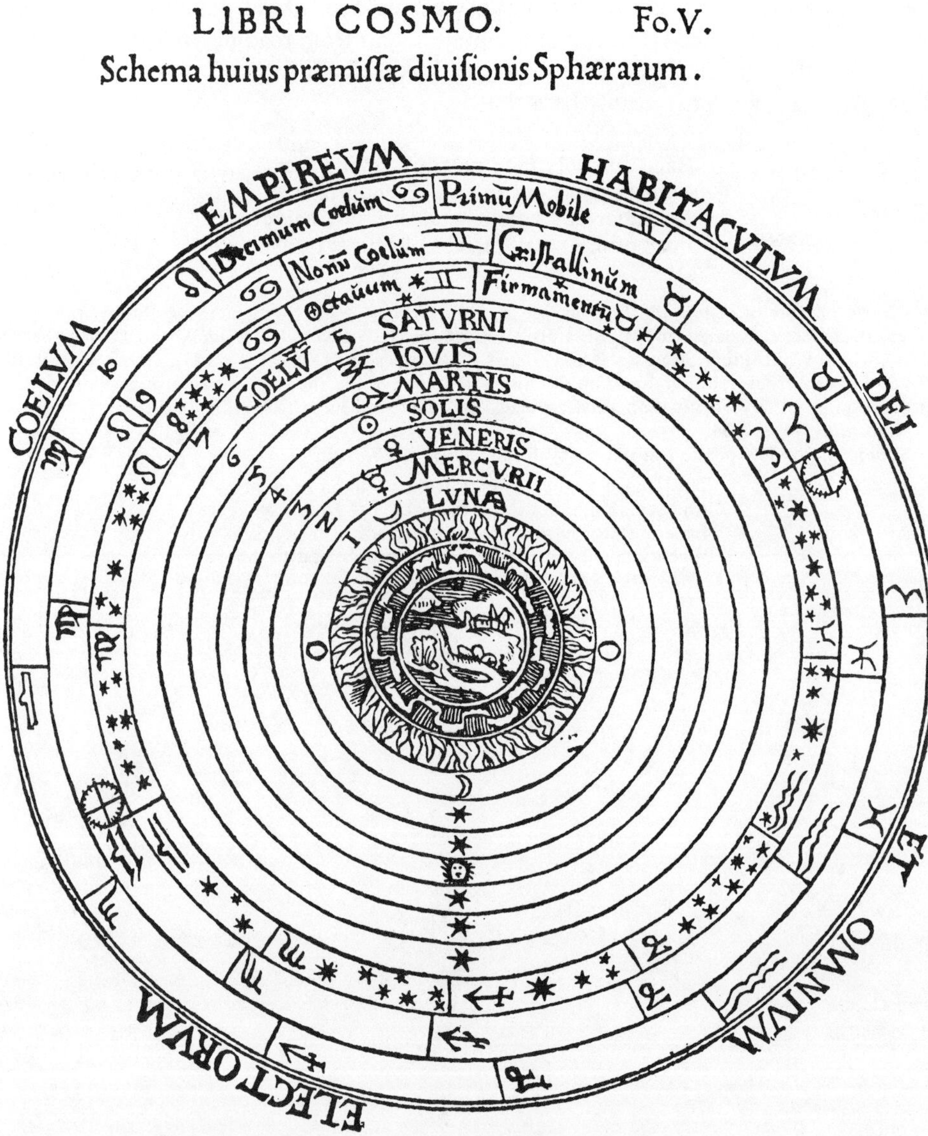

The celestial orbs as depicted in Peter Apian's Cosmographia *(Antwerp, 1539). Courtesy of the Rare Book Division, New York Public Library.*

Celestial Orbs in the Latin Middle Ages

By Edward Grant

MODERN VIEWS OF MEDIEVAL ORBS

HOW DID MEDIEVAL ASTRONOMERS and natural philosophers understand the celestial orbs and spheres that surrounded their immobile earth? Modern scholars have presented a few interpretations. Undoubtedly the most widely held opinion is that the orbs were believed to be solid, where *solid* is taken to be synonymous with *hard* or *rigid*.[1] Here the image is one of transparent glass or crystalline globes.

A second view assumes that medieval astronomers and natural philosophers followed Aristotle in thinking of the celestial orbs as composed of the celestial ether and adhered to Aristotle's dicta about that substance. According to this interpretation the orbs or spheres could be neither solid nor fluid because Aristotle had argued that contrary qualities such as hard-soft, dense-rare, and so on, were inapplicable to the incorruptible celestial ether of which they were composed. Nicholas Jardine observes that to pose a question about the hardness or softness of celestial spheres would have been to make a "category mistake."[2] Hardness and softness are qualitative opposites characteristic only of terrestrial matter. Since pairs of opposite qualities are the source of all terrestrial change, they must of necessity be absent from the celestial region, where change is impossible. Thus to inquire about the possible hardness or softness of celestial orbs is to ask a senseless question.

I am grateful to the National Science Foundation for its generous support of my study on the medieval cosmos, 1200–1687, of which this article forms a part.

[1] See C. S. Lewis, *The Discarded Image: An Introduction to Medieval and Renaissance Literature* (Cambridge: Cambridge Univ. Press, 1964), p. 96; Herbert Butterfield, *The Origins of Modern Science* (rev. ed., New York: Macmillan, 1959), pp. 20–21; A. R. Hall, *The Scientific Revolution, 1500–1800: The Formation of the Modern Scientific Attitude* (London: Longmans, Green, 1954), p. 13 (no similar statement appears in the third edition, 1983); Marie Boas, *The Scientific Renaissance, 1450–1630* (The Rise of Modern Science, 2) (London: Collins, 1962), pp. 42, 45; Thomas S. Kuhn, *The Copernican Revolution: Planetary Astronomy in the Development of Western Thought* (1957; New York: Random House, n.d.), pp. 80, 111, 113; Stephen Toulmin and June Goodfield, *The Fabric of the Heavens* (1961; Harmondsworth, U.K.: Penguin Books, 1963), p. 207; A. C. Crombie, *Medieval and Early Modern Science*, 2 vols. (Garden City, N.Y.: Doubleday, 1959), Vol. I, p. 75; Nicholas H. Steneck, *Science and Creation in the Middle Ages: Henry of Langenstein (d. 1397) on Genesis* (Notre Dame, Ind./London: Univ. Notre Dame Press, 1976), p. 62; and Noel Swerdlow, "Pseudoxia Copernicana: or, Enquiries into Very Many Received Tenents and Commonly Presumed Truths, Mostly Concerning Spheres," *Archives Internationales d'Histoire des Sciences*, 1976, *26*:108–158, esp. pp. 116–118. Swerdlow's comments apply only to medieval astronomers, however, since he excludes natural philosophy and astrology from his discussion (p. 114, n. 11).

[2] Nicholas Jardine, "The Significance of the Copernican Orbs," *Journal for the History of Astronomy*, 1982, *13*:168–194, on p. 175.

In order to avoid posing such questions, William Donahue supposes that peripatetics conceived of the heaven and its orbs as "immaterial or quasi-material" entities, whatever these terms may signify.[3] A similar position was recently adopted by the late Edward Rosen, who denied that Aristotle held the notion of solid celestial orbs.[4] Rosen insisted that because Aristotle denied the existence of corruptible terrestrial matter in the incorruptible and unchanging ethereal heaven, it followed that his celestial ether was not material and could therefore be neither solid nor hard, as Pierre Duhem had claimed. Indeed by similar reasoning (though Rosen does not draw the inference), it could not be fluid either. Not only does this argument misunderstand Aristotle's intent, but when applied to the Middle Ages it distorts medieval opinion. It will therefore be well to eliminate this second interpretation before proceeding.

Although Aristotle may have denied that alterable terrestrial-type matter could exist in the heavens, his ether may be construed as a fifth kind of substance, or element—a *quinta essentia,* as many commentators would call it—with properties radically different from those of the four sublunar elements. Whatever Aristotle may have thought about the properties of the celestial ether, there is no doubt that in *De caelo* he assumed the corporeality, and therefore the physicality, of the heavenly orbs.[5] As nonspiritual, corporeal, and therefore three-dimensional physical entities composed of ether, celestial orbs had to be something akin to hard or soft, even though Aristotle himself was committed to a formal denial of contrary qualities. It is no doubt for this reason that he completely ignored the physical nature of celestial spheres and so provided no helpful clues as to their possible hardness, softness, or fluidity; indeed this may well explain why his medieval scholastic commentators also neglected the problem. But, just as many scholastic authors ignored Aristotle's famous dictum that neither place, nor void, nor time could exist beyond the outermost sphere of the physical world and began to inquire what indeed might lie beyond, so, to a lesser extent, did some of those same authors reveal an opinion or judgment, usually

[3] See William H. Donahue, "The Solid Planetary Spheres in Post-Copernican Natural Philosophy," in *The Copernican Achievement,* ed. Robert S. Westman (Berkeley/Los Angeles: Univ. California Press, 1975), pp. 251, 256–257, 258, 259, 275.

[4] Edward Rosen, "The Dissolution of the Celestial Spheres," *Journal of the History of Ideas,* 1985, *46*:13–31, on p. 13.

[5] See Aristotle, *De caelo* 2.12, 293a8, trans. W. K. C. Guthrie (Loeb Classical Library) (London: Heinemann; Cambridge: Harvard Univ. Press, 1960): "The last sphere moves round embedded in a number of spheres, and each sphere is corporeal." Cf. the Latin translation of the second phrase in Averroes's commentary on *De caelo,*" et omnis orbis eorum est corpus: *Aristotelis opera cum Averrois commentariis* (Venice, 1562–1574), Vol. V, Bk. 2, text 70, fol. 70r, col. 1. Robert Grosseteste reports that "John Damascene also implies in his book of *Sentences* that the existence of an immaterial body, that which is called a fifth body among the wise men of the Greeks, is impossible," but he declares that Aristotle and his followers did assume the existence of a "fifth body" in addition to the four elements: Robert Grosseteste, *Hexaemeron,* ed. Richard C. Dales and Servus Gieben (London: Oxford Univ. Press for the British Academy, 1982), p. 106. (Unless otherwise indicated, all translations are mine.) John may have had in mind the passage from Aristotle cited above. Aristotle presents his theory of the celestial ether in *De caelo.* The ether moves naturally with perfect, uniform, circular motion; it is eternal, ungenerable, and incorruptible; and, because it is incorruptible, it lacks the various contrary qualities that ordinarily cause change in the four elementary bodies of the sublunar world and the bodies compounded of them. Indeed Aristotle argues that the celestial ether is neither heavy nor light. By implication, it is neither hot nor cold, nor rare nor dense, and so on. For a detailed discussion see Friedrich Solmsen, *Aristotle's System of the Physical World: A Comparison with His Predecessors* (Ithaca, N.Y.: Cornell Univ. Press, 1960), pp. 287–303.

indirectly, about the hardness or softness of the celestial orbs, which they all assumed to be physical bodies.[6]

To my knowledge no medieval natural philosopher rested content to depict the celestial orbs as immaterial entities devoid of physical properties, although many denied—in the abstract—that the celestial ether could possess terrestrial attributes such as hot-cold and dense-rare. When confronted with specific problems about the spheres themselves—that is, about their arrangement and the relationships between successive surfaces—scholastic natural philosophers speak in a quite different vein, revealing, often inadvertently, a concern about real, physical spheres. Indeed, numerous medieval discussions about possible physical problems that might affect eccentric orbs—for example, whether vacua can occur between successive celestial surfaces or whether two orbs can overlap and occupy the same place simultaneously—provide ample evidence that the spheres were conceived of as physical bodies.[7] I am aware of no instance in which physical considerations were dismissed because the celestial orbs were deemed immaterial or quasi-material. Because those orbs were judged to be physical, it was difficult to avoid the attribution of some physical properties to them.

With the second interpretation eliminated from further consideration, we shall now attempt to determine whether, during the late Middle Ages, the celestial orbs were conceived of as hard and rigid or fluid and soft.

WHAT SOLID SPHERES SIGNIFIED IN THE SEVENTEENTH CENTURY

Because his world system required an intersection between the orbits of Mars and the sun, which would have been impossible if hard spheres existed, Tycho Brahe used his knowledge that the comet of 1577 was moving in the celestial region beyond the moon to deny the existence of solid celestial orbs and to suggest instead that the heavenly region was composed of a fluid substance.[8] The solid celestial orbs whose existence Tycho denied were, of course, of the hard and rigid variety. He explains that he "first showed and clearly established that by the motions of comets [the heaven] is fluid and that the celestial mechanism is not a hard and impervious body filled with various real orbs, as has been believed by many up to this point, but that it is very fluid and simple with the orbits of the planets free and without the efforts and revolutions of any real spheres."[9]

[6] On the problem of extramundane space see Edward Grant, *Much Ado About Nothing: Theories of Space and Vacuum from the Middle Ages to the Scientific Revolution* (Cambridge: Cambridge Univ. Press, 1981), Pt. 2.

[7] For a description and discussion of these arguments see Edward Grant, "Eccentrics and Epicycles in Medieval Cosmology," in *Mathematics and Its Applications to Science and Natural Philosophy in the Middle Ages, Essays in Honor of Marshall Clagett,* ed. Edward Grant and John E. Murdoch (Cambridge: Cambridge Univ. Press, 1987), pp. 189–214.

[8] See Victor E. Thoren, "The Comet of 1577 and Tycho Brahe's System of the World," *Arch. Int. Hist. Sci.,* 1979, *29*:53–67.

[9] "Ubi per Cometarum motus prius ostensum et liquido comprobatum fuerit, ipsam Coeli machinam non esse durum et impervium corpus varijs orbibus realibus confertum, ut hactenus a plerisque creditum est, sed liquidissimum et simplicissimum, circuitibusque Planetarum liberis, et absque ullarum realium Sphaerarum opera aut circumvectione": Tycho Brahe, *De mundi aetherei recentioribus phaenomenis,* in *Tychonis Brahe Dani opera omnia,* ed. I. L. E. Dreyer, Vol. IV (Copenhagen, 1922), p. 159. On p. 222 of the same work, Tycho says much the same thing, emphasizing that "very many modern philosophers . . . distinguished the heaven into various orbs made of hard and impervi-

From Tycho's assertions that "many" or "very many" contemporaries believed that the heaven was composed of hard, celestial orbs we learn that this was apparently the commonly held opinion of his day. Tycho's influence was so great that, not surprisingly, seventeenth-century astronomers and natural philosophers who mentioned solid spheres followed him in assuming spheres that were hard and rigid.

Almost from the first formulation of Tycho's radical interpretation scholastic natural philosophers found themselves divided. As a direct reflection of that division of opinion scholastic authors introduced a new question into their commentaries on Aristotle's *De caelo,* one that was unknown to the Middle Ages. They asked whether the heavens are solid or fluid. Although some would side with Tycho and assume a fluid heaven without spheres, while others would defend the existence of hard spheres, almost all were agreed that a solid sphere signified a hard sphere.[10] Giovanni Baptista Riccioli, the important seventeenth-century Jesuit astronomer, underscores the powerful association of solidity with hardness when he explains that although the term *soliditas* means three-dimensional, it also "has associated with it hardness [as] opposed to softness, as we say that marble is solid, and metal, as long as it does not liquefy; and even ice before it melts."[11]

THE MEANING OF THE TERM *SOLIDUM* IN THE MIDDLE AGES

The virtual synonymity between solid spheres and hard or rigid spheres in the late sixteenth and the seventeenth centuries became fixed in the subsequent literature of the history of astronomy and cosmology and was, as we saw, eventually applied retrospectively to the Middle Ages. Thus when modern scholars speak of references to solid orbs in the Middle Ages, they almost invariably assume orbs that are hard and rigid.[12] But was this the medieval understanding of a celestial orb? To ascertain whether, during the late Middle Ages, the description of an orb as solid also implied its hardness or rigidity, we must examine the meanings that were assigned to the term *solid* (*solidum*).

At the beginning of his famous thirteenth-century treatise *On the Sphere* (*De spera*), John of Sacrobosco defined a sphere by citing Euclid and Theodosius,

---

ous matter" ("et recentiores etiam Philosophos quamplurimos, qui Coelum ex dura et impervia materia Orbibus varijs distinctum"); see also p. 223. Tycho's *De mundi aetherei* was reprinted in 1603 and 1610 and was thus widely known. For these and other references to Tycho I am grateful to my colleague Victor E. Thoren.

[10] E.g., see Roderigo de Arriaga, *Cursus philosophicus* (Antwerp, 1632), pp. 499, cols. 1–504, col. 1 ("Whether the heavens are incorruptible and solid"); Sigismundus Serbellonus, *Philosophia Ticenensis,* 2 vols. (Milan, 1663), Vol. II, p. 25, cols. 1–28, col. 1 ("Whether the celestial bodies are solid or liquid"); Thomas Carleton Compton, *Philosophia universa* (Antwerp, 1649), pp. 398, cols. 2–399, col. 2 ("Whether the heaven is solid or fluid"); and Melchior Cornaeus, *Curriculum philosophiae peripateticae uti hoc tempore in scholis decurri solet* (Herbipoli, 1657), pp. 494–500 ("Whether the heavens are hard and solid").

[11] "Controversia igitur est de soliditate presse sumpta quae praeter trinam dimensionem habet adiunctam duritiem mollitiei oppositam, quomodo solida dicimus esse marmora et metalla quamdiu non liquescunt et ipsa glacies antequam dissolvatur": Giovanni Baptista Riccioli, *Almagestum novum astronomiam veterem novamque complectens,* Bk. 9, Section 1, Ch. 7 (1651; Bologna, 1653), p. 238, col. 2.

[12] E.g., Jardine assumes that solidity is the opposite of fluidity; therefore it is equivalent to hardness: "Significance of Copernican Orbs" (cit. n. 2), p. 175.

both of whom considered it a solid body.[13] As a consequence it became common, in commentaries on this work, to inquire about the nature of a sphere and occasionally to ask about the sense in which a sphere was a solid. In a commentary ascribed to Michael Scot we learn that the term *solid* (*solidum*) is spoken of in three ways: "In one way, it is the same as hard, just like earth; in another way solid is the same as continuous, and thus the elements and supercelestial bodies are called solid; in a third way it is like a three-dimensional thing, and thus it is the same as a body. Therefore it is not superfluous to say that *a sphere is a solid body*."[14]

Although this significant passage poses serious problems, it is striking that Michael—we shall, for convenience, assume that Michael Scot was the author—invokes the earth as an illustration of what is meant by a *hard solid,* but mentions the celestial bodies (and the elements) as illustrations of what is meant by a *continuous solid.*[15] Does this signify that Michael thought of the celestial bodies as continuous, but not hard? This may depend on whether the term *elements* (*elementa*), in the second sense of *solid,* includes or excludes the earth. Was Michael, in effect, dividing the elements into hard (earth) and soft (water, air, and fire), with only the latter assumed to be continuous? If so, the celestial bodies would also be continuous *and soft,* just like water, air, and fire.

But it is also possible that Michael had something else in mind: to signify that solid bodies possessed all three attributes—hardness, continuity, and tridimensionality. This interpretation seems less plausible because, in a sentence immediately following the one proclaiming his threefold sense of the term *solidum,* Michael provides a clue that he may have intended that there were three quite distinct senses of *solid* rather than that a solid possessed all three attributes. For he says that "It is also known that a surface (*superficies*) is threefold: it is plane, as in a wall; it is concave, as in a tub; [and it is] convex, as in a mountain. And it is by such a surface [i.e., the convex surface] that a round solid is contained because it includes everything within itself, leaving nothing outside."[16]

Is there a parallel between the respective threefold senses of the terms *solid* and *surface?* Since Michael obviously intended three senses of the term *surface,* may we also infer that he intended three distinct senses of the term *solid?* If so, then he may also have intended that celestial bodies be conceived of as continuous, but not hard, or at least not necessarily hard.

This interpretation gains credence from an examination of a discussion by Robertus Anglicus in his *Commentary on the "Sphere" of Sacrobosco,* written around 1271. In a passage that he may have drawn, and perhaps even copied, from Michael Scot, Robertus describes the same three senses of the terms *solidum* and *superficies.* He assumes the existence of nine celestial orbs and also proclaims the immutability of the material from which they are composed. The

[13] See Lynn Thorndike, ed. and trans., *The "Sphere" of Sacrobosco and Its Commentators* (Chicago: Univ. Chicago Press, 1949), pp. 76–77 (Latin), p. 118 (English).

[14] "Item nota quod solidum dicitur tribus modis: uno modo est idem quod durum, sicut terra; alio modo solidum idem quod continuum, et sic elementa et corpora supercelestia solida dicuntur; tertio modo, id est, quod trina dimensio, et sic idem est quod corpus, unde non est ibi negatio, *Spera est corpus solidum*": *ibid.,* p. 256.

[15] By the expression *corpora supercelestia* I assume that Michael means all celestial bodies, i.e., both planets and spheres.

[16] Thorndike, *"Sphere" of Sacrobosco* (cit. n. 13), p. 256.

orbs are distinguished as being larger and smaller orbs by means of "greater and smaller intelligible [i.e., imaginary] circles." That is, an orb, or sphere, is the space that is cut off between two such circles and is the place where each planet is carried.[17] Robertus illustrates the arrangement of the nine celestial orbs by imagining nine wheels of such sizes that they can be arranged concentrically. These nine nested wheels are assumed to be in the air and to move around the same center. Robertus now explains that the quantity or volume of air between any two wheels is like a celestial orb that carries the planet around that lies within it. By choosing air enclosed by wheels as his analogy with celestial orbs, Robertus leaves the impression that he conceived of the celestial orbs as somehow fluid in nature.

This interpretation gains support when Robertus later considers whether the celestial spheres are continuous (*continue*) or contiguous (*contigue*). He decides that they are continuous, which means that the convex surface of one sphere is identical with the concave surface of the next superior orb.[18] But if the successive surfaces of successive orbs are continuous, there is a problem: "Since the orbs are moved by contrary motions, . . . then one and the same [surface] would be moved by contrary motions, which is impossible. Also, it would then follow that, if one orb were moved by some motion, all the other orbs would be moved by the same motion, which, nevertheless, we know to be impossible."[19]

Replying to this difficulty, Robertus indicates that orbs are fluid. For he says:

> We suppose the outer edge of the orb immobile and the middle of the orb to be moved, just as we see that the center of water is moved, yet at its sides the water is still. And it seems much more likely that this can be done in the orbs, which are much simpler than water. Nor, as is now clear, need all orbs be moved when one orb moves, although they are continuous, just as it is not necessary that, when a part of the water is moved, all the water should be moved, although the water is continuous.[20]

Whatever Robertus may have meant by these examples, the fact that he uses water to illustrate the continuity and motion of celestial orbs suggests that he thought of those orbs as continuous and fluid rather than as continuous and hard. Moreover, his description is of great interest, for he seems to say that the surface of an orb can be assumed to be immobile while the part toward its center is in motion. Thus the planet itself is somehow carried by the fluid part of the orb lying within its immobile surfaces. But what is the nature of an orb's immobile surfaces? The water analogy indicates that these are fluid, since they are in no way differentiated from their mobile content.

By adopting an approach in which only the inner part of an orb was assumed to move while its outermost surface lay immobile, Robertus avoided the seemingly

[17] Robertus uses the terms *spera* and *orbis* interchangeably. For the Latin see *ibid.*, p. 145 (trans. p. 200). Robertus substitutes the expression "corpora celestia" for Michael Scot's "corpora supercelestia." E. J. Aiton mentions Robertus's discussion of the term *solid* and correctly explains that "the Earth was solid in the first sense, while the celestial bodies were solid in the second and third senses, but were not necessarily hard"; Aiton, "Celestial Spheres and Circles," *History of Science*, 1981, *19*:75–114, on p. 90.

[18] "Ad primam questionem dicendum quod omnes orbes novem sunt continui": Thorndike, *"Sphere" of Sacrobosco* (cit. n. 13), p. 147. If the surfaces were distinct, the orbs would be contiguous. This discussion can be found *ibid.*, pp. 146–147 (Latin), pp. 202–203 (English).

[19] *Ibid.*, p. 202. I have added the bracketed word.

[20] *Ibid.*, pp. 202–203.

impossible dilemma that would have resulted from an assumption of the continuity of the celestial orbs. For on that assumption two separate, but successive and immediately proximate, orbs would necessarily move in the same direction because the convex surface of the inner sphere would be continuous—that is, identical—with the concave surface of the next superior orb. Despite the assumption of continuity, however, Robertus could now declare that although the convex surface of the planetary orb of Mars and the concave surface of the sphere of Jupiter, for example, were one and the same, those two planetary spheres could, nonetheless, move in different directions because only the middling parts of each sphere or orb—and not the surface itself—would actually move.

Although the evidence is stronger in the case of Robertus Anglicus, both he and Michael Scot appear to have thought of the celestial orbs as soft rather than hard. Whatever the merits of that claim, however, one strong inference may be drawn from the discussions of our two thirteenth-century commentators on the *Sphere* of Sacrobosco: when the term *solid* (*solidum,* or any of its variants) occurs in the context of a discussion on celestial orbs, it may, in the absence of other decisive criteria, refer to either hard or soft spheres. The modern interpretation, which always equates solidity with hardness, is untenable. In the course of the sixteenth century, and certainly by the seventeenth century, the earlier ambivalence seems to have vanished: *fluid* was then opposed to *solid,* with the latter clearly taken as equivalent to *hard,* as when John Punch (Johannes Poncius, O.F.M., 1599–1661), declared that "some say that the heaven is a continuous body and fluid, like air . . . ; and other moderns think that the firmament is a solid [i.e., hard] body."[21] For Michael Scot and Robertus Anglicus, and perhaps for many other natural philosophers during the thirteenth to fifteenth centuries, both of these interpretations formed part of the concept of a solid body.

THE HEAVENS

Because of the profound interrelationship between Greek, and especially Aristotelian, thought on the one hand and Christianity on the other, concepts of the physical nature of the celestial orbs, and the heavens in general, were strongly influenced by theological considerations. The term *caelum,* or *heaven,* was broadly conceived. During the Middle Ages some interpreted it as representative of the entire heavens, while others, like Thomas Aquinas, distinguished three physical heavens: the empyrean, the crystalline (or aqueous), and the sidereal, which embraced all seven planetary spheres as well as the sphere of fixed stars. In the same century Bartholomaeus Anglicus spoke of seven heavens (*septem sunt coeli*), namely, the empyrean, the aqueous (or crystalline), the firmament, the fiery heaven, the olympian heaven, the ethereal heaven, and the airy heaven.[22]

[21] Johannes Poncius (John Punch), commentary on John Duns Scotus, *Sentences,* in *Opera omnia,* 12 vols. (Lyon, 1639; Hildesheim, FRG: Georg Olms, 1968), Vol. 6, Pt. 2: *R.P.F. Ioannis Duns Scoti, Doctoris Subtilis, Ordinis Minorum, Quaestiones in Lib. II. Sententiarum . . . Cum commentariis R.^{mi}P.F. Francisci Lycheti . . . Et supplemento R.P.F. Ioannis Poncii, eiusdem Ordinis, in collegio Romano Hibernorum Theologiae primarii Professoris,* p. 727, col. 1. The bracketed word is my addition. Punch observes that the solidity—i.e., hardness—of the heavens is an opinion "more common" to Peripatetics (p. 727, col. 2).

[22] See Edward Grant, "Cosmology," in *Science in the Middle Ages,* ed. David C. Lindberg (Chicago: Univ. Chicago Press, 1978), p. 275; and *Bartholomaei Anglici De genuinis rerum coelestium*

More important than the medieval concept of heaven, or heavens, is the fact that natural philosophers and astronomers divided the celestial region into as few as eight and as many as eleven major units. From the innermost to the outermost, the eleven units were as follows: (1) the moon, (2) Mercury, (3) Venus, (4) the sun, (5) Mars, (6) Jupiter, (7) Saturn, (8) the sphere of the fixed stars, often identified with the biblical firmament and occasionally also embracing the planetary orbs below, (9) the crystalline, or aqueous, or indeed ninth sphere, (10) the primum mobile, or first movable sphere, and (11) the empyrean sphere. Except for the empyrean sphere, which was commonly interpreted as an immobile sphere enclosing the world, each of these celestial units was assigned one (units 8 to 10) or more (units 1 to 7) orbs to account for its motion or motions. Not all would assign a separate astronomical existence to units 9 to 11.

If we leave aside the empyrean sphere, which was a purely theological, though not a biblical, concept—the abode of God and the elect—the major spheres or zones of the celestial region that had both traditional astronomical (or cosmological) and theological relations were the crystalline sphere and the firmament. We must now determine whether these entities were conceived of as hard or soft.

THE CRYSTALLINE SPHERE

In its theological aspect the idea of the crystalline sphere developed from commentaries on Genesis 1:6–7, which spoke of a division of waters into those above the firmament and those below. From the time of the church fathers the meaning and significance of the waters above the firmament were much debated. Because the Bible itself spoke of waters above the firmament, Christian authors, following Augustine, were generally agreed on the necessity for a literal interpretation and were therefore committed to the existence of waters of some kind above the firmament.[23] All else was seemingly arguable. Indeed, the debate hinged on the interpretation of the terms *waters* (*aquae*) and *firmament* (*firmamentum*), the meaning of the latter largely determining the meaning of the former.[24]

From the time of the church fathers to the end of the Middle Ages one finds a variety of interpretations of the waters above the firmament. The interpreters divide essentially into two camps: those who thought of them as fluid and those

---

terrestrium et inferarum proprietatibus libri XVIII . . . (Frankfurt, 1601; rpt. Frankfurt am Main: Minerva, 1964), Bk. 8, Ch. 2 ("De coelorum distinctione"). After assuming and describing seven heavens (*ibid.*, p. 372), Bartholomew the Englishman (Bartholomaeus Anglicus) mentions that the philosophers assume only one heaven ("Nam philosophi non ponunt nisi solum coelum unum"): *ibid.*, p. 373.

[23] An exception is William of Conches (fl. 1120–1149), who in his *Philosophia mundi* denied that waters could exist above the firmament and insisted that the scriptural passage in which this is asserted must be interpreted allegorically: see Helen R. Lemay, "Science and Theology at Chartres: The Case of the Supracelestial Waters," *British Journal for the History of Science*, 1977, *10*:226–236, on p. 231.

[24] To see how the meaning of *firmament* determined the meaning of the *waters* above that firmament, and to obtain an excellent sampling of the different interpretations given to both of these terms, see Thomas Aquinas, *Summa theologiae*, ed. and trans. William A. Wallace, Vol. X: *Cosmogony* (New York: McGraw-Hill: London: Eyre & Spottiswoode, 1967, Pt. 1, question (quest.) 68, article (art.) 1 ("Was the firmament made on the second day?"), pp. 71–77, art. 2 ("Are there any waters above the firmament?"), pp. 77–83, and art. 3 ("Does the firmament separate some waters from others?"), pp. 83–87. Aquinas explains (p. 79) that "we maintain that these waters are material. Just what they are must be explained in different ways depending on various theories about the firmament." He then offers a number of interpretations.

who considered them to be solid and hard. Among the former, who probably made up the larger group during the Middle Ages, we may mention Ambrose, John Damascene, Robert Grosseteste, Richard of Middleton, Bonaventure, and Vincent of Beauvais.[25] In this group some, like Richard of Middleton and Bonaventure, provide little or no description of the alternative that the supraheavenly waters might be hard like a crystal. They were agreed, however, that although these waters were unlike ordinary elemental water they shared with it a few important properties, namely, transparency (*perspicuitas*), coldness (*frigiditas*), and, for Richard, wetness (*humidum*). Vincent of Beauvais, who assumed that the waters were immutable, believed they were luminous (*luminosum*), transparent (*perspicuum*), and subtle (*subtile*). These authors characterized the suprafirmamental waters as crystalline not because of their hardness but because of their immutability, transparency, and luminosity.[26]

Among those who thought of the supraheavenly waters as solid and hard we may include Jerome, Basil, and Bede, who likens the waters to "the firmity of a crystalline stone" ("cristallini lapidis quanta firmitas"). Peter Lombard was probably aware of Bede's opinion and seems to approve of it in his famous and widely used *Sentences,* composed in the twelfth century, when he declares that the waters above the heaven are not in a vapory form but suspended by icy solidity (*glaciali soliditate*) to prevent their fall.[27] In all these instances the justification

[25] Ambrose, *Hexameron,* third homily (Bk. 2: The Second Day), Ch. 3, pp. 51–59, in *Saint Ambrose: Hexameron, Paradise, and Cain and Abel,* trans. John J. Savage (The Fathers of the Church: A New Translation, 42) (New York, 1961). John Damascene, *The Orthodox Faith,* Bk. 2, Ch. 9, p. 224 in *Saint John of Damascus: Writings,* trans. Frederic H. Chase, Jr. (The Fathers of the Church: A New Translation, 37) (New York, 1958). Grosseteste, *Hexaemeron* (cit. n. 5), Pt. III, Ch. 3, ¶4, p. 104. Richard of Middleton, *Clarissimi theologi Magistri Ricardi de Media Villa . . . Super quatuor libros Sententiarum Petri Lombardi quaestiones subtilissimae* (*Commentary on the Sentences*), 4 vols. (Brescia, 1591); rpt. as *Richardus de Mediavilla* (Frankfurt: Minerva, 1963), Vol. II, pp. 167–168 (Bk. 2, dist. 14, quest. 1: "Utrum coelum crystallinum dictum sit naturae aquae"). Bonaventure, *Commentaria in quatuor libros Sententiarum Magistri Petri Lombardi: In secundum librum Sententiarum,* in *Opera omnia,* edita studio et cura PP. Collegii a S. Bonaventurae (Quaracchi, 1882–1901), Vol. II, pp. 335–338 (Bk. 2, distinction (dist.) 14, art. 1, quest. 1: "Utrum caelum crystallinum sit natura aquae"). Vincent of Beauvais, *Speculum naturale,* Bk. 3, in *Vincentii Burgundi, ex Ordine Praedicatorum venerabilis episcopi Bellovacensis, Speculum quadruplex, naturale, doctrinale, morale, historiale,* 2 vols. (Douai, 1624; rpt. Graz, Austria: Akademische Druck- u. Verlaganstalt, 1964), Vol. I, cols. 221–229, esp. col. 224.

[26] Richard is wholly silent, whereas Bonaventure mentions only that Bede believed that the waters above the heaven rested, and were sustained, by virtue of their solidity: "Et ibidem aquae illae quiescunt et sustentantur vel sua soliditate, sicut videtur Beda dicere, vel sua subtilitate, vel etiam Dei virtute, quae sic ordinavit": *Opera omnia,* Vol. II, p. 337, col. 2. For Vincent see *Speculum naturale* (cit. n. 25), col. 224. Vincent also describes the alternative opinion that the waters are congealed like a crystal and rejects it because he can find no cause that would congeal the waters; *ibid.,* col. 221. Grosseteste described the opinion that the waters above the firmament were like a hard, crystalline stone (*cristallus lapis*), but rejected it: *Hexaemeron* (cit. n. 5), Pt. III, Ch. III, ¶¶ 3, 4, p. 104.

[27] On the strength of Thomas Aquinas's declaration that Basil held that "the waters that are above the heavens are not necessarily fluid, but rather are crystallized around the heavens in a state similar to ice," I have placed Basil among those who believed that the waters above the firmament are hard; see Thomas Aquinas, *Summa theologiae* (cit. n. 24), Pt. I, quest. 68, art. 2, p. 81. Inspection of Basil's third homily, however, where he discusses the suprafirmamental waters, suggests a belief in waters that are fluid rather than hard. Thus Basil explains that the term *firmament* "has been assigned for a certain firm nature which is capable of supporting the fluid and unstable water": *Saint Basil: Exegetic Homilies,* trans. Sister Agnes Clare Way, C.D.P. (The Fathers of the Church: A New Translation, 46) (Washington, D.C.: Catholic Univ. of America Press, 1963), p. 43. In the first homily Basil asserts that "concerning the substance of the heavens we are satisfied with the sayings of Isaia [Isa. 51:6, Septuagint version, per Sister Way], who in simple words gave us a sufficient

was undoubtedly the conviction that only frozen or congealed waters could remain suspended above the firmament. In liquid form the waters would surely flow downward.

That there was no ready consensus on such matters is established by William of Conches, who not only denied the existence of waters above the firmament but was particularly incensed at those who thought they were frozen. Congealed waters, he claimed, would be so heavy that "they must either descend to the earth by their natural heaviness or they must be moved. They cannot be moved, for motion cannot exist without heat. Thus, if they are moved, their movements generate heat in them, and if this happened they would be dissolved by that heat."[28]

Although terms like *crystalline* and *icy solidity* seem to imply hardness, they could be interpreted otherwise. In commenting on Peter Lombard's passage on the icy solidity of the waters above the heaven, Bonaventure insists that the sense of *solidity* that implies that those waters are heavy and held in position above the firmament by violence is contrary to the order of the universe. We should rather understand that "those waters compare with icy solidity (*glaciali soliditati*) not because of heaviness, but because of continuity and stability because they do not ebb or flow, nor do they descend downward."[29] Bartholomew the Englishman is even more explicit when he explains that the water above the heaven is called "crystalline not because it is hard (*durum*) like a crystal, but because it is uniformly luminous and transparent. Moreover it is called watery in so far as water is moved by virtue of its subtlety and mobility."[30] Thus it is more than likely that when medieval authors spoke of the crystalline sphere, they had in mind those properties of a crystal such as luminosity, transparency, and even a quasi-immutability, rather than hardness.

THE INFLUENCE OF THE NEW LEARNING ON THE CONCEPT OF THE WATERS ABOVE THE HEAVENS

Until the thirteenth century most authors were concerned with the need to explain the meaning of the biblical waters. But with the introduction of the cosmo-

---

knowledge of its nature when he said: 'He established the heaven as if smoke,' that is, He gave the substance for the formation of the heavens a delicate nature and not a solid and dense one": p. 14.

For the relevant passages from Jerome and Bede see *Campanus of Novara and Medieval Planetary Theory: "Theorica planetarum,"* ed. and trans. Francis S. Benjamin, Jr., and G. J. Toomer (Madison: Univ. Wisconsin Press, 1971), pp. 393–394, n. 54. For Peter Lombard, see his *Sentences,* Bk. 2, dist. 14, Ch. 4, ¶ 1, in *Sententiae in IV libris distinctae* (3d ed., Grottaferrata [Rome]: Collegio San Bonaventura, 1971), Vol. I, Pt. 2, Bks. 1 and 2, p. 396.

[28] William of Conches, *Glosae super Macrobii In somnium Scipionis,* trans. in Lemay, "Science and Theology at Chartres" (cit. n. 23), pp. 229–230 (for the MSS used by Lemay see *ibid.*, p. 235, n. 19). William's irritation with those who upheld the interpretation of frozen waters is obvious from these further remarks: "This opinion does not suit me, nor is there any reason why it would be true, nor indeed is it even possible"; "there are many other reasons by which it can be proved that what they say is worth nothing" (pp. 230, 231).

[29] Bonaventure, *Commentary on the Sentences,* Bk. 2, dist. 14, pt. 1, art. 3, quest. 2, *dubium* 1, in *Opera omnia* (cit. n. 25), Vol. II, p. 350, col. 1.

[30] "Et ideo in summo dicitur crystallinum non quia durum sicut crystallus, sed quia uniformiter est luminosum et perspicuum. Aqueum autem dicitur quemadmodum aqua ex sua subtilitate et mobilitate movetur": Bartholomew, *De genuinis rerum . . . proprietatibus* (cit. n. 22), Bk. 8, Ch. 3 ("De coelo aquaeo sive crystallino"), p. 379. Vincent of Beauvais held a similar view about the crystalline sphere (see. n. 26).

logical and astronomical works of Aristotle, Ptolemy, and many others in the late twelfth and the thirteenth centuries, we find clear attempts to accommodate the supraheavenly waters to the new cosmology of nested orbs as well. Thus for Thomas Aquinas the waters above the sphere of the fixed stars are not like any waters we know but are most like Aristotle's ether, or fifth essence, since both exhibit the similar properties of lucidity and transparency.[31]

Vincent of Beauvais, showing the influence of Aristotelian and Ptolemaic cosmology, argues that because no vacuum can exist, the space between the empyrean heaven and the firmament, which Vincent identifies with the fixed stars, must be filled with body, a body that he equates with the supracelestial waters. From the implied assumption that both the empyrean heaven and the fixed stars are spherical in shape, Vincent infers that the waters that lie between them must also be spherically shaped. Thus did the "crystalline heaven" become the "crystalline orb."[32]

And, finally, Vincent of Beauvais, Thomas Aquinas, and many others identified the crystalline sphere with the ninth sphere that had been introduced into Ptolemaic astronomy as a starless orb.[33] It was assumed to possess a single proper motion, usually one of the following three: the daily motion, motion of precession, and motion of trepidation. For some it also served to conserve existing things.[34]

Because of its identification with the ninth sphere, the crystalline orb became an integral feature of the medieval cosmos. But most natural philosophers seem to have thought of it as in a fluid rather than a solid and hard state. Since the biblical account specifies waters, those who discussed the waters above the

[31] See Thomas's commentary on Peter Lombard's *Sentences,* Bk. 2, dist. 14, quest. 1, art. 1 ("Utrum aquae sint super caelos") in *S. Thomae Aquinatis Scriptum super libros Sententiarum Magistri Petri Lombardi Episcopi Parisiensis,* ed. R. P. Mandonnet, 4 vols. (Paris: P. Lethielleux, 1929[?]–1947), Vol. II, p. 347. Campanus of Novara also insisted that all celestial orbs, including the crystalline, were composed of a fifth essence (*quinta essentia*); see Benjamin and Toomer, *Campanus of Novara and Medieval Planetary Theory* (cit. n. 27), p. 183.

[32] Vincent, *Speculum naturale* (cit. n. 25), col. 226. That the crystalline heaven would not have been perceived as obviously spherical may be gleaned from a problem raised by Basil and repeated by Ambrose. In the latter's *Hexameron,* day two, he replies to those who inquire how waters above the heavens can remain in position even as the world rotates on its axis. Would the waters not be thrown off? Ambrose replies that just as some buildings have spherical ceilings but flat roofs where water may collect, so also may the heaven possess a concave surface as viewed from below, but a flat convex surface where waters may collect and remain: see Ambrose, *Hexameron* (cit. n. 25), pp. 52–53. See also Thomas Aquinas's report of this opinion in *Summa theologiae* (cit. n. 24), Pt. 1, quest. 68, art. 2, p. 81 and note g.

[33] For Vincent see *Speculum naturale* (cit. n. 25); cols. 227, 228; for Aquinas see *Scriptum super libros Sententiarum* (cit. n. 31), Vol. II, p. 348 (Bk. 2, dist. 14, quest. 1, art. 1). In *Theorica planetarum* Campanus of Novara hesitates between identifying the crystalline heaven with the ninth or tenth sphere and concludes that "it is apparent that the heavenly spheres are 11 in number if the ninth sphere and the crystal heaven are different, or only 10 if they are identical": *Campanus of Novara and Medieval Planetary Theory* (cit. n. 27), p. 183.

[34] For mention of a few scholastic authors who assumed one or the other of these motions see Grant, "Cosmology" (cit. n. 22), p. 278 and the relevant notes on p. 297. For the sphere as conserving existing things see Vincent, *Speculum naturale* (cit. n. 25), col. 228. Benjamin and Toomer (*Campanus of Novara and Medieval Planetary Theory* [cit. n. 27], p. 393, n. 53) explain that "the necessity for a 'ninth sphere' was astronomical: precession was regarded as the motion of the sphere of the fixed stars ('the eighth sphere') eastward with respect to a stationary ninth sphere." I have not yet found any scholastic natural philosopher who described the ninth sphere as stationary. Although most late medieval natural philosophers assumed the existence of a ninth crystalline orb, Nicole Oresme did not; see Oresme, *Le livre du ciel et du monde,* ed. Albert D. Menut and Alexander Denomy, trans. Albert D. Menut (Madison: Univ. Wisconsin Press, 1968), p. 491.

heaven but did not mention the possibility that they might exist in the form of solid ice were probably supporters of the fluid theory. Even where icy hardness is mentioned, it may not reflect an author's own opinion but may have been included merely for the record.[35] Or, as with Bonaventure, icy solidity could be explained in terms that effectively denied its hardness. It seems a reasonable conclusion, then, that the crystalline orb of the Middle Ages was more often conceived of as fluid and soft than as hard and rigid.

THE FIRMAMENT

Genesis 1:6–7 says that God created the firmament (*firmamentum*) on the second day to divide the waters above from the waters below. What could such a firmament signify? Not surprisingly, a variety of interpretations emerged. Augustine set the tone for such discussions. In his commentary on Genesis he observed that although much subtlety and learning had been expended on explicating the nature of the firmament, he himself had "no further time to go into these questions and discuss them, nor should they have time whom I wish to see instructed for their own salvation and for what is necessary and useful in the Church." But he went on to say that those who do consider the meaning of the firmament ought to "bear in mind that the term 'firmament' does not compel us to imagine a stationary heaven: we may understand this name as given to indicate not that it is motionless but that it is solid and that it constitutes an impassable boundary between the water above and the waters below."[36] Without choosing between them, Augustine thus explained the "firmity" of the firmament in two ways: because it is motionless, or because it is solid and prevents the passage of waters from above or below. Augustine gives no indication whether those who assumed a "solid" and impenetrable firmament meant also to signify that it was hard, although that seems a reasonable inference.

In contrast with Augustine, Basil unequivocally denied the hardness of the firmament when he wrote in his *Hexaemeron:* "Not a firm and solid nature, which has weight and resistance, it is not this that the word 'firmament' means. In that case the earth would more legitimately be considered deserving of such a name. But, because the nature of superincumbent substances is light and rare and imperceptible, He called this firmament, in comparison with those very light substances which are incapable of perception by the senses."[37]

During the Middle Ages most authors were vague and noncommittal. Few were ever as explicit about the hardness of the firmament as Hartmann Schedel,

[35] See, e.g., Thomas Aquinas, *Summa theologiae* (cit. n. 24), Pt. 1, quest. 68, art. 2, p. 81; he cites it as an opinion of Basil.

[36] *St. Augustine: The Literal Meaning of Genesis,* trans. John Hammond Taylor, S.J., 2 vols. (Ancient Christian Writers: The Works of the Fathers in Translation, 41) (New York/Ramsey, N.J.: Newman Press, 1982), Vol. I, pp. 60–61.

[37] Basil, *In Hexaemeron* (homily 3) (cit. n. 27), p. 47. As it stands, Basil characterizes as imperceptible both the firmament and "those very light substances" to which the firmament is compared. It would seem that the firmament should have been described as perceptible in contrast to "those very light substances which are incapable of perception by the senses." Whatever the case, it does not affect the import of the passage, which proclaims that the firmament is not a hard substance. Earlier Basil had denied that the firmament could be compared to water that is "like either frozen water or some such material which takes its origin from the percolation of moisture, such as is the crystalline rock which men say is remade by the excessive coagulation of the water, or as is the element of mica which is formed in mines" (p. 43).

who in his *Liber chronicarum* of 1493 declared that God "made [the firmament] solid, out of water congealed like a crystal."[38]

Perhaps because Augustine thought a Christian's interpretation of the meaning of *firmament* was relatively unimportant, some who subsequently discussed its meaning did not think it essential to arrive at a definitive judgment or to consider the problem in any great detail.[39] But certain questions about the firmament were frequently considered and play a role in our deliberations. The most obvious were, Where was it located? What was it composed of? and With what in the heavens was it to be identified?

The location of the firmament created on the second day could be understood in two ways, as sublunar or as celestial. Basil seems to have conceived of it as in the airy part of the world "due to the density and continuity of the air which clearly comes within our vision and which has a claim to the name of heaven from the word 'seen,' namely where the Scripture says: 'The birds of the heavens,' and again, 'the flying creatures below the firmament of the heavens.' "[40] Although few would follow Basil, we have already seen that his identification of the firmament with the air's "density and continuity" could not imply a hard firmament.

From the church fathers to the scholastic natural philosophers and theologians of the Middle Ages and the early modern period, almost all authors located the firmament in the celestial region. Most described, and often chose one of, two opinions about its composition, namely, that it was either constituted of one or more of the four elements or from a fifth element wholly distinct from the other four. In his *Hexaëmeron* Robert Grosseteste presents a typical description of the two opinions but declines to choose between them:

> The philosophers write mutually contrary statements about these things. Some affirm it [the firmament] to be the four elements and that the whole thing that is called the firmament is fiery in nature, which is what Plato seems to feel in the *Timaeus*. Augustine also seems to accept this opinion in several places. John Damascene also implies in his book of *Sentences* that the existence of an immaterial body, that which is called a fifth body among the wise men of the Greeks, is impossible. Others, however, like Aristotle and his followers, say that a fifth body exists besides the four elementary bodies. To include here the controversies and arguments of these people would be too prolix and tedious for our audience and does not seem very necessary for our purpose.[41]

[38] Hartmann Schedel, *Liber chronicarum,* Pt. I (Nuremberg, 1493), in *The Story of the Creation of the World,* ed. Bern Dibner, trans. Edward Rosen (New York: Burndy Library, 1948). The book is unpaginated; see "On the Work of the Second Day," ll. 4–5. On the opposite page the Latin text reads: "Ex aquis congelatis in modum cristalli solidavit et in eo fixa sidera." Schedel seems to have conflated the crystalline sphere with the firmament. As we saw in note 37, Basil mentioned crystal and the firmament in the same passage but sought to dissociate the two.

[39] Judging from his discussion of the firmament in the *Summa theologiae* (cit. n. 24), pp. 71–77, Thomas Aquinas was certainly one of these. Robert Grosseteste thought it would be tedious and prolix to present a detailed analysis of the nature of the firmament; see Grosseteste, *Hexaemeron* (cit. n. 5), p. 106, ¶1.

[40] Basil, *In Hexaemeron* (homily 3) (cit. n. 27), pp. 49–50. I am ignorant of the Greek word rendered by "seen," as well as of its precise relationship to the scriptural passages. In his *Summa theologiae,* Pt. 1, quest. 68, art. 1 ("Was the firmament made in the second day?"), Thomas Aquinas declares that the region where the clouds are formed is "called a *firmament* because of the density of the air in that part, since the dense and solid is said to be *a firm body, to distinguish it from a mathematical body,* as Basil observes" (p. 75; italics in Wallace's translation).

[41] Grosseteste, *Hexaemeron* (cit. n. 5), p. 106. On pp. 112–117 (Chs. 14, 15), Grosseteste gives numerous allegorical interpretations of the firmament.

There is no doubt that the church fathers, including Augustine, favored the opinion that the firmament was composed of the fiery element. Since fire was considered a fluid-like substance, the firmament was obviously not conceived of as hard. Moreover, since the element fire was corruptible, some of the saints inferred that the heavens were corruptible.[42] With the introduction of Aristotle's natural philosophy in the universities of the thirteenth century, however, almost all natural philosophers adopted his assumption that the heavens were composed of a fifth, incorruptible element. As Bonaventure explained it, the philosophers were right because the fifth element had no contrary, unlike fire, and therefore did not suffer change. Indeed, the contrary of fire was earth, and because the latter moved naturally with a rectilinear motion, so did fire, but in an opposite direction.[43] Because rectilinear motion was inapplicable to the celestial region, where all motions were circular, Bonaventure concluded that the firmament was composed of a fifth element, which moved naturally with a circular motion.

But if most scholastic authors followed Aristotle and assumed a celestial ether—that is, a firmament composed of a fifth element—with what did they identify the firmament? As with the crystalline sphere, there was much uncertainty, but two basic interpretations held sway. One identified the firmament with the entire region between the lunar sphere and the sphere of the fixed stars, a position defended by Robert Grosseteste and Bonaventure, while the other assumed that the eighth sphere of the fixed stars alone was the firmament.[44] Vincent of Beauvais cites both interpretations without choosing between them, but says that the first is assumed in Genesis 1:14–16. Since God had placed the sun and moon in the firmament and these luminaries occupied lower places among the circles of the planets, it followed that the firmament embraced all the inferior planets as well as the sphere of the fixed stars.[45]

All these factors played a role in shaping an understanding of the nature of the firmament. The very name *firmamentum,* with its implications of strength, power, and stability, seemed to invite an explanation and thus to provide an occasion for the expression of opinions about its possible hardness or softness. Few, however, chose to take the opportunity to explain why, in Genesis, the term *firmamentum* was used for the heaven created on the second day.[46] Two

[42] Bonaventure in his *Commentary on the Sentences,* Bk. 2, dist. 14, art. 2, quest. 2, "Whether the firmament is the same as the element fire," affirms that it is: *Opera omnia* (cit. n. 25), Vol. II, p. 340, col. 1. John Damascene, who believed that the celestial region was made of fire, is one who assumed the corruptibility of celestial bodies: see his *On the Orthodox Faith* in the Latin translation of Burgundio of Pisa: *Saint John Damascene, De fide orthodoxa, Versions of Burgundio and Cerbanus,* ed. Eligius M. Buytaert, O.F.M., S.T.D. (Franciscan Institute Publications, Text Series, 8) (St. Bonaventure, N.Y.: Franciscan Institute, 1955), p. 95, ¶17.

[43] Bonaventure, *Opera omnia* (cit. n. 25), Vol. II, p. 340, col. 1. For similar arguments see Richard of Middleton, *Commentary on the Sentences* (cit. n. 25), Vol. II, pp. 169–170: Bk. 2, dist. 14, art. 1, quest. 3 ("Whether the firmament is of a fiery nature").

[44] Grosseteste, *Hexaemeron* (cit. n. 5), p. 106; and Bonaventure, *Commentary on the Sentences,* Bk. 2, dist. 14, pt. 2, art. 1, quest. 1 ("Whether all the celestial luminaries are located in one continuous body"), in *Opera omnia* (cit. n. 25), Vol. II, p. 352.

[45] Vincent concludes his discussion with these words: "Porro firmamentum dicitur secundum duplicem modum. Uno modo dicitur octava sphaera tantum. Alio modo dicitur tota natura corporis quinti, et sic accipitur in Genesi ubi dicitur quod fecit Deus luminaria in firmamento coeli": *Speculum naturale* (cit. n. 25), col. 230.

[46] Although Bonaventure, e.g., discussed the firmament in a few questions in his *Commentary on the Sentences,* Bk. 2 (see *Opera omnia* [cit. n. 25], Vol. II, pp. 338–341, 351–352), he nowhere considers why the term *firmamentum* was used to describe the one or more heavens embraced by it.

who did were Vincent of Beauvais and Campanus of Novara. Vincent declared that the term *firmament* was used because that heaven is ungenerated and incorruptible rather than because it is immobile with respect to place. It is indeed indissoluble, because it lies beyond the action and passion of contraries. Campanus explains that the firmament is so called because "its motion always seems to be firm and uniform and because the fixed stars seem to be firmed in it."[47] Nowhere, however, does either associate solidity or hardness with the term *firmamentum*.

Perhaps at this point the reader might wonder whether the stellar sphere had to be rigid in order to maintain the relative positions of the numerous fixed stars that are allegedly embedded in it and carried around by it.[48] In Aristotelian cosmology, however, a fluid ethereal sphere could perform the same function. A fluid ether carrying the stars would rotate with a uniform circular motion and carry each star immersed in it with a uniform motion. Moreover, because the stars are neither light nor heavy, they would have no inclination to move up or down. Consequently, the stars would retain their positions relative to each other.

THE PLANETARY ORBS

Although the crystalline sphere was probably more commonly regarded as fluid or soft than as hard and icy, the biblical *firmamentum* was rarely identified as either hard or soft during the Middle Ages, whether one limits the firmament to the eighth sphere of the fixed stars or includes all the inferior planetary spheres. Few described the firmament in ways that reveal their opinions. Indeed this is true whether the planetary orbs are discussed as part of the medieval concept of the firmament, taken together as a collection of spheres, or considered one by one.[49]

This silence on its character is somewhat surprising, because a line from the text of Job (37:18) provided an excellent opportunity for those who sought support for a belief in a solid *and* hard heaven, where *heaven* is understood as the

---

The same may be said about Richard of Middleton in *Commentary on the Sentences* (cit. n. 25), Bk. 2.

[47] Vincent: "Nos autem dicimus ad primum quod firmamentum dicitur a firmitate naturae quia non generatur, nec corrumpitur et non ab immobilitate secundum locum. . . . Et propter hoc dicitur firmamentum quia indissolubilis est concensus ille cum extractus sit extra actionem et passionem contrariorum"; *Speculum naturale* (cit. n. 25), col. 230. Campanus identifies the firmament with the eighth sphere and says: "Et dicitur firmamentum quoniam ipsius motus semper videtur esse firmus et uniformis et quia stellae fixe videntur firmari"; Campanus, *Tractatus de sphaera,* printed with a large number of astronomical treatises in a work titled *Spherae tractatus* (Venice, 1531), fol. 196r.

[48] This seems an appropriate place to express my gratitude to an anonymous reader who offered some valuable suggestions on this point and saved me from a few embarrassing errors.

[49] One possible exception was Peter of Abano (1257–ca. 1315), who, according to Pierre Duhem, rejected Aristotle's assumption that rigid orbs carried each planet around and assumed instead that each planet is self-moved within a fluid substance that constitutes the heaven, or orb, of that planet; see Pierre Duhem, *Le système du monde; Histoire des doctrines cosmologiques de Platon à Copernic,* 10 vols. (Paris: Hermann, 1913–1959), Vol. IV, p. 253. Nancy Siraisi repeats this claim in *Arts and Sciences at Padua: The "Studium" of Padua Before 1350* (Toronto: Pontifical Institute of Mediaeval Studies, 1973), p. 86. Peter's unpublished *Lucidator astronomiae* is the treatise in which he is alleged to have declared in favor of fluid spheres. Duhem's summary and discussion are based on a single Paris manuscript (Bibliothèque Nationale, fonds latin, MS 2598). Unfortunately he does not cite the Latin text. Although Duhem's assertion is prima facie evidence that Peter probably believed in fluid spheres, we cannot determine whether he considered the surfaces, or shells, of those spheres to be hard or soft.

sky with all its orbs. In mentioning the many wondrous things that God can do, Elihu asks Job whether, like God, he could fabricate the heavens as if they were made of molten metal.[50] The reference to metal would at least have implied a hard surface for those who wished to invoke this text as a description of the firmament. In fact it was rarely used, and even when it was an author might himself be noncommittal. Thus Thomas Aquinas cited the Job passage in his commentary on Boethius's *De trinitate,* but the context of his discussion provides no clue as to whether he believed in hard spheres.[51] Although the text of Job 37:18 was largely ignored during the Middle Ages, it was often invoked in the seventeenth century by scholastic authors who used it to defend the notion of solid and hard spheres against Tycho Brahe and his supporters, who rejected the existence of spheres of any kind and argued for a fluid celestial medium.[52] By that time, however, the term *solid sphere,* as we have seen, almost invariably implied hardness as opposed to fluidity and softness.

HARD SPHERES IN THE MIDDLE AGES

Although few scholastics explicitly considered whether the celestial spheres were hard or fluid, a few, like Themon Judaeus and Henry of Hesse, did. In his *Questions on the Meteorology* Themon debated whether the sky or heaven is of a fiery nature and rejected the possibility. He argued that if the heaven had an elemental nature, it would be like earth and water, rather than fire. This is because "every heaven [i.e., orb] is a hard (*durum*) body without capacity for flowing." But "fire in matter proper to it is not hard and lacking in the capacity to flow, as is obvious by experience, as we see in flames. [Experience] also shows [that it is quite otherwise with] water, ice, and earth. For earth [and water] can be made hard and even transparent (*perspicua*), as is obvious from glass and ice."[53] Thus fire cannot be hardened and so cannot be the material from which the celestial orbs are composed.

Around 1390, in a commentary on Genesis (*Lecturae super Genesim*), Henry

[50] In the Vulgate the text (Job 37:18) reads: "Tu forsitan cum eo fabricatus es caelos, qui solidissimi quasi aes fusi sunt." The Douay-Challoner translation of the Vulgate (ed. Rev. John P. O'Connell, Chicago: Catholic Press, 1950) renders "qui solidissimi quasi aes fusi sunt" as "which are most strong, as if they were of molten brass." A recent translation from the Hebrew text provides a more graphic version to describe the rigid heavens: "Can you beat out the vault of the/skies as he does,/ hard as a mirror of cast metal?" *The New English Bible with the Apocrypha, Oxford Study Edition* (New York: Oxford Univ. Press, 1976).

[51] Nor indeed does Thomas give an unequivocal indication in his other writings. Thomas Litt includes the section citing Job in his *Les corps célestes dans l'univers de Saint Thomas d'Aquin* (Louvain: Publications Universitaires; Paris: Béatrice-Nauwelaerts, 1963), p. 348.

[52] See, *e.g.,* Punch, commentary on Duns Scotus, *Quaestiones in lib. II Sententiarum* (cit. n. 21), p. 727, col. 2–p. 728, col. 1; and Raphael Aversa, *Philosophia metaphysicam physicamque complectens quaestionibus contexta,* 2 vols. (Rome, 1625, 1627), Vol. II, p. 67 (quest. 32, sect. 6).

[53] Themon, *Questions on the Meteorology,* in *Questiones et decisiones physicales insignium virorum: Alberti de Saxonia in octo libros Physicorum; . . . Thimonis in quatuor libros Meteororum; . . . Recognitae rursus et emendatae summa accuratione et iudicio Magistri Georgii Lokert Scotia quo sunt Tractatus proportionum additi* (Paris, 1518), Bk. 1, quest. 3 ("Utrum coelum sit nature ignis"), fol. 157v, col. 2. The full text of Themon's second conclusion reads: "Secunda conclusio: Si celum esset nature elementalis potius esset nature aque vel terre quam ignis. Probatur conclusio: quia celum est corpus durum influxibile, alias fieret permixtio astrorum et stellarum nimium irregularis propter divisionem eius. Sed ignis in propria sibi materia non est durus et influxibilis, ut patet per experientiam, videmus enim de flammis. De aqua autem et glacie et terra patet. Terra enim potest indurari etiam perspicua fieri, ut patet de vitro et glaciebus. Ergo celum potius esset nature aque vel terre, quod est propositum."

of Hesse presented an unusual interpretation of the celestial orbs. According to Nicholas Steneck, Henry argued that the firmament created on the second day was composed of "a series of concentric shells or spheres that stretch from the region of the moon to the region of the fixed stars. They are clear, firm, impenetrable, and have thickness. . . . In fact, the image of glass globes spinning on fixed axes around the central earth, so commonly used to describe the medieval conception of the celestial orbs, seems to fit quite nicely the discussion in the *Lecturae*."[54] The various orbs had congealed like water or lead. Henry rejected Aristotle's postulated celestial ether, or fifth element, and insisted that the heavenly region was composed of matter similar to that of the earth. He further argued that on the fourth day of creation the planets were formed from mixtures of elemental matter that rose up through the various hard celestial orbs. Because Henry believed that the movement of such relatively coarse matter through the hard celestial orbs was physically impossible, and since he was not prepared to abandon his interpretation, he chose to explain it by miraculous intervention.

Few were as explicit as Themon Judaeus and Henry of Hesse. Only occasionally—and then indirectly—can we determine the apparent opinions of others about the hardness of the spheres. In this connection Nicole Oresme's discussion in his French translation of and commentary on Aristotle's *De caelo* is of interest. Here Oresme describes the surfaces of all celestial spheres as perfectly polished and smooth. Because no vacua can exist between any two celestial surfaces,

> it follows necessarily that the concave surface of the sovereign heaven and the convex surface of the second or next heaven below must be absolutely spherical, with no roughness or humps, and that these heavens must move one inside the other without any friction. Rather, the passage of one surface above the other must be as smooth, as gentle, and as effortless as possible. The same holds for the second and third heavens and thus through all of them in descending order down to the concave surface of the lunar sphere, which is concentric with the earth and with the heavenly body which contains or comprises or is composed of all the partial heavens; otherwise, all this body taken together would be thicker in one part than in another, which is neither probable nor reasonable. Therefore the concave surface of the lunar heaven must be perfectly spherical.[55]

The perfect circularity of the concave lunar surface causes the convex surface of the sphere of elemental fire to assume a perfect circular shape. Ordinarily, none of the imperfect four elements could do this. But the convex surface of fire, which Oresme describes as "perfectly polished and spherical," is an exception. But "this is not due to the element of fire itself, but to the concave surface of the lunar sphere which contains the fire and which is perfectly spherical, being everywhere in contact with the fire without intermediate plenum or vacuum."[56] On the basis of Oresme's discussion, it seems reasonable to assume that he judged the celestial surfaces to be hard. Otherwise the lunar concavity could not have shaped the outermost surface of grosser fire into a perfectly circular surface.[57]

[54] Nicholas H. Steneck, *Science and Creation in the Middle Ages* (cit. n. 1), pp. 61–62.

[55] Oresme, *Le livre du ciel et du monde* (cit. n. 34), pp. 317, 387 (quotation).

[56] *Ibid.*, p. 399.

[57] Although he speaks of celestial bodies, it is likely that the author of an anonymous fourteenth-century treatise on natural philosophy meant celestial orbs when he declared that "celestial bodies do

CONTINUITY AND CONTIGUITY

Those in the Middle Ages who chose to declare outright, or to provide some indication of, their belief in hard or fluid celestial orbs were few indeed. Why? Surely not because they consciously wished to avoid the issue. Rather, it appears that the issue never really surfaced. An excellent illustration of the tendency to describe the relationships between orbs without indicating whether the orbs' surfaces were hard or soft appears in al-Biṭrūjī's *De motibus celorum,* translated from Arabic to Latin in 1217 by Michael Scot. Here al-Biṭrūjī explains that

> 5. It is well known by all men that the whole heaven is composed of mutually different spheres and that one touches another in perfect contact. And because one [sphere] is moved inside another, there is a finiteness of the rotation and an equality of surfaces. And these [orbs] are continuous with each other because no other body lies between them.
>
> 6. And it is known that the concave surface of a higher [orb] is the place of the [orb] next below it and between them there is neither a plenum of another extraneous body, nor is there a vacuum, but one [orb] touches the other [orb] with its whole surface.[58]

In this passage al-Biṭrūjī gives no indication of his opinion concerning the hardness or softness of the celestial orbs. And despite his statement that the orbs "are continuous with each other" (*continuantur unum cum altero*), it is not even clear whether he assumed the surfaces of the successive orbs to be continuous or contiguous. That is, it is unclear whether he assumed that the concave surface of a superior sphere is identical with the convex surface of the next inferior sphere so that they are one and the same, or continuous; or whether those surfaces are distinct and separate but in contact at every point, that is, contiguous. However, numerous natural philosophers, and at least one astronomer, considered the problem and decided on one or the other of these options.[59] But does an author's opinion about the relationships between successive surfaces of celestial orbs—that is, whether they are continuous or contiguous—provide a clue to his view on the hardness or softness of the surfaces of the celestial orbs?

---

not rub together in their local motions because they are highly polished. Nor is there any friction between them that could generate heat" ("dicendum est quod corpora celestia in suis motibus localibus non confricantur quia sunt corpora politissima. Nec inter ipsa sit aliqua confricatio talis ex qua possit gigni calor"): MS Bibliothèque Nationale, Paris, fonds latin, 6752, fol. 214v. For a description of the contents of the treatise see Lynn Thorndike, *A History of Magic and Experimental Science,* 8 vols. (New York: Columbia Univ. Press, 1923–1958), Vol. III, pp. 568–584.

[58] *Al-Biṭrūjī De motibus celorum,* critical edition of the Latin translation of Michael Scot, ed. Francis J. Carmody (Berkeley/Los Angeles: Univ. California Press, 1952), p. 82. The words "there is a finiteness of rotation" are an uncertain rendition of the Latin *ideo ipse est in fine rotationis.* For an English translation based on Arabic and Hebrew versions of Biṭrūjī's treatise see Al-Biṭrūjī, *On the Principles of Astronomy,* an edition of the Arabic and Hebrew versions with translation, analysis, and an Arabic-Hebrew-English glossary by Bernard R. Goldstein (New Haven, Conn./London: Yale Univ. Press, 1971), pp. 65–66.

[59] The definitions of *continuity* and *contiguity* were derived from Aristotle, *Physics* 5.3.227a.9–12, 21–23. For discussions of the problem see Michael Scot, *Commentary on the "Sphere" of Sacrobosco,* and Robertus Anglicus, *Commentary on the "Sphere" of Sacrobosco,* both in Thorndike, *"Sphere" of Sacrobosco* (cit. n. 13), pp. 279–283, 201–203; Bonaventure, *Commentary on the Sentences,* Bk. 2, dist. 14, pt. 2, art. 1, quest. 1, in *Opera omnia* (cit. n. 25), Vol. II, pp. 351–352; Albert of Saxony, *Questions on De celo,* in *Questiones et decisiones physicales insignium virorum* (cit. n. 53), fols. 88v, cols. 1–89v, col. 2; Richard of Middleton, *Commentary on the Sentences* (cit. n. 25), Bk. 2, art. 3, quest. 1, p. 184, col. 1; and Oresme, *Le livre du ciel et du monde* (cit. n. 34), pp. 385–387.

If an author simply opted for continuity or contiguity without providing any further clues about hardness or softness, we would have no good independent reasons for assuming a belief in either hard or soft orbs. This will be obvious from an examination of these two important concepts.

The view that successive celestial surfaces were continuous was popular because it enabled astronomers and natural philosophers to present a simple formula for the distances of planetary spheres. The distance of the convex surface of one planetary sphere, say Mars, was set equal to the distance of the concave surface of the next higher planet, Jupiter. As Campanus of Novara expressed it, "the highest point of the lower [sphere] coincides with the lowest point of the higher." For if this were not true, two successive orbs would either overlap and their parts would occupy the same place simultaneously, or there would be a void between them, both of which are impossible. On the basis of the first alternative, "which is contrary to Aristotelian thinking," the editors of Campanus's *Theorica planetarum* inferred that Campanus "supposes the spheres to be solid," that is, hard.[60] But this is an unwarranted inference. If the spheres were composed of a single fluid substance, the same objection would obtain: the overlapping parts of the two spheres could not possibly occupy the same place simultaneously. Thus Campanus's spheres may indeed be solid, but whether that solidity is to be associated with hard or soft surfaces and orbs cannot be specified.

The crucial additional clues that Campanus and most scholastics omitted were supplied by another supporter of the continuity interpretation, who made it clear that he considered the orbs fluid. In order to permit the different orbs to have independent motions and to have continuous and identical outer surfaces Robertus Anglicus, as we saw earlier, assumed that the outer surfaces of celestial orbs were immobile but that their inner parts, which he likened to a fluid, were capable of motion.[61]

The contiguity theory of distinct and separate celestial surfaces that are everywhere in perfect contact to avoid the possibility of void space or extraneous matter between them may seem at first glance to indicate that the surfaces are hard. If the convex surface of Mars's planetary sphere were contiguous to the concave surface of Jupiter's orb, and if those two distinct but touching surfaces moved with different speeds and perhaps even in different directions, one might argue that they would retain their separateness and move independently of each other only because their surfaces are hard. But because of the attributes traditionally assigned to the celestial ether, the conditions just described do not by themselves warrant the inference of hard spheres. They are equally compatible

[60] See *Campanus of Novara and Medieval Planetary Theory* (cit. n. 27), p. 331; see also pp. 181–182 and 356–363 for a table of distances and dimensions. Although Campanus was aware that the planetary orbs moved in different directions and at different rates, he failed to realize that the continuity, or identity, of the surfaces of successive planetary orbs was incompatible with the independent motions of each planetary orb. As Roger Bacon explained, continuous surfaces would cause all orbs to "be moved with equal velocity, even with the same motions, which is contrary to experience"; see Bacon, *Opus tertium,* in Pierre Duhem, *Un fragment inédit de l'Opus tertium de Roger Bacon précedé d'une étude sur ce fragment* (Quaracchi, 1909), p. 123. Despite Campanus's clear assumption of the identity and therefore continuity of successive celestial surfaces, Benjamin and Toomer attribute the contiguity interpretation to him (*Campanus of Novara and Medieval Planetary Theory,* p. 412 n. 47), as indeed does E. J. Aiton, "Celestial Spheres and Circles" (cit. n. 17), p. 90.

[61] See Robertus, *Commentary on the Sphere of Sacrobosco,* in Thorndike, *"Sphere" of Sacrobosco* (cit. n. 13), pp. 147 (Latin) and 202–203 (English).

with soft and fluid spheres. After all, not only is the celestial ether unalterable and incorruptible, but it was usually assumed to be more subtle than air and fire.[62] With such properties the ethereal orbs could be contiguous and fluid. Moreover, because of their presumed unalterable and incorruptible natures, fluid orbs could be in contact at every point and also retain their perfect spherical shapes. Indeed perfect, incorruptible, contiguous fluid surfaces, no less than hard surfaces, could be polished and move without resistance or friction. If I earlier attributed to Nicole Oresme an implicit belief in hard spheres, it was not because of his assumption of polished surfaces moving without resistance, but rather because of his declaration that the lunar concavity shaped the contiguous, convex surface of the fiery sphere below it. On the basis of the properties of the celestial ether and in the absence of an explicit assertion in favor of hard spheres, or indirect independent evidence indicating rigid spheres, it is more plausible to assume that a medieval author considered the celestial orbs solid and soft rather than solid and hard. After all, the celestial ether was traditionally assigned properties that made it appear rare and fluid-like. Those scholastics who believed in a celestial ether and also in hard spheres must somehow have lost sight of this.

CONCLUSION

Medieval natural philosophers and astronomers rarely discussed the hardness or softness of the spheres. This was less a conscious effort to avoid the issue than an innocent omission. Like many other problems that we wish had been explicitly debated in the Middle Ages, the hardness or softness of the spheres never surfaced as a genuine issue. For most medieval authors a solid sphere signified a material sphere, without implying any obvious commitment to hardness or softness. And yet, in view of the arguments presented here, if these authors had been asked whether they thought the orbs hard or fluid, I believe most would have opted for fluid orbs, largely because they would have had great difficulty reconciling hard, rigid orbs with the usual properties assigned to the celestial ether. It therefore seems a plausible inference that the firm and almost invariable association of the properties hard and rigid with solid celestial orbs occurred during the sixteenth century, and not during the Middle Ages. Although it is very likely that Copernicus believed in the existence of solid spheres, it by no means follows that he, "like every other astronomer of his time, envisioned planetary models to be composed of nonintersecting, rigid spheres."[63] Nothing that Copernicus said or implied in the *De revolutionibus* enables us to say whether he assumed hard or fluid spheres. Copernicus fits the pattern that obtained in the Middle Ages, when explicit opinions about the rigidity or fluidity of the orbs were rarely presented. In this regard Copernicus is no different from two of his important predecessors, Ibn al-Haitham (Alhazen) and Georg Peurbach, whose treatises played instru-

[62] Aristotle nowhere says this, but it seems an implication of his ordering of the four elements—namely, earth, water, air, and fire—which, as we move up from the earth, become rarer and more subtle. Since the celestial ether extends beyond fire, it should exceed the latter in rarity and subtlety. In the strict sense, of course, Aristotle denied that contrary qualities such as rarity and density could be applied to the ether. But because physical attributions were sometimes unavoidable, the divine celestial ether was likely to be considered pure, rare, and subtle. It was certainly never considered impure, dense, and gross.

[63] Swerdlow makes this claim in "Pseudoxia Copernicana" (cit. n. 1), pp. 108–109.

mental roles in disseminating knowledge about solid spheres during the late Middle Ages and the Renaissance. There can be no doubt that both spoke about real, material orbs, but neither gave any indication whether those orbs were rigid or fluid.[64] Indeed, the silence on this issue during the Latin Middle Ages is partially attributable to an earlier silence in the Arabic and Greek sources on cosmology and astronomy that were rendered into Latin during the twelfth and thirteenth centuries.

If what has been said here is a reasonably accurate reflection of medieval opinions and attitudes, it was not until the sixteenth century—not long before they would be forever banished from the heavens—that the celestial orbs were hardened and subsequently opposed as the primary alternative to a heaven devoid of orbs and filled with a fluid substance. Our conclusion gives rise to a number of important questions. Were the celestial orbs hardened in the period between Copernicus and Tycho Brahe?[65] In view of the fact that the hardness or softness of spheres was rarely discussed in the late Middle Ages, from what source or sources did Tycho learn of their alleged hardness? And, in view of Tycho's claim that many of his contemporaries believed in hard spheres, when and why did mention of hard celestial orbs become explicit and commonplace?[66] Who made *solidity* and *hardness* virtually synonymous even before Tycho insisted that hard, solid orbs could not exist? Was it the astronomers? The natural philosophers? Both? Were the spheres hardened as a result of new interpretations of medieval or ancient texts?

The time has come to seek answers to these, and other, questions about the emergence of an explicit concept of hard and rigid celestial spheres, a concept that appears to have emerged with the Copernican challenge to Aristotelian-Ptolemaic cosmology and was destined to endure but a short while.

[64] Swerdlow, who assumes that solid orbs or spheres were rigid, implies that in both the Latin translation of Ibn al-Haitham's "On the Shape of the Universe," which Swerdlow claims introduced "solid sphere planetary models into medieval European astronomy" (*ibid.*, p. 117), and in Georg Peurbach's *Theoricae novae planetarum*, the solid spheres were assumed to be hard and rigid (*ibid.*, pp. 109, 110, 116–117). I have examined both treatises and have found only the usual silence on the issue of hardness or softness. For Ibn al-Haitham's treatise see José Maria Millás Vallicrosa, ed., *Las traducciones orientales en los manuscritos de la Biblioteca Catedral de Toledo*, App. II: "Tratado de Astronomia de Al-Hasan ben Al-Haytam, en Traducción Latina Desconocida," pp. 285–312; for Peurbach I have used the edition of the *Theoricae novae planetarum* included in the *Spherae tractatus* (cit. n. 47), accompanied by the commentary of Joannis Capuan.

[65] Before the comet of 1577 led Tycho to deny the existence of hard spheres in the heavenly region, he had assumed their existence.

[66] Perhaps the answers to some of these questions may be found in the various commentaries that were written in the late fifteenth and the sixteenth centuries on Peurbach's *Theoricae novae planetarum*.

Theory and Observation in Medieval Astronomy

By Bernard R. Goldstein

THERE IS NO MODERN STUDY of the relationship between theory and observation in medieval astronomy, as far as I am aware. A number of modern astronomers, foremost among them Simon Newcomb, have discussed medieval observations with regard to their accuracy and usefulness for the determination of certain astronomical parameters.[1] Moreover, historians of astronomy have devoted much effort to the elucidation of various medieval modifications of ancient theories.[2] But the impression one gains from reading these many studies is that theory and observation developed in isolation from one another. In the course of my researches I began to realize that this picture was seriously in need of revision and that some medieval astronomers had a much better understanding of the procedures and aims of ancient scientists than they have been given credit for. They not only held the view, already prevalent in antiquity, that observations should be used to test and construct theories, but they succeeded in applying it as well. One difficulty they faced was that the formidable achievements of Ptolemaic astronomy seemed to allow little room for improvement. Indeed, the theoretical advances introduced by Kepler rested on the observations of Tycho Brahe, which were clearly more accurate than those available to medieval astronomers.

This paper is not intended to exhaust all known discussions of theory and observation in medieval astronomy, but simply to call attention to some illustrative passages that I have recently uncovered. That the passages are largely drawn from Hebrew and Arabic sources is not entirely a reflection of my own interests, for the evidence seems to suggest that these questions were not as actively pursued in the Latin West. One may note that observations of phenomena not discussed in the *Almagest*, such as planetary brightness, transits, and even comets, were introduced to resolve problems arising out of Ptolemaic astronomy. Moreover, at least one lunar theory introduced in the Middle Ages was based on a new set of observed lunar positions, and Levi ben Gerson, who constructed it, had reason to believe it was superior to Ptolemy's lunar theory. Finally, solar eclipses were used to test and modify astronomical theories, although Ptolemy did not record or use any such observations in the *Almagest*. In effect, the ancient texts served as a point of departure for subsequent developments in astronomy and were not understood as setting firm limits on the range of topics worthy of consideration.

Throughout the history of astronomy, beginning with the Babylonian astronomers,

This study was supported by a National Science Foundation Research Grant.

[1] *Cf.* S. Newcomb, *Researches on the Motion of the Moon*, Pt. 1 (Washington, 1878).

[2] See E. S. Kennedy, "A Survey of Islamic Astronomical Tables," *Transactions of the American Philosophical Society*, 1956, Vol. 46.

observations of celestial phenomena have been used to construct theories whose goal has been the prediction of the positions of heavenly bodies as well as the times of certain phenomena such as eclipses. In a series of texts, now called diaries, we have the observational records of the Babylonians which in their original form covered the period from about 750 B.C. to the beginning of the Christian era.[3] Another group of texts, mathematical (but not geometrical) in character, list successive planetary positions at certain specified times.[4] These predictions could be, and were, confirmed by subsequent observations. The procedure of basing astronomical theories on observation, constructing mathematical theories which produce quantitative predictions and then allowing subsequent observations to confirm or refute the theory, has remained an essential feature of this field.

The Greek astronomers owed many debts to their Babylonian predecessors, well beyond the relatively few observations which are cited in the ancient literature. One notable change, however, was the insistence by the Greeks that the scheme for predicting planetary positions be capable of geometric representation, a restriction not imposed by the Babylonians. This approach is related to, but not identical with, the feeling that the predictive scheme ought to reflect physical reality.

One of the most important steps in the development of Greek astronomical theory was taken by Apollonius (second century B.C.) when he proved that the eccentric and epicyclic hypotheses were equivalent.[5] In modern terms he proved that vector addition is commutative (see Fig. 1). Ptolemy (second century, A.D.) made extensive use of

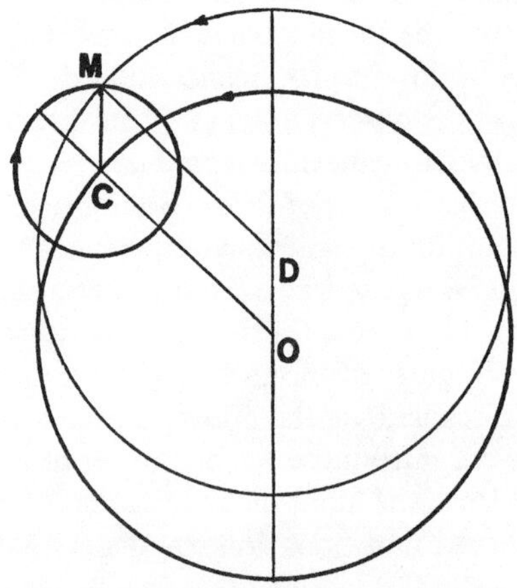

Figure 1. *The eccentric and epicyclic hypotheses.* In the epicyclic model the center of the epicycle, *C*, moves uniformly about the observer at *O*, while point *M* moves at the same rate in the contrary direction about *C*. Thus *CM* maintains a fixed direction. In the eccentric model *M* moves uniformly on a circle whose center *D* is some distance from the observer at *O*. We set the eccentricity *OD* equal to the radius of the epicycle *CM*. Then the equivalence of these two models follows from the commutativity of vector addition, that is, **OD** + **DM** = **OC** + **CM**, or from the fact that *ODMC* is a parallelogram.

these two hypotheses in his planetary models and realized that their equivalence meant that the two were not distinguishable on those physical grounds which relate to planetary positions.

Medieval scholars often used Aristotelian arguments to refute one or the other of these hypotheses. So we learn from Maimonides (late twelfth century):

[3] See A. Sachs, "A Classification of the Babylonian Astronomical Texts of the Seleucid Period," *Journal of Cuneiform Studies*, 1948, *2*:271–290.

[4] See two works by O. Neugebauer: *Astronomical Cuneiform Texts* (London: Lund Humphries, 1955) and *The Exact Sciences in Antiquity* (2nd ed.; New York: Dover, 1969).

[5] *Almagest*, III, 3; Ptolemäus, *Handbuch der Astronomie*, trans. K. Manitius (2nd ed.; Leipzig: Teubner, 1963), Vol. I, pp. 148 ff.

> I have heard that Abū Bakr [Ibn Bājja] discovered a system in which no epicycles occur, but eccentric spheres are not excluded by him. I have not heard it from his pupils; and even if it be correct that he discovered such a system, he has not gained much by it, for eccentricity is likewise contrary to the principles laid down by Aristotle. . . . I have explained to you that these difficulties do not concern the astronomer, for he does not profess to tell us the existing properties of the spheres, but to suggest, whether correctly or not, a theory in which the motion of the stars and planets is uniform and circular, and in agreement with observation.[6]

The astronomical systems of Ibn Bājja (or Avempace, d. 1138) has not otherwise survived, but al-Biṭrūjī (or Alpetragius, fl. *c.* 1200), who may be counted among his successors in Muslim Spain, found epicycles and eccentrics equally repugnant and wished to avoid both of them in constructing his new system.[7] In the opinion of Maimonides one could not properly distinguish between these two mathematically equivalent hypotheses by introducing Aristotelian principles. Moreover, he clearly stated the view that astronomical models are to be formulated in geometric terms without regard to physical reality.

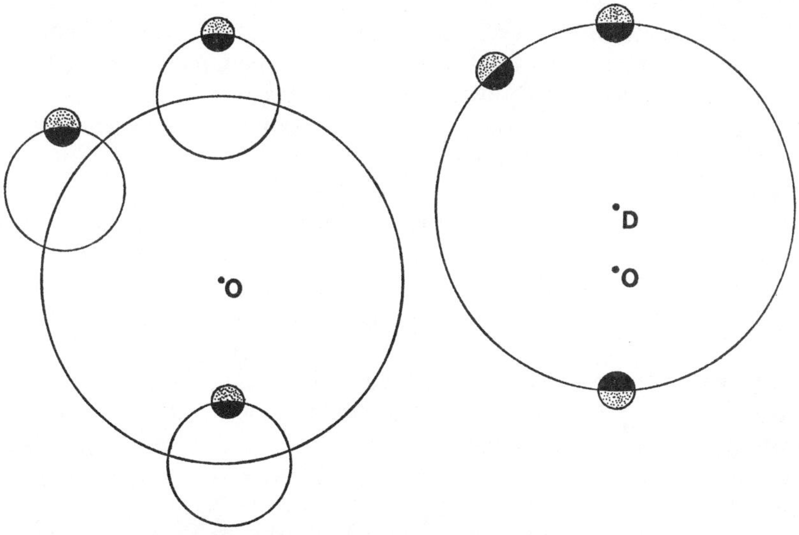

Figure 2. *The apparent disk of the moon according to the epicyclic and eccentric models. Note that the sun may shine on both the solid and the shaded portions of the moon.*

But in the fourteenth century Levi ben Gerson believed he had found an observation which allowed him to distinguish properly between these two hypotheses. To do so, he depended on the Aristotelian principle, universally accepted by astronomers and philosophers in his day, that a planet is fixed rigidly on its sphere—that is, that a planet has no motion of its own but only moves with the motion of its sphere. This principle allowed him to affirm that according to the epicyclic hypothesis in certain circumstances we should see one side of the moon, and in other circumstances the

[6] *The Guide for the Perplexed*, II, 24, trans. M. Friedländer (New York: Dover, 1956), p. 196.

[7] *Cf.* B. Goldstein, *Al Biṭrūjī: On the Principles of Astronomy* (New Haven, Conn.: Yale Univ. Press, 1971).

other side of the moon. But according to the eccentric hypothesis we should always see the same side of the moon, roughly speaking. However, the markings on the lunar disk prove that we always see the same side of the moon, and therefore the epicyclic hypothesis must be rejected as contrary to the phenomena (see Fig. 2). By extension he argued that the same must hold for the planets, and for this reason he eliminated epicycles entirely.[8]

In the *Almagest* Ptolemy restricted his attention to planetary positions (generally stated in longitude and latitude) and specifically avoided the question of a planet's distance from the observer. To be sure, we have long known of a system of planetary distances and sizes generally called the Ptolemaic system, but the textual basis for attributing it to Ptolemy was only recently uncovered. In the first instance, Willy Hartner collected a variety of evidence, mainly from medieval sources, that clearly pointed to Ptolemy as the author of this system.[9] He then suggested that the original Greek text of Ptolemy's *Planetary Hypotheses* (most of which is not extant) contained his description of this system, even though the published version of this text, largely based on the surviving Arabic translation, had no such passage in it. In following up Hartner's paper, I had occasion to check the Arabic text and found that the missing passage (perhaps one-quarter of the entire text) was in the manuscripts of the Arabic translation and had simply been omitted in the published version. Though I found it difficult to understand how such a gross error could go unnoticed, there are a number of mitigating circumstances, including the death of the principal collaborator in this work prior to its completion.[10]

Fundamental to Ptolemy's geocentric system is the nesting hypothesis: that the maximum distance of one planet from the earth is equal to the minimum distance of the planet above it. Moreover, the ratio of a planet's maximum distance to its minimum distance is equal to the ratio of maximum to minimum distance in the model that is used to represent its motion. Thus the geometric models were converted into physical models, and the nesting principle served to exclude a vacuum between the planetary spheres. Accordingly, the values for solar and lunar distances in the *Almagest*, derived from observations of parallax, could now be used to fix the distances of the planets.[11] Ptolemy indicated that another question had to be resolved before he could proceed with the determination of these distances—namely, the order of the planetary spheres—and he tells us that he was unable to decide this with certainty. Ptolemy, as well as most of his successors, found no difficulty in placing spheres of the moon, Mercury, Venus, Mars, Jupiter, Saturn, and the fixed stars in ascending order from the earth, but he wondered where the sun belonged in this order, whether any planet's sphere lay entirely below the sphere of the sun. He described his views as follows in *Planetary Hypotheses*:

[8] Levi ben Gerson, *Wars of the Lord*, Bk. V, Pt. 1, Ch. 20 (unpublished). *Cf.* MS Paris, *hebr.*, 724, fol. 31[a]. On Levi, see B. Goldstein, "Preliminary Remarks on Levi ben Gerson's Contributions to Astronomy," *Proceedings of the Israel Academy of Sciences and Humanities*, 1969, *3*:239–254.

[9] W. Hartner, "Mediaeval Views on Cosmic Dimensions," *Mélanges Alexandre Koyré* (Paris: Hermann, 1964), pp. 254–282.

[10] *Cf.* B. Goldstein, "The Arabic Version of Ptolemy's *Planetary Hypotheses*," *Trans. Am. Philos. Soc.*, 1967, Vol. 57.

[11] *Almagest*, V, 13–15; trans. Manitius, Vol. I, pp. 299–312. The method for finding the solar distance depends on certain assumptions concerning solar and lunar eclipses and cannot be applied to planets.

> The distances of the five planets are not as easy to determine as those of the two luminaries, for the distances of the two luminaries were determined, mostly, on the basis of combinations of eclipses. A similar proof cannot be invoked for the five planets, because no phenomenon allows us to fix their parallax with certainty. Moreover, up to this time we have not seen an occultation of the Sun (by any of the planets), and therefore it is possible for one to assert that all five planetary spheres lie above the sphere of the Sun. But the argument so far does not permit one, whose intention is to seek the truth, to draw a conclusion. Firstly, the occultation of a large body (the Sun) by a small one (a planet) may not be perceptible on account of the remainder of the solar body which would still be exposed, for when the Moon eclipses part of the Sun equal to, or even somewhat greater than, the diameter of one of the planets, the eclipse is not perceptible. Moreover, such events could only take place at long intervals, for (an inner planet) is closest to the Sun (in longitude) when it is at the apogee and perigee of its epicycle; but (the planet) is found in the plane of the ecliptic only twice in every revolution on the epicycle, when it passes from the north to the south, and when it passes from the south to the north. When the center of the epicycle is at one of the nodes, and the planet is also at that node, and the planet is also at the apogee or perigee (of its epicycle), then the planet may hide (part of the Sun). According to those who report observations and examine them carefully, a long time (*miqdār al-zamān*) must elapse before the return (of the center) of the epicycle and the planet in conjunction (with the Sun) above the earth. With these conditions, it is clear that one cannot judge with certainty for the two (inner) planets, nor even for the planets on which it is agreed that they lie above the sphere of the Sun, i.e. Mars, Jupiter and Saturn.[12]

Ptolemy clearly stated that a planet observed to transit the sun would be demonstrably below it, but he knew of no such observation. In the absence of conclusive evidence he depended on the nesting hypothesis to resolve the issue. In the *Almagest* he found the maximum lunar distance from the earth to be about 64 terrestrial radii. Assuming the sphere of Mercury to lie directly above that of the moon, and the sphere of Venus to lie directly over that of Mercury, he then found the maximum distance of Venus to be 1,079 terrestrial radii.[13] But from the data given in the *Almagest* the minimum solar distance is 1,160 terrestrial radii; and the closeness of these results was taken by him as confirming the order: moon, Mercury, Venus, sun, Mars, Jupiter, Saturn, fixed stars. Though Ptolemy did not state it explicitly, it is clear that this near coincidence in the distances of Venus and the sun also served to confirm the nesting hypothesis.

With rare exceptions medieval astronomers supported the nesting principle, but they considered the order of the planetary spheres to be an open question. There were three views: Mercury and Venus both lie beneath the sun; Mercury is below the sun and Venus above it; Mercury and Venus both lie above the sun. I think it is fair to say that a modern scientist confronted with this situation would try to predict the time of a forthcoming planetary conjunction with the sun and then would observe to see if a transit occurred, for Ptolemy himself had indicated this method of resolving the question. Only a positive result would be conclusive, because Ptolemy had already suggested that the body of a planet might be too small to be seen against the sun. We may then ask if such a procedure was technically feasible for a medieval astronomer. The first part of his work would be to predict the conjunction for a planet near one of its nodes. Although Ptolemy did not arrange his planetary tables to facilitate this

[12] *Planetary Hypotheses*, Bk. I, 2,2; Goldstein, "The Arabic Version," pp. 6–7. Parentheses indicate that a word or phrase in the translation does not appear in the text and has been added for clarity.

[13] For a more detailed discussion of these computations, see *ibid.*, pp. 9–11.

computation, it is possible to use them for this purpose without an excessive amount of labor. Since most of the medieval discussions concerned Venus, one may also ask if a Venus transit is, in fact, visible to the naked eye. This may be answered affirmatively, as I learned from an old newspaper clipping left between the leaves of a book in the Yale Astronomy Library.[14] One further point of difficulty arises from the medieval view that the heavens are incorruptible. Thus the observation of a sunspot might well be interpreted as a transit in order to maintain this philosophical tenet.

Let us consider one medieval report of a transit in the light of this analysis. In his *Compendium of the Almagest*, Avicenna stated that he saw Venus as a spot on the surface of the sun, from which he concluded that Venus lies below the sun.[15] Avicenna died in 1037, and a Venus transit took place on May 24, 1032. Since he mentioned no date, it is not clear whether he was claiming to have seen this transit or whether, as so often was the case, he had mistaken a sunspot for a planet passing in front of the sun.

A consequence of Ptolemy's planetary models is that each planet must vary in size inversely as its distance from the observer. Thus Venus and Mars should vary in size enormously in the course of their orbits. In the *Almagest* planetary sizes are not considered, and in the *Planetary Hypotheses* sizes are stated for mean distances only—the variation in size is not discussed. Several medieval scholars, however, noted that Venus did not appear to vary much in size, and that this was contrary to Ptolemy's model for Venus.[16] An anonymous treatise in Arabic, known to me from a thirteenth-century manuscript, indicates that if one accepts Ptolemy's value for the size of Venus at mean distance—that it is one tenth the size of the sun—then at its closest approach to the earth Venus would appear to be two-fifths the size of the sun, which is grossly contrary to observation. Therefore, this anonymous author argues, Venus must be one-tenth the size of the sun at its minimum distance rather than at its mean distance. Since Ptolemy says that the observations of apparent sizes were made by Hipparchus, this author claims that he is merely reconstructing what Hipparchus had actually done.[17] Here we have a case of an observation being used to refute the implications of a theory and then the theory being modified to conform to the observation. Yet the author gives no indication that he tried to check the observational data himself; rather, he was satisfied with textual criticism.

Some medieval scholars recognized another difficulty in the assumption that the

[14] The relevant passage is quoted in B. Goldstein, "Some Medieval Reports on Venus and Mercury Transits," *Centaurus*, 1969, *14*:49 f.

[15] *Ibid.*, pp. 52 f.

[16] For example, *cf.* Levi ben Gerson, *Wars of the Lord*, V, 1, 17. Medieval astronomers frequently considered planetary sizes, but direct observations of their apparent sizes played only a minor role in most of these discussions. The generally accepted values for the apparent sizes were taken from Ptolemy's *Planetary Hypotheses*, and these values are all too large from a modern point of view. One difficulty in naked-eye observations that they failed to recognize is the limited resolving power of the eye. But another difficulty lay in their implicit assumption that apparent brightness is a measure of apparent size. In the case of Venus the size of its apparent disk varies enormously, but its apparent brightness varies very little. The principal reason for this phenomenon is Venus' phases which, however, are not visible to the naked eye. In fact, when Venus is seen near its closest approach to the earth, the sun illuminates only a narrow crescent of its apparent disk; but when it is farthest from us and the sun shines on most of the disk facing the earth, its apparent diameter is very much smaller. As a result, its apparent brightness varies relatively little. For a modern discussion of apparent planetary brightness, see *Explanatory Supplement to the Astronomical Ephemeris*, issued by H. M. Nautical Almanac Office (London, 1961), pp. 310 ff.

[17] *Cf.* B. Goldstein and N. Swerdlow, "Planetary Distances and Sizes," *Centaurus*, 1970, *15*: 135–170.

sun's sphere lay entirely above that of Venus. If Venus received its light from the sun, then, like the moon, it should have phases; yet they are not observed, and similarly for Mercury. Al-Biṭrūjī resolved this difficulty by affirming that the planets are self-luminous.[18]

One may infer from these reports that the absence of Venus' phases in pre-telescopic observations afflicted Ptolemaic astronomy as much as it did Copernican astronomy. When discovered, however, the pattern of these phases confirmed that the orbit of Venus enclosed the sun, as Copernicus had asserted,[19] and proved that Venus was not always closer to the earth than the sun, as Ptolemy's system required.

Levi ben Gerson discussed the variation in the sizes of the planets with the goal of refuting Ptolemy's planetary models, and, unlike most other medieval scholars, he preferred to rely on his own observations. He first tells us that Ptolemy's model for Venus implies a sixfold variation in its size. But in comparing its size with that of first- and second-magnitude stars, he failed to see any variation in Venus' size. He also observed Venus during the day to compare its size with that of the sun, noting that Venus is visible in sunlight when it is more than twenty degrees away from the sun. He made other observations on moonless nights using a pinhole camera to measure the size of Venus.[20]

Levi also considered the size of Mars and argued that it seemed to vary only twofold instead of the more than sixfold that Ptolemy's model required.[21] Levi introduced several observations of Mars near opposition to the sun when the planet is in retrograde motion and where Levi expected its size to be more or less constant. In comparing its size to that of Saturn, he found that on two occasions (in different years) Mars was larger than Saturn, but once it was not larger than Saturn. He could find nothing in the planetary motions to account for this phenomenon and showed considerable resourcefulness in producing an explanation. He argued that on this anomalous occasion Mars had generated a comet directly below it from thick vapors in the upper reaches of the sublunary world and added that this comet lasted for three months and passed by the north pole. Levi did not mention the year in which he made this observation, but from the description he was clearly alluding to the comet of 1337, which was seen and reported both in Europe and China.[22]

Levi's report is noteworthy on two accounts. First, the theory of comet formation on which Levi depended goes back to Aristotle and was widely accepted in the Middle Ages, but, as far as I know, no one else suggested that the planet would appear dimmer at the time the comet was formed. Second, I know of no other medieval author who discussed the observation of a comet in an astronomical context. Comets were

[18] Goldstein, *Al-Biṭrūjī*, Vol. I, p. 33.

[19] Ptolemy's system implies that Venus is crescent shaped at inferior and superior conjunction, but Galileo notes that this phenomenon took place only at inferior conjunction. *Cf.* Galileo, *Dialogue Concerning the Two Chief World Systems*, trans. S. Drake (Berkeley/Los Angeles: Univ. of California Press, 1962), pp. 324 and 339; and S. Drake, *Discoveries and Opinions of Galileo* (New York: Doubleday Anchor Books, 1957), pp. 93–94.

[20] *Cf.* Levi ben Gerson, *Wars of the Lords*, V, 1, 17, MS Paris, *hebr.*, 724, fol. 27^b.

[21] See *ibid.*, fol. 28^a. The MSS agree on "more than sixfold," although Ptolemy's model clearly implies more than a sevenfold variation. See Goldstein, "The Arabic Version," p. 11. For Levi's parameter, *cf.* MS Paris, *hebr.*, 724, fol. 28^a:6, where it appears as a word, and MSS Paris, *hebr.*, 696, fol. 129^b:6 and Paris, *hebr.*, 725, fol. 19^a:9, where it appears as a numeral.

[22] *Cf.* L. Thorndike, *Latin Treatises on Comets* (Chicago: Univ. of Chicago Press, 1950), pp. 219 ff.; Ho Peng Yoke, "Ancient and Mediaeval Observations of Comets and Novae in Chinese Sources," *Vistas in Astronomy*, 1964, *5*:196.

frequently noted in chronicles and astrological treatises, but as sublunary phenomena they did not belong to the domain of astronomy. This report by Levi mentioned the comet for purely scientific reasons, and he did not consider its significance for prognostication, in marked contrast to the practice of his contemporaries.

Comets did not formally enter the realm of astronomy until Tycho Brahe's famous discovery in the sixteenth century that they were clearly more distant than the moon. Nevertheless, they were treated in the same way as astronomical phenomena beginning sometime in the fifteenth century, as witnessed by Regiomontanus' discussion of methods for finding cometary distances and his presentation of cometary observations alongside other astronomical phenomena without distinguishing between them.[23] I do not have any textual evidence to account for this new attitude toward comets in the fifteenth century, but I am tempted to suggest that non-Aristotelian ancient works played a significant role. In particular Seneca argued that the regular movement of comets, as contrasted with meteors, implied that they were celestial phenomena, and that astronomers should attempt to discover whether any comet returned after it had disappeared.[24]

Ptolemy's lunar model was occasionally criticized for its inadequate representation of the observed phenomena. Though Ptolemy succeeded very well in accounting for the lunar positions in longitude and latitude, his model seriously distorted the lunar distance, with the result that by his reckoning the moon ought to appear twice as large in diameter at quadrature (at half-moon phase) than at opposition (or full-moon phase). Even though the most casual observation would indicate the error of this consequence of Ptolemy's lunar model, Ptolemy and most of his medieval successors passed over it in silence.[25] There were two notable exceptions, both of whom lived in the fourteenth century but worked quite independently of each other: Ibn al-Shāṭir[26] and Levi ben Gerson. Levi described two kinds of observations by which he attempted to determine precisely the variation in apparent lunar diameter.[27] One method was to use a pinhole camera for measuring the image of the moon on a screen, and the other was to make two observations of the moon's position in longitude separated by an interval of four or five hours and then to fix the difference in parallax from the moon's motion. In this way its distance and hence its size could be computed.

The most striking difference between Levi and Ibn al-Shāṭir in their discussions of lunar theory is that Ibn al-Shāṭir started with Ptolemy's lunar observations and model and wished to find a new model that agreed with them as closely as possible—that is, he accepted Ptolemy's lunar positions in longitude and latitude and merely wished to repair the distances—but Levi preferred to depend on his own observations and dis-

[23] Regiomontanus, *De cometae problemata XVI* (Nuremberg, 1531); "Ioannis de Monteregio, Georgii Peurbachii, Bernardi Waltheri, ac aliorum, Eclipsium, Cometarum, Planetarum ac Fixarum observationes," *Scripta Clarissimi Mathematici* (Nuremberg, 1544), fols. 43^a–43^b.

[24] Seneca, *Quaestiones naturales* VII, 2; trans. Clarke, p. 274.

[25] For example, one of the most prominent astronomers of the 13th century, Naṣīr al-Dīn al-Ṭūsī, described a new lunar model in which Ptolemy's extreme values for the lunar distances were preserved. *Cf.* W. Hartner, "Naṣīr al-Dīn al-Ṭūsī's Lunar Theory," *Physis*, 1969, *11*:287–304.

[26] Ibn al-Shāṭir, *Nihāyat al-Sūl*, MS Bodleian, *arab.*, Marsh 139, fol. 3^a:13. I wish to thank Victor Roberts for allowing me to consult his unpublished translation of this work. *Cf.* V. Roberts, "The Solar and Lunar Theory of Ibn ash-Shāṭir," *Isis*, 1957, *48*:428–432.

[27] Levi ben Gerson, *Wars of the Lord*, V, 1, 16.

carded both Ptolemy's observations and his model. Levi then found that Ptolemy's model, but for small errors, worked well at syzygy (when the moon is in a straight line with the sun and earth) and at quadrature, but that at octant it was not as satisfactory. He therefore constructed a new model (devoid of epicycles) which agreed very closely with Ptolemy's model at syzygy and quadrature, but which departed from Ptolemy's results at the octant points.[28] He computed that his model produced closer agreement with his own observations of the moon than could be obtained from Ptolemy's model. Levi remarked, "No argument can nullify the reality that is perceived by the senses; for true opinion must follow reality, but reality need not conform to opinion."[29]

Solar eclipses were considered to be proper tests for astronomical theories, but their very sensitivity meant that the theory in question was most likely to be found inaccurate. Johannes de Muris observed a solar eclipse on March 3, 1337, which appeared to begin when the sun had an altitude of ten degrees above the horizon. He then stated that true first contact must have taken place when the sun's altitude was nine degrees, but he claimed that according to the Alfonsine Tables the eclipse should not have begun until the sun had reached an altitude of thirteen degrees. He reckoned this to be an error in the tables of about one-third hour, which he considered excessive.[30] As far as I know, there is no evidence to suggest that Johannes de Muris sought to rectify this error produced by the theory. One may also note Johannes de Muris' careful distinction between true first contact and visible first contact. Since the theory produced the moment of true first contact, it could not be compared directly with observation. Very few medieval astronomers concerned themselves with this awkward problem, but among those who did there was a wide range of views.[31] But most often the time cited for the beginning of an eclipse was that of true first contact without any information concerning its derivation from visible, or observed, first contact.

Levi ben Gerson was one of those astronomers who failed to make this distinction in reporting his observations of solar eclipses, but on the other hand he altered his solar theory after finding a discrepancy of one-half hour between the predicted and observed times of this very solar eclipse of March 3, 1337.[32] Medieval scientists lacked a systematic theory of observational error and often considered any perceptible discrepancy intolerable.

In sum, medieval scientists had a variety of attitudes concerning the relationship of astronomical theory and observation; indeed many of these scientists were prepared to challenge or modify significant elements in the tradition that they received. In particular, new theories were proposed and quantitative criteria continued to be invoked in discussions of the adequacy of these theories. To be sure, the transmission of ancient doctrines demanded a significant part of a medieval scientist's efforts, but it did not exclude him from making positive contributions in his field.

[28] *Cf.* B. Goldstein, "Levi ben Gerson's Lunar Model," forthcoming in *Centaurus*.

[29] MS Paris, *hebr.*, 724, fol. 75[a].

[30] MS Escorial, latin, O.II.10, fol, 93[b]. L. Gushee, "New Sources for the Biography of Johannes de Muris," *Journal of the American Musicological Society*, 1969, *22*:3–26. I am grateful to Professor Gushee for bringing this passage to my attention.

[31] *Cf.* J. Ali, *Al-Bīrūnī's Determination of the Coordinates of Cities* (Beirut: American Univ. of Beirut, 1967), p. 130.

[32] Levi ben Gerson, *Wars of the Lord*, V, 1, 100, MS Paris, *hebr.*, 724, fol. 186[a].

Situating Arabic Science

Locality versus Essence

By A. I. Sabra

LOCALITY AS A FOCUS OF HISTORIOGRAPHY

I trust that no one would wish to contest the proposition that all history is local history—whether the locality is that of a short episode or of a long story. All history is local, and the history of science is no exception. There can be no history of science that is not concerned with a localized episode or a sequence of such episodes. Philosophers of science, or some of them, may restrict themselves to analyzing the formal or logical and timeless structure of a piece of scientific thought; or they may substitute an emphasis on growth as a basic feature of scientific inquiry that requires for its logical elucidation the addition of a temporal dimension that retains an abstract character. But historians of science have a different task. For while they can ignore the cognitive core of scientific practice only at the cost of forfeiting their claim to a distinctive problematic and a distinctive discipline, they are especially concerned with science as a process that takes place in actual time or science as a series of phenomena that, owing to their special character of chronological and geographical locality, we call "historical"—this "special character" being due to the fact that the phenomena in question are not merely in space and time but events associated with, and indeed produced by, individuals acting in what we broadly call "cultural settings." The thesis outlined in these few abstract sentences is but a generalization of a weak version of the familiar contextualist thesis in scientific historiography; or, to put it another

This lecture was delivered at the annual meeting of the History of Science Society, Minneapolis, Minnesota, 28 October 1995. I am greatly honored to have been invited to give this talk. I have since made a few additions and changes, some of them in response to comments, questions, or criticisms by the anonymous referees, to whom my thanks are due.

way, contextualism is but an obvious consequence of the simple, undeniable fact of the local character of all events, including historical events. Since a historical event is where and when and how it is, inseparably tied to all the circumstances that combine to define it for us as historians, then, to be genuinely historical, all *history* of science must be contextual, because all historical events are local.

I said "weak version" because I do not wish to subscribe to a stronger, reductionist version that seems to me to misinterpret the local character of cognitive expression and behavior by appearing to deprive them of objective import. At the risk of being much too brief, let me indicate what I mean with just a sentence or two before I get on to my subject proper. I am persuaded, as I hope you are, that locality is not a property of a piece of scientific thought—say, the Pythagorean theorem. In other words, the Pythagorean *theorem,* not being an event in the spatiotemporal world, has no space-time coordinates. But *recognizing* it as a theorem in a given geometrical system with certain definable features was the achievement of someone or other who certainly had such coordinates. From this it would seem to follow that the historical study of scientific thought is, strictly speaking, not concerned with the thought content of science as such (that would be to engage in science, logic, or philosophical analysis), but with the *historical occurrence* of the recognition of thought. I am aware that all of this has been said before and in various ways by philosophers and historians to serve rather different agendas. So, to avoid misunderstanding, I will just add the observation that recognition of thought always involves, among other things, a cognitive context that itself is location-bound, in the sense of being part of a problem situation that depends on the existing state of knowledge at a given time and place. And this means that engaging in science, logic, and philosophical analysis may all be involved in the *historical* study of scientific thought.[1]

My purpose here is to try to illustrate the advantages of a strict adherence to the axiom of locality in situating the tradition of Arabic science with reference both to the place that this tradition occupies in the general history of science and to its place in the civilization where it emerged and developed. My discussion will thus be concerned with two contexts that, though distinct from one another, are, as I shall suggest, intimately connected with each other. Obviously what I shall present to you can only be the bare sketch of what might be described as a framework for research, but, I hope, a framework that is not irrelevant to other scientific traditions and that may even propose a correction to other historiographies that seem to pay little or no attention to the intercultural transmission of scientific knowledge.

Let us begin with an apparently neutral and innocent definition of Arabic, or what may also be called Islamic, science in terms of location in space and time: the term *Arabic* (or *Islamic*) *science* denotes the scientific activities of individuals who lived in a region that roughly extended chronologically from the eighth century A.D. to the beginning of the modern era, and geographically from the Iberian Peninsula and North Africa to the Indus valley and from southern Arabia to the Caspian Sea—that is, the region covered for most of that period by what we call Islamic civilization, and in which the results of the activities referred to were for the most part expressed in the Arabic language. We need not be concerned over the refinements that obviously need to be introduced even into this seemingly neutral definition. But what about the term *scientific* in it? What does it mean, and

[1] For ideas underlying the theoretical structure of this essay I am indebted to the writings of Gottlob Frege (on the distinction between thought and recognition of thought), Karl Popper (on methodological individualism, situational logic, interaction of Worlds 1, 2, and 3), and Alfred North Whitehead (all about events).

can it be regarded as in any way "innocent"? To me it seems clear that the only correct answer to this last question must be an unequivocal No. *Science* and *scientific* are *our* own terms and they express *our* own concepts (which, by the way, does not mean that they are sharply defined or unproblematic); and, therefore, the study of any past intellectual activity can be relevant to what we call "history of science" only to the extent that such an activity can be shown to help us understand the modes of thought and expression and behavior that *we* have come to associate with the word *science.* This is not anachronism, presentism, whiggism, or any of the other objectionable isms, but a consequence of the fact that we who are writing the history also have a location of our own that defines our perspective and, hence, the questions we pose from our vantage point and the terms in which these questions are framed. Nor should this admission to a definite point of view discourage or detract from investigating past modes of thought and expression and behavior under other categories deemed suitable for elucidating these modes "in their own terms," as the phrase goes. But, without aiming to replace other approaches that put the emphasis on certain concerns of sociology or anthropology or cultural history, our historiography of science will always change as a function of our changing position, being ourselves forever located at the end point of the process that is continually shaping and reshaping what we call "science." And so I am led to combine a self-evident proposition with another that seems no more, and no less, than a corollary of it: that all history of science is local, and no history of science can ever be neutral.

The character of Arabic science, its strengths and failings, the course of its development, and its ultimate fate have all been variously explained in terms of language as a matrix of thought and expression, of religion as an inexorable shaping force, of natural aptitudes or inclinations of a certain race or inherent mentality, or as one inevitable expression of a world culture of which Islamic civilization was a late embodiment. A perceived emphasis on algebra in the Arabic tradition has been attributed to certain features of Semitic languages that make these languages or their native users prone to "algebraization," as opposed to Greek "geometrization." The persistent attempts of Islamic astronomers to construct kinematic models primarily designed to save the principles and the logical consistency of Ptolemaic astronomy have been seen as a sign of poverty of imagination or of the tendency of the "Semitic mind" toward things it can easily perceive by the senses. Islamic religion has been cited both as the origin and source of vigor of medieval Islamic science and as the major cause of its final demise. And the "spirit of culture," in this case a Magian culture already at work in "so-called" late antiquity, has been invoked to account for every aspect of Islamic civilization, including its scientific products.[2]

It is not difficult to expose the weaknesses from which such explanations suffer. One

[2] Roger Arnaldez and Louis Massignon, "Arabic Science," in *History of Science: Ancient and Medieval Science, from the Beginnings to 1450,* ed. René Taton, trans. A. J. Pomerans (New York: Basic, 1963), Vol. 3, Ch. 2, pp. 385–421, esp. pp. 402–405; Massignon, "Réflexions sur la structure primitive de l'analyse grammaticale en arabe," *Arabica,* 1954, *1:*3–16; Pierre Duhem, *Le système du monde,* Vol. 2 (Paris: Vrin, 1954), pp. 117–179, esp. pp. 117–119 (on "le réalisme des arabes"); Duhem, *To Save the Phenomena: An Essay on the Idea of Physical Theory from Plato to Galileo,* trans. Edmund Dolan and Chaninah Maschler (Chicago/London: Univ. Chicago Press, 1969), pp. 25–35, esp. p. 26; Bernard Carra de Vaux, "Les sphères célestes selon Nasîr-Eddîn Attûsî," in Paul Tannery, *Recherches sur l'histoire de l'astronomie ancienne* (Paris: Gauthier-Villars, 1983), App. 6, pp. 337–361; G. E. von Grunebaum, "Muslim World View and Muslim Science," in *Islam: Essays in the Nature and Growth of a Cultural Tradition* (London: Routledge & Kegan Paul, 1964), pp. 111–126; von Grunebaum, "Islam and Hellenism," in *Islam and Medieval Hellenism: Social and Cultural Perspectives* (London: Variorum, 1976), Ch. 1; J. J. Saunders, "The Problem of Islamic Decadence," *Journal of World History,* 1963, *7:*701–720; and Oswald Spengler, *The Decline of the West,* trans. Charles Francis Atkinson, 2 vols. (London: Allen & Unwin, 1959), esp. Vol. 1, pp. 71–73, 207–216 (and see the index under "Arabian Culture").

can refer, for example, to the considerable and highly successful efforts of Islamic mathematicians in the fields of geometry and trigonometry. One can relate the theoretical program of Islamic astronomers to the work of Ptolemy himself and to earlier ideals of Greek astronomy. One can point out the great complexity of the relationship between science and religion throughout Islamic history and in various parts of the Islamic world. And one can easily show the vacuousness of theories born of the spirit-of-culture approach. And, in fairness to those who have advanced explanations of these sorts, it must be said that they tend to be poorly informed (or worse) about Arabic science and, in many cases, about Islamic civilization—a fact that, unfortunately, does not seem to have discouraged their influence on minds that seek ready-made and perhaps comforting explanations.

What is wrong with these explanations, and others like them, is not their consideration of language, religion, and culture as factors in the formation of a scientific enterprise that consciously adopted earlier traditions with markedly different languages and religious and cultural values but, rather, their essentialist character, which has tended to prejudice or obstruct historical research. Now locality—that is, the character of being local—is an ineradicable or, if you like, an essential property of all historical events, but the actual where and when and how of any such events are happenings created by human effort. With the sure perceptiveness of a true historian, Richard Southern once described the process of acquisition and adaptation of Greek learning in Islam as "the most astonishing event in the history of thought."[3] The event is astonishing because it strikes us as unexpected, and the best way I know to explain the unexpected in history, insofar as it can be explained at all, is to try to understand it, not in terms of essences or spirits or inevitabilities, but as the outcome of choices by individuals and groups responding to their situations as *they* perceived and experienced them. Let me illustrate.

THE INTERSECTION OF ISLAMISM, ARABISM, AND HELLENISM IN NINTH-CENTURY BAGHDAD

The powerful drive that eventually led to the transfer of the bulk of Greek science and philosophy (as well as elements of the scientific thought of India and Persia) to Islam was launched as a massive translation effort that took place in the context of empire and under the patronage of the confident Abbasid court in Baghdad. Translations into Arabic had been made earlier, and these had been preceded in the Middle East by translations from the Greek into Syriac and Persian, but it was the Abbasids who mounted a concentrated translation effort soon after they came to power in the middle of the eighth century and who further organized and intensified their support during the ninth century. Under their predecessors, the Umayyads who ruled from Damascus (661–750), the Islamic empire already encompassed large areas—including Egypt, Syria, and Persia—that had come under the influence of Hellenism from the time of Alexander; and before the ninth century was over Islamic rule had reached Kashmir in the east and Khwarazm to the north. In the early Abbasid period the higher administration of the court itself was in the hands of cultivated Persians who had gained much favor and influence with the Abbasid rulers and whose intellectual interests inclined them to various forms of secular learning and to a rationalizing approach for understanding matters of religious belief. Some of these Persian officials acted as translators, especially from Persian, and in general they constituted an

[3] R. W. Southern, *Western Views of Islam in the Middle Ages* (1962; Cambridge, Mass./London: Harvard Univ. Press, 1978), pp. 8–9. These lectures were delivered at Harvard in April 1961.

important, politically influential part of Baghdad's intellectual elites. Two other groups within the empire (and concentrated mainly in Syria, Iraq, and Persia) had maintained a long-established tradition of Hellenized Syriac learning. One consisted of Christian physicians and Christian theologians, who continued to pursue their interests in Greek logic and philosophy in scattered monastic schools; the other was the pagan Ṣābians of Ḥarrān, in northern Mesopotamia, an ancient Semitic group whose astral religion connected them to Hellenistic astrology and astronomy and to Hermeticism. It was from these two last groups that the Abbasids were able to recruit the scholars who carried out the translations of Greek medical, philosophical, and mathematical works into Arabic, either from preexisting Syriac versions or directly from the Greek.

The survival of these pockets of Hellenic learning during the first centuries of Islamic rule in the Middle East and Asia, although scattered and limited at first in scope and appeal, ensured a certain continuity with the classical tradition—a continuity that was largely lacking, for example, in the case of the "Renaissance of the Twelfth Century," when European scholars first had to journey to the edges of Western Christendom to acquire Arabic and Greek learning from across the borders with Islam and Byzantium. In the earlier Middle Eastern episode, this continuity meant the immediate availability of texts, in Greek or Syriac versions, and of translators already conversant with these languages and with Greek thought itself in a number of scientific, medical, and philosophical disciplines. And although much additional Greek material was later to be brought over the borders with Byzantium, the continuity with the Greco-Syriac tradition helps to explain the high level of competence, even sophistication, that characterized scientific writings in Arabic from an early period that overlapped the translation movement.

One might then say, and with much justification, that the stage was set, at a certain place and time, for the translation movement that quickly acquired unprecedented proportions—unprecedented not only in the Middle East but in the world at large. But in order to explain the momentum, scope, and multiple dimensions of that movement, it is necessary to go beyond the availability of favorable conditions, and even beyond the important consideration of practical expectations that must have loomed large at least in the minds of the Muslim patrons. Islamic religion had introduced a new ideology with sweeping and universalist claims. Already during the swift expansion of Islamic conquests, that ideology had come into direct contact with a large variety of creeds (Jewish, Christian, Zoroastrian, Mazdian, Manichaean, etc.) with which it inevitably collided and against which it had not merely to defend but—much more importantly—to define itself, often in terms borrowed from its opponents. The result was a huge intellectual ferment, centered especially in multicultural Iraq, to which the movements of Islamic theology, philosophy, and science owed their birth.

Or, should we not rather say, more accurately, that the creation of these fields of thought represented the *responses* of so many groups of individuals to aspects of what was, in the context of religion and politics and power and the variety of competing ways to salvation, a very complex and live intellectual atmosphere? With regard to the creation of the tradition of science and philosophy in Islam, I am tempted to borrow an obsolete term, *aspecting,* in order to refer to the way in which individuals in a given culture *aspect* another culture as they direct their gaze to the other from their own location. Aspecting in this sense is conditioned both by the interests, aspirations, and aptitudes of the aspecting individuals and by the accessible aspects of the viewed culture, that is to say, the aspects that happen to be disclosed to them by the accidents of history or by their further, determined effort. Thus, for example, through the Ṣābians of Ḥarrān, Muslim thinkers were able to view facets of Hellenistic thought that might not have been available to them by way of the Christian theologians, who had already made their own choices from their own standpoints. And, as has been plausibly suggested, the absence of Greek literature and Greek historiography from the translated corpus may be attributable to a lack of acquaintance or serious interest on the part of the Christian translators.[4] In a similar way, the twelfth- and thirteenth-century Arabo-Latin translations in northern Spain were understandably limited to the types and the levels of the learning that was currently available in Al-Andalus, with all the features of that learning that had undoubtedly been shaped by a combination of circumstances peculiar to Al-Andalus.[5]

The scholars of eighth- and ninth-century Iraq looked east to Persia and India and west to Greece and especially to Alexandria. Both looks deeply affected the character of Arabic science, especially in mathematics and astronomy, in both of which we find combinations of identifiable elements from the East and the West. But it was the westward gaze that proved most enticing and, as it turned out, most consequential. Some years ago I used the term *appropriation* to characterize the attitude of Muslim scholars and patrons who made it their business to get hold of and make their own what they called "the sciences of the ancients," an expression that clearly revealed a sense of distance in time between themselves, as "the moderns" (*al-muta'akhkhirūn*), and the appropriated legacy of "the ancients" (*al-mutaqaddimūn*), even as the appropriators set about gaining possession of the ancient legacies with great energy.[6] Without aiming here to unfold the full meaning of that sense of distance (which has frequently been misinterpreted and misused in modern scholarship), let me indicate briefly how it was understood and evaluated by some of those who promoted or participated in the appropriation drive of the ninth century.

I will not expand, not even briefly, on the role of the Abbasid caliph al-Ma'mūn, whose contributions as a patron of astronomical research and as the one who turned the library of Greek philosophical sciences collected by his immediate predecessors into an organized center of translation are well known. But I must emphasize the significance of his deep involvement in a theological dispute that prompted him to initiate the "inquisition" against the conservative opponents of the Muʿtazilite school of *kalām* (or "theology") that he favored. It is natural to think of a connection between al-Ma'mūn's support of the Muʿtazilite emphasis on the role of reason in elucidating religious dogma and his championing

[4] Paul Kunitzsch, "The Two Movements of Translation into and from Arabic and Their Importance in the History of Thought" (in Arabic), *Zeitschrift für Geschichte der Arabisch-Islamischen Wissenschaften,* 1987/1988, *4:*93–105.

[5] Guy Beaujouan, "The Transformation of the Quadrivium," in *Renaissance and Renewal in the Twelfth Century,* ed. Robert L. Benson and Giles Constable (Cambridge, Mass.: Harvard Univ. Press, 1982), pp. 463–487, on p. 465: "Having found its principal Arabic sources in Spain, medieval Western science obviously reflected, at the start, the choices of the Hispano-Moslem culture. The lack of interest in abstract mathematics, the predominance of astronomy and astrology in the early translations, the relatively late date of the Arabo-Latin versions of Aristotle's natural philosophy, the failure to use important works by eastern Arabic scholars: all are explained by the evolution of Arabic science in the Iberian peninsula, with its peculiarities of history and geography, its particularist pride within the Islamic world, its conditioning by the oppressive domination of the Malikite fakihs." In addition to the references cited by Beaujouan see A. I. Sabra, "The Andalusian Revolt against Ptolemaic Astronomy: Averroes and al-Biṭrūjī," in *Transformation and Tradition in the Sciences: Essays in Honor of I. Bernard Cohen,* ed. Everett Mendelsohn (Cambridge: Cambridge Univ. Press, 1984), pp. 133–153, rpt. in Sabra, *Optics, Astronomy, and Logic* (Aldershot: Variorum, 1994); Julio Samsó, *Islamic Astronomy and Medieval Spain* (Aldershot: Variorum, 1994), esp. Chs. 1 and 12; and R. S. Avi-Yonah, "Ptolemy vs. al-Biṭrūjī: A Study of Scientific Decision-Making in the Middle Ages," *Archives Internationales d'Histoire des Sciences,* 1985, *35:*124–147.

[6] A. I. Sabra, "The Appropriation and Subsequent Naturalization of Greek Science in Medieval Islam," *History of Science,* 1987, *25:*223–243, rpt. in Sabra, *Optics, Astronomy, and Logic.*

of Greek science and philosophy. We are even given a hint of a contemporary interpretation of such a connection in a story that circulated in tenth-century Baghdad, if not earlier. The story reported a dream that the caliph was alleged to have had, in which none other than the pagan Aristotle appeared as an instructor to the "overawed" Commander of the Believers. In answer to al-Maʾmūn's question about the nature of "the good" (*al-ḥasan*), the Greek philosopher identifies its authoritative source in the first place with reason (*al-ʿaql*), in the second place with "the (religious) law," and in the third place with what is accepted by "the majority of people" (i.e., the majority of legal scholars). In one version of the dream Aristotle ends by urging his Muslim student to uphold the unity of God (*wa ʿalayka biʾl-tawḥīd*)![7] Affirmation of the unity of God (*tawḥīd*) was the first article in the Muʿtazilite credo, and it was in the name of divine unity, *as interpreted by reason* (*ʿaql*), that the Muʿtazila maintained the doctrine, which al-Maʾmūn and his two immediate successors sought to impose on the conservative legists, that the Qurʾān—as the speech of God—had not existed from all eternity but was created. The suggested connection between Greek rationalism and Islamic Muʿtazilism is impossible to overlook.

Turning now from the patron al-Maʾmūn to leading Muslim intellectuals of the period, one quickly detects an attitude of openness and gratefulness to the recently imported wisdom, mingled with a feeling of high optimism and a certain trust in humanism that was quite pronounced. The major Arabic prose writer of the period, and one of the most influential figures in the history of Arabic literature, al-Jāḥiẓ (d. 869), was also the leader of a prominent branch of Muʿtazilism named after him (al-Jāḥiẓiyya) and a reader of Aristotle. In one of his most important works, the monumental *Book of Animals*, which owed much to Aristotle and which he dedicated to the famous *wazīr* Ibn al-Zayyāt (who served the Abbasid court in the reigns of al-Maʾmūn's successors, al-Muʿtaṣim and al-Wāthiq [833–842, 842–847]), he thanks "the ancients" (i.e., the Greeks) for their considerable intellectual contribution to the scanty possessions of the people of his time and place and culture. He wrote, in a passage that is characteristic for its utter lack of inhibition or ambiguity: "Our share of wisdom would have been much reduced, and our means of acquiring knowledge weakened, had the ancients not preserved for us their wonderful wisdom, and their various ways of life, in writings which have revealed what was hidden from us and opened what was closed to us, thereby allowing us to add their plenty to the little we have, and to attain what we could not reach without them."[8]

An exact contemporary of al-Jāḥiẓ, and arguably the single most important figure in this phase of appropriation, the celebrated Muslim philosopher, scientist, and mathematician al-Kindī (d. ca. 870), was a member of the Arab nobility (his grandfather, we are told, was *amīr* of al-Kūfa in southern Iraq). As a tutor to a son of Caliph al-Muʿtaṣim, he was much closer to the Abbasid court than al-Jāḥiẓ. In his work *On First Philosophy*, dedicated to al-Muʿtaṣim, we find strong acknowledgment of the accomplishments of ancient Greece that is combined with the assertion of truth as the universal good that must be sought out wherever it may be found, in addition to a clear concept of the growth of knowledge as a process of accumulation that requires the cooperative effort of different peoples and successive generations.[9] These were the deep convictions of a true devotee,

[7] Ibn al-Nadīm, *al-Fihrist*, ed. Gustav Flügel (Leipzig, 1871–1872), Vol. 1, p. 243 (also in an edition by Riḍā Tajaddud [Tehran, 1971], pp. 303–304).

[8] Al-Jāḥiẓ, *Kitāb al-Ḥayawān*, ed. ʿAbd al-Salām Hārūn, 2nd ed., Vol. 1 (Cairo: Maktabat M. al-Bābī al-Ḥalabī, 1965), p. 85.

[9] Alfred Ivry, *Al-Kindī's Metaphysics: A Translation of Yaʿqūb ibn Isḥāq al-Kindī's Treatise "On First Philosophy"* (Albany: State Univ. New York Press, 1974); Al-Kindī, *Fī al-sināʿa al-ʿuẓmā*, ed. ʿAzmī Ṭāhā al-Sayyid Aḥmad (Cyprus: Dār al-Shabāb, 1987); and Al-Kindī, "Fī ḥudūd al-ashyā wa rusūmihā," in *Rasāʾil al-Kindī al-falsafiyya*, ed. M. Abū Rīda, Vol. 1 (Cairo: Dār al-Fikr al-ʿArabī, 1950), pp. 163–180.

and they carried with them an entire Hellenic or Hellenistic world view, a distinct concept of wisdom in both the theoretical and practical senses (also borrowed from the Greeks), and a sense of mission on the part of the author to do his utmost to disseminate the ancient and especially Greek heritage in his milieu. Both this world view and this concept of wisdom, as well as the universalist character of al-Kindī's mission, are recognizable Hellenistic themes. But we all know better than to push the continuity thesis too far—especially not in the direction of blatant essentialism: to look on al-Kindī as a latter-day Hellene or Hellenist would be as helpful as the affirmation, found in Spengler's *Decline of the West,* that the Pantheon built by Hadrian in the second century was the first mosque ever. Such a view of al-Kindī and other Muslims who shared his outlook in the ninth century would tear them away from their unique position in history and from the role they consciously and freely chose to assume in the context of their own culture. Al-Kindī's mission, as he understood it, clearly reveals his adoption of the humanistic theme, his avowed intent being, as he said, "the perfecting of our [human] species."[10] But as one who lived in an Arabic culture, his immediate role, as he also conceived it, was to introduce and, he hoped, to convert his Arabic-reading contemporaries (*ahl lisāninā:* the people who speak in our tongue) to the Greek wisdom that had captivated him—a task that he actually undertook to achieve by producing a huge number of Arabic epitomes and adaptations, with supplementary clarifications and additions when necessary, of a very large number of Greek disciplines of science and philosophy. This was a preposterously optimistic project for anyone to envisage; but not only was al-Kindī able to carry it out, whether singlehandedly or with the help of others, it proved to be remarkably successful—so successful, in fact, as to make him truly worthy of the reputation he quickly gained as one of the founders of the Arabic tradition in philosophy and science.

With these three pivotal figures in mind (and there are others that can be brought into consideration along with them), I am inclined to portray that crucial phase in the *appropriation* process as the accomplishment of individuals who experienced the intersection, at a certain place and time, of three major movements at work—namely those of Hellenism, Arabism, and Islamism. By viewing ninth-century Baghdad as a point of intersection in the manner I have tried to outline, I hope to render useless such questions as whether the scientific tradition then being established was essentially Arabic, Islamic, or Greek[11] and to open the way for empirical research aimed at identifying the actual workings of these movements as revealed in the writings and records of those individuals who experienced and responded to them. No doubt there is reason here to celebrate the creative genius of a moment in the history of civilization, but my overriding aim is to direct attention to the complexity and richness of that extraordinary moment and away from the misleading "essences."

THREE LOCI OF SCIENTIFIC ACTIVITY IN ISLAM: THE COURT, THE COLLEGE, AND THE MOSQUE

Abundant richness also awaits the empirical investigator into the subsequent course of scientific activity in Islamic civilization. What I have just described is the attitude of certain

[10] "idh kunnā ḥirāṣan ʿalā tatmīmi nawʿinā": *al-Falsafa al-ūlā,* in *Rasā'il,* ed. Abū Rīda, Vol. 1, p. 103. For the translation see Ivry, *Al-Kindī's Metaphysics,* p. 58. See also Spengler, *Decline of the West* (cit. n. 2), Vol. 1, p. 211.

[11] Ernest Renan, *L'islamisme et la science,* 2nd ed. (Paris: Calmann Lévy, 1883).

individuals who were favorably disposed to the imported knowledge and who played an active part in bringing about what later proved to be a long-lasting tradition. But, of course, a few first steps, even significant ones, might not have been followed by others in the same direction or with the same determination and vigor. And, indeed, there were other, contemporary individuals and groups whose markedly different or contrary attitudes and intellectual commitments, then and in later periods, did much to shape the course of Arabic science. As I turn now to later developments, our story acquires a degree of complexity that I cannot hope to convey in a lecture. But I shall try to give you a sense of it. It is a complexity that further illustrates the usefulness of the methodological concept of locality, and it will lead me at the end of my talk to pose the general question of whether, and in what sense, Arabic science should be investigated as a single enterprise. The story unfolds in at least three distinct but by no means isolated loci whose different structures and modes of operation and interaction have yet to be explored from the standpoint of our subject. These loci are the college or institution of higher learning, the royal or princely court, and the mosque. I shall arbitrarily ignore the hospital as a result of excluding medicine from my present account.[12]

I will begin with the college, and in order to bring you closer to an unfamiliar situation I will start with two general observations by way of comparison with more familiar episodes. The first is this: as far as science and philosophy are concerned, the European Renaissance of the sixteenth century was in part a reaction, which became more pronounced in the seventeenth century, against patterns of thought and argument associated with medieval "scholasticism." In Islamic history, events followed the reverse order: the "renaissance" (if that is the right word) came first, in the ninth and tenth centuries, and a form of scholasticism followed, though not immediately and not uniformly in all parts of the Muslim world. My second observation points to another contrast between Islam and medieval Europe that is crucially important but more difficult to describe briefly. In Islam, whether in ninth- and tenth-century Baghdad, eleventh-century Egypt and central Asia, twelfth-century Spain, thirteenth-century Marāgha in northwestern Iran, or fifteenth-century Samarkand, the major scientific work associated with the names of those who were active at those times and places was carried out under the patronage of rulers whose primary interests lay in the practical benefits promised by the practitioners of medicine and astronomy and astrology and applied mathematics. Many of these practitioners were also prolific writers on "philosophy," a mode of thinking known by the Arabicized term *falsafa* and characterized to a large extent by a mixture of Aristotelian and Neoplatonic doctrines and forms of argument—the kind of mixture we find, for example, in the works of al-Kindī, al-Fārābī, and Avicenna. In those circumstances science and "philosophy," or *falsafa,* were secular activities that were practiced, developed, and propagated as rational inquiries completely independent of any religious authority—which, of course, did not prevent the proponents of this autonomous, self-legitimizing mode of thinking from offering their own rationalistic (i.e., Hellenic) interpretations of religious doctrines such as revelation or prophecy or providence and of religious institutions such as law. After all, *falsafa* was an all-embracing world view that claimed the right to scrutinize and account for everything within the sphere of human experience, including religious experience. In

[12] The only "excuse" for this exclusion is to avoid further complicating an already complex picture. No story of Arabic/Islamic science or philosophy is complete without taking into account their relation to medical thought, the effect of medical patronage, the place of science and natural philosophy and logic in the institutions of medical education and practice, and the role of Galenic writings as a source of ideas and doctrines that shaped the minds and attitudes of Islamic philosophers *and* scientists, as well as physicians.

one case among the prominent devotees of *falsafa,* that of al-Kindī, a serious compromise was made by renouncing the Greek doctrine of the eternity of the world in favor of *creatio ex nihilo.* But unlike most of their Christian counterparts in medieval Europe, Islamic philosophers (the self-styled *falāsifa*) and philosopher-scientists in that Greek sense were not "theologians" or members of religious orders. The one major exception, of sorts, is the twelfth-century Andalusian Averroes, who came from a celebrated traditional family of Mālikite jurists and practiced the Mālikite version of Muslim law as a judge, but who nevertheless believed himself to have inherited the mantle of Aristotle. I shall come back to him later.

Now, Islamic "theology," or what has come to be known in Western scholarship by this name, followed a different course—with important, indeed far-reaching consequences for the development of both science and *falsafa.* It began to make a conspicuous appearance in the eighth century (the second Islamic century), well before the patronized translation movement got under way, as the activity of spontaneously sprouting groups of Muslim intellectuals in the urban centers of Baṣra and Baghdad who immersed themselves in probing discussions (*kalām:* speech, discussion, argument), obviously driven by their interest in current religious and political controversies. They gradually developed somewhat varied and sometimes seemingly indecisive but sophisticated and sophisticatedly argued doctrines concerning a comprehensive array of subjects that ranged all the way from God and his relation to man and the world, to questions of epistemology and morality and political leadership, to subtle and difficult speculations about the ultimate constitution of all created being, which they characteristically proposed to understand in atomistic terms. The *falāsifa* later dubbed these *kalām* practitioners as religious apologists, thereby seeking to downgrade their rivals or, if possible, to circumscribe their role by subsuming their enterprise under the authority of *falsafa.*[13]

There can be no doubt that the early practitioners of *kalām,* the *mutakallimūn,* were influenced by a multiplicity of pre-Islamic traditions in ways that still remain mostly veiled in obscurity. But whatever the remote sources of their ideas, and despite their fundamental concern with the elucidation and critique of religious tenets, it is my conviction (which I share with a few others) that the discourse of the early "school" of the Muʿtazila, the one favored by al-Maʾmūn, and of the later and subsequently dominant Ashʿarites, represents an important turn in the history of philosophical thought—one that gave rise to new styles of thinking that seriously challenged the Aristotelianism and Neoplatonism of *falsafa* by proposing a thoroughgoing atomism that viewed the world as a creative process. It was this new philosophy, the "philosophy of the *kalām,*" as Harry Wolfson called it,[14] that, in the Ashʿarite version, later found its way into the colleges of higher education, the so-called *madrasas* that ultimately spread wide and far over the Islamic world as endowed or charitable institutions, having been first introduced on a large scale in the eleventh century by the Sunnite Saljūqs in Iraq and Persia as part of a political agenda and in response to the Ismāʿīlī propaganda emanating from Fāṭimid Egypt and Syria.

The *madrasas,* it should be noted, were first conceived of as primarily schools of law,

[13] A. I. Sabra, "Science and Philosophy in Medieval Islamic Theology: The Evidence of the Fourteenth Century," *Z. Gesch. Arab. Islam. Wiss.,* 1994, *9:*1–42.

[14] Harry Austryn Wolfson, *The Philosophy of the Kalam* (Cambridge, Mass./London: Harvard Univ. Press, 1976). The view of *kalām* as religious apologetics has been prevalent in modern literature but is currently being revised. See Richard M. Frank, "The Science of *Kalām,*" *Arabic Sciences and Philosophy,* 1992, *2:*7–37; and Sabra, "Science and Philosophy in Medieval Islamic Theology."

an emphasis that they retained throughout their history.[15] But, as creations of private endowments, they generally enjoyed a degree of informality that allowed for a variable range of intellectual pursuits that depended on local circumstances and the interests of their professors and their sponsors. Many, perhaps a large number, of the *madrasas* included some teaching in arithmetic, algebra, astronomy, and logic as part of the intellectual equipment of the practicing jurist, along with the indispensable disciplines of language and rhetoric. *Kalām,* as a study of the "fundamental tenets of religion" (*uṣūl al-dīn*), performed the dual function of supplying a superstructure of theory for the rest of the "religious sciences" as well as a substitute for Greek metaphysics and natural philosophy.

Combining these two general observations should now help us to appreciate the following result. The sciences of the Greeks, which were first welcomed in Islam along with Greek theories of cosmology and epistemology and metaphysics (be they Hellenic or Hellenistic), eventually came to be confronted in the *madrasas* by a homegrown religious philosophy that claimed to develop viable alternatives to the Greek paradigms. That much we can say in light of what we already know. But there is no end to the questions that have yet to be examined. When and where and in what circumstances did that process take place? Was it one process or many? How did *kalām* manage to subdue *falsafa,* given the tentative beginnings of the former and the originally strong and full-grown nature of the latter? How much, if anything, of *falsafa* doctrines and forms of argument was absorbed into *kalām?* How did *falsafa* react to the assault of the *mutakallimūn,* given that *falsafa* continued to pursue its activities long after the advent of the *madrasas,* at least in some parts of the Muslim world? And—the question of special importance for the historian of science—what was the effect of the *kalām* point of view on the dissemination and development of scientific disciplines such as cosmology and astronomy, about which the *mutakallimūn* had a lot to say as an integral part of their own world view?

None of these questions can be answered a priori. They are all empirical questions that require empirical research. Some of my colleagues, I am happy to say, are now beginning to tackle them in earnest. Others are reluctant to embrace them, being afraid of the possible danger of diverting too much attention from the vast quantities of scientific texts that remain to be edited and analyzed. The skeptics have a point, and I share their concern. But this is not an either/or matter. As for the argument that "we do not yet know enough to ask the big questions," my answer is this: it is only by attempting to formulate appropriate questions that can be fruitfully examined in light of what we now know that we make it possible for others to come up with deeper and more probing questions in the future. We do not know much (that is for certain), but the day when we know "enough" will never come. On the other hand, by altogether abandoning all programs of full-fledged historical research, we only tempt others to fill the vacuum with easy and useless essentialist generalizations.

The *madrasas* were not, therefore, in general a locus where scientific research was promoted for its own sake, but one in which science was interpreted and judged and

[15] The literature on the *madrasa* is growing rapidly, but see especially the wide-ranging studies of George Makdisi: "Muslim Institutions of Learning in Eleventh-Century Baghdad," *Bulletin of the School of Oriental and African Studies,* 1961, *24:*1–56 (rev. by A. L. Tibawi, *ibid.,* 1962, *25:*225–238), and *The Rise of Colleges: Institutions of Learning in Islam and the West* (Edinburgh: Edinburgh Univ. Press, 1981). Most of the recent publications are concerned with the Mamlūk period (for which there is abundant material), but they have not yet directed special attention to the question of the place of science and philosophy in the *madrasas.* This must be due in part to the long-held assumption that science and philosophy had no place in the *madrasas,* which is not quite true, as is now being realized.

presented to a large group, indeed the vast majority, of educated Muslims. When we talk of scientific advance in Islam, whether in mathematics, astronomy, or experimental science, we usually have in mind the contributions of men who carried out their work outside of the *madrasas* with the support of kingly patrons—men like al-Khwārizmī, al-Khayyāmī, Ibn Yūnus, Ibn al-Haytham, al-Bīrūnī, al-Ṭūsī, and al-Shīrāzī. As always, there were exceptions—sometimes important ones, especially in the later period—and in some cases the systems of patronage and *madrasa* even seem to merge, either partially, or even completely, as for example under Ulugh Beg's initiative in fifteenth-century Samarkand. Such exceptions should direct our attention to the significant overlap among all three loci of court patronage, college, and mosque. But the general picture of scientific advance in Islam as a patronized activity holds.

What we are still far from understanding is how patronage worked in contexts that obviously differed from one center of activity to another. Let me give an example with reference to two or three situations about which we know a little more than about the rest. Consider again the pagan sage, Aristotle, who was said to have inspired Caliph al-Maʾmūn and with whom the Mālikite Muslim jurist Averroes identified himself. In both cases we encounter acceptance of the authority of an alien thinker and of the intellectual values he represented. In both cases a Muslim theological context was involved: Muʿtazilite *kalām* in the earlier episode, and the religious ideology of the Almohad dynasty in the later Andalusian episode. But the theological commitments were different in the two cases, with rather different implications for the place and authority of the law, and so were the patterns of patronage. The differences were, moreover, enhanced by an emphatic self-conscious Andalusian identity vis-à-vis the rest of the Islamic world. It is not surprising that Averroes, as the intellectual who responded most powerfully to this Andalusian situation, should develop a new theory of religious authority, a totally negative attitude to *kalām,* and a new valuation of his adopted Aristotelianism that set him against his Peripatetic predecessors *and* earlier mathematicians in the eastern part of the Islamic world. It is, therefore, with reference to this special context in all its geographical, political, and intellectual particularities that we should try to gain a historical understanding not only of Averroes' rebuttal to the attack launched by the Eastern al-Ghazālī against the *falāsifa,* but also of his important and explicit divergences from fellow-*faylasūfs* like Avicenna or from a recognized mathematical authority like Ibn al-Haytham and of his ultimate rejection of the hitherto-dominant Ptolemaic astronomy.[16]

Similar, or greater, contrasts with regard to attitudes, patterns of relations between patron and client, and implications for the practice of science are what we should expect to find as we turn our attention to later periods. For example, soon after the Mongol Īlkhāns had captured Baghdad in 1258, thus bringing the Abbasid caliphate to an end, their leader Hūlāgū was persuaded to establish an observatory at Marāgha in northwestern Iran, an event that marked the beginning of one of the longer-lasting and important episodes in the history of Arabic science. Most of the scholars who were soon to be gathered at Marāgha were Muslims (there are reports of one or more Chinese scholars). The man put in charge of organizing the new enterprise was Naṣīr al-Dīn al-Ṭūṣī, a Persian from Ṭūs with serious interests in Shīʿite theology and Avicennan philosophy. (See Figure 1.) At age fifty-five when he surrendered himself to the Mongols upon their capture of the Ismāʿīlī stronghold of Alamut, he was already famous as a scholar and known to the Mongols as a competent astronomer and astrologer. Two other scholars, both of them Sunnīs, were brought over

[16] Sabra, "Andalusian Revolt against Ptolemaic Astronomy" (cit. n. 5).

***Figure 1.** Naṣīr al-Dīn al-Ṭūṣī and colleagues at the Marāgha observatory. (MS F1418, early sixteenth century, Istanbul University Library.)*

from Syria. One of them, Mu'ayyad al-Dīn al-'Urḍī, had a reputation as a building engineer and instrument-maker. The other, the mathematician Muḥyī al-Dīn al-Maghribī al-Andalusī, was captured by the Mongols during their campaign in Syria in 1259–1260. He managed to save his life only by presenting himself to his captors as an astrologer who could be of use to "the lord of the earth," the great Mongol Khān.[17] The Mongol patron of the astronomical enterprise at Marāgha was not Muslim, and his interest in the work of the observatory was undoubtedly astrological. The immediate goal was to produce a new

[17] The story of al-Maghribī's capture is told by Barhebraeus (in *Ta'rīkh mukhtaṣar al-duwal*), who heard it from al-Maghribī in Marāgha; see A. I. Sabra, "Simplicius's Proof of Euclid's Parallels Postulate," *Journal of the Warburg and Courtauld Institutes,* 1969, *32:*1–24, esp. pp. 13–14, rpt. in Sabra, *Optics, Astronomy, and Logic* (cit. n. 5). On al-'Urḍī see George Saliba, ed., *The Astronomical Work of Mu'ayyad al-Dīn al-'Urḍī: Kitāb al-Hay'a* (Beirut: Center for Arab Unity Studies, 1990), pp. 27–30.

***Figure 2.** Giant armillary set up in the open air at the sixteenth-century Istanbul observatory. (MS 1404, late sixteenth century, Istanbul University Library.)*

set of astronomical tables, based on new observations, of the type that had been used by Arabic astronomers and astrologers for planetary predictions since the time of al-Maʾmūn. The *Zīj-i Īl-Khānī,* as the new Persian handbook came to be known, was not completed until 1272, after Hūlāgū's death. But in the meantime, and for several decades afterward, the scholars at Marāgha and nearby Tabrīz were able to pursue their individual interests in theoretical astronomy and in various branches of mathematics. It was in this singular situation that the aporetic research in planetary theory, which was initiated by Ibn al-Haytham before the middle of the eleventh century and had attracted the attention of a few individual scholars in Asia and Syria, first found a sustaining atmosphere; and it was from here that this type of research later spread further east, south, and west, where it was carried on in different terms or with different emphases as scholars with different commitments responded to changing contexts. (See Figure 2.)

The phenomenon of the mosque as a significant locus of scientific activity came into being at about the same time as the establishment of the Īlkhānid rule in Iran and Iraq, being associated with the ascension to power of the Mamlūks in Egypt and Syria in 1250. Most of what is now known about this important phenomenon is due to David King, whose work over the last twenty years has been responsible for putting this phenomenon on the map of Arabic science.[18] It was the longest-lasting episode within the tradition of Arabic/Islamic science, having continuously endured in some of the major mosques all the way up to the nineteenth century, and it possessed interesting features that distinguish it in several ways from court patronage and the *madrasas,* the two other loci of consequence.[19]

Though primarily a place of worship, the mosque, from its inception, and as distinguished from the *madrasa* that was sometimes attached to it, often served as a forum for propagation and discussion of subjects related to Arabic language, grammar, and rhetoric, as well as the vital issues of law, religion, and politics. Through the introduction, apparently for the first time under the Mamlūks, of the office of *muwaqqit,* the timekeeper in charge of regulating the times of the five daily prayers, a place was created for the utilization of one form of scientific knowledge in a permanent religious institution.

Strictly speaking, it would be wrong to consider the *muwaqqit* a "professional" astronomer. His institutional role in the mosque was not to pursue the goals of astronomy as these had been defined and elaborated by Arabic astronomers since the ninth century but, as is clearly indicated by his status title, to offer reliable guidance to his local Islamic community with regard to definite religious observances (mainly prayer times) as specified by religious law. This function the *muwaqqit* nonetheless performed in his capacity as an expert in what was called "the science of reckoning time" (*ʿilm al-mīqāt*) by means of exact astronomical computations, and this distinguished him from the traditional *muʾezzin* (the man who called for prayer), who relied on traditional prescriptions. The main task of the *muwaqqit* was therefore to use the methods of spherical astronomy in order to construct tables, usually computed for a certain locality or latitude, that would enable anyone who could operate a simple observation instrument (such as an astrolabe or a quadrant) to determine the time of day or night from the altitude of the sun or a star. A *muwaqqit* might also possess the skill to construct such instruments. And some distinguished *muwaqqits* in the thirteenth and fourteenth centuries accomplished the impressive feat of providing universal solutions of timekeeping problems (indeed, all problems of spherical astronomy) for all latitudes. One *muwaqqit,* the fourteenth-century Ibn al-Shāṭir (d. ca. 1375), who was attached to the Umayyad mosque in Damascus, ventured into the area of theoretical astronomy to produce the most complete solution to the equant problem, which Ibn al-Haytham had forcefully pointed out as a threat to the principles of Ptolemaic astronomy and which was diligently pursued by mathematical astronomers in the thirteenth century. These were all accomplishments that must be regarded as accomplishments in astronomy proper, regardless of their institutional setting. And the same can be said of other equally impressive investigations aimed at determining the direction of Muslim prayer. As in the case of timekeeping, these investigations also culminated in universal solutions for all latitudes.

And yet it is noticeable, as King has pointed out, that the legal scholars and interpreters

[18] David King, "The Astronomy of the Mamluks," *Isis,* 1983, *74:*531–555; and King, *Astronomy in the Service of Islam* (Aldershot: Variorum, 1993).

[19] The distinction has to be maintained despite occasional or even frequent overlappings, as, for example, when a local ruler was responsible for the appointment of a favored professor in a *madrasa.*

of the religious law continued to apply the simpler considerations, based on observations of twilight and horizon phenomena (rising and setting) or of shadow lengths, leaving alone the sophisticated mathematical treatises, which they "generally considered to be too complicated or even completely irrelevant."[20] This is not really surprising (mathematical precision need not be considered a prerequisite of religious piety!), but it does render problematic the concept of the *muwaqqit*'s mathematical work as "service to religion." On the other hand, the theoretical triumph of Ibn al-Shāṭir in planetary theory does not seem to have elicited serious attention from other contemporary *muwaqqits,* and this appears to be the result of the fact that their institutional position did not demand or encourage theoretical ventures for their own sake. Such paradoxes may simply reflect our present, inadequate knowledge of the circumstances in which a new institutional structure brought together mathematical and religious interests. But whatever the correct understanding of these paradoxes might be, it would be gratuitous to regard the work of the *muwaqqit* in aiding religious ritual as constituting "the essence of Islamic science" (as King puts it)[21] or even as the most revealing aspect of scientific activity in Islam. To propose such a view may have the advantage of highlighting the *uniqueness* to Islamic civilization of a certain emphasis on some programs of astronomical research. But the disadvantages of this proposal are also glaringly conspicuous. It disregards the full extent of scientific research in Islam, and it ignores the characteristic complexity of Islamic civilization itself by neglecting the variety of religious attitudes with regard to the status, the function, and the value of scientific knowledge. And it might appear to equate "Islamic science" with narrowly circumscribed programs that largely developed within the confines of an institution with no commitment to "science" as such, and this alone would tend to obstruct or prejudice vital questions about scientific practice in Islam by identifying a single locus of activity with a widespread and extremely complex phenomenon. And, of course, it would again open the way into the trap of essentialism.

To come finally and very briefly to the general question formulated earlier: Was Arabic science one or many? A similar question has sometimes been asked with reference to Islamic art, where manifest varieties of styles and functions are displayed in the artifacts and architectural monuments of the vast Islamic world. As far as science is concerned, it seems to me that important considerations lead us to say that we have to do with a single, unitary *tradition.* These are considerations of language, which—for science and philosophy—was for the most part one language (Arabic), and of Islamic religion as an ever-present point of reference though not always a point of departure, in addition to considerations of the dominance of dynastic rules over large regions for extended periods of time and the remarkable ease of movement and communication all through the Muslim world—a feature itself connected to religion and law and language. And, with regard to communication of learning, we must also keep in mind that crucial Chinese invention, paper, which took the whole Islamic world by storm from the moment of its appropriation in the middle of the eighth century.

One example will have to suffice as an illustration of what I mean by these remarks. Writing in fourteenth-century Damascus, Ibn al-Shāṭir linked his studies in theoretical astronomy to those of earlier mathematicians, four of whom had worked in thirteenth-century Marāgha, one in eleventh-century Egypt, two in twelfth-century Spain, and two

[20] David King, "Science in the Service of Religion: The Case of Islam," in *Astronomy in the Service of Islam* (cit. n. 18), Ch. 1, p. 246.

[21] *Ibid.*, p. 245.

who originated in thirteenth-century Syria and North Africa. All had written in Arabic, the language in which Ibn al-Shāṭir also wrote. The example is representative of situations that existed before and after Ibn al-Shāṭir, though there were cases in which Persian and Turkish were the languages of composition, especially in later times. But once we direct our attention to *situations,* as distinguished from *tradition,* our picture and our problematic will change with every case, as we turn from one set of circumstances to another in which individual choices are made with reference to specific problems proposed by specific contexts. Not, of course, that tradition and individual response are separable: on the contrary, the former provides an inseparable part of the intellectual context in which the other must take place. When I started to write this talk I hoped to be able to illustrate and perhaps also to characterize in some general terms the interplay of tradition and individual response with reference to one or two episodes of Arabic science. In the end I am forced to leave that subject for another time and place.

The Solar and Lunar Theory of Ibn ash-Shāṭir

A Pre-Copernican Copernican Model

By Victor Roberts

1. Introduction.

THE purpose of this note is to present the solar and lunar theory developed by the Damascene astronomer, Ibn ash-Shāṭir (1304–1375/6 A.D.) [1] in his *Kitāb Nihāyat as-Sūl fi Taṣḥīḥ al-Uṣūl* (A Text of Final Inquiry in Amending the Elements). Since he dispenses completely with the Ptolemaic eccentric deferent and introduces a second epicycle, both his solar and lunar models are non-Ptolemaic. What is of most interest, however, is that his lunar theory, except for trivial differences in parameters, is identical with that of Copernicus [2] (1473–1543).

Ptolemy assumed a circular path for the sun, whereas the orbit of Ibn ash-Shāṭir's sun deviates slightly from circular motion. The major fault of the Ptolemaic lunar model is in its exaggeration of the variation in lunar distance. The major Copernican contribution to the lunar theory lies in the elimination of this Ptolemaic fault.

Five copies of Ibn ash-Shāṭir's work are known to exist, four of which are at the Bodleian Library. They are MSS Marsh 139, Marsh 290, Marsh 501, and Hunt 547. The fifth copy is at Leiden, Arabic MS 1116.[3]

In the preparation of this paper a microfilm of the first of the above-mentioned copies was used, made available through the courtesy of the Keeper of Oriental Books at the Bodleian. This manuscript of sixty-four folios was copied in 768 A.H. (1366 A.D.), i.e., within the lifetime of the author. In the introduction Ibn ash-Shāṭir claims that in his treatise he is presenting a planetary theory of finest achievement, and that he has traced and explained all astronomical uncertainties in his *Ta'līq al-Arṣād* (Comments on Observations). Unfortunately, this work is apparently non-extant.

According to our source the inventor's name is Abū'l-Ḥasan 'Alī ibn Ibrāhīm ibn Maḥmalain al-Humām ibn Muḥammad ibn Ibrāhīm 'Abdur-Raḥmān al-Anṣārī. Other versions are: Abū'l-Ḥasan 'Alī bin Ibrāhīm bin Muḥammad al-Muṭ'im al-Anṣāri,[4] and 'Alā' ud-Dīn 'Alī bin Ibrāhīm Muḥammad.[5] In all

[1] Suter, H., *Die Mathematiker und Astronomen der Araber und ihre Werke*, Abhandlungen zur Geschichte der Mathematik . . . (Leipzig, 1900), X. Heft, p. 168.

[2] This was pointed out by Professor O. Neugebauer.

[3] Brockelmann, C., *Geschichte der arabischen Litteratur* (2d ed., Leiden, 1943), vol. 2, p. 156.

[4] Suter, *op. cit.*, p. 168.

[5] Brockelmann, *loc. cit.*

versions he is known by the name of Ibn ash-Shāṭir. Besides being an astronomer, he held the post of *muwaqqit* — one who determines the time of prayer — at the Umayyad Mosque in Damascus.

2. *The Solar Theory.*

The mean sun is represented by the center a (Figure 1) of an epicycle moving ing from west to east above the horizon with a daily mean motion of 0;59,8,9,51,-46,57,32,3° [6] around the center of the universe E on a deferent of radius 1,0;0.

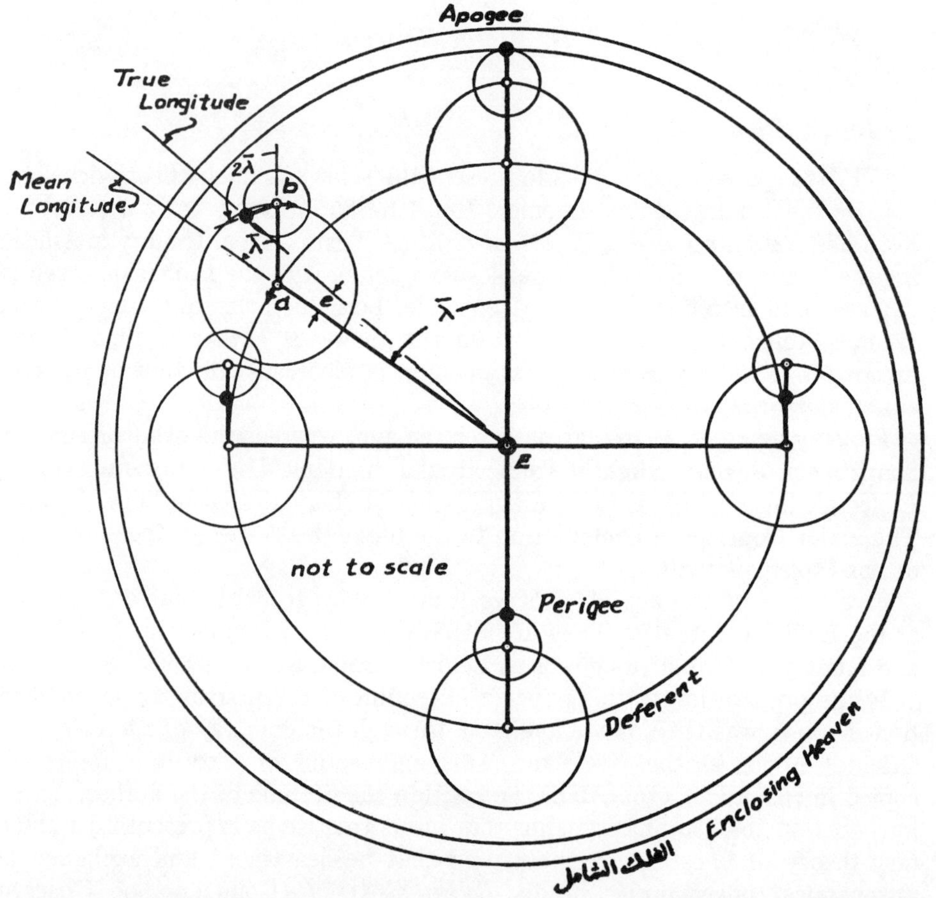

Fig. 1.

The radius of this epicycle, in length 4;37 units, rotates with respect to the deferent radius, at the same speed but in the opposite sense, carrying the center b of a second epicycle of radius 2;30. This latter epicycle, with the true sun of radius 0;10 on it, rotates with respect to the first epicycle radius

[6] The value is expressed in the sexagesimal system. For instance, in our notation 1,2;32,59, stands for $1\cdot60^1 + 2\cdot60^0 + 32\cdot60^{-1} + 59\cdot60^{-2}$.

at double the daily mean motion and in the same direction. The *enclosing heaven* (*al-falak ash-shāmil*), has an inner radius equal to the sum of the three given radii, or 1,7;17. Its width is 0;43, apparently in order to round off its outer radius to an integer (1,7;17 + 0;43 = 1,8;0). The enclosing heaven moves from west to east at the rate of 0;0,0,9,51,46,51° per day. This is the motion of the solar apogee, and comes to one degree in sixty Egyptian years of 365 days each. Figure 1 shows the initial and subsequent positions of the true sun. With the given values of the parameters the solar distance varies between 52;53 and 1,7;7. Accordingly, the apparent solar diameter has a mean value of 0;32,32°, varying between extreme values of 0;29,5° and 0;36,55°.

The maximum solar equation *e* given by Ibn ash-Shāṭir is 2;2,6°, and occurs at a mean longitude λ of 97° or 263°, measured from apogee. It disappears at the apogee and perigee. These values have been verified by independent computation by the present author. The following method, expressed in modern symbolism, is given in the manuscript for computing *e* and the solar distance, *ρ*.

$$7;7 \cos \bar{\lambda} = A, \text{ say,}$$

(The manuscript has sin $\bar{\lambda}$, an obvious error.)

$$1,0;0 + A = B, \text{ say}$$

(The manuscript says add *A* if 97° 7′ > $\bar{\lambda}$ > 262° 53′, otherwise, subtract *A*. Such special rules to take care of negative functional values are typical of mathematical exposition prior to the use of signed numbers. The limits of $\bar{\lambda}$ represent the mean longitudes which yield the mean solar distance.)

$$2;7 \sin \bar{\lambda} = c, \text{ say,}$$

therefore $$\rho = \sqrt{B^2 + c^2},$$

and $$\sin e = \frac{c}{\rho}.$$

With simple geometry this method can easily be verified by noting that 7;7 and 2;7 are respectively the sum and difference of the radii of the epicycles.

The only observation mentioned in this connection is one made in Damascus on Tuesday, the first day of the year 701 of Yazdigerd, or 24 Rabīʿ al-Awwal 732 A.H. (25 December, 1331). On the mid-day of this date, according to the inventor, the mean sun and its apogee showed longitudes of 5,9;0° and 1,19;12° respectively.

Ibn ash-Shāṭir gives no motivation for his adoption of two epicycles, and it is difficult to see how his model constitutes an improvement over the Ptolemaic one.

3. *The Lunar Theory.*

The orbit of the moon is inclined at an angle of 5° to the parecliptic [7] (of radius 1,9;0), and moves from east to west at the rate of 0;3,10,38,27° per day.

[7] The parecliptic (*al-falak al-mumaththal*) is defined as a circle in the plane of the ecliptic with the earth as its center.

The deferent, of radius 1,0;0, moves from west to east around the center of the universe (R in Figure 2) at the daily mean latitudinal rate of 13;13,45,39,40°. Therefore, the mean value of the moon is the sidereal motion, or 13;10,35,1,13°, which is the difference between the two values given above. The first epicycle radius ab, in length 6;35, rotates with respect to Ra at a daily anomalistic rate a of 13;3,53,46,18°, and in a direction opposite to the mean motion. To explain the "second inequality" of the moon, Ibn ash-Shāṭir places the true moon on a second epicycle bc of radius 1;25 whose daily motion from west to east is equal to double the elongation ϵ or 24;22,53,23° (double the difference between the mean sun 0;59,8,20°, and the mean moon). It is of interest to compare the radii of the epicycles with the Copernican values, which are 6;34,55,12 and 1;25,19, 12 [8] respectively. Figure 2 illustrates the motion of the moon, starting from a mean conjunction in which R, a, and b are collinear.

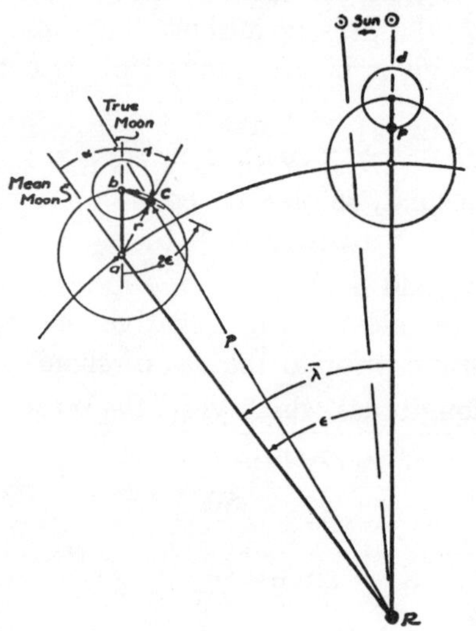

FIG. 2.

As a consequence of the resultant motion, the moon will always be in perigee p of the second epicycle at mean syzygies, and in its apogee d at quadratures. Hence, the apparent epicycle of radius 5;10 (difference of the epicyclic radii) at the syzygies accounts for the equation of the center, and the gradual increase in its apparent radius (maximum 8;0) as it approaches the quadratures accounts for the evection. The maximum values of the first and second inequalities are 4;56 and 2;44 respectively. Thus Ibn ash-Shāṭir has retained the ancient sum of the two inequalities, which is obtained at the quadratures when the line from the earth R to the true moon is tangent to the apparent epicycle.

It follows from the given values of the parameters that the lunar distance

[8] *De revolutionibus*, IV, 8. See also Dreyer, J. L. E., *History of the Planetary Systems* (Cambridge, 1906; republished in 1953 under the title, *A History of Astronomy from Thales to Kepler*), p. 336.

varies between 54;50 and 1,5;10 at the syzygies, and between 52;0 and 1,8;0 at the quadratures. Accordingly, the apparent lunar diameter, with a mean value of 0;32,54,33°, varies between extremes of 0;29,2,15 and 0;37,58,20°. (The last two values are computed from the first, the only one given in the text.)

The manuscript gives the following method, expressed in modern symbolism, for computing *e*, the lunar equation.

$$1;25 \sin 2\epsilon = \text{A, say,}$$
$$1;25 \cos 2\epsilon = \text{B, say,}$$
$$6;35 + \text{B} = \text{C, say.}$$

(The manuscript says, subtract *B* if $90° \geqq 2\epsilon \geqq 0°$ or $360° \geqq 2\epsilon \geqq 270°$, otherwise add.)

$$\sqrt{\text{C}^2 + \text{A}^2} = \text{r, the radius of the apparent epicycle.}$$

Therefore $\sin \gamma = \frac{\text{A}}{\text{r}}$, or $\gamma = \sin^{-1} \frac{\text{A}}{\text{r}}$.

$$5;10 \sin (\alpha + \gamma) = \text{D, say,}$$
$$5;10 \cos (\alpha + \gamma) = \text{E, say,}$$
$$1,0;0 + \text{E} = \text{F, say.}$$

(Add *E* if $90° \geqq (\alpha + \gamma) \geqq 0°$ or $360° \geqq (\alpha + \gamma) \geqq 270°$, otherwise, subtract *E*.)

$$\sqrt{\text{D}^2 + \text{F}^2} = \rho.$$

Therefore $\sin e = \frac{\text{D}}{\rho}$, where *e* is the first inequality. When 5;10 is replaced by 8;0, *e* will represent the sum of the two inequalities.

4. *Concluding Remarks.*

The same manuscript gives full parameters and descriptions of the models of the other planets. These await further study. Suffice it to say at the present that these, like the solar model, are quite different from those of Copernicus.

The question which immediately arises is that of possible influence of the earlier astronomer on the later. There seems to be no way at present of resolving the problem either in one way or the other.

Ibn ash-Shāṭir's work was fairly well-known in the Near East, as numerous copies and recensions of his astronomical handbook (*Az-Zīj al-Jadīd*) and other works testify.[9] There is no evidence for any of his books having been translated into Latin,[10] and it is possible that the two individuals hit upon the same solution independently. On the other hand, Copernicus utilized the work of other Islamic scientists, notably that of Thābit bin Qurra. *Allāhu aʿlim* (God knows better).

[9] Dr. D. J. Price suggests that Ibn ash-Shāṭir may have been the maker of two astrolabes preserved in Paris collections, Nos. 6 and 142 (in Gunther, *The Astrolabes of the World*, Oxford, 1932).

[10] For instance, Ibn ash-Shāṭir is not mentioned by F. J. Carmody in his *Arabic Astronomical and Astrological Sciences in Latin Translation. A Critical Bibliography*, University of California Press, 1956.

The True Place of Astrology in the History of Science

By Lynn Thorndike

THE true place of astrology in the history of science is a vast subject with countless ramifications which it would take a long time to pursue and many pages to relate. My present purpose is to emphasize a single point, but one which is the most important and fundamental of all. Briefly stated, it is that, during the long period of scientific development before Sir Isaac Newton promulgated the universal law of gravitation, there had been generally recognized and accepted another and different universal natural law, which his supplanted. And that universal natural law was astrological.

One may, of course, employ the words "astrology" and "astrological" in a variety of senses and meanings, and they have often been so employed in times past. One may restrict the terms to the art of, or attempt at, prediction of the entire life of any and every human individual from the hour of his nativity, and dismiss this as an idle superstition. But nativities were only a single branch or department of astrology in the broad sense, and their validity depended upon the underlying assumption that the entire world of nature was governed and directed by the movement of the heavens and the celestial bodies, and that man, as an animal naturally generated and living in the world of nature, was also naturally under their rule. Astrological medicine was an obvious sequel of this assumption and was free from the objection that prediction of man's fate violated freedom of the will.

Huillard-Bréholles interpreted the word, "astrology," in a narrow sense, when he wrote of the emperor, Frederic II, "He believed in astrology to the last hour of his life," although he accounted for that emperor's penchant for astrology by his insatiable thirst for knowledge.[1] But he went on to say that Frederick's astrological speculations were only a branch of mathematics, which larger subject he cultivated "with a sort of passion," and loved to relieve the cares of government by study of the exact sciences.

Aristotle and the astrologers [2] were agreed that the heavens and celestial bodies were incorruptible and unchanging, that their motion was regular and

[1] *Historia diplomatica Friderici Secundi,* VII (1859), dxxxi–dxxxiii.

[2] The friction between the faculties of arts and of theology at Paris, of which Professor Mary Martin McLaughlin treated in a paper on "Medieval University Masters and Ideas of Intellectual Freedom," at the sixty-ninth annual meeting of the American Historical Association on 28 December 1954, was not merely a matter of difference between a rational and a dogmatic mode of approach, or of how the works of Aristotle should be taught, but a more fundamental divergence between the supernatural and this universal natural law which was astrological. Of the 219 opinions which were condemned at Paris in 1277, a large number was astrological in one or another sense of that word. This would have proved to be even more true in the case of universities like Padua and Bologna, where the arts course was pre-medical rather than leading on to the study of theology, as it did at Paris.

eternal, circular and perfect, that they were a superior fifth essence distinct from the inferior elemental world of our earth and its atmosphere, where generation and corruption and alteration prevailed, changing seasons of the year and daily weather permutations, geological formation and dissolution, changing sea-coasts and river-beds, earthquakes, volcanoes and floods, growth and decay of vegetation, birth and death of animals.

Moreover, all these processes corresponded to the movements and positions of the heavenly bodies. The stars were not themselves affected by their movement and light, since they were eternal and incorruptible. But their motion and rays had to have some effect, and an outlet for this vast store of energy was found in our elemental world, whose changes and fluctuations and variations paralleled the shifting pattern of the eternal heavens and the varying projection of rays of light and influence thence. Furthermore, the earth was thought of as the center and bottom of the universe, and it was fitting that inferiors should be ruled and governed by superiors — the heavenly bodies. As the concluding sentence of a philosophical manual in a manuscript of the fifteenth century [3] put it, "Certain therefore is the influence of the heavens on these inferior bodies." The heavenly bodies were responsible for all meteorological phenomena in the region or regions of air; they caused the tides in the sphere of water; they affected all generation upon earth, and sometimes produced spontaneous generation without sexual intercourse; they formed gems and strata of rocks underground, so that it was the most logical thing in the world to relate seven metals to the seven planets, as was done for centuries.

A special appeal might be made to celestial influence in order to explain the operation of occult virtue, such as that of the magnet. But this should not blind us to the fact that celestial influence was the general and universal cause of all inferior nature. "Coelum per formam suam agit," quoth Duns Scotus.

The most scientific form of weather prediction was astrological, and an abundant literature on the subject, derived from India as well as from the Arabic, instructed one how to forecast rain, winds, and changes of temperature from the movements of the planets — direct, retrograde or stationary — through the twelve signs of the zodiac and the twenty-eight mansions of the moon.

In other cases, it might be possible to arrive at roughly correct and satisfactory conclusions from the observation of terrestrial phenomena as immediate causes, but in such cases too the heavens were the remote and primary cause. This had been the law of nature and the fate of the ancient Stoics. This general rule of the heavens over the world of nature was held by Albertus Magnus and Thomas Aquinas, by Tycho Brahe and Kepler.

More than thirty years ago I wrote with regard to Albertus Magnus:

> This general law that the world of nature and of life on this earth is governed by the movements of the stars is expressly repeated again and again in Albert's works, and its truth is assumed even oftener.[4]

I may add now two further specific illustrations from his works of the applica-

[3] Vatican Barberini Latin MS 343, fols. 66–80.

[4] *A History of Magic and Experimental Science*, II (1923), 583.

tion of this law. The figure of a plant imitates the pyramid of celestial light,[5] and, although living bodies are more highly organized than inanimate things, they also deviate less from the norm and are more closely related to celestial nature than are other material bodies. By their equality of complexion they participate by analogy in the principle of celestial life. That celestial principle has more power over the matter of the body than the body's own corporal form has. Hence the influence of the stars has more effect upon animate creatures than their corporal nature has, and moves them to forms which are not of the elements, nor are their compounds consequences of the elements, but the celestial force works in them not one but many impressions, none of which their corporal nature could effect.[6] Even in his *Summa theologica*, in discussing the sixth day of creation and the divine command, "Let the earth bring forth the living creature," Albert asked how that could be, "since the power to produce animals is not in the earth, but, according to astronomers, is in the heavens." Ultimately Albert reached the conclusion that, when God said, "*Producat terra*," He designated the earth merely as the material principle from which animals are formed, but that the active principle is the heavens.[7]

This rule of the heavens should be kept constantly in mind by every student of the history of science before Newton in evaluating any aspect of scientific or, for that matter, human activity. Most past critics of what they were pleased to call astrology never questioned this assumption, which was its very basis. They began to do so in the sixteenth and seventeenth centuries, as the distinction between earth and heavens was gradually obliterated, and this led on to Newton's new development. But to hold that natural or physical law was a concept then first inaugurated is to do astrology, in the sense that I have been using the term, and also the previous period, a grave injustice.

As an example of this common misconception may be quoted a passage from Bidney's book on Spinoza in the year 1940:

> The characteristic features of Renaissance philosophical thought were the revival of the atomic theory in conjunction with mathematics, and the consequent stress upon the primary importance of the efficient cause as a principle of definition and explanation. Beginning with the experiments of Galileo and his success in measuring the rate of acceleration of falling bodies, the conviction gradually arose that nature as a whole was subject to immutable laws which could be mathematically expressed — a process which Newton brought to a brilliant culmination. In other words, there were universal laws of becoming, laws of change which governed the correlation between particular events. This was an entirely new and seminal idea which was destined to revolutionize the natural sciences.[8]

But surely astrology had for centuries before believed in universal laws which governed particular events, and that nature as a whole was subject to immutable laws.

[5] *De veget. et plantis*, II, i, 5.

[6] *Ibid.*, I, i, 1.

[7] *Op. cit.*, II, xi, 61.

[8] David Bidney, *The Psychology and Ethics of Spinoza*, 1940, pp. 13–14. Edgar Zilsel, "The Genesis of the Concept of Physical Law," *Philosophical Review*, 61 (1942), 245–79, dismissed astrology in two or three sentences such as (p. 252) "In the astrological literature of late antiquity sometimes laws of nature are mentioned in an entirely magical sense."

Modern historians of science have been strangely blind to the fundamental and universal importance of this sweeping, all-inclusive hypothesis, that all operations of the inferior world of nature spring from and are controlled by the eternal movement of the incorruptible celestial bodies. Strangely blind also to its supreme significance in the appreciation and comprehension and evaluation of pre-Newtonian scientific thought and activity! A few, however, have come close to the realization of this. George Sarton, in one passage in his monumental *Introduction to the History of Science*, comes close to recognizing it, when he writes of Roger Bacon:

> He was deeply interested in astrology, in which he believed implicitly. . . Astrology itself he divided into two kinds; a legitimate kind, and a forbidden one, mere superstition. This was very sound, in spite of the fact that much if not all of Bacon's legitimate astrology was nothing but error and superstition from our point of view.[9]

In another passage Sarton says:

> However, Bacon's thoughts did not dwell so much on statics as on dynamics. He was pondering on the nature of force, especially on force or action at a distance. Curiously enough, these thoughts, earnest as they were, were partly astrological. For among the forces or actions considered by him were light and gravity, but also astrological influences, the reality of which were beyond doubt. How were these astrological influences transmitted across the open spaces? How were these distant causalities propagated? It was very remarkable to ask such questions, and we must not blame him for failing to solve them.[10]

On this second passage our comment must be that there was nothing curious about Bacon's thought; it was quite logical and natural. Nor was it very remarkable to ask such questions. Rays of light and of astrological influence were propagated from the same heavenly bodies in accordance with the natural law of the universe, the causation of inferiors by superiors.

Mark Graubard has recently written of astrology not as a past superstition but as a fossil science.[11] Yet he does not seem to realize that it depended upon a once generally accepted hypothesis of universal natural law.

Such a law is tacitly or virtually admitted in many a passage in A. C. Crombie's *From Augustine to Galileo: the history of science, A.D. 400–1650.* At page 9 he speaks of "astrological interpretation of the world of nature as a whole"; at pages 35–36, he says, " 'Celestial virtue' was admitted as a cause by nearly all the Latin writers of the 13th century"; at page 40 we read, "From the four elements were produced, under the influence of the celestial spheres, plants, animals and man himself"; at page 95 he states that Ristoro d'Arezzo "was very astrological. He attributed the elevation of dry land above the sea to attraction by the stars"; at page 122 we learn that Arabic writers and Latin schoolmen "generally supposed that such forces" (of generation) "were supplied by celestial virtue." But the basic generalization towards which these scattered passages point is not brought out, and sometimes astrology is spoken of as a superstition which the more enlightened made fun of.[12]

[9] *Op. cit.*, II, 956.

[10] *Ibid.*, II, 763–64.

[11] *Astrology and Alchemy, Two Fossil Sciences*, New York, 1953.

[12] *Op. cit.*, pp. 8–9.

Bally-hooers for the uniqueness of modern science have further repeatedly insisted that mathematical method was then first applied to the investigation of natural phenomena. Francis Bacon, with his inductive method, has been sharply criticized for his neglect of mathematical method. I hold no brief for poor old Francis. But give me leave to say that astrology certainly applied mathematical method to natural phenomena. Geometry and trigonometry, sines and chords, were needed to trace the courses and to find the positions of the planets. The *Almagest* of Ptolemy was primarily mathematical in its method, nor were its mathematics of an easy sort. In astrology, furthermore, timing was of utmost importance. Tables were carried not merely to minutes and seconds, but thirds and even farther. The projection of rays and problems of reflection and refraction were as much the concern of astronomy and astrology as they were the concern of perspective and optics. Regiomontanus has been called the first modern mathematician. But he was not the first to draw up tables of astrological directions. All the observations and measurements of the stars, all the devising of astronomical instruments, all the calculation of astronomical Tables, were motivated by the urge to implement the universal law of nature in the service of humanity.

In previous periods astrology and astronomy had regarded themselves as far superior to physics and mechanics. Newton's *Principia* destroyed the age-long distinction between superiors and inferiors. The astrological wings of high-flying science melted; it fell back to earth and became terrestrial. But this change came about slowly. The Jesuit, Menestrier, writing in 1694 seven years after the publication of Newton's *Principia*, while attacking engraved astrological images and the belief in the stars as particular causes, yet had no doubt that they were general causes of all inferior nature. Moreover, the new Newtonian physics did not affect biology and medicine, which had to wait for the doctrine of evolution, whereas the astrological hypothesis had been a universal law for them too.

Herbert Dingle, in a paper read last year at the Seventh International Congress of the History of Science at Jerusalem, upon "The Essential Elements in the Scientific Revolution of the Seventeenth Century," held that the object of all science is

> . . . the detection of regularities in our experience and the expression of those regularities in the simplest and most comprehensive rational form.

Such regularity had been the aim, the ideal, the boast and the assumption of astrology. Dingle further argued that

> It is reasonable to look forward to the time when all the sciences will be fused into a single science, using a single set of concepts, though that time may yet be distant.

If we may look forward to such a unified science, we may also look backward to it, before the seventeenth century, to the time when all change and all phenomena in the elementary world were believed to be governed by the radiations of the eternal and incorruptible, yet moving, celestial spheres.[13]

[13] It should hardly be necessary for me to state that I do not share this belief or have any faith in the methods of astrology. I further recognize that the rule of all terrestrial nature by the movements and influence of the heavenly bodies was not a law in the modern sense of being mathematically demonstrable. My purpose has been merely to emphasize that this belief was generally held by scientists and by mankind at large for centuries, and should be taken into account by every historian of that period.

This paper was first presented at the annual meeting of the History of Science Society, on 30 December 1954.

The Stars and Human Sexuality: Some Medieval Scientific Views

By Helen Lemay

Astrology was an integral part of the Arabic scientific corpus which was so decisive in transforming Western intellectual life in the twelfth and thirteenth centuries. Along with purely Aristotelian writings, astrological works expanded considerably the sources of scientific information available to the Latin West. Astrology was referred to by the Arabs as the "science of the stars," and this science had as its object the examination of all aspects of man in his social context, including the study of the human reproductive system. Sexual behavior is given extensive consideration in Arabic astrological treatises.

Western scientists did not merely receive Arab astrology as part of the baggage that contained Aristotelian science: they eagerly embraced the study of the stars and their effects on man. Arabic astrological works were widely translated and copied; Latin authors composed their own summaries of astrology; and astrologers played an important part in the cultural life of the West. For example, Michael Scot served as court astrologer for Frederick II, and Giovanni Pontano, the renowned Italian humanist, composed a major astrological treatise. Both of these men treat human sexuality in detail in their writings.

Astrology played a particularly significant role in the field of medicine. Astrological works were formally incorporated in the curricula of medieval medical faculties from at least the fourteenth century, and even from the moment of their reception in the West they were studied avidly by doctors. The pseudo-Ptolemaic *Centiloquium,* for example, an astrological handbook designed primarily for the use of physicians, was translated into Latin six times during the twelfth century by many of the central figures of this period, and over 150 manuscripts have been identified so far.[1] Physicians wrote treatises on

[1]The *Centiloquium* (*Kitāb al-Tamara* or *Liber fructus*) was composed by pseudo-Haly (Abū Ja'far Aḥmad ibn Yūsuf ibn Ibrāhīm al-Kātib al-Tūlūni) in the late 9th or early 10th century in Cairo. It was used in Paris (1348), Bologna (1405), and Cracow (mid 1400s). See Heinrich Denifle and Emile Châtelain, *Chartularium Universitatis Parisiensis, Auctarium* (Paris: Delalain, 1889), Vol. I, p. 235; Carlo Malagola, *Statuti delle Università e dei Collegi dello Studio Bolognese* (Bologna: Zanichelli, 1888), pp. 276 and 214; Aleksander Birkenmajer, *Études d'histoire des sciences en Pologne, Studia Copernicana,* 1972, *4*:469–495. The *Centiloquium* is being studied at present by a research group at the CUNY Graduate Center under the direction of Richard Lemay. See Lemay, "Origin and Success of the Kitāb Tamara of Abū Ja'far Aḥmad ibn Yūsuf ibn Ibrāhīm from the Tenth to the Seventeenth Century in the World of Islam and the Latin West," *Aleppo University, Proceedings of the First International Symposium for the History of Arabic Science, 5–12 April 1976* (Aleppo: University of Aleppo Institute for the History of Arabic Science, 1978), pp. 91–107. The author of the *Centiloquium* will henceforth be referred to as Aḥmad ibn Yūsuf.

ISIS, 1980, *71* (127–137)

astrology, and astrologers wrote on medical topics or incorporated medical findings in their astrological writings. Michael Scot's treatise on physiognomy, which is really an account of human generation, and Albubather's treatment of fetal development in his *Book on Nativities* are among the more prominent examples of the overlap of the two disciplines in society and of the interest medieval astrologers had in the human reproductive system.

Astrologers, doctors, and natural philosophers of the period share a basic assumption that the stars influence terrestrial events. They are not, however, in accord over the degree of this influence: medical and natural-philosophical writers grant a much more limited power to the planets than do the astrologers. Many medical works do not mention extraterrestrial control over human sexuality, and Latin treatises on human generation limit their discussion of stellar influence to the section on embryological development. For example, the thirteenth-century author of the *De secretis mulierum* and the fifteenth-century humanist Peter Candidus Decembrius state that during the fifth month of fetal life Venus is in control and takes on responsibility for the configuration of the exterior members—ears, nose, fingers, and sexual organs. Venus bestows upon the individual the capacity for physical desire, which is ruled over by the sign of the zodiac Scorpio.[2] When these authors discuss sterility, determination of virginity, or other aspects of adult sexual conduct, however, they turn first to terrestrial phenomena for their explanations. Hot or cold complexions, foreign matter in the urine, diet, and climate are some of the factors that physicians and natural philosophers list as determinants of sexual behavior.

The astrologers' approach is very different. When seeking to understand particular events, astrologers look first at the configuration of the heavens, and they are very precise about how information can be derived from the sky. Astrological writings are divided into a number of branches, each serving a specific purpose in the practitioner's attempt to understand human sexual behavior. The *Introductoria* or Judgments are general works, setting down rules and principles; the Nativities apply these rules to individuals, explaining the proper manner for casting horoscopes; the study of Interrogations uses the stars to find the answers to questions; and Elections concentrate on choosing the right moment to act. Thus, a general writing will tell us that the sperm receives its power directly from the firmament;[3] a Nativities will spell out precisely the position of Venus, Mars, and the moon that will cause large genitals or sickness in these organs;[4] a treatise on Interrogations will give instructions on how to determine if a woman is a virgin;[5] and a work on Elections will indicate when an individual is most likely to experience great desire.

This neat classification of the astrological sciences is found in the Arab writers and was retained by Western astrologers. The two groups disagreed, however, on the role actually played by the stars in causing an individual's sexual makeup. The Latin astrologers in particular were extremely wary of falling into the trap of determinism, and long apologies for astrology such as the one found in the beginning of Guido Bonatti's thirteenth-century Latin astrological treatise provide us with clear evidence of con-

[2]Pseudo-Albertus Magnus, *De secretis mulierum* (Lyons: Quadratus, 1580), pp. 52, 66; cf. Petrus Candidus Decembrius, *De genitura hominis* (Rome: Stephan Plannck, c. 1487). On the authorship of the *De secretis mulierum* see Lynn Thorndike, "Further Consideration of the *Experimenta, Speculum astronomiae,* and *De secretis mulierum* ascribed to Albertus Magnus," *Speculum,* 1955, *30*:413–443. On Peter Candidus Decembrius see Lynn Thorndike, *History of Magic and Experimental Science,* Vol. IV (New York: Columbia University Press, 1934), pp. 398–399.

[3]E.g., 'Alī ibn Ridwān, *Commentarium super Ptolomaei Quadripartitum* (Venice, 1493), fol. 54rb.

[4]E.g., Albubather, *De nativitatibus* (Venice, 1492), fol. c3.

[5]E.g., Zahel, *De interrogationibus* (Venice, 1493), fol. 130rb.

temporary theological opposition to this science.[6] Although Bonatti does offer the argument from authority that astrology was used by the holy fathers and therefore should not be shunned, he stresses planetary control over natural phenomena in his discussion of determinism. Treating stellar influence as a kind of physical force that acts upon men, Bonatti states that the stars are noble, perfect beings who rule over creatures in the corruptible world of the four elements. But they do not rule a man or a horse or a ship in the same way that a man would; rather they govern them by controlling the elements of which corruptible beings are composed.[7] Like other Latin astrologers, he accepts the Arabic astrological interpretation of Aristotle's statements in his *Treatise on Generation and Corruption* and his *Meteorology* that inferior things are subject to superior ones.[8] Thus, on the subject of human sexuality, we find Bonatti repeating the statement of the ninth-century Arab astrologer Albubather that the stars act directly on the seed, at times corrupting the womb or semen, as well as Ahmad ibn Yusuf's assertion in the *Centiloquium* that masculine planets cause males, and feminine planets females.

Arabic discussion on the question of determinism is more developed. Although the problem of the freedom of the will versus divine power was not so easily solved for Muslim theologians, the astrologers resolved the issue by concluding that the heavens control human sexuality. They considered that the stars were a natural force directly dependent upon God, just as individual human beings also derived their powers directly from Allah. Thus, according to the Arab astrologers, one's sexual behavior is to a large degree beyond his control. The astrological texts underline this notion of divine omnipotence by constantly repeating *inshā'llāh,* "if God wills it," rendered into Latin as *si deus voluerit,* or *nutu Dei.*

The *Centiloquium,* composed in Cairo in the late ninth or early tenth century, provides us with perhaps the clearest formulation of Arabic astrological thinking on the issue of determinism. Verba 1 and 2 state that the stars are animate beings, of a higher order than man. They operate according to immutable laws, and it is up to man to discern these laws, to reach the level of perception where his soul becomes one with the intelligences.[9] The relation between man and these higher powers is one of relative subservience, however. Although an individual can come to know the rules by which the stars operate, he can do nothing to change them. At best he can work around the inevitable, choosing favorable moments for his acts, or fabricating amulets in the shape of a constellation to capture a certain position of the sky and prolong its effect.[10]

Yet not all the astrological writings state as clearly as the *Centiloquium* that the stars are actually the causes of an individual's sexual makeup. Many Arabs and their Latin interpreters are more ambivalent. Abu Ma ʿshar, for example, the ninth-century Arab whose *Introductorium maius in astronomiam* was the most influential astrological writing for Western scientific development, alternates between this position and the stance that the heavens are merely indicators of what already exists; indeed, the Latin manuscripts of the *Introductorium* clearly divide into two groups, one using *facere* and the other *significare quod.*[11]

[6]See Thorndike, *History of Magic,* Vol. II, p. 830.

[7]Guido Bonatus, *Decem tractatus astronomie* (Venice, 1506), fol. 147.

[8]For a listing of Aristotelian statements on this point see Richard Lemay, *Abu Ma'shar and Latin Aristotelianism in the Twelfth Century* (Beirut: Publication of the Faculty of Arts and Sciences of the American University, 1962), p. 60, n. 4.

[9]Aḥmad ibn Yūsuf, *Centiloquium,* trans. Plato of Tivoli, Verba 1 and 2 in *Liber Quadripartiti Ptolomaei* (Venice, 1484), fol. 7a. See also ed. Venice, 1493, fol. 107.

[10]*Ibid.,* 1493 ed., Verba 5 and 9, fols. 107va, vb.

[11]See Richard Lemay, "Fautes et contresens dans les traductions arabo-latines médiévales: *L'Introductorium in Astronomiam* d'Abou Ma'shar de Balkh," in *Colloques: Textes des rapports, XII*ᵉ

The same disagreement is found among the Latin interpreters of Alchabitius, the tenth-century summarizer of Abū Ma ʿshar. John of Seville, the twelfth-century translator of Alchabitius' treatise on astrological judgments, renders the Arabic by referring to the planet Venus as the *significatrix* of women and wives and mothers; a later commentator, however, perhaps John of Saxony, adds the remark that Venus has an influence over the humors and especially the genitalia and private parts of women.[12] Thus, although Arabic and Latin astrologers proceed from the assumption that ultimate responsibility for physical phenomena lies in the supercelestial world, they never entirely agree upon the scope of this planetary influence. The differences among astrologers, however, as the examples below show, are less important than the wider gap between astrologers and medical writers; the former, regarding their discipline as heuristic, look for signs, while the latter, regarding theirs as potentially therapeutic, look for cures.

VIRGINITY

A question that preoccupied Arabs and Latins, astrologers, doctors, and natural philosophers, was how to determine if a woman's virginity had been corrupted. This aspect of sexuality obviously possessed a social significance that was not limited to the Middle Ages, and the topic was consequently well developed in the scientific sources. For astrologers it fell logically under the category of Interrogations, the use of the planets as indicators of all things hidden to men, and Zahel in his ninth-century treatise *De interrogationibus,* translated from the Arabic by John of Seville, includes a section on how to determine from the stars whether a woman is a virgin.[13] Abenragel's eleventh-century treatise on Judgments, translated into Latin about 1256, is also concerned with this problem. If the ascendent and the moon are in fixed signs, he tells us, the woman is a virgin and free from all suspicion; if they are in mobile signs, she is corrupt. The planets can tell us as well whether she was propositioned, what she answered, and whether she really sent her beguiler away.[14]

Latin astrologers were just as eager to make use of the stars to inquire into a girl's purity. John of Saxony in the fourteenth century copies Abenragel practically word for word on this question.[15] The fullest treatment by a Latin astrologer of how to detect the corruption of virginity, however, occurs a century earlier in Guido Bonatti. Bonatti was determined to establish by looking at the stars exactly how far toward final penetration by the male the woman had progressed; while the Arabs gave indications of how one could make judgments, he worked out the full details on his own. Thus he states that if the lord of the ascendent and the moon are in fixed signs and the angles are in mobile signs, she was tempted and did not give in. If the moon is joined to Venus or Mercury, another woman delivered the man's message but she still remained steadfast in virtue. If the ascendent and the moon are in mobile signs she was tempted and is still thinking it over. Further refinements of these planetary positions indicate that

Congrès international d'histoire des sciences, Revue de Synthèse, 3rd Ser., Nos. 49–52 (Paris: Albin Michel, 1968), p. 122.

[12] "Additio. Venus habet influentiam super . . . humores superfluitates maxime circa genitalia et mulierum pudibunda." Alchabitius, *Astronomie iudiciarie principia,* trans. John of Seville (Lyons, 1520), fol. xli. On Alchabitius see Manfred Ullmann, *Die Natur- und Geheimwissenschaften im Islam* (Leiden/Cologne: Brill, 1972), p. 332.

[13] "Questio de muliere si sit virgo vel corrupta." Zahel, *De interrogationibus,* fol. 130rb. On Zahel see Ullmann, *Die Natur- und Geheimwissenschaften im Islam,* pp. 309–312.

[14] Abenragel (Albohazen), *Liber in iudiciis astrorum* (Venice, 1485), fol. 27. See Carlo A. Nallino, *Raccolta di scritti editi e inediti* (Rome: Istituto per l'Oriente, 1944), Vol. V, pp. 340–341; Francis J. Carmody, *Arabic Astronomical and Astrological Sciences in Latin Translation* (Berkeley/Los Angeles: University of California Press, 1956), pp. 150–154.

[15] Iohannes Saxonii, *Commentarium,* in Alchabitius, *Astronomie iudiciarie principia,* fol. xxxii.

the man touched the female subject's private parts; that he placed his organs on top of hers, causing her to believe mistakenly that her virginity had been lost; or that he ejaculated without penetration.[16]

Bonatti's discussion of virginity gives us evidence of how the practitioner of the science of the stars could have functioned as an arbiter of sexual disputes within the society. Having set down detailed information about how to determine the truth about a girl's corruption from the configuration of the heavens, he continues his exposition by giving instructions on how this information should be presented. If you find that a woman is truly a virgin, Bonatti declares, tell the questioner that nothing can be said against her. However, he continues, corruption of virginity is a complex matter. A female's virginity can expire without her engaging in sexual intercourse with a man. She can be devirginized by her own hands or someone else's; titillation with the fingers can cause her to ejaculate (or emit her own seed) and therefore leave the condition of purity. In this case, the astrologer can rationalize that although the girl may have laughed or played with another, she was not known by a man in the way she has been accused, and therefore he should not tell his client the full truth. He should say simply that the stars indicate that she is excused before her interrogator, because if he gives the full information, the girl will be taken as corrupt, as if she had actually been with a man.[17]

This sympathy for the female is also present occasionally in the medical works, and physicians sometimes provide "cures" to make a corrupt woman appear to be a virgin. In attempting to determine whether a woman has lost her virginity, doctors and natural philosophers concentrate most on the urine—its color and the manner in which it is passed are two of the more important indicators. Physical examination also plays a role in finding the answer to this question, although it is less emphasized than urinalysis. Whether the reason for this is that women did not easily submit to this sort of scrutiny, or that gynecological examination was practiced so infrequently that doctors were unable to develop sufficient expertise to make an accurate assessment on this topic is a matter of conjecture.[18]

[16]Bonatus, *Decem tractatus astronomie,* fols. 203v–204.

[17]*Ibid.,* fol. 204ra.

SEXUAL PROCLIVITIES

Arab astrologers took a great deal of interest in the varieties of sexual expression, and in their discussions of sexual proclivities they grant a large degree of control to the stars. Indeed, they are convinced that the precise manner in which the individual will perform the carnal act is fixed at the moment of conception. They rely on the outline in the basic work of Greek astrology, Ptolemy's *Quadripartitum,* composed approximately 150 A.D. in Alexandria, which was an important source for Arabic understanding of the science of the stars. Ptolemy states that if the luminaries are alone in the masculine signs (for certain parts of the sky are considered to be masculine and others feminine), women possess an exceeding amount of unnatural quality. This stellar configuration increases the virility of their souls and makes them lustful for unnatural congresses, when they act as if their female friends were their wives. Depending upon the position of Mars and Venus they may perform these acts either secretly or openly. The stars can also cause women to become harlots, and they can produce similar aberrations in men.[19]

[18]The author of the *De secretis mulierum* states, e.g., that the signs of virginity can be feigned, including among them a tight vaginal opening. He asserts that the only sure sign is found in the urine. Pseudo-Albertus Magnus, *De secretis mulierum,* p. 119.

[19]Ptolemy, *Tetrabiblos* III 14, ed. and trans. F. E. Robbins (London/Cambridge, Mass: Harvard University Press, 1971), pp. 368–369.

In his eleventh-century commentary on the *Quadripartitum,* ᶜAlī ibn Riḍwān, relying most probably on the pseudo-Ptolemaic *Centiloquium,* interprets Ptolemy to mean, when he states that women act as if their female friends were their wives, that they rub one another as if they were men.[20] Similarly detailed accounts of the effect of the stars on sexual proclivities are found in Albubather, Ibn Ezra, Albohali, Abenragel, and Alchabitius. For example, Albohali tells us that the planets can cause a woman to be sexually aggressive;[21] Ibn Ezra relates that Venus in the twelfth or seventh house leads to daily sexual intercourse;[22] Albubather asserts that certain signs produce sodomy, excessive sexual appetite, impotence, sterility, and hermaphroditism. (Albubather is so detailed on stellar influences that he even informs his reader when the individual will suffer hemorrhoids or bad breath.[23]) According to Abenragel, it is even possible to satisfy one's curiosity about a neighbor's sex life by means of astrology; he outlines a method for determining whether the people next door will have sexual intercourse on a certain night.[24]

A personal testimony of the consequences of planetary determination of sexual makeup for the life of a ninth-century individual is found in Verbum 95 of the *Centiloquium,* where we are privy to the confession of a man who suffers from secret desires. "Although you see me dressed in clean clothes," he tells us, "I do this for the eyes of men . . . I am in great tension, for my inclination is otherwise." His desire, it turns out, is to spend the night with a negress or negro, preferably the latter, wrapped in a coarse blanket of goat's hair. This urge springs directly from the constellation Aries which determined his nativity, or horoscope, and he is powerless to change it. Although the poor man swears by God that he has always hated homosexuality, this desire continues to torment him. He does not act on it, however, nor does he reveal it to his soldiers and familiars.[25] For this individual, his sexual complexion is predetermined by the heavens, and he can conform to the mores of society only with a good deal of inner turmoil.

Latin astrologers do not provide us with a similarly detailed case history, although Giovanni Pontano does cite a number of individuals by name to demonstrate the effects of planetary conjunctions on sexual behavior. Pontano maintains that not only do the planets cause perversions, but they cause sex change as well. He names a number of women in Italy who changed into men—one after twelve years of marriage; and another, forty years old, who was so ridiculed that she joined a monastery.[26]

For the most part, however, Western authors follow the Arabs on this subject. Guido Bonatti quotes the Arab writers extensively. For example, he cites the late eighth- or early ninth-century astrologer Aomar, who was translated in the twelfth century, to the effect that the stars can cause a woman to be insatiable, virile, or, presumably, both.[27] Leopold of Austria draws from the *Centiloquium* and its commentaries in explaining this

[20] ʿAlī ibn Ridwān, *Commentarium,* fol. 84ra. On ʿAlī ibn Ridwān see Manfred Ullmann, *Die Medizin im Islam* (Leiden/Cologne: Brill, 1970), p. 158.

[21] Albohali, *De iudiciis nativitatum,* trans. John of Seville (Nuremberg, 1546), fol. k2v. See Nallino, *Raccolta,* Vol. V, p. 330; Carmody, *Arabic Astronomical and Astrological Sciences,* p. 49; Ullmann, *Die Natur- und Geheimwissenschaften im Islam,* pp. 312–313; Lynn Thorndike, "John of Seville," *Speculum,* 1959, *34*:20–38, esp. p. 32.

[22] Abraham ben Meir Ibn Ezra, *Abrahe Avenaris Judei Astrologi peritissimi in re iudiciali opera,* trans. Petrus de Abano (Venice, 1507), fol. D2v. This writing was composed during the 12th century and translated by Peter of Abano in 1293. See Thorndike, *History of Magic,* Vol. II, p. 917.

[23] Albubather, *De nativitatibus,* fols. c–c4.

[24] Abenragel, *Liber in iudiciis astrorum,* fol. 27v: Cap. 24, "In questione si habuerit rem cum muliere hac nocte vel non."

[25] Aḥmad ibn Yūsuf, *Centiloquium,* 1493 ed., Verbum 95, fols. 115vb–116ra.

[26] Iohannes Iovianus Pontanus, *De rebus coelestibus* in *Opera omnia* (Venice, 1519), Vol. III, fol. 250.

[27] "Et dixit Aomar . . . siquidem fuerit mulier et fuerit in signo feminino erit insaturabilis et insaturata coitu et meretrix fiet . . . Et dixit Aomar . . . si vicerint in supradictis locis in nativitatibus feminarum planete masculini puella que nata tunc fuerit erit virilis quasi masculinus." Bonatus, *Decem tractatus astronomie,* fol. 312rb.

phenomenon: masculine tendencies are determined by humidity, feminine by dryness, and the oriental stars control the former and the occidental the latter.[28] Bonatti often cannot resist adding extra details, such as that if Venus is in a certain position when a male child is conceived, his future sexual intercourse will be in the woman-above position.[29]

The physicians and natural philosophers do not spell out as clearly the variations of sexual activity, and they emphasize organic causes in their discussions of sexual proclivities. Arnald of Villanova, for example, following Averroës, states that a man's sexual appetite is dependent on the complexion of his testicles; if they are hot, he will have a great desire for sexual intercourse, beget many males, and grow an abundance of pubic hair.[30] Michael Scot describes women with hot natures in similar terms.[31] Physicians are also apt to prescribe therapy—for example, corrective diet—for sexual disorders, where the astrologers counsel resignation.

Illustrative of the different approaches of the two disciplines are two passages dealing with a desire for foul odors during sexual intercourse. In the *Centiloquium* Aḥmad ibn Yūsuf states that if Mars is in the ascendent at a person's conception, he will take delight in terrible odors, and if Saturn should also be in his nativity, he will want to smell feces during coitus. "I once knew a rich man," he continues, "who swore that when he was with beautiful and pleasant-smelling women he was unable to have an erection." When he was with a vile, stinking one, however, this problem was solved.[32] The Arab offers no remedy for his impotence; in the *Centiloquium* it is clearly stated that one should simply follow one's natural inclinations.[33] The fifteenth-century physician Anthonius Guainerius takes an entirely different approach to this situation. Chapter 24 of his *Treatise on the Womb* states that a desire for foul-smelling odors can prevent delectation in sexual intercourse, and therefore impede conception. In contrast to Aḥmad ibn Yūsuf, Guainerius attempts to cure this condition by natural remedies, prescribing a number of ointments to be applied to the penis to produce long-lasting erections.[34]

MARITAL FIDELITY AND ILLEGITIMATE PROGENY

Sexual loyalty to one's spouse is another topic treated in detail in the astrological literature. According to practitioners of the science of the stars, the heavens are divided into twelve houses, and the seventh of these controls marriage. Leopold of Austria outlines all sixteen parts of this house, which include the marriage of men, the marriage of women, male cunning used against women, and sexual intercourse. The fifth part is of particular importance; it controls licentiousness and coitus of men. If this part is in a good position, states Leopold, the marriage will be successful; if not, it will fail.[35]

The tenth-century Arab Alchabitius gives us more specific information on how to determine the character of a future husband, in particular, his sexual behavior. In his treatise on astrological judgments, translated during the twelfth century by John of Seville, Alchabitius states that we must determine the position of the sun at the hour of a woman's nativity. If it is in the two eastern quarters her husband will either be very young himself, or after a few days he will

[28]Leopold of Austria, *Compilatio de astrorum scientia* (Venice/Augsburg, 1489), fol. h6v.

[29]Bonatus, *Decem tractatus astronomie,* fol. 312rb.

[30]Arnaldus de Villanova, *De coitu* in *Opera* (Lyons, 1509), fol. 273. See Averroës, *Colliget* (Venice, 1549), fol. 68v.

[31]Michael Scotus, *Liber physiognomiae* (Venice, 1477), fol. b2v. See also ed. Lyons, 1580 (entitled *De secretis naturae*), p. 254.

[32]Aḥmad ibn Yūsuf, *Centiloquium,* 1493 ed., Verbum 40, fol. 110va.

[33]"Et ego dixi et ostendi ei: 'Nullum est dedecus id operari quod eius natura exquirit; sed tunc est maximum dedecus cum mala consuetudo naturam superat.'" *Ibid.,* Verbum 95, fol. 115vb.

[34]Antonio Guainerius, *Tractatus de matricibus,* Cap. 24 in *Opera omnia* (Pavia, 1481).

[35]Leopold of Austria, *Compilatio,* fol. 3 (numbered 31 in the Columbia University Library copy).

copulate with a young boy. If it is in the two western quarters, either he will begin to have sexual intercourse late in life, or he will choose an old woman as a sexual partner soon after the marriage. If the sun is joined with Saturn, he will be a man of the law who likes to work; if it participates with Jupiter he will be prone to abstinence and have a good and noble heart; and if it is linked with Mars he will be brave in heart but bear her no love.[36] The twelfth-century Jewish astrologer Abraham ibn Ezra, whose book on Nativities was translated in 1293 by Peter of Abano, adds that if Venus is in the twelfth or seventh house of a man's nativity, he will have sexual intercourse with his wife daily.[37]

Almost all the Arabic astrological treatises include a description of the configuration of the heavens that will produce an adulterous spouse. Ibn Ezra states that if Jupiter does not face Mars the individual will be adulterous,[38] and the ninth-century Arab Albubather outlines the position of Venus, Mercury, and Mars that indicates whether a newborn child will become a fornicator and whether children will result from his illicit sexual activity.[39] Albubather's contemporary Albohali also considers that Venus can cause an individual to be unstable in marriage, with straying sexual desires.[40] Abenragel in the eleventh century devotes a chapter of his book on Judgments to the method for determining whether a girl child will be faithful to her husband, and Zahel, another ninth-century Arab, gives instructions on how to tell from the planets whether a woman has a lover.[41] These statements became part of Western scientific literature in the twelfth and thirteenth centuries and were adopted by the Latin astrologers. For example, Giovanni Pontano in his fifteenth-century treatise *On Celestial Matters* also stresses the importance of Venus in causing a woman to become adulterous.[42]

The question of marital fidelity is directly related to legitimacy of children, and the astrologers also believed that they could determine paternity by the position of the planets. Thus Zahel and Abenragel each devote a chapter of their astrological works to the topic of how to tell whether a pregnancy is legitimate or a result of fornication.[43] I have seen no attempt in these writings to decide this question after the child has been born, nor have I seen any mention in the medical or natural-philosophical literature of how to tell whether a particular man is the father of a child. The topics of hereditary characteristics and length of gestation are treated in a general manner in the thirteenth- and fourteenth-century treatises on the development of the embryo, however, and these discussions did have a certain legal influence in paternity cases.[44]

Another topic that interested the astrologers was whether a spouse was guilty of incest. The three ninth-century Arabs Zahel, Aomar, and Alkindi, whose writings were translated into Latin in the twelfth century, give directions on how to tell whether suspicion of incest is well founded.[45] According to Alkindi, the position of the sun and moon with respect to Mars tells us whether an individual has an incestuous

[36] Alchabitius, *Astronomie iudiciarie principia*, fol. xxxi v.

[37] Abraham ben Meir ibn Ezra, *Liber nativitatum et revolutionum eorum*, in *In re judiciali opera*, fol. D2v.

[38] *Ibid.*

[39] "Quando Venus in occidente vel medio celi et Mercurius cum ea in termino Martis fuerit, natus erit fornicator ac prolem in fornicatione habebit." Albubather, *De nativitatibus*, fol. b6v.

[40] Albohali, *De iudiciis nativitatum*, fol. Kv.

[41] Cap. 30, "pro muliere si erit fidelis viro suo de corpore vel non." Abenragel, *Liber in iudiciis astrorum*, fol. 28v. Zahel, *De interrogationibus*, fol. 130va.

[42] "Quod si Venus quoque et ipsa foemineo in signo constituta fuerit, eadem haec mulier et adultera erit. . . ." Pontanus, *De rebus coelestibus*, fol. 291.

[43] "Quaestio de pregnante si fit ex fornicatione vel non." Zahel, *De interrogationibus*, fol. 130va. Cap. 28, "in questione si pregnatio est legitima vel de fornicatione." Abenragel, *Liber in iudiciis astrorum*, fol. 28.

[44] See M. Anthony Hewson, *Giles of Rome and the Medieval Theory of Conception: A Study of the* De formatione corporis humani in utero (London: Athlone, 1975), Ch. X, esp. pp. 212–216.

[45] *Liber novum judicum in judiciis astrorum* (Venice, 1509), fols. 34v–35.

relationship, and variations in the places of these planets will indicate whether the incest has taken place publicly or privately. Incest is another topic that does not find a place in the medical or natural-philosophical literature, and indeed the only comment I have seen on marital sexual relations in this group of writings is a statement by Michael Scot in his treatise on physiognomy that some spouses always hate one another because they are extremely dissimilar in complexion—for example, one choleric and another sanguine in temperament.[46]

CONCEPTION AND FETAL DEVELOPMENT

The area of greatest overlap between astrological, medical, and natural-philosophical writings is in their discussion of conception and fetal development. Just as the natural philosophers considered that the planets controlled the various stages of embryonic growth, so did the astrologers manifest an intense interest in biological details when discussing development of the fetus. The reason for this is to be found perhaps in the crucial importance of this period of life for the reception of planetary influence. The basic Arabic astrological theory was that stellar determination of an individual's makeup took place at the moment of conception, and that the planets continued to affect a person during the nine months spent in the womb. The difficulty of determining the exact moment of conception led to extended discussion on how to achieve greater precision in this matter. The position of the moon at birth was the most important factor, and Verbum 51 of the *Centiloquium, Locus luni,* is heavily interpolated in most copies, demonstrating the importance of this topic for Arabic astrology.[47]

The thirteenth-century Latin astrologer Leopold of Austria includes medical material in his discussion of this fundamental astrological problem. The time of birth informs us of the moment of conception, and vice versa, he tells us, and conception can be noted by certain biological signs. Leopold then lists the standard indicators of pregnancy in his *Compilation on the Science of the Stars*—changes in the breast, eyes, and body tone, and he even includes a discussion of how to tell whether a woman is carrying a male or female child.[48]

Leopold is really compiling material here from various sources, as the title of his work indicates, and the result is therefore basically a juxtaposition of quotations. We have a true integration of medical and astrological material in the *Book on Nativities* by Albubather, translated during the twelfth century by Plato of Tivoli and again in 1218 by Salio at Padua.[49] In his discussion of fetal development, Albubather tells us that the root of feminine matter is cold and humidity, and this is warmed by the natural heat of the womb during sexual intercourse. At this point matter descends to this seven-chambered matrix from all parts of the body. When the male seed fertilizes one or more of these chambers, the newly conceived child will be nourished by the feminine substance, which is really menstrual blood.

This account, which is standard in the medical works, is followed by a detailed explanation of planetary direction of the growing fetus. For the first month, according to Albubather, the male and female seed do not mix, but remain as separate as the white and yolk of an egg, under the rule of the planet Saturn. In the second month, Jupiter, the planet of wind, takes over, and causes the seeds to become one, producing nausea in the woman, whose whole body is transformed by the wind of Jupiter and the seed within her. The child who has much of this wind will be wise, for it is even subtler than the soul, and produces the

[46] Michael Scotus, *Liber physiognomiae,* fol. dv.
[47] Lemay, "Origin and Success," pp. 91–107.

[48] Leopold of Austria, *Compilatio,* fols. g2v–g3. Leopold takes this information from Abū Ma 'shar's *Introduction to Astrology.*
[49] See Ullmann, *Die Medizin,* p. 158.

child's thoughts, just as the soul gives the five senses.[50]

Albubather continues his description through the nine months, but what is significant here is that he links astrological and medical theory. That is, he does not merely repeat the statement of Aristotle that superior things act on inferiors, but he outlines for us precisely how the planets directly cause biological phenomena such as nausea in early pregnancy as well as functions of the brain. His month-by-month treatment of fetal development also demonstrates how important the period of evolution in the womb was considered to be in terms of an individual's future character.

The importance accorded to conception is parallelled in the astrological writings by the detailed treatment of the problem of failure of conception, or sterility. According to Albubather, the sterile signs are Gemini, Leo, and Virgo, and if Venus and the moon should be situated in these signs in a newborn's horoscope, and if Mars and Saturn are in a bad position with relation to them, he or she will never have any offspring.[51] Ibn Ezra states that the sperm is universally controlled by Venus, and judgments concerning the seed should therefore be made according to its strength or depression. He quotes the astrologer "Hermes" to the effect that if Venus is in Leo or Scorpio the individual will not have sons because of an impediment in his member.[52] The Latin astrologer Giovanni Pontano also stresses the importance of Venus in causing male sterility, and he notes as well that certain stars make a woman sterile and cause a man to lose interest in coitus. Pontano also lists some physical causes of sterility normally cited in the medical and natural-philosophical writings; he states, for example, that extreme obesity will make a woman incapable of bearing children.[53]

Sterility is also given detailed treatment in the medical and natural-philosophical literature, although here both causes and remedies are considered to be found in the natural world. Indeed, Averroës, who provides a detailed medical treatment of this problem, largely in terms of the complexions of both the male and female partner, states categorically that he does not agree with those who hold that these complexions can be caused by the disposition of the celestial bodies.[54] Doctors prescribe diets, medicines, and sometimes surgery to cure infertility problems, once again in contrast to the astrologers who assume that nothing can alter what the heavens have disposed.

CONCLUSION

The Arabic astrological writings, translated into Latin during the twelfth and thirteenth centuries, contain detailed discussion of the human reproductive system, and the precise manner in which our sexual functioning is controlled by planetary conjunctions. These astrological texts were highly influential among Western scientists, and their importance to Latin doctors is widely documented. Yet, although astrologers, physicians, and natural philosophers all accept the Aristotelian statements that inferior things are subject to superior ones, the disciplines remain distinct in their approaches to sexual matters. Astrologers believe that an individual's sexual complexion is determined by the conjunction of the planets at the moment of his conception, and he can do little during his lifetime but resign himself to what the stars have wrought. Doctors and natural philosophers, on the other hand, look to the natural world for explanations of sexual characteristics and cures for sexual problems. Further, although all three groups of writers give detailed treatment

[50]Albubather, *De nativitatibus,* fol. a3v.

[51]*Ibid.,* fol. c3v.

[52]Ibn Ezra, *Liber nativitatum,* fol. D. On "Hermes" see Carmody, *Arabic Astronomical and Astrological Sciences,* pp. 52–70.

[53]Pontanus, *De rebus coelestibus,* fol. 248v.

[54]Averroës, *Colliget,* fols. 62–62v.

to the topics of virginity, conception, and fetal development, astrologers are more interested in aspects of sexual behavior that have a direct impact on society. Thus, they concentrate on homosexuality, marital fidelity, illegitimacy, and incest, topics which doctors and natural philosophers generally omit. The medical/natural-philosophical tradition, on the other hand, examines all aspects of the reproductive act, and consequently contains lengthy treatment of the proper manner for performing sexual intercourse for greatest enjoyment and maximal probability of conception. Bedroom details of ordinary married couples are not part of the astrologers' purview; most likely the social structure of the Islamic world, where the "science of the stars" was most highly developed, is a factor in explaining the different foci of these disciplines.

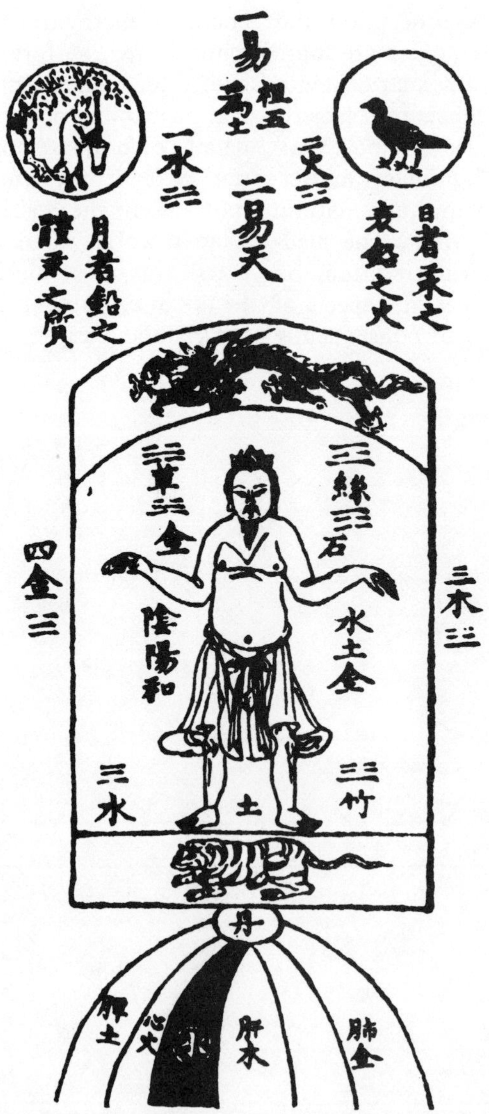

*A Chinese version of the alchemical Great Work, portrayed as an array of correspondences. The adept is surrounded by correlates of yin and yang (the moon containing a hare pounding the elixir in a jade mortar; the sun containing a three-legged raven; tiger and dragon), the Five Phases, and the trigrams of the Book of Changes. Below the tiger in the circle is the elixir of immortality, and issuing from it the essences characteristic of the Five Phases and four of the visceral spheres of function within the body. From a text of internal alchemy (*Chen-yuan miao tao hsiu tan li-yen ch'ao,[au] *probably between the seventh and tenth centuries, in* Yun chi ch'i ch'ien), *72:34a.*

Chinese Alchemy and the Manipulation of Time

By N. Sivin

For the dedication of Willow House

TOWARD WHAT ENDS did the Chinese alchemists master chemical manipulations? On what principles did they shape their processes? Their goals, I claim, were not in any significant sense chemical—a point easily overlooked if we restrict our attention, as we usually do, to those isolated accomplishments that entitle the alchemist to credit in the light of modern chemistry.

Alchemy (*wai tan*)[a,1] was the art of making elixirs of immortality, perfected substances which brought about personal transcendence and eternal life—which, in other words, communicated their own perfection to human beings when ingested (or, as I will show, even without being ingested). Their power could also be applied to the "maturation" of metals into silver and gold, to the cure of sickness, and other matters of practical benefit. But almost every alchemical treatise used its manipulations as concrete metaphors for cosmic and spiritual processes.

Even so routine and preliminary a process as the extraction of silver from lead by cupellation was assimilated in many-layered imagery reflecting the separation of the positive and negative macrocosmic energies and the emergence of the perfected self when a human is metamorphosed into an immortal:

Refine silver within lead
And the spiritual being is born of itself.
In the ash reservoir, melting in the flame,

This essay summarizes as concisely as possible parts of an extensive and considerably more technical study of the theoretical foundations of laboratory alchemy, completed in 1970 and due for publication c. 1978 in Joseph Needham, Ho Peng Yoke, Lu Gwei-djen, and N. Sivin, *Science and Civilisation in China*, Vol. V, Pt. 4 (Cambridge: Cambridge University Press). Many more examples and citations are provided there for each step in my argument, but this version incorporates five years' further reflection. I intend only to exhibit the general character of alchemy over the period in which it flourished (second to eleventh century A.D.). Because few texts can yet be dated precisely, I am not prepared to trace its origin or evolution without excessive resort to guesswork. The reader should be aware that given the enormous limitations of understanding and control of the literature, even a rough sketch of the sort I offer can have little value except to encourage further study. For additional orientation and bibliography, see Needham and Lu, *Science and Civilisation*, Vol. V, Pt. 2 (1974), and Sivin, *Chinese Alchemy: Preliminary Studies* (Harvard Monographs in the History of Science, 1) (Cambridge, Mass.: Harvard University Press, 1968).

All translations below are my own. I acknowledge with gratitude support from the National Science Foundation and the National Library of Medicine and hospitality from the Sinologisch Instituut, Leiden, Gonville and Caius College, Cambridge, and the Research Institute for Humanistic Studies, Kyoto.

[1] Superscript letters refer to the list of Chinese characters at the end of the article.

Lead sinks down, silver floats up.
Pristine white the Treasure appears
With which to make the Golden Sprout.[2]

A necessary first step toward understanding the character of Chinese alchemical thought is to look at how alchemists drew on concepts of natural change, especially temporal change, that practically all Chinese thinkers shared.

TIME IN NATURAL PHILOSOPHY

Scientific thought began, in China as elsewhere, with attempts to comprehend how it is that although individual things are constantly changing, always coming to be and perishing, nature as a coherent order not only endures but remains conformable to itself. In the West the earliest such attempts identified the unchanging reality with some basic stuff out of which all the things around us, despite their apparent diversity, are formed.

In China the earliest and in the long run the most influential scientific explanations were in terms of time. They made sense of the momentary event by fitting it into the cyclical rhythms of natural process. The life cycle of every individual organism—its birth, growth, maturity, decay, and death—followed essentially the same progression as those more general cycles which went on eternally and in regular order: the cycle of day and night, which regulated the changes of light and darkness, and the cycle of the year, which regulated heat and cold, activity and quiescence, growth and stasis. It was the nested and intermeshed cycles of the celestial bodies that governed the seasonal rhythms and, through them, the vast symphony of individual life courses.

This pattern held for minerals as well as for flora and fauna. The Chinese shared what seems to have been a traditional belief among miners everywhere that earths matured within the terrestrial womb. The metallurgist was merely accelerating a natural progress of metals toward perfection—a perfection that, except in the cases of silver and gold, nature's corrosion would eventually undo.[3]

The concepts that molded this organismic world-view were above all functional and relational. They specified what was significant about each thing (its "way" or *tao*)[b] in terms of what it did in relation to what others did, its role in the cosmic system (the greater *Tao*) rather than its static qualities. Yin and

[2] *Chin pi ching*[t] (Gold and Cerulean Jade Canon, c. A.D. 200?), cited in *Tan lun chueh chih hsin chien*[u] (Heart-Mirror of Mnemonics and Explanations from Writings on the Elixir, probably prior to the tenth century; in *Cheng-t'ung Tao tsang*[v] [hereafter TT], Vol. 598), p. 7a; excerpted from an unpublished complete translation of the Heart-Mirror, based on a text edited from three printed versions. Although this quotation refers to a chemical operation, the Heart-Mirror contains a great deal of material on the physiological analogue of alchemy (*nei tan*),[w] which uses alchemical language to describe the pursuit of immortality through breathing or meditative disciplines. This "internal" (*nei*) form and "external" (*wai*) alchemy were so closely complementary, and were so regularly practiced in conjunction before the eleventh century (when external alchemy began to die out), that it is often impossible to distinguish which sort of operation a text is concerned with. Their theoretical basis was largely identical. The Golden Sprout is an intermediary in the preparation of the elixir.

[3] This view of the miner, the smith, and the metallurgist in all the major civilizations has been documented by Mircea Eliade in *Forgerons et alchemistes* (Paris: Flammarion, 1956), translated by Stephen Corrin as *The Forge and the Crucible* (New York: Harper, 1962); for more recent references see Mircea Eliade, "The Forge and the Crucible: A Postscript," *History of Religions*, 1968, *8*:74–88.

yang were the passive and active phases through which any cycle must pass. Although used analytically, to break down change into its parts, the aim of these concepts was as often as not synthetic, to explain the relation of parts in the whole. Yin and yang could also be used to distinguish aspects of configurations in space—down and up, back and front, inside and outside. Here too the *Oxford English Dictionary* sense 2 of "phase" ("any one aspect of a thing of varying aspects; a state or stage of change or development") is not out of place; yin and yang were almost always applied to spatial relations that are either changing or in dynamic equilibrium. The Five Phases (*wu hsing,*[c] often mistranslated as Five Elements) were a very closely analogous division of cycles or configurations into five functionally distinct parts (not, in scientific explanations, into five types of ultimate constituents). The phases Fire and Wood, for instance, made possible a finer analysis of the yang, or active, aspect of change. Wood is the name of the phase of growth and increase, and Fire is the maximal flourishing phase of activity, when the yang is about to begin declining and yin must once again reassert itself. In the cycle of the year Wood and Fire thus correspond to spring and summer respectively.[4] A third set of concepts of the same kind, much used in alchemy, were the eight trigrams and a set of twelve hexagrams from the Book of Changes, each defined with a function that played a due part in the preparation of elixirs.

All these concepts belong to the most general level of early Chinese thought about nature. The various fields of science, such as medicine and alchemy, simply applied them to different classes of phenomena, redefining and supplementing them with technical concepts as necessary.

THE NATURAL ELIXIR

Gold was an obvious endpoint for the evolution of minerals, for it was exempt from decay. In China a second line of development, more important in alchemy, led toward cinnabar (HgS). Chinese gave its large, translucent, nearly vitreous crystals, the color of fresh blood, a special place among semi-precious stones. Cinnabar was, like gold, associated with vitality and immortality from very early times. Alchemists associated it with the maximal phase of yang, Fire, and adapted *tan*[d] (cinnabar) as a term for elixirs whether or not mercuric sulfide was among their ingredients.[5]

The alchemist's elixir, although certainly the product of artifice, reproduced the work of nature. Here is a theoretical text, so far undatable but before

[4]The aspects of yang that correspond to Wood and Fire are conventionally distinguished as young or minor yang (*shao yang*)[x] and mature, old, or major yang (*t'ai yang*).[y] Metal and Water are the corresponding subphases of yin, and Earth is the neutral point of balance between yin and yang. I capitalize the names of the Five Phases to avoid confusion with ordinary wood, fire, and so on. The first philologically adequate explication of yin-yang and the Five Phases in any language is that of Manfred Porkert, *The Theoretical Foundations of Chinese Medicine* (MIT East Asian Science Series, 3) (Cambridge, Mass.: MIT Press, 1974), pp. 9–54. Porkert is concerned only with their use in certain aspects of medicine.

[5]A typical scheme of cinnabar evolution is given in *Huang ti chiu ting shen tan ching chueh*[z] (The Yellow Emperor's Canon of the Nine-Vessel Spiritual Elixir, with Explanations, Chs. 2–20 apparently of the late tenth century; *TT*584–585), 14:1a. Some early writers, e.g. those cited c. 320 by Ko Hung,[aa] make cinnabar a stage in the evolution of gold; see *Pao p'u tzu nei p'ien*[ab] (The Inner Chapters of the Philosopher Pao p'u tzu; *P'ing chin kuan ts'ung-shu*[ac] ed.), 16:5a, translated by James R. Ware in *Alchemy, Medicine, and Religion in the China of A.D. 320. The Nei P'ien of Ko Hung (Pao-p'u tzu)* (Cambridge, Mass.: MIT Press, 1966), p. 268.

A.D. 900, on the elixir that the cosmic order produces in its own time. It corresponds to the alchemist's two-ingredient process in which silver extracted from lead and mercury recovered from cinnabar are conjugated to bring about a perfect balance of yin and yang function. The natural elixir, unattainable by mere mortals, is called "cyclically transformed" because it is matured by the alternations of the seasons, which we will see the alchemist modeling by alternations of his fire:

> Natural cyclically transformed elixir is formed when flowing mercury, embracing Squire Metal[e] [i.e., lead], becomes pregnant. Wherever there is cinnabar there are also lead and silver. In 4320 years the elixir is finished. . . . It embraces the *ch'i*[f] of sun and moon, yin and yang, for 4320 years; thus, upon repletion of its own *ch'i*, it becomes a cyclically transformed elixir for immortals of the highest grade and for celestial beings. When in the world below lead and mercury are subjected to the alchemical process for purposes of immortality, [the elixir] is finished in one year. . . . What the alchemist prepares succeeds because of its correspondence on a scale of thousandths.[g,6]

The number 4320 implies the "scale of thousandths" that relates the natural and artificial elixirs. There were twelve "hours" (*shih*)[h] in a Chinese day and 4320 in a round year of 360 days. An hour in the laboratory thus recapitulated a year in the terrestrial matrix.

Just as the diversity in color and other physical properties of native gold suggested that there could be red and purple varieties which excelled the normal, cinnabar was believed to constitute a range of substances. The rarest and most evolved kinds (again unattainable by mortals) were in effect natural elixirs. Ch'en Shao-wei's[i] great monograph Arcane Teachings on the Alchemical Preparation of Numinous Cinnabar (written perhaps c. 712) first discusses the best cinnabar at the alchemist's disposal (similar to that shown in Fig. 1), then its eventual maturation into an elixir, and finally the endpoint of cinnabaric evolution which the alchemist tries to approximate:

> Now the highest grade, lustrous cinnabar,[j] occurs in the mountains of Ch'en-chou and Chin-chou [in modern Hunan] upon beds of white toothy mineral. Twelve pieces of cinnabar make up one throne. Its color is like that of an unopened red lotus blossom, and its luster is as dazzling as the sun. There are also thrones of 9, 7, 5, or 3 pieces or of one piece. . . . In the center of each throne is a large pearl [of cinnabar], ten ounces (*liang*)[k] or so in weight, which is the "monarch." Around it are smaller ones . . . ; they are the "ministers." They surround and do obeisance to the great one in the center. About the throne are a peck (*tou*)[l] or two of various kinds of cinnabar, encircling the "jade throne and cinnabar bed." . . . If lustrous cinnabar is further taken in the sevenfold-recycled or ninefold-cyclically-transformed state, then without ado the yin soul is transformed and the outer body destroyed, the spirit made harmonious and the constitution purified. The yin *ch'i* [i.e., the material configuration organized by energy of yin type] dissolves and the persona floats up, maintaining its shape, to spend eternity as a flying immortal of the highest grade of realization . . . .

[6] *Tan lun chueh chih hsin chien*, p. 12b. *Ch'i* refers to the active energy (in the colloquial, purely qualitative sense of the word) that organizes matter into configurations, causing change, or that maintains the organization of configurations and thus resists change. The word is also applied to configurations of matter so organized. Such configurations are generally defined by their functions rather than by their constituents. See Porkert, *Theoretical Foundations*, pp. 167–168. The earliest senses of the term are surveyed in Kuroda Genji,[ad] "*Ki*," *Tōhō shūkyō*, 1953, No.3:1–40; 1955, No.7:16–44.

Figure 1. "Throne" formations of large cinnabar crystals, growing on beds of what is apparently drusy quartz. From a 1957 Peking reproduction of the Chinese Materia Medica printed in 1249 (Ch'ung hsiu Cheng-ho ching shih cheng lei pei yung pen-ts'ao),[ar] 3:2b.

After imbibing the active energy of the cosmos for twenty-two thousand years, lustrous cinnabar is spontaneously

> . . . transformed into celestial throne cinnabar, in which the throne is jade-green. There are nine pieces in the center, growing in layers, pressed closely about by 72 [smaller] pieces. It floats in the midst of the Grand Void, constantly watched over by one of the spirits of the Supreme Unity. On a Superior Epoch day[m] [the full moon of the first lunar month] the Immortal Officials descend to collect it. The mountain [on which it is found] suddenly lights up as if illuminated by fire. This celestial throne cinnabar is collected [only] by Immortal Officials; people of the world can have no opportunity to gather it.[7]

ALCHEMICAL PROCESSES

The alchemist's simulation of the hidden operations of nature was founded upon a much more ancient simulation, that which made possible the accomplishments of the metalworking artisan. In extracting a metal from its ore or in making strong steel from brittle cast iron, the metallurgist was demonstrating that humans can imitate natural processes, can stand in the place of nature and bring about natural changes at a rate immensely faster than nature's own time. People in traditional societies did not know of any way to speed up the growth cycle of plants or animals. Only the life rhythms of minerals could be manipulated.

Upon the craftsman's mastery of metalline growth alchemists built processes to fit their own purpose. They constructed in their laboratories working models of the cosmic order as manifested in time. They not only shrank the dimensions of the universe to fit their four walls but also compressed time to make the

[7] *Ta tung lien chen pao ching hsiu fu ling sha miao chueh,*[ae] (*TT*586), pp. 5b–7b; excerpted from a complete draft translation based on collation of two printed texts. The first six characters of the title indicate that this book (like the one cited in n. 11) provides supplementary instructions for the process given in the Great Void Canon on Making the Realized Treasure, an otherwise unknown work. "Realization" and "realized" throughout this essay are my rather literal translations of *chen,*[af] which denotes the authenticity and perfection attained by the elixir, or by the adept who attains a high grade of immortality. The text mentions Immortal Officials and grades because immortals were given appointments in the hierarchy of the gods, bureaucratically organized like the imperial government.

duration of their manipulations feasible. Thus through point-by-point correspondence the artificial circumstances of the laboratory were made profoundly natural and responsive to the operation of the cosmic *Tao*, to the rhythms of the great order outside: "When earth mixes with water to form mud, and is kneaded [by subterranean forces] below a mountain, gold will be formed, and generally cinnabar above it. When this [cinnabar] is ceaselessly metamorphosed and cyclically treated [by the alchemist], and once again forms gold, this is merely a reversion to the root substance, and not something to be wondered at."[8] How could the alchemists be sure that what went on within their reaction vessels was identical with the work of nature? If they erred they would be overruled by the *Tao*, and their elixirs would wither.

They depended upon numerological correlations, correspondences, and resonances—in a word, upon analogy.[9] There were three points at which analogies could be applied to cosmic process: materials, apparatus and other aspects of spatial arrangement, and control of time and temperature in combustion. Space permits examination only of the last, but it is important to keep in mind that all were used in conjunction.

FIRE PHASING

Since the heat of the flame stood for the active cosmic forces, to recreate those forces in the laboratory required that fire be bound by time. "Fire phasing" (*huo hou*)[n] was the gradual increase and decrease of fire intensity by using precisely weighed increments of fuel. It is true that a constant increment in the weight of charcoal burnt in the furnace does not cause a constant increase in temperature. But it was well known by craftsmen that each weight of fuel burnt in the same way yields a fixed heat and a predictable product. What the alchemists did was to make this concept of fire phasing dynamic, varying the weight of fuel in a regular way over a very long period of heating. They were bringing their processes under the control of one of the few exact measuring instruments at their disposal, the balance. The profile of the flame's intensity was thus governed by correspondences that tied it to the seasonal cycles. An anonymous early author calls into play yin and yang phases and other divisions of the annual cycle and cites two of the most canonic sources of alchemical theory: "The amounts of fuel to be weighed out are increased and decreased in cyclical progression according to the proper order of yin and yang. They must conform with the signs of the Book of Changes and the Threefold Concordance, tally with the 4, 8, 24, and 72 seasonal divisions of the year, and agree with the implicit correspondences and *ch'i* phases of the year, month, day and hour—all without a jot or tittle of divergence."[10]

[8] *Huang ti chiu ting shen tan ching chueh*, 13:2a.

[9] On correlative thinking see Joseph Needham and Wang Ling, *Science and Civilisation in China*, Vol. II (1956), pp. 279–291 and elsewhere. The most important source on numerology in China is Marcel Granet, *La pensée chinoise* (L'évolution de l'humanité, 5.3) (Paris: La Renaissance du Livre, 1934), pp. 149–299.

[10] Cited in *Chu chia shen p'in tan fa*[ag] (Wonderful Elixir Formulas of the Masters, tenth century or later; *TT* 594), 4:1b. On this collection, see Sivin, *Chinese Alchemy*, pp. 75–76. The Threefold Concordance is the *Chou i ts'an t'ung ch'i*,[ah] traditionally dated c. A.D. 140, which would make it the earliest extant alchemical text. Even though the present version is probably much later, its putative antiquity gave it great status in the eyes of later writers. See *Chinese Alchemy*, pp. 36–40. There is a very rough English translation, Wu Lu-ch'iang, "An Ancient Chinese Treatise

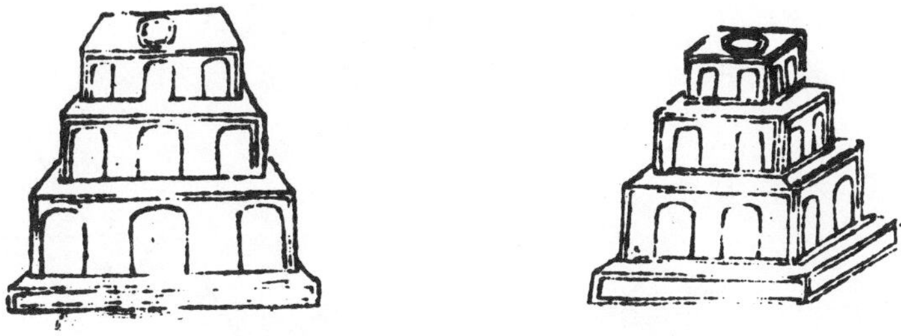

Figure 2. Furnaces similar to the one described by Ch'en Shao-wei, from Wu Wu,[as] Tan fang hsu chih[at] (Indispensable Knowledge for the Alchemical Laboratory, 1163; in TT588). Ch'en specified nine openings in the top level, twelve in the middle, and eight in the lower. His furnace was fired only through the median doors, each of which was equipped with tuyeres. In Wu's book these structures were no longer used as furnaces; they merely supported a small stove which enclosed the reaction vessel.

Consider a single fire-phasing scheme, from Arcane Teachings on the Ninefold Cyclically Transformed Gold Elixir (c. 712?), a companion work to the book by Ch'en Shao-wei cited earlier. Ch'en repeats a sixty-day cycle six times over a round year to make his elixir. The furnace is built in three tiers (see Fig. 2), which correspond to sky and earth and man centered between them. The central tier has twelve doors, which correspond to hours in the day and months in the year. Ch'en fires through each door for five days to complete a cycle.

> As for the time of firing the furnace, the fire should be applied at a midnight which is also the first hour of a sixty-hour cycle, on the first day of a sixty-day cycle, in the eleventh month [i.e., the month which contains the winter solstice]. Begin by firing through door A for 5 days, using 3 oz of charcoal. There must always be 3 oz of well-coked charcoal, neither more nor less, in the furnace. Then open door B and start the fire, firing for 5 days, using 4 oz of charcoal. Then open door C and start the fire, firing for 5 days, using 5 oz of charcoal.

This scheme proceeds until 8 oz of charcoal have been fed through the sixth door for 5 days. Then the alchemist is told to move on to what Ch'en calls the yin doors, reducing the fuel by 1 oz per day at each door, until at the twelfth the charge has been reduced to 4 oz. The cycle is not quite closed, for the weight of fuel has not been reduced to the original amount. The second sixty-day cycle begins with 5 oz of fuel and goes up to 10 oz. The third cycle begins with 7 oz and goes up to 12 oz. The sixth cycle, which completes the elixir, begins at 17 oz, rises to 22 oz, and falls again to 16 oz.[11]

---

on Alchemy Entitled 'Ts'an T'ung Ch'i,' Written by Wei Po-Yang about 142 A.D. with an Introduction and Notes by Tenney L. Davis," *Isis*, 1932, *18*:210–289.

"*Ch'i* phases" are phases of the regular temporal variations in energy and activity (see n. 6). In other words, the cycles which yin-yang and the Five Phases analyze are cycles of *ch'i*, whether manifested as heat, light, or some other factor that brings about change or maintains a configuration.

[11] *Ta tung lien chen pao ching chiu huan chin tan miao chueh*[ai] (*TT* 586), pp. 12a–13a, excerpted from a complete translation based on two texts. The first sentence quoted places the epoch of the process at the moment when calendrical cycles begin. For certain purposes connected with imperial ritual—including astronomical computation—the year was considered to begin in the eleventh month, the month which contained the winter solstice. See N. Sivin, "Cosmos and

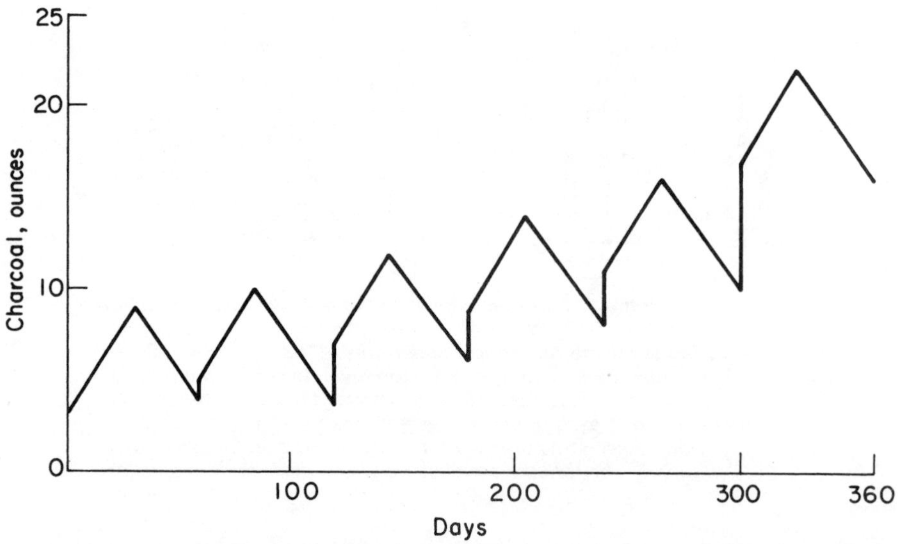

Figure 3. The cyclical pattern of precisely controlled "fire phasing."

As graphed in Figure 3, the zigzag time-weight profile represents most reasonably the notion of a change in the alchemical ingredients at once cyclic and progressive. If, mentally, we were to superimpose the six cycles, which in principle repeat each other, it would be very tempting to call this a helical fire-phasing schema.

Some fire-phasing schemata are a good deal simpler than this one, and some are more elaborate—for instance, a thirteenth-century "two-variable" procedure in which the weight of charcoal below the reaction vessel and that of water in a cooling apparatus in its upper part are varied regularly in such a way that the sum of this yin-yang pair remains constant.[12]

THE SIGNIFICANCE OF FIRE PHASING

From the documents I have quoted and many others like them, one can see that number and measure were being used in a way not much like their

Computation in Early Chinese Mathematical Astronomy," *T'oung Pao* (Leiden), 1969, *55*:10–11. There is an excellent discussion of furnaces and other apparatus in Ho Ping-yü and Joseph Needham, "The Laboratory Equipment of the Early Mediaeval Chinese Alchemists," *Ambix*, 1959, 7:57–115.

[12] A system in which the increase in weight of charcoal is linear and repeated in each cycle is given in *T'ai shang pa ching ssu jui tzu chiang wu chu chiang sheng shen tan fang*[aj] (Method for Making the Eight-Radiance Four-Stamens Purple-Fluid Five-Pearl Incarnate Numinous Elixir, a Most High Scripture, probably late fourth century; in *Yun chi ch'i ch'ien*,[ak] The Seven Bamboo Tablets of the Cloudy Satchel, c. 1023; *TT*691), 68:4b–5a. The two-variable system appears in *Chin hua ch'ung pi tan ching pi chih*[al] (Confidential Instructions on the Canon of the Heaven-Piercing [?] Golden Flower Elixir, preface dated 1225; *TT*592), 2:20a–21b. Although Meng Hsun,[am] who recorded this treatise, claims that he carried out the process, its great complexity suggests a thought experiment. On the origins of this treatise, see Needham and Lu, *Science and Civilisation*, Vol. V, Pt. 2, p. 316. Note that Needham and Lu translate the last word of the title as "manual." It is true that although *ching* usually designated books of philosophic or religious doctrine, it

use in modern science, where time-temperature specifications also control chemical preparations. What makes alchemical operations resemble those of modern chemistry is that the quantities that controlled the process were derived, by way of theory, from observations and measurements. While they served as models of cosmic change, the alchemical processes also had to transform one substance into another, and so at some point certain chemical and physical conditions had to be satisfied. Ch'en Shao-wei was well aware, for instance, that 1 lb of cinnabar would yield 14 oz of mercury (a figure he almost certainly learned from mercury smelters).

Alchemy and modern chemistry differ in the ways number served theory, as well as in the links between theory and the observations on which it was based. The foundation on which chemical theories rest is essentially mathematical; Chinese alchemical theories were essentially numerological. The latter used numbers, not as measures, but as a means of ranking phenomena into a qualitative order, an order that reflects their subjective values. It is not at all unusual for measured quantities to be combined with symbolic numbers in such schemes, nor for the results of numerological manipulations to be treated as though they were quantitative. (These complexities may be seen in the grading system prevalent in American universities, the most influential survival of numerology in modern society.)

Fire phasing was one of many means of constructing in the laboratory a system analogous part by part to another system—the cosmic order. Alchemists drew upon all the numerological and emblematic correspondences available to describe the activity of the cosmos, especially those correspondences related to the Five Phases, yin-yang, and the symbols of the Book of Changes. But because these concepts had already been extended to so many areas of thought, their connotations were too rich and diverse to be entirely integrated into alchemical theory (or any other sort of early scientific theory). They were quite adequate for some kinds of explanation, but not for others. By weaving a new fabric of associations to supplement conventional natural concepts adapted to their own concerns, the alchemists were in effect picking and choosing what meanings would be included in the overtones of their discussions. Fire phasing was a way of creating special structures embodying number—patterns of times and weights—which conveyed no more and no less than what alchemists wanted to teach their disciples.

SPATIAL AND MATERIAL CORRELATIONS

The alchemist's model could be complete only if it were aligned with the universal order in space as well as time. In spatial correlations an enormous variety of verbal and numerological associations, mostly analogous in kind to those I have already mentioned, were called into play. The laboratory was oriented to the cardinal points of the compass, the furnace centered in it, and the reaction vessel centered in the furnace to make it the axial point of change. The designs of furnaces and vessels were precisely specified to

---

was occasionally applied to what moderns would consider technical manuals; but this was done precisely to assert their status as canons of craft traditions, and there is no reason not to translate *ching* literally.

make them concrete metaphors for sky and earth, with the work of man centered between the two and in perfect concord with both.

The more theoretically inclined alchemists tended to use processes in which two ingredients represented the alternating conjugation and separation of yin and yang. One of the two most usual processes began with cinnabar and lead, representing yang and yin, from which were extracted mercury, representing yin emergent from yang, and silver, representing yang emergent from yin. The other combined mercury and sulfur to form vermilion (artificial cinnabar, finely divided), which could be repeatedly decomposed and resynthesized. As Ch'en Ta-shih[o] (Sung period) put it, "That cinnabar should come out of mercury and again be killed by mercury: this is the mystery within the mystery."[13]

Despite the many types of modeling that it incorporated, alchemy was not model-making for its own sake, nor was it the pursuit of chemical knowledge for its own sake, nor even an alternative way of doing natural philosophy. It had its own goals, and we are now ready to look at them.

THE GOALS OF ALCHEMY

I have already mentioned that in a certain sense alchemy grew out of the metalworking artisan's ability to accelerate mineral maturation. The alchemist's discovery was that the life courses of minerals could be accelerated by man in the interest not only of metal production but of understanding and even of wisdom. No one could wait 4,320 years to experience nature's production of an elixir, and in any case her womb was not open to contemplation. An alchemist who set out to fabricate an elixir in a few months or a year was creating an opportunity to witness the cyclical sweep of universal change.

The alchemists' enterprise, as they themselves defined it, was not chemistry. They were not primarily motivated by curiosity about the properties and reactions of specific substances (which they generally learned about from craftsmen). Those properties and reactions fascinated some alchemists, but they were no more intrinsically important than the characteristics of pigments that painters must know about in order to set down images. In fact the response of alchemists to their materials was at least as esthetic as it was manipulative. "Charming" (*k'o-ai*)[p] is one of the most common adjectives in alchemical writing for mineral specimens.

Chemical knowledge and concepts of a chemical kind, in other words, were to some extent indispensable means and to some extent byproducts (as were physical relations and ritual patterns). Alchemists recorded a great deal of their chemical knowledge as well as their attempts to make sense of it. Their documents are chemically far richer and more explicit than those of the West before the Renaissance.[14] But the aims of Chinese alchemy were of a different sort than those of modern science.

[13] *Pi yü chu sha han lin yü shu kuei*[an] (On the Cerulean Jade and Cinnabar Jade-Tree-In-a-Cold-Forest Process, eleventh to fourteenth century?; *TT*587), p. 1a.

[14] The chemical knowledge reflected in alchemical practice is being surveyed systematically in Needham and Lu, *Science and Civilisation,* Vol. V, Pts. 2 (1974) and 3 (1976), and I discuss chemical aspects of theory in Vol. V, Pt. 4 (see n. *). Chinese alchemy, no less than the traditions of other civilizations, depended on recondite language and symbolic images (see the Frontispiece) both to multiply levels of meaning and to exclude people unprepared spiritually and morally to use its teachings responsibly. But the degree of mystification has been badly overestimated

The old books reflect many different aims and many kinds of curiosity. Some interests were on the whole pragmatic; it was the product, the elixir, that mattered, and one way of making it was in principle as good as another. The emphasis was on benefits elixirs yielded: health, wealth, above all immortality. This was especially true of certain medical men such as Sun Ssu-mo[q] (fl. 673), who described his alchemical processes in much the same way as his techniques for preparing drugs.[15] But the aim that runs through most of the roughly one hundred treatises on laboratory alchemy that still exist,[16] the aim that makes them a coherent literature, that conditioned every step in the design of the processes, was to construct a model of the *Tao*, to reproduce in a limited space on a shortened time scale the cyclical energetics of the cosmos.

This goal values contemplating the process over using the product. In the Threefold Concordance, the most influential of the early alchemical canons, the product was practically ignored. There are no instructions for compounding or ingestion. There are a mere couple of cursory descriptions of that immortal beatitude which to pragmatic alchemists was the highest reward of their art.[17] Among the posterity of the Threefold Concordance we find such a concern with gnostic rapture over the process that the steps between understanding and becoming an immortal are occasionally skipped altogether. This gap is apparent in the Most Ancient Canon of Earth and *Tui* (between sixth and ninth centuries?): "If [the devotee] attains a clear and penetrating understanding of these Five Phases, one can proceed to a discussion of fire-subduing, and can then talk to him about the *tao* [i.e., the art] of projection. When he has comprehended every aspect of the Five Phases, he will be a man of balanced Realization, and the Three Worms[r] will leave his body."[18] In other words, one can understand the preparation of elixirs by rendering volatile minerals invulnerable to the fire, and the use of these elixirs in transforming lesser substances into gold, only after comprehending exactly how the Five-Phases theory governs these changes. The adept who thus has mastered the alchemical process is invulnerable to the Three Corpseworms (demonic presences who

---

by a few historians who, because they could not trouble themselves to understand the conceptual language of yin-yang and the Five Phases, assumed that it was also occult.

[15] His *T'ai ch'ing tan ching yao chueh*[ao] (Essential Formulas from the Alchemical Classics, a Most Pure Scripture; in *Yun chi ch'i ch'ien*, Ch. 71) was edited and translated with commentary in Sivin, *Chinese Alchemy*.

[16] Needham and Lu, *Science and Civilisation*, Vol. V, Pt. 2, Bibliography A and Concordance for *Tao tsang* Books and Tractates, lists over two hundred alchemical texts. One can only estimate how many are concerned with external alchemy (see n. 2). For a list of thirteen published translations, see Sivin, *Chinese Alchemy*, pp. 322–324.

[17] The extent to which the Threefold Concordance (see n. 10) was originally concerned with laboratory procedures remains problematic. It contains elements of external, internal, and sexual alchemy (i.e., disciplines for immortality through the most literal union of yin and yang), but was interpreted by external alchemists as primarily concerned with their discipline. See R. H. Van Gulik, *Sexual Life in Ancient China. A Preliminary Survey of Chinese Sex and Society from ca. 1500 B.C. till 1644 A.D.* (Leiden: E. J. Brill, 1961), pp. 80–84.

[18] *T'ai ku t'u tui ching*[av] (*TT* 600), 1:4b. *T'u* refers to the phase Earth (in which opposed tendencies reach balance), and *tui* to one of the eight trigrams of the Book of Changes, but there is no indication in the text of what association of the latter the author meant to invoke. "Balanced Realization" describes the perfect complementarity of yin and yang energies that characterizes immortals; see n. 7.

normally live in the body and are responsible for natural death) and is therefore immortal.

The content, tone, and balance of the evidence strongly suggest that the dominant goal of Chinese alchemy was contemplative, and even ecstatic. A Taoist book of the early sixth century has this to say about the use of a small amount of elixir to turn lead to gold:

> You may also first place the lead in a vessel, heat it until it is liquefied, and then add one spatula of the Scarlet Medicine [i.e., the elixir] to the vessel. As you look on, you will see every color flying and flowering, purple clouds reflecting at random, luxuriant as the colors of Nature—it will be as though you were gazing upward at a gathering of sunlit clouds. It is called Purple Gold, and it is a marvel of the *Tao*.[19]

This is a splendid description of what a metallurgist sees on a lead button as it oxidizes and the oxide is moved by surface tension. The richness and vividness of the details suggest a state of heightened awareness. That is perhaps not surprising, since meditative practices were normally part of the alchemist's discipline. If Purple Gold was that fugitive iridescence on the surface of molten lead, as the text clearly says it was, it was not a gold for selling or spending, only for seeing and wondering.

The alchemists constructed their intricate art, made the cycles of the cosmic process accessible, and undertook to contemplate them because they believed that to encompass the *Tao* with their minds—or, as they put it, with their hearts and minds (comprised in one word, *hsin*)[s]—would make them one with it. That belief (by no means confined to Taoists) was not at all unlike one of the central convictions that underlay the birth of physics in the West before the time of Plato: the idea that to grasp the unchanging reality that underlies the chaos of experience is to rise above that chaos, to be freed at least for the moment from the limits of personal mortality.

CONCLUSION

In the great metaphysical and religious significance given to alchemical processes the Chinese art was remarkably similar to that of the Hellenistic tradition, although the dominant metaphors were constructed out of very different world-views and beliefs. The idea of immortality did not enter European alchemy until the twelfth century and may have been received from China (predominantly via Islam), but it would be arbitrary to deny that the tradition of Zosimos was true alchemy because it lacked this notion.[20] The Stoic and Gnostic scheme of spiritual death and rebirth played much the same role

[19] *T'ai ch'ing chin yeh* (or *i*) *shen tan ching*[ap] (Classic of Liquefied Gold and Divine Elixir, a Most Pure Scripture; in *Yun chi ch'i ch'ien*), 65:15b. A more extended version appears in *TT*582. The part of the text cited almost certainly dates from the early fifth century. See Henri Maspero, "Une texte taoïste sur l'Orient Romain," pp. 97–98 in *Études historiques (Mélanges posthumes sur les religions et l'histoire de la Chine,* 3) (Paris: Civilisations du Sud, 1950).

[20] Needham has taken the position that for this reason "the Hellenistic proto-chemists ought not to be called 'alchemists'" (*Science and Civilisation,* Vol. V, Pt. 2, p. 12); for early criticism see the review by Robert P. Multhauf in *Ambix,* 1975, *22*:218–220. A view of Alexandrian alchemy broadly similar to my understanding of the Chinese tradition is taken by Allen G. Debus, e.g., in "The Significance of the History of Early Chemistry," *Journal of World History,* 1965, *9*:39–58, and by H. J. Sheppard in "The Origin of the Gnostic-Alchemical Relationship," *Scientia,* 1962, *97*:146–149 and other writings.

in thought about transcendence as the Chinese belief that the adept's perishable body would be left behind as a lifeless husk when his new, imperishable self (embodying his old but "purified" personality) was ready to hatch out and join the ranks of the immortals. The religious depth as well as the chemical ingenuity of alchemists at both ends of Asia deserves serious consideration. There is evidence that the arts of both civilizations were means toward self-perfection through disciplined contemplation of manual operations (sometimes the patterns of manipulation were so idealized that they could not possibly have been carried out). It misses the point to assume that either tradition was dedicated to the search for abstract knowledge.

In China the operative alchemy of the laboratory, no less than the physiological and introspective disciplines that borrowed its language and symbols, was a form for self-cultivation, a means toward transcendence.

> Necessary that the maturing come within man,
> Due to the maturing of his heart and mind.
> If heart and mind have reached divinity, so will
> the Medicine;
> If heart and mind are confused the Medicine will be
> unpredictable.
> The Perfect *Tao* is a perfect emptying of heart
> and mind.
> Within the darkness, unknowable wonders.
> When the wise man has attained the August Source
> In time he will truly reach the clouds.[21]

CHINESE CHARACTERS

a. 外丹
b. 道
c. 五行
d. 丹
e. 金公 = 鉛
f. 氣
g. 象而成之大千之數
h. 時
i. 陳少微
j. 光明砂
k. 兩
l. 斗
m. 上元
n. 火候
o. 陳大師
p. 可愛
q. 孫思邈
r. 三尸
s. 心
t. 金碧經
u. 丹論訣旨心鑑
v. 政統道藏
w. 内丹
x. 少陽
y. 太陽
z. 黃帝九鼎神丹經訣
aa. 葛洪
ab. 抱朴子内篇

[21] From the Arcane Memorandum of the Red Pine Master (*Ch'ih sung tzu hsuan chi*,[aq] probably T'ang or earlier), cited in *Tan lun chueh chih hsin chien* (see n. 6), p. 14a. "Medicine" is a conventional synonym for the elixir.

ac. 平津館叢書

ad. 黑田源次

ae. 大洞鍊真寶經修伏靈砂妙訣

af. 真

ag. 諸家神品丹法

ah. 周易參同契

ai. 九還金丹妙訣

aj. 太上八景四蘂紫漿五珠降生神丹方

ak. 雲笈七籤

al. 金華沖碧丹經秘旨

am. 孟煦

an. 碧玉朱砂寒林玉樹匱

ao. 太清丹經要訣

ap. 太清金液神丹經

aq. 赤松子玄記

ar. 重修政和經史證類備用本草

as. 吳悞

at. 丹房須知

au. 真元妙道修丹歷驗抄

av. 太古土兌經

Technology and Alchemical Debate in the Late Middle Ages

By William Newman

THE MEDIEVAL ATTITUDE toward technology is one of the more interesting topics available to historians concerned with the transition from antiquity to modernity. Considerable debate has focused on the sweeping claims made by Lynn White for the social and cultural impact of technological changes wrought in the Middle Ages.[1] Others, such as Guy Beaujouan and James Weisheipl, have looked at Scholastic classifications of the arts and sciences and found a higher appreciation for the role of the artisan in society than earlier sources betray.[2] Despite the growing consensus that the Middle Ages provided a fertile seedbed for technological development, one significant contemporary debate has been largely overlooked, namely, the late medieval dispute over the importance of alchemy—whether it fit into the legitimate fields of knowledge and whether its claims were possible or even legal.[3] Why should we consider this topic within the context of medieval technology? As I shall show in this essay, alchemy provided a natural focus for the issue of man's artisanal power in the natural world.

The medieval world view was marked by a deep division between art and nature. Stemming partly from Aristotle, and partly from other Greek, Latin, and

The research for this article was made possible through the generous funding of the Belgian American Educational Foundation, the Sheldon Traveling Fellowship and the Niles G. Whiting Fellowship of Harvard University, and the National Science Foundation. I would also like to thank John E. Murdoch of Harvard University and Robert Halleux of the Université de Liège. An earlier version of the article won the Schuman Prize in 1986.

[1] Lynn White, jr., *Medieval Technology and Social Change* (Oxford: Oxford Univ. Press, 1962). A generally positive assessment of White's work can be found in Brian Stock, "Science, Technology, and Economic Progress in the Early Middle Ages," in *Science in the Middle Ages,* ed. David C. Lindberg (Chicago: Univ. Chicago Press, 1978), pp. 1–51. For general treatments of medieval technology, see Bertrand Gille, *Histoire générale des techniques,* ed. Maurice Daumas, 5 vols. (Paris: Presses Universitaires de France, 1962–1979), Vol. I, pp. 427–598, and Vol. II, pp. 2–139; and Jean Gimpel, *The Medieval Machine* (New York: Holt, Rinehart & Winston, 1976). For the aura of magic surrounding technology in the Middle Ages see William Eamon, "Technology as Magic in the Late Middle Ages and the Renaissance," *Janus,* 1983, *70*(3–4):171–212. See also Bert Hanson, "Science and Magic," in Lindberg, *Science in the Middle Ages,* pp. 483–506.

[2] Guy Beaujouan, *L'interdépendence entre la science scolastique et les techniques utilitaires (XIIe, XIIIe, et XIVe siècles)* (Conférence faite au Palais de la Découverte, ser. D, 46) (Paris: Université de Paris, 1957); James Weisheipl, "The Nature, Scope, and Classification of the Sciences," in Lindberg, *Science in the Middle Ages,* pp. 473–474, 480; and Weisheipl, "Classification of the Sciences in Medieval Thought," *Mediaeval Studies,* 1965, *27*:54–90. See also Jan Legowizc, "Le problème de la theorie dans les *artes illiberales* et la conception de la science au moyen âge," in *Arts libéraux et philosophie au moyen âge: Actes du quatrième congrès internationale de philosophie médiévale* (Montreal: Institut d'Etudes Médiévales, 1969), pp. 1057–1061.

[3] On the issue of legality in this context see Johannes Chrisippus Fanianus, *De iure artis alchemiae . . . ,* in J. J. Manget, *Bibliotheca chemica curiosa,* Vol. I (Geneva, 1702), pp. 210–216.

Arabic sources, this view placed strict boundaries on the conceptual limits of technical innovation. The twelfth-century monastic writer Hugh of Saint Victor, famous for his influential inclusion of technology in the field of the sciences, wrote that "the products of artificers, while not nature, imitate nature, and in the design by which they imitate, they express the form of their exemplar, which is nature." Here Hugh is merely echoing the conviction of ancient Greek philosophy that the various branches of the "mechanical arts," what we would call technology, were originally learned by copying natural processes. As he also writes, "The human work, because it is not nature but only imitative of nature, is fitly called mechanical, that is adulterate."[4] The pejorative view that the mechanical arts derived their name from the Greek word for adultery (*moicheia*), because of their trickery, was of course widespread in the Latin West.[5] Although man could copy nature by means of art, his products—however cunning—could never be identical to their natural models. Hugh of Saint Victor was one of the more appreciative writers to consider the "adulterine arts," among which he grouped "fabric making, armament, commerce, agriculture, hunting, medicine, and theatrics,"[6] but he too considered them to be born of—and limited to—the mimicry of nature.

Latin alchemists did not generally strive to vitiate the principle enunciated by Hugh of Saint Victor; on the contrary, few medieval alchemical works fail to pay homage to the notion that "art imitates nature." But alchemical writers, unlike those in the mainstream of the Scholastic tradition, were willing to argue that human art, even if it learned by imitating natural processes, could successfully reproduce natural products or even surpass them. In so doing the alchemists of the Middle Ages developed a clearly articulated philosophy of technology, in which human art is raised to a level of appreciation difficult to find in other

[4] *The Didascalicon of Hugh of Saint Victor,* ed. and trans. Jerome Taylor (New York: Columbia Univ. Press, 1961), pp. 51, 55–56 (all quotations here from Hugh's *Didascalicon* are from Taylor's English translation). Taylor suggests that Hugh's depiction of the relationship between art and nature is drawn from Chalcidius's early medieval commentary on the *Timaeus.* See also, e.g., Aristotle, *Meteorologica,* 381b6–7; *Physics,* 199a15–17. Further sources are explored in Kurt Flasch, "Ars imitatur naturam," in *Parusia: Studien zur Philosophie Platons und zur Problemgeschichte des Platonismus* (Frankfurt: Minerva, 1965), pp. 265–306. Stock has found Hugh of Saint Victor's inclusion of seven mechanical arts among the traditional liberal ones "in an early *Ysagoge in theologiam,* in an exposition of Martianus of Capella, and in the *Aeneid* commentary attributed to Bernard Silvester": "Science, Technology" (cit. n. 1), pp. 45–48. See also George Ovitt, Jr., "The Status of the Mechanical Arts in Medieval Classifications of Learning," *Viator,* 1983, *14*:89–105, and Lynn White, jr., "Cultural Climates and Technological Advance in the Middle Ages," *Viator,* 1971, *2*:171–201, esp. pp. 196–197, where the positive influence of Hugh's *Didascalicon* is discussed.

[5] Lynn White's belief that this spurious etymology "was to have small influence in the West" ("Cultural Climates," cit. n. 4, pp. 192–193) is contradicted by the many examples collected by Peter Sternagel, *Die artes mechanicae im Mittelalter,* Vol. II of *Münchener Historische Studien, Abteilung Mittelalterliche Geschichte* (Kallmünz über Regensburg: Michael Lassleben, 1966). Among these we find the 9th-century monastic writer Martin of Laon, who very possibly originated this derivation of *mechanica,* the 9th–10th-century figure Remigius of Auxerre; the anonymous Bamberger classification of the sciences dating from the early 12th century; the 12th-century writers Richard of Saint Victor, Bernardus Silvestris, and the canonist Hugutio; an anonymous commentary on Aristotle's *Perihermeneias;* and such 13th-century authors as Radolfus de Longo Campo (c. 1216), Vincent of Beauvais, Albertus Magnus, Johannes Balbi (c. 1286), and Engelbert of Admont (fl. 1250–1331); see pp. 45–46, 89–91. Sternagel maintains that learned society gradually devalued the mechanical arts between the beginning of the 12th century and the middle of the 13th; the artisans themselves were of course making real technological progress.

[6] Hugh, *Didascalicon* (cit. n. 4), p. 74.

writings until the Renaissance. The degree to which medieval alchemists and their supporters were forced to develop their positive views about the power of technology in order to salvage their art from the increasingly hostile audience of the late Middle Ages is truly remarkable. This essay will attempt to trace the alchemical debate from its inception in the second half of the twelfth century up to a definite crisis reached in the first quarter of the fourteenth century.[7]

ALCHEMICAL DEBATE IN THE THIRTEENTH CENTURY

Alchemy first made its appearance in the Latin West around the mid twelfth century, when Robert of Ketton translated the *De compositione alchemiae* of Morienus from Arabic into Latin. Between the time of Robert's translation and the end of the fourteenth century, a massive quantity of alchemical literature appeared in Latin, much of it original in character. Yet the university curricula of the Middle Ages did not choose to incorporate alchemy, nor did any institutes of higher learning teach it until the early seventeenth century.[8] Although many medieval alchemical works were apparently written by Scholastic authors, using their characteristically dry, orderly style of exposition, the Scholastics effectually relegated alchemy to the category of marginality by denying it university status.

The reasons for this are complex. It is not enough to say that the medieval universities were bastions of Aristotelianism and that Aristotle had nothing to say about alchemy. Although both halves of this statement are from a modern perspective true, they are misleading. First, some Scholastic writers steeped in Aristotle, such as Albertus Magnus and Roger Bacon, definitely believed in the possibility of alchemical transmutation, as I shall presently elaborate. In fact, the alchemy of the late Middle Ages was a perfectly reasonable and sober offshoot of Aristotle's theory of matter. In this we must carefully distinguish medieval alchemy from the eclectic, Neoplatonic alchemy of the Renaissance, suffused with theosophy and cabalism.

Second, it was commonly believed in the late Middle Ages that Aristotle himself had written about alchemy. At least eighteen different pseudonymous works on alchemy attributed to Aristotle in the late Middle Ages survive in modern libraries. One alchemical work of the late twelfth or early thirteenth century

[7] This topic has been touched on in a brief and rather ahistorical fashion by Reijer Hooykaas, *Religion and the Rise of Modern Science* (Edinburgh: Scottish Academic Press, 1972), pp. 55–58. A more useful synopsis ignores the influence of Avicenna: Chiara Crisciani, "La 'quaestio de alchimia' fra duecento e trecento," *Medioevo: Rivista di Storia della Filosofia Medievale,* 1976, *2*:119–168. Crisciani's useful study of an alchemical systematist of the 1330s falls outside the chronological scope of this essay: "The Conception of Alchemy as Expressed in the *Pretiosa Margarita Novella* of Petrus Bonus of Ferrara," *Ambix,* 1973, *20*:147–162.

[8] For Robert of Ketton see Robert Halleux, *Les textes alchimiques: Typologie des sources du moyen âge occidental* (Turnhout, Belgium: Brepols, 1979), fasc. 39, pp. 49, 70; this work contains by far the best available treatment of the historiography of medieval alchemy as a whole. Some controversy still lingers as to the genuineness of the Morienus text and its ascription to Robert; see Julius Ruska, "Zwei Bücher de compositione alchemiae und ihre Vorreden," *Archiv für die Geschichte der Mathematik, der Naturwissenschaften, und der Technik,* 1928, *11*:28–37; and Lee Stavenhagen, "The Original Text of the Latin *Morienus,*" *Ambix,* 1970, *17*:1–12. On the integration of alchemy into the German universities during the Scientific Revolution see Owen Hannaway, *The Chemists and the Word: The Didactic Origins of Chemistry* (Baltimore, Md.: Johns Hopkins Press, 1975).

attributed to Aristotle exists today in over thirty-five manuscripts, many of them medieval.[9] Similarly, a long section of Avicenna's *Book of the Remedy* that attacked alchemy was habitually attributed to Aristotle by medieval writers. Some, confronted by Aristotle's seeming to support alchemy in one text while attacking it in another, went so far as to devise a developmental hypothesis in which the young Stagyrite was harshly critical but changed his views in the wisdom of old age.[10]

Nor is it sufficient to argue that alchemy was denied university status because of its characterization as a technology (mechanical art). In fact, alchemy occupied a medial position between the arts and the sciences, a position also occupied by medicine. Like medicine, alchemy consisted of a body of theory about certain aspects of the natural world; this theory was then used to support a plethora of manual practices. Thomas Aquinas, to cite one example, variously calls alchemy an "operative science," a "mechanical art," and an "operative art." In the first case, he ranks "medicine, alchemy, and moral [philosophy]" together, since they have a practical use and pertain more to specific subjects than do such fields as metaphysics, physics, and mathematics. In the second and third cases, Thomas groups alchemy with agriculture and medicine as technological pursuits subordinate to physics.[11] But the medieval university curriculum frequently included such subjects as medicine and moral philosophy, despite Thomas's classification of them as, respectively, "mechanical" or "operative." We cannot therefore view the learned disdain for the practical and technological as sufficient cause for alchemy's exclusion from the medieval university, though it may have been a contributing factor.

Far more convincing explanations for alchemy's lack of institutional success than such general ones as the genuine Aristotle's silence or the learned disdain for technology can be found by examining specific medieval documents. It then becomes clear that between the time of alchemy's inception in the mid-twelfth century and the end of the thirteenth century a general backlash against this discipline gradually developed, with mainstream scientific and religious authorities coming to agree in its denunciation. In such an atmosphere it would have been academically unprofitable, to say the least, for a university master to teach alchemy publicly. The result is that alchemical writers went "underground." Anyone who seriously consults the alchemical bibliography of the Latin Middle Ages cannot fail to be impressed by the large number of pseudepigrapha.[12] The

[9] Charles B. Schmitt and Dilwyn Knox, *Pseudo-Aristoteles Latinus: A Guide to Latin Works Falsely Attributed to Aristotle before 1500* (Warburg Institute Surveys and Texts, 12) (London: Warburg Institute, 1985), entry 58; for texts primarily or exclusively of alchemical content, see entries 1–5, 10, 21–22, 25–26, 54–56, 58, 73–74, 85, and 93. Other texts, such as the *Secretum secretorum,* contain substantial sections on alchemy.

[10] Petrus Bonus, *Margarita pretiosa,* in Jean Jacques Manget, *Bibliotheca chemica curiosa,* Vol. II (Geneva, 1702), pp. 76, 80.

[11] See Thomas Aquinas, in his commentary on Boethius's *De trinitate*: "Et ideo etiam quanto alia scientia magis appropinquat ad singularia, sicut scientiae operativae, ut medicina, alchimia, et moralis, minus possunt habere de certitudine." *S. Thomae Aquinatis opera omnia, curante Roberto Busa S.I.* (Stuttgart: Frommann-Holzboog, 1980), Vol. IV, p. 536, col. 3. On p. 532, col. 1, Thomas classes agriculture, alchemy, and "aliae artes quae dicuntur mechanicae" together, again as "operative arts," and at p. 533, col. 1, he explicitly subordinates medicine, alchemy, and agriculture to physics.

[12] Over 30 medieval alchemical works are attributed to Albertus Magnus alone: Pearl Kibre, "Alchemical Writings Ascribed to Albertus Magnus," *Speculum: A Journal of Medieval Studies,* 1942, *17*:499–518. I have recently proved that the *Semita recta,* long considered to have the greatest claim

following discussion will not limit itself to the recapitulation of Scholastic viewpoints concerning alchemy but will also examine some of this pseudonymous literature. We shall find that, in the process of justifying this discipline before its opponents, the alchemists and their supporters gave a conscious and articulate defense of technology, indeed, one of the earliest and most thorough to be found in Latin Christendom. The texts to be discussed make up a disputation literature that may justly be called the "alchemical debate" of the late Middle Ages, although this debate was not really resolved until the university of the Scientific Revolution incorporated chemistry as a part of its curriculum.

THE EARLY THIRTEENTH CENTURY

Our story begins with the English translator Alfred of Sareshel, who around 1200 translated a meteorological section of the Persian philosopher Avicenna's (980–1037) *Kitāb al-Shifāᵓ (The Book of the Remedy)* and inserted it into the fourth book of Aristotle's *Meteorologica,* already translated by Henricus Aristippus.[13] This short text, which came to be known in Latin as *De congelatione et conglutinatione lapidum,* immediately acquired the authority of a genuine Aristotelian production, since it appeared to be the conclusion of the *Meteorologica*'s fourth book.[14] It became thereby the locus classicus for all subsequent attacks on alchemy, and virtually any alchemical writer—whether philosophically sophisticated or not—felt obliged to respond to the arguments of "Aristotle" (i.e., Avicenna). In the process, the *De congelatione* of Avicenna became a focal point for the discussion of human artisanal power in general.

The *De congelatione* contains a description of geological processes, including the formation of the known metals—gold, silver, copper, tin, lead, and iron. Following the doctrines of Arabic alchemy, Avicenna asserts that these six are composed of mercury (mercury is considered not a metal but a component of metals) and sulfur in varying quantities and degrees of purity. It therefore comes as something of a shock when he proceeds to denounce the doctrine of metallic transmutation, upon which alchemical practice is based. Avicenna's main points may be summarized under two heads:

1. Artificial and natural products are intrinsically different, for art is inherently inferior to nature and cannot hope to equal it. Therefore artificers cannot change an inferior metal to a better one, although they can produce passable imitations of the precious metals by inducing superficial characteristics.
2. The true species-determining characteristics of metals cannot be known, since they subsist beneath the level of sense. Since these specific differences are unknown, it will be impossible to bring about the

to authenticity of any Albertine alchemical work, is a forgery: William Newman, "The Genesis of the *Summa perfectionis,*" *Archives Internationales d'Histoire des Sciences,* 1985, *35*:240–302, esp. pp. 246–260.

[13] James Otte, "The Life and Writings of Alfredus Anglicus," *Viator,* 1972, *3*:275–291.

[14] *Avicennae de congelatione et conglutinatione lapidum,* ed. and trans. E. J. Holmyard and D. C. Mandeville (Paris: Geuthner, 1927), pp. 1–11. Unfortunately, these editors have not supplied a critical Latin text. I have prepared a temporary working edition of the section known as *Sciant artifices* (discussed below) in William Newman, "The *Summa Perfectionis* and Late Medieval Alchemy," 4 vols. (Ph.D. diss., Harvard Univ., 1986), Vol. I, pp. 59–62.

> transmutation of one metal into another, for the alchemist cannot manipulate what he does not know.

Avicenna's argument may seem prima facie self-evident to the modern reader, but the terms "species" and "specific difference" are somewhat nuanced. Avicenna's terminology has a logical basis. When he used the term *nau*ᶜ, rendered in Latin by Alfred of Sareshel as *species,* he meant above all to refer to the group of characteristics defining a particular kind of thing. To Avicenna there are six such species among metals: gold, silver, copper, iron, tin, and lead. All six belong to the more general *genus* of metals, which he informally defines as "malleable," "fusible," "mineral" (that is to say "mined") bodies.[15] Hence each type of metal shares the set of properties that define the genus: anything that is a malleable, fusible body found in mines will be a metal. But the metals are not all identical: gold, silver, copper, iron, tin, and lead, while metals, also have their own specific differences that make each of them belong to a particular species. The thrust of Avicenna's conclusion is that the specific differences that make metals fall into different species are not such easily perceived properties as melting point, malleability, specific gravity, and color. Instead, the specific differences are really underlying and imperceptible: we cannot know them, and therefore we cannot change them.

It may be tempting for the modern reader to view Avicenna's rejection of alchemy as a forward-looking event that foreshadowed the weaning of chemistry from the "irrational" or "pseudoscientific" doctrines of alchemy. A closer look will reveal, however, that it was Avicenna, and not the alchemists, who held reactionary views. Avicenna begins his attack with the "self-evident" assertion that natural products are intrinsically superior to their artificial counterparts and that the latter cannot possibly match up to the naturally occurring exemplars of which they are copies. As two modern commentators on the *De congelatione* have remarked, Avicenna would have been on the side of "the general public [today], who usually imagine that synthetic indigo, for example, is not veritable indigo, but only a very good imitation."[16]

Avicenna, though basing himself on an ancient prejudice, in fact takes a considerably stronger position about the schism between natural and artificial products than did Aristotle. In the *Physics* (2.8, 199a) the latter allows art either to mimic nature or to carry some of her works to a greater state of perfection than they would otherwise have: "One sort of art perfects that which nature cannot complete, while another sort imitates nature."[17] One almost gets the impression that a personal experience with alchemical counterfeiters led Avicenna to his disdain for human art as expressed in the *De congelatione*. Whatever the sources

[15] Avicenna, *De congelatione,* pp. 32–33. *Nau*ᶜ is an Arabic rendering of the Greek *eidos,* used by Aristotle to denote either "species" or "form." In the *De congelatione* Avicenna uses the Latin *species* primarily to distinguish the individual types of metal (such as lead and tin) from the *genus* of metals in general. See Ibrāhīm Madkūr, *L'organon d'Aristote dans le monde Arabe: Ses traducions, ses études, et ses applications* (Etudes Musulmanes, 10) (Paris: Vrin, 1969), pp. 70, 299. For the term itself see Avicenna, *De congelatione,* p. 24. My discussion of this topic owes a debt to conversations with John Murdoch and A. I. Sabra.

[16] Holmyard and Mandeville, in Avicenna, *De congelatione,* p. 41, n. 5.

[17] Aristotle, *De physico auditu,* in *Aristotelis opera cum Averrois commentariis,* Vol. IV (Venice, 1562), fol. 78r, col. 2: "Et omnino ars alia quidem perficit que natura non potest efficere, alia vero imitatur."

for his view, the universal proposition that art is inferior to nature, coupled to the belief that natural species are intransmutable, constituted an attack not on alchemy alone but on the totality of technology and applied science. Avicenna's point was not merely that human technology cannot *outdo* nature but that man cannot even hope to *imitate* nature in a truly successful fashion. Avicenna thus first clothes the ancient philosophical disdain for technology in the form of an "authoritative" enunciation, then spells out the specific reasons for alchemy's failure in terms of Aristotelian natural philosophy. Later we shall find such varied claims as the impotence of demons to work miracles and the inability of horticulturalists to produce new breeds of plants supported by reference to Avicenna's dictum that alchemists cannot transmute species. The effects of the *De congelatione* were by no means restricted to alchemy but served to crystallize an antitechnological bias in many areas.

In response to Avicenna's dictum that species are intransmutable—which came to be referred to in abbreviated form by the incipit *Sciant artifices*—the alchemists developed counterarguments adopting a radical view of technology in which man assumed extraordinary power over nature. Centuries before Francis Bacon's philosophy of nature with its Draconian decree to "put nature to the rack," we find protagonists of alchemy asserting that man's ability to transform the natural world is virtually unbounded. Their justification of human art was not based on vague optimism, however; it was supported by practical observation, analogical reasoning, and a Neoplatonizing Aristotelianism.

One of the earliest sets of counterarguments to the *De congelatione* can be found in a pseudonymous *Book of Hermes* written in the first half of the thirteenth century or before. This work contains a series of elliptical attacks on alchemy, each with its matching rebuttal. The first argument, that metals are natural products and hence may not be replicated by artificial means, implicitly contains Avicenna's axiom that natural products are always better than artificial ones. The author of *Hermes* rebuts this by saying that human technology frequently succeeds better than nature herself, since artificial verdigris, vitriol, zinc oxide, and sal ammoniac are all better than the naturally occurring forms, "which [anyone] who knows about minerals does not contradict."[18] Similarly, the horticulturist improves on nature by making successful graftings. We have here a sort of manifesto proclaiming the power of technology in general and chemical technology in particular. Interestingly, "Hermes" does not deny that art learns by mimicking nature: in order to eviscerate Avicenna's proposition that art is *weaker than* nature, it is sufficient for him to point to the empirical fact that certain products have greater efficacy when prepared artificially.

When the author of *Hermes* comes to Avicenna's assertion that species cannot be transmuted, he adopts the approach of logic: he replies that the metals belong

[18] *Liber Hermetis,* quoting from the partial working edition in Newman, "The *Summa perfectionis*" (cit. n. 14), Vol. I, pp. 63–67, on p. 65, ll. 36–40: "Sal vero viride et dragantum et thutia et sal armoniacus et naturalia et artificialia sunt. Immo et artificialia naturalibus potiora sunt, quod qui de mineriis sciunt non contradicunt." The *Liber Hermetis* has never been printed, or for that matter analyzed. I have found it in the following 13th–14th-century MSS: Cambridge, Trinity College 1400, fols. 131r–133r; Oxford, Bodleian Library (BL), Bodley 679, fols. 20r–21r; London, British Museum (BM), Add. 41486, fols. 218r–222r; Paris, Bibliothèque Nationale (BN), Latin (Lat.) 6514, fols. 135r–v; and in the following 14th-century MSS: London, BM, Sloane 1754, fols. 60r–62r; Palermo, Biblioteca Communale, 4QqA10, fol. 37v (incomplete).

to a single definition, any metal being "a composite, fusible, incombustible, malleable body."[19] Logically, there is no compelling reason why this should be called a genus rather than a species, since such differentiation is merely a matter of degree (a genus is merely comprehended by a more general definition than a species). In providing a single definition for all the metals, the author of *Hermes* can therefore argue that they all belong to a single "species," and that the "species" of which Avicenna speaks are by implication only "more specific species" (*species specialiores*). Thus *The Book of Hermes* does not need the transmutation of species. This purely logical approach to undermining the *Sciant artifices* soon gave way in the West to a more hylomorphic tendency. As we shall see, Albertus Magnus—among others—took Avicenna's *species* to mean a form that "inheres" physically in the substance of a metal in order to determine its particular set of characteristics. Although permissible within the framework of Aristotelian philosophy (where *eidos* means either "species" or "form"), Albert's interpretation would have the effect of turning Avicenna's discussion of *genera* and *species* into an argument about matter and form.

THE MID-THIRTEENTH CENTURY: VINCENT OF BEAUVAIS, ALBERTUS MAGNUS, AND ROGER BACON

The *Book of Hermes,* although it offered a succinct and early defense of alchemy, does not seem to have been known to the three Scholastic authors of the mid thirteenth century most concerned with alchemy, Vincent of Beauvais, Albertus Magnus, and Roger Bacon. The works of these three authorities provide a gauge of the degree of controversy alchemy aroused at that time.

I will dispense with Vincent briefly, since he gives a rambling account of alchemy, devoid of originality. Vincent wrote his *Speculum doctrinale* and *Speculum naturale* between 1244 and 1250.[20] Since these two works contain much the same material on alchemy and on mineralogy in general, I shall focus on the somewhat more orderly *Speculum doctrinale.* The *Speculum doctrinale* places alchemy among the mechanical arts. Alchemy, unlike the sciences per se, is merely useful from a practical point of view—to the metalworker, since it teaches "the examination, intermixture, separation, and transmutation" of the metals, and to the physician, because it aids in the isolation of healthy from harmful components, which "are often found mixed together in simple medicines." Vincent adds that alchemy is descended from the "science of minerals" (*ab illa parte naturalis philosophie que est de mineris*) in the same way that agriculture is derived from the "science of plants." To Vincent, therefore, alchemy "is properly the art of transmuting mineral bodies, such as metals and the like, from their own species to others."[21]

[19] *Ibid.*, p. 66, ll. 57–61: "De speciali vero differentia cui dubium est metalla specialiter non differe, cum in una diffinitione conveniant. Verbi gratia, corpus compositum, in igne fusibile, non combustibile, sub malleo extendibile."

[20] William A. Wallace, "Vincent of Beauvais," *Dictionary of Scientific Biography,* ed. Charles C. Gillispie, 16 vols. (New York: Scribners, 1970–1980) (hereafter *DSB*), Vol. XIV, p. 35.

[21] Vincent of Beauvais, *Speculum doctrinale* (Venice, 1494), Book 11, Ch. 105: "Ad fabrilem quidem propter metallorum examinationem, commixtionem, disgregationem, transmutationem. Ad medicinam itidem propter substantiarum vel qualitatum salubrium a noxiis que frequenter etiam in medicinis simplicibus permixte sunt separationem. . . . Alkimia proprie est ars transmutandi corpora mineralia a propriis speciebus ad alias, ut sunt metalla et huiusmodi."

So far, Vincent is reasonably consistent: he considers alchemy to be a simple practical art, entirely devoid of theoretical content. But the *Speculum doctrinale* here begins to contradict itself, for the introductory passage is directly followed by passages from alchemical writers giving long theoretical descriptions of the generation of minerals from sulfur and mercury within the earth. Clearly this is a speculative sort of alchemy, not just a mechanical art. A parallel confusion reigns in Vincent's description of Avicenna's attack on alchemy. He quotes Avicenna's broadside without giving his own point of view, then replies with an extract from pseudo-Avicenna containing a number of garbled arguments *in favor* of alchemy.[22]

Vincent's rather complacent and confused account of alchemy is followed chronologically by the *De mineralibus* of Albertus Magnus. Here we find a considerably more coherent assessment. Between 1250 and 1254 Albert took on the task of writing a comprehensive study of mineralogy as part of his endeavor to explain the totality of natural science.[23] Since Albert could find no Aristotelian book on minerals to comment on, he had to turn to the texts of the alchemists.

In the course of his investigation, Albert therefore felt the need to respond to the arguments of the *De congelatione,* which he knew to be a work of Avicenna's. Albert begins his analysis of transmutation with an attack on previous authors who have proposed that all metals share one form, that of gold, in varying states of completion. Arguing from sense, he says that the metals appear to be "stable" (*permanens*); under normal circumstances they do not become other metals. Therefore they must each have their own substantial form by which they are "perfected." Similarly, each metal has its own peculiar set of properties, so their accidents are not common. As a result, "the substances and specific form [*species*] [of different metals] must be different."[24]

Given that Albert believes the metals to differ in their species, we might expect him to uphold the viewpoint of the *De congelatione*. This is not the case, however. In a special chapter he directly attacks the pronouncement *Sciant artifices,* where Avicenna had argued that alchemists could not transmute species. In this chapter it becomes clear that Albert has understood the Latin *species* to mean "specific form."[25] This substitution of "specific form" for "species" allows Albert to circumvent the *Sciant artifices,* since he can now draw on a well-defined Scholastic theory concerning the physical corruption of a preexistent form followed by the induction of a subsequent form. Thus Albert believes that *species*

[22] *Ibid.,* Book 8, Chs. 42, 84.

[23] Robert Halleux, "Albert le grand et l'alchimie," *Revue des Sciences Philosophiques et Théologiques,* 1982, *66*(1):58.

[24] *Albertus Magnus: Book of Minerals,* trans. Dorothy Wyckoff (Oxford: Clarendon Press, 1967), pp. 172–173. Here as elsewhere Wyckoff uses the expression "specific form" for the Latin *species.* Wyckoff's translation, however, obscures the fact that Albert's argument is motivated by the *Sciant artifices* of Avicenna. Like Avicenna, Albert uses the Latin term *species* rather than *forma specifica,* though what he means is closer to the latter.

[25] *Ibid.,* pp. 177–179. For the original Latin see Albertus Magnus, *Mineralium libri quinque,* in *B. Alberti Magni . . . opera omnia,* ed. Auguste Borgnet, Vol. V (Paris, 1890), pp. 70–71. For further evidence that Albert conflates *species* and *forma specifica* see the following passages: "Experimenta autem alchimicorum graves duas nobis hic ingerunt dubitationes. Videntur enim illi dicere quod sola auri species est forma metallorum" (68a); and "Quod si forte concederetur quod substantiam auri inducat, adhuc non est sufficiens probatio ad hoc quod non sit nisi una species metallorum: quoniam calcinando et sublimando et distillando et caeteris operationibus quibus elixir per materiam metallorum faciunt penetrare, corrumpere potest species metallorum quae primitus infuerunt materiae metallorum" (69a).

can indeed be transmuted, inasmuch as one specific form can be destroyed and replaced by another.

Albert's interpretation, however, slightly distorts Avicenna's use of the Arabic term *nau*ᶜ, or *species*, in the *De congelatione*. By *species* Avicenna meant primarily a logical entity, in the same way that the term is contrasted to *genus* by logicians. In the *De congelatione* Avicenna does not speak of species as "inhering" in matter, or as being corrupted and induced. Instead, his species are above all abstract categories that existed in the Creator's mind when he fashioned the natural world. To say that such logical species are transmutable would be fatuous indeed, since they represent the distinct underlying concepts by which God created separate metals.[26]

At any rate, armed with his hylomorphic interpretation of "species," Albert says that honest alchemists act toward metals just as physicians do toward their patients.[27] The alchemist first cleans and purifies the old metal, just as a doctor employs emetics and diaphoretics to purge his patient. Then he strengthens the "elemental and celestial powers" in the metal's substance, apparently by adding druglike components and observing astrological "judgments." As a result, the purged metal receives a new and better specific form from the celestial virtues of the stars. Hence the alchemist has not *transmuted* any species: he has only removed one specific form and prepared the way for another to be received.

Albert's benign view of alchemy does not bear witness to a heated debate on this subject. He is not responding to any *moderni* but only to Avicenna and other Arabic authors. The equanimity of his tone, furthermore, seems to reflect a period in which alchemical transmutation was not yet a general subject of irascible dispute. When we turn to Roger Bacon, the atmosphere changes radically.

Roger wrote his *Opus tertium* around the year 1266, as a part of the trilogy also comprising his *Opus maius* and *Opus minus*. The three books were intended as an advertisement for reform, and as such they were sent by special courier to Roger's friend Clement IV.[28] In the *Opus tertium* Roger proposes that alchemy should form the primary means of reforming Scholastic science. He asserts that alchemy teaches things of which Aristotle was completely ignorant, such as the precise generation of minerals, pigments, precious stones, and humors from the elements. Furthermore, since alchemy is the science of the elements per se, while natural philosophy and medicine concern things made out of the four elements, such as the four humors, alchemy is the most basic of the sciences. Hence Roger's approbation of alchemy far exceeds that of Vincent or Albert: whereas they see alchemy as primarily a practical art whose masters have provided empirical examples for real philosophers to explain, he wants to make it the wellspring of all medical and natural knowledge. Although modern historians have stressed Roger's mathematics (to the ultimate misfortune of the poor friar),

[26] Avicenna attacked specific transmutation in at least one other text: the *R. fi ibṭāl aḥkām al-nujūm* or *R. al-Ishāra ilā ᶜilm fasād aḥkām al-nujūm*. Georges C. Anawati paraphrases it as follows: "Ce sont des absurdités [specific transmutation]; car pour tout ce que Dieu a créé moyennant la force de la nature, l'imitation artificielle est impossible; comme au contraire les productions artificielles et scientifiques n'appartiennent d'aucune manière à la nature." Anawati, "Avicenne et l'alchimie," in *Oriente e Occidente nel Medioevo: Filosofia e scienze* (Rome: Academia Nazionale di Lincei, 1971), pp. 285–341, on pp. 300–301.

[27] Albertus Magnus, *Book of Minerals* (cit. n. 23), pp. 178–179.

[28] A. C. Crombie and J. D. North, "Roger Bacon," in *DSB,* Vol. I, p. 378.

we must note that in the passage above he explicitly lauds alchemy as "greater than all the foregoing [sciences]," of which the science of mathematics was one.[29]

It is equally significant that Roger nowhere mentions the attack of the *De congelatione* in the three *Opera* addressed to Clement IV. Why, after all, should he have interjected the views of a doubting Thomas into his promotional broadsides? But Roger clearly knew the *Sciant artifices* from an early date, since he supported its point of view in his commentary of about 1245 on the pseudo-Aristotelian *De plantis*.[30] There he universalizes the proposition that "species cannot be transmuted," taking this supposedly Aristotelian view to apply to plants as well as metals. Clearly Roger underwent a serious change of mind between 1245 and the 1260s. If we examine his *Communium naturalium* of 1266, the reason for his new disregard of the *Sciant artifices* appears. Here he attacks its attribution to Aristotle.[31] Despite the view of "fools," he says, the *Sciant artifices* is only a second-rate commentary by Alfred of Sareshel. By replacing the authority of Aristotle with that of Alfred, Roger makes it an easy matter to dismiss the proposition that species may not be transmuted. In the *Communium naturalium* Roger goes far beyond his predecessors in rejecting the theoretical validity of the *Sciant artifices*. Whereas Vincent chose to take no position, and Albert circumvented the issue by interpreting species to mean specific form, Roger simply says that the proposition "species cannot be transmuted" is not true.

Furthermore, he adds that "fools" abuse the authority of Aristotle by attributing this position to him, apparently in attacking alchemy. Since Islamic authors did not attribute the *De congelatione* to Aristotle, it follows that the fools to whom Bacon refers must have been Latin fools. Albert's dispute was limited to Arabs, and Vincent found no need to take sides at all. It is therefore evident that the alchemical debate had grown in magnitude since the 1240s and 1250s, when Vincent and Albert were concerned with mineralogy. Further evidence is the great space given to disputation in alchemical works written after 1250. One such work was written by a practicing alchemist, probably in the last third of the thirteenth century.

ALCHEMICAL DEBATE IN THE LATE THIRTEENTH CENTURY: PAUL OF TARANTO

I have recently shown that Paul of Taranto, a thirteenth-century Franciscan, was the probable author of a famous alchemical text, the *Summa perfectionis,* spuriously attributed to the Arab Jābir ibn Ḥayyān.[32] More pertinent to our present argument, however, is the defense of alchemy that Paul presents in another text, the *Theorica et practica*.

[29] Roger Bacon, *Opus tertium,* in *Opera quaedam hactenus inedita,* ed. J. S. Brewer, Vol. I (London, 1859), pp. 39–40.

[30] "Item, quarto *Metheororum,* 'sciant artifices alkimie species metallorum transmutari non posse,' quare similiter nec species plantarum": Roger Bacon, *Questiones supra de plantis,* in *Opera hactenus inedita Rogeri Baconi,* ed. Robert Steele et al. (Oxford: Clarendon Press, 1911–1940), Vol. XI (1932), ed. Steele and F. M. Delorme, p. 251; see also p. 241.

[31] Roger Bacon, *Communium naturalium,* in *Opera hactenus inedita,* Vol. II (1911), ed. Steele, p. 7.

[32] See William Newman, "New Light on the Identity of 'Geber,' " *Sudhoffs Archiv,* 1985, *69*:76–90; and Newman, "Genesis of the *Summa perfectionis*" (cit. n. 12).

The *Theorica et practica,* as its title implies, is a didactic work comprehending both the theoretical and the practical bases of alchemy. It contains a long defense of this discipline, in which Paul attempts to defuse the *Sciant artifices* of pseudo-Aristotle. Paul prefaces his specifically alchemical comments, however, with a general defense of human art. He begins with a proem, heavily dependent on the pseudo-Aristotelian *Liber de causis,* in which he tries to justify the power of man over nature. He does this by identifying the Plotinian hypostasis *intellectus* with the human intellect, a not-uncommon conflation among thirteenth-century thinkers. Because nature is inferior and subject to intellect, man must therefore be in a position to manipulate and rule nature. Paul then proceeds to a special chapter on the relation of art to nature (see Appendix). Since human intellect rules over nature, he says, artisans such as "sculptors, painters, horticulturists and physicians" have nature subjected to themselves "as matter and instrument." Drawing on Aristotle's *Physics,* Book 2, Paul divides human art into two categories—that which generates an "extrinsic" form, as in the case of painting and sculpting, and that which terminates in an "intrinsic form," such as medicine or agriculture.[33] Those arts that produce an intrinsic form use the Aristotelian primary qualities—hot, cold, wet, and dry—as instruments. Arts inducing an extrinsic form rely on secondary qualities, such as colors and tastes.

Arts such as agriculture and medicine that act on the primary qualities can actually transmute substance, whereas sculpture, painting, carpentry, and other arts that work only on secondary qualities can only induce accidents into their subject. The genuine physician, horticulturist, or alchemist, therefore, produces real changes in essence and substance, because he manipulates the first qualities of matter. False artisans, on the other hand, produce only the appearance of change; they attack the symptom rather than the cause. When Paul comes to the *Sciant artifices,* he uses this bifurcation of the arts to remove its force, saying, "We do not consider the opinion of Aristotle which he writes at the end of the *Meteorologica*—'The alchemists should know that species cannot be transmuted'—to be true unless it be understood in the foresaid way, [that is, as occurring] through purely artificial agents."[34] In other words, the *Sciant artifices* holds only if the artisan employs secondary, "artificial" qualities, since these do not affect the substance of a given subject. Otherwise, if he uses primary qualities, it is indeed possible to induce substantial change and thus to alter species.

Paul of Taranto's argument, although directed mainly toward alchemy, is undeniably a justification of technology in general, since it upholds the power of those arts capable of manipulating primary qualities to induce real change in natural products. In other words, Paul consistently affirms the power of man really to alter and improve natural products. At the same time, his reasoning implicitly contains more than an apology for technical skill per se. By dividing the "arts" into two categories he implicitly distinguishes between pure technol-

[33] Paul of Taranto, *Theorica et practica,* Paris, BN, Lat. 7159, fols. 1r–55r, on fols. 1r–v, 2v, ll. 17–18. Since the *Theorica et practica* has never been printed, quotations are from an edition I have prepared, available in Newman, "The *Summa perfectionis*" (cit. n. 14), Vol. III, pp. 1–237. (See also the Appendix, for a partial edition and translation and the other manuscripts.)

[34] Paul of Taranto, *Theorica et practica,* fol. 4r, ll. 12–15: "Non putamus igitur esse verum verbum Aristotelis quod in fine sui libri scribit in methauris—'*Sciant artifices* species se transmutare non posse'—nisi hoc predicto modo per pura artificialia intelligatur."

ogy (e.g., sculpting, painting, and carpentry) and applied science (e.g., medicine, agriculture, and alchemy).

The difference between these two categories hinges on the second argument of the *Sciant artifices,* where Avicenna, remarking that the differences between metals are not known, asks: "When the [specific] difference is not known, how will it be possible to know whether it is taken away or not or even how it could be taken away? Removal of the accidents within, however, such as taste, color, and weight, or at least their lessening, is not impossible, for reason is not opposed to this."[35] Avicenna's argument relies on the impossibility of the artificer's manipulating that which he cannot recognize, namely, the hidden essential differences that make one metal different from another. The theoretical part of Paul of Taranto's *Theorica et practica,* therefore, is precisely an attempt to become acquainted with these essential principles of the different metals. These are perceived by means of tests performed in the laboratory or foundry. We know, for example, that metals contain sulfur for the following reason: when metallic ores are "calcined"—oxidized by intense heating—they give off a sulfurous, stinking smoke.[36] Some metals, however, contain more sulfur than others. Lead, for example, contains more sulfur than tin and of a more entrenched sort. Paul determines this from the fact that two calcinations of lead leave a yellow, sulfurous calx, while two such firings of tin leave a white calx, although the metal's smoke is still yellow.[37]

The presence of mercury in the metals is proved by similar means. Mercury readily forms an amalgam with gold, silver, copper, tin, or lead, simply by remaining in intimate contact with any of the said metals. The medieval writers even claimed to be able to amalgamate mercury with iron, though this claim was probably the result of observational error on their part. As Paul says, this striking affinity between quicksilver and the then-known metals "is due to a similarity of substance."[38] But he later describes a procedure by which mercury subjected to

[35] Avicenna, *De congelatione,* ed. Newman (cit. n. 14), pp. 61–62: "Differentie metallorum enim non sunt cognite et cum differencia non sit cognita, quomodo poterit sciri utrum tollatur nec ne, vel quomodo tolli possit? Sed expoliacio intus accidentium ut saporis, coloris, ponderis, vel saltem diminucio non impossibilis, quia contra hoc ratio non est."

[36] Paul of Taranto, *Theorica et practica* (cit. n. 33), fol. 8r, ll. 20–24: "Demonstrat etiam sulphuris cum mercurio secundum naturam esse admixtionem pro substantia et tinctura cum ad calcinationem metalla funduntur, maxime si fuerint imperfecta, mollia vel dura, fumus eorum et fetor sulphureus et etiam color in calcibus eorum."

[37] *Ibid.,* of lead, fol. 25r, 11. 24–30: "Probatur [autem] in eo sulphureitas esse duplex, quarum una est adustiva et fixa parum[que], altera⟨que⟩ magis fixa ex eo—quod una eius sulphureitas cito in fumum resolvitur, cuius probatio est sulphureus eius fetor et color; et facit deperditionem corporis in se ipso. Alia vero sulphureitas eius non nisi per difficilem maximum magisterium transit, que quidem remanet etiam in calce eius, cuius etiam probatio est sulphureus fetor et color citrinus." Of tin, fol. 26v, ll. 20–22: "Nam licet post primam calcinationem iovis et reductionem ad ignem magne sue ignitionis, apparet adhuc fumus eius citrinus, quod etiam sicut in saturno est necesse contingere ex resolutione in evolatione partis sulphuree in eo non fixe, tamen quoniam remanet eius calx alba, scilicet in saturno citrina, manifestum est ex hoc sulphureitatem in iove non ita profundatam esse in intimo suo substantie, nec tantam esse, nec taliter fixam sicut est in saturno." In quotations from the original language (notes and Appendix), editorial deletions are in square brackets and editorial interpola sections in angle brackets. (In the translations interpolations are in square brackets.)

[38] *Ibid.,* fol. 7v, l. 32–fol. 8r, l. 8: "Sed et sensus probat in plumbo, et stagno, et luna, et mercurio subtilitatem esse per similitudinem rei, magis autem fit hoc evidens cum ipse mercurius congelatur in massam, licet frangibilem, ad odorem ipsius saturnus vel iovis, cum scilicet liquatis ad ignem his, iam cum incipiunt frigefieri extra ignem, et facta in eis foveola cum forma ligni alicuius, proiicitur intus

the vapor of molten silver or lead is congealed per se "without the admixture of any other."[39] From the modern point of view, the product of this experiment would again be an amalgam. To the alchemist, however, it appeared that the hot, arid vapor of lead served to "dry out" the excess humidity of the mercury, thus allowing it to become a full-blown metal. Hence Paul says that this procedure, by which mercury is "independently" converted from a "spirit" to a metal, shows better than any other demonstration that the "subtlety [of the substance] of the metals" comes from mercury.

Paul supports these experimental demonstrations of the metals' components by an attempt to explain the nature of the metallic principles, sulfur and mercury, in terms of the four primary qualities, hot, cold, wet, and dry. Without going into details, I will point out the thrust of his argument. By arriving at the composition of the two metallic principles in terms of the four qualities, then showing how the two principles can be manipulated to form the six known metals, Paul manages to satisfy Avicenna's objection that the alchemist cannot manipulate that which he does not recognize. Using experimental demonstrations such as those described above and the philosophical framework provided by Aristotle's *De generatione et corruptione* and *Meteorologica,* Paul responds that properly educated alchemists really can recognize and manipulate the primary qualities directly. This ability to recognize qualities is precisely what distinguishes alchemists and physicians on the one hand from carpenters and painters on the other. The latter do not attempt to understand the nature of their material at its most fundamental level. They are technically skilled, but their work operates only on the level of appearance, not fundamental change. In other words they are scientifically uninformed artisans—pure technicians—whom Paul is careful to distinguish from those who derive their skill from a direct knowledge of the four Aristotelian qualities. I can think of no better term for the latter than *applied scientist,* a category already employed in this sense by Lynn Thorndike.[40]

In effect, therefore, Paul has made a distinction between the applied scientist, who understands and employs the true causes of things, and the simple artisan, who works to produce an effect without true knowledge of its causes. As Avicenna said in the *De congelatione,* such an artisan cannot change species, for if the specific difference—the cause of the species—is not known, "how will it be possible to know whether it is taken away or not, or how it could be taken away?"[41] The applied scientist, according to Paul, does understand the causes of species and can therefore change them.

Of the three Scholastics examined earlier, only Roger Bacon matches the pre-

mercurius. Tunc enim frigefacto tali plumbo vel stagno plene, mercurius invenitur densatus in massam argenteam plumbo vel stagno consimilem."

[39] *Ibid.,* fol. 20v, ll. 12–17: ". . . subtilitas scilicet metallorum ex mercurio fiat potissime sicut probatur ad sensum, quod maxime patet cum per artificium mercurius figitur et fit per se solum verum argentum, aut etiam aurum per tincturam sulphuream, sine alterius alicuius admixtione."

[40] See the term *applied science* in Lynn Thorndike, *History of Magic and Experimental Science,* 8 vols. (New York: Columbia Univ. Press, 1923–1958), *passim;* e.g., in Vol. I, Ch. 5, of the ancient mechanical engineering of Vitruvius and Hero of Alexandria; in Vol. II, p. 81, of pseudo-Farābī's *scientia de naturis*; and in Vol. II, p. 663, of Roger Bacon's *Epistola de secretis operibus.* Attempts like those made in these works to apply philosophical or scientific principles for technological ends must be distinguished from nonscientifically oriented craft traditions in which no attempt to utilize scientific first principles is made.

[41] Avicenna, *De congelatione,* ed. Newman (cit. n. 14), p. 61: ". . . cum differencia non sit cognita, quomodo poterit sciri utrum tollatur nec ne, vel quomodo tolli possit?"

mium put on technology, or rather applied science, by Paul of Taranto. It is possible, however, that the very success of such arguments as Paul's and Roger's led to the condemnation of their views. In the *Theorica et practica* Paul goes to the limit by insisting that "anything short of the animated and the soul itself can be made naturally from anything else with regard to elementary form . . . such as bodies composed of the four elements, as for example stones and metals."[42] In other words, the powers of art are limited only by the human inability to make and infuse another soul. Roger Bacon, in a similar fit of hubris, goes so far as to say that alchemical gold, because it contains the four elements in an even better proportion than natural gold, can restore the human body to a condition of elemental equality like that of Adam and Eve and the resurrected at the end of time. Roger's enthusiastic views may have contributed to his apparent imprisonment during the last fifteen years of his life.[43]

ANTI-ALCHEMICAL SENTIMENT IN THE LATE THIRTEENTH CENTURY AND THE RESULT OF THE DEBATE

What then was the opposite camp doing, while Paul and Roger were concocting their alchemical manifestos? The first direct counterattack by a Latin author that I have been able to locate is contained in a work by the Thomist Giles of Rome (Aegidius Romanus), written between 1286 and 1291.[44] Before turning to Giles, however, it will be useful to summarize briefly the views of his teacher, Thomas Aquinas.

Determining Thomas's opinions on alchemy is not as straightforward a task as one might hope, since his genuine works were sometimes completed posthumously by others. This appears to have been the case with his commentary on the *Meteorologica* of Aristotle: the portions of the text that give a positive portrayal of alchemical transmutation were actually written by another author. The *Summa theologiae,* finished or broken off in 1272, refers several times to alchemy, but only in passing.[45] Fortunately, Thomas's commentary on the *Sentences* of Peter Lombard, probably written between 1252 and 1256, contains a revealing treatment of demonology in which alchemy—though not the main topic—is discussed.[46]

When commenting on Book 2 of the *Sentences* Thomas asks "whether demons can induce a true corporeal effect into corporeal matter." He then lists five authoritative opinions that deny the possibility of such demonic power. The last of

[42] Paul of Taranto, *Theorica et practica* (cit. n. 33), fol. 6v, ll. 18–20: ". . . probaverimus in prefato de libello de causatis et causis naturaliter fieri posse ex quolibet quidlibet citra animatum et animam, scilicet quantum ad formam elementarem, sive mixtorum sive simplicium, ut sunt quatuor elementorum corpora ac lapides et metalla."

[43] Roger Bacon, *In libro sex scientiarum . . . ,* in *Opera hactenus inedita,* Vol. IX, ed. A. G. Little and E. Withington (1928), pp. 183–184. See also Thorndike's comments in *History of Magic and Experimental Science* (cit. n. 40), Vol. II, pp. 628–629.

[44] Aegidius Romanus, *Quodlibeta revisa, correcta, et varie illustrato, studio M. F. Petri Damasi de Coninck* (Louvain, 1646), pp. 147–149 (= *Quaestio* 3, *Quodlibeti* 8, *Membri* 3).

[45] Aquinas, *In meteorologicorum continuatio,* in *Opera* (cit. n. 11), Vol. VII, p. 627, cols. 1 and 2; and Aquinas, *Summa theologiae, ibid.,* Vol. II, p. 623, cols. 2, 3; p. 873, cols. 1, 2. See also Francesco Migliorino, "Alchimia lecita e illecita nel Trecento: Oldrado da Ponte," *Quaderni Medievali,* 1981, *11*(June):33.

[46] Thomas Aquinas, *In quatuor libros sententiarum,* in Aquinas, *Opera,* Vol. I, p. 145, cols. 1–3. On the probable date of composition see William O. Wallace and James Weisheipl, "Thomas Aquinas," in *The New Catholic Encyclopedia,* Vol. XIV, p. 104.

these is the *Sciant artifices*: "Demons cannot work except through the method of art. But art cannot give a substantial form, whence it is said in the chapter on minerals that the authors of alchemy should know that species cannot be transformed. Therefore neither can demons induce substantial forms."[47]

Thomas next defines legitimate art as a procedure that merely joins passive natural products to active natural powers in order to produce a required effect. A good example, Thomas says, is the lighting of a fire. The artisan conjoins the form of fire (the natural agent) and the wood (the passive material) in order to draw forth the effect of fire. Demons act in the same way: they "cannot produce new effects by creation," like God himself; they can only apply natural agents to natural patients. When demons appear to raise the dead or perform other supernatural acts, they are acting by means of illusion only; the effects of such illusion are false and ephemeral.

Thomas returns to alchemy at the end of this distinction, in order to buttress his earlier comments.

> Art by its own power cannot confer a substantial form, but it can do this by means of a natural agent, as is clear in the following [*hoc*], that the form of fire is produced in logs through art. There are some substantial forms, however, which art cannot induce by any means, since it cannot find the proper active and passive subjects. Even in these art can produce a similitude, as when alchemists produce something similar to gold as to exterior accidents. But it is still not true gold, since the substantial form of gold is not [induced] by the heat of fire—which alchemists use—but by the heat of the sun in a determinate place where the mineral power flourishes. Hence such [alchemical] gold does not operate according to the species [of real gold], and the same is true for the other things that they [alchemists] make.[48]

Thus it is impossible for the alchemist to join the form of a precious metal to the substance of a base one in the way that the form of fire is joined to wood, because this must be done deep within the earth, where the mineral power or *virtus* is subjected to a special strengthening. For the same reason, Thomas adds, "the other things that they [alchemists] make" must also be deficient when compared with their naturally occurring counterparts. Thomas therefore rejects not only the alchemical creation of metals but the artificial synthesis of any chemical product. Such "alchemical" substances as ammonium chloride produced by the destructive decomposition of hair, or copper acetate made with vinegar left in a copper flask, are implicitly rejected as "fake" because they were not generated in the bowels of the earth, "where the mineral power flourishes." A form of this

[47] Aquinas, *In quatuor libros sententiarum*, p. 145, col. 1: "Utrum daemones possint inducere in materia corporali verum effectum corporalem. . . . Praterea, daemones non operantur nisi per modum artis. Sed ars non potest dare formam substantialem; unde dicitur in cap. de *numeris*: sciant auctores alchimiae, species transformari non posse. Ergo nec daemones formas substantiales inducere possunt." The text used by Busa contains a manifest error: *numeris* should be corrected to *mineris*.

[48] *Ibid.*: "Ad quintum dicendum, quod ars virtute sua non potest formam substantialem conferre, quod tamen potest virtute naturalis agentis; sicut patet in hoc quod per artem inducitur forma ignis in lignis. Sed quaedam formae substantiales sunt quas nullo modo ars inducere potest, quia propria activa et passiva invenire non potest, sed in his potest aliquid simile facere; sicut alchimistae faciunt aliquid simile auro quantum ad accidentia exteriora; sed tamen non faciunt verum aurum: quia forma substantialis auri non est per calorem ignis quo utuntur alchimistae, sed per calorem solis in loco determinato, ubi viget virtus mineralis: et ideo tale aurum non habet operationem consequentem speciem; et similiter in aliis quae eorum operatione fiunt."

argument had already been rebutted by the *Book of Hermes,* where the pseudonymous author relied on the empirical testing of artificial reagents to confirm their equivalence to the natural forms. Furthermore, the *Book of Hermes* used the artificial incubation of chicks to disprove directly the necessity of a special *virtus loci* (a power linked to a certain place).[49]

In the work of Giles of Rome these rather incidental comments of Thomas are fleshed out to become a full-fledged attack on alchemy. Like Thomas, Giles relies on the *Sciant artifices* and the argument that the generation of metals requires a specific *virtus loci,* a mineralizing power found only deep within the earth. Similarly, Giles, like Thomas, does not consider alchemy in the context of natural philosophy—although he too wrote commentaries on *De generatione et corruptione* and the *Meteorologica*—but in his *Quodlibeta,* a treatise concerned primarily with theology. Furthermore, the question "whether man can make gold" belongs to the subsection of the *Quodlibeta* devoted to the subject of man, and here man is being treated "in relation to his art," not "in relation to nature."[50]

Giles's *quaestio* actually contains two questions: first, "whether man can make true gold by art," and second, "given that he can make gold, whether it be permissible to sell such gold." Relying on the *Sciant artifices,* Giles paraphrases Avicenna's argument that nature is better than art, saying that art is only a principle of artificial things, whereas gold is not artificial but natural.[51]

Then Giles introduces the argument of the *virtus loci.* Admitting that some creatures, such as the bees generated spontaneously from dead cattle, do not need a specific place of generation but only a "material principle" (putrefying matter), he argues that other things, such as wine made from grapes, need both this material principle and a specific place of generation, for wine is produced only "in the depth of the grape" (*in ventre vitis*). Similarly, Giles says, "it is also believable" that metals must be generated deep within the earth.

The second question, "given that man can make gold, whether it is permissible to sell such gold," Giles refuses to entertain seriously, since he is unequivocally convinced that artificial gold cannot be made. At this point he reveals the true nature of his argument, saying that even if gold that would withstand the assayer's test of cupellation could be made, it would still not be legal tender, since it would not have all the medical properties of natural gold.[52] It follows that such a product would not be real gold, despite the assayer's judgment. No doubt Giles would have said the same even if such artificial gold had the same specific weight as natural gold, for to him, mineral gold and artificial gold can *never* be the same, regardless of their properties. Like Avicenna, Giles has adopted the immutable principle that artificial products can never be the same as their natural models.

The last three decades of the thirteenth century witnessed an increasingly hostile attitude by religious authorities toward alchemy that culminated eventually in the denunciation *Contra alchymistas,* written by the well-known inquisitor Nicholas Eymeric in 1396. Giles's attack was preceded, for example, by a number

[49] *Liber Hermetis* (cit. n. 18), p. 66: "Loci oppositio cassatur quia sicut ex ovo in ventre animalis nascitur, sic etsi sub mamilla vel in fumario ponitur, animalis nascetur."
[50] Aegidius Romanus, *Quodlibeta* (cit. n. 44), p. 147.
[51] *Ibid.*
[52] *Ibid.*, p. 149.

of interdictions issued by the religious orders; the Dominicans alone propounded condemnations of alchemy in 1272, 1287, 1289, and 1323. The movement to prohibit alchemy was given papal authority in 1317, when John XXII issued his well-known bull "Spondent quas non exhibent,"—after he had held a public disputation between alchemists and their detractors, according to Eymeric. This papal document is directed specifically against alchemists who employ their artificial gold for counterfeiting; it was motivated by purely fiscal reasons, for the debasement of coin by counterfeiters, alchemical or otherwise, presented a serious problem to the medieval commonwealth. It thus contains little theoretical justification. Nonetheless, the bull does say that the alchemists feign "that which is not in the nature of things," indicating that John did not believe alchemical transmutation to be physically possible.[53]

What was the reason for this great backlash against alchemy that seems to have begun around the time Paul of Taranto was writing his *Theorica et practica*? We have already suggested that the alchemical proponents were themselves in part responsible, by arrogating too much power to the claims of their art. Yet deeper causes were at work. These are evident even in a well-known *consilium* by the consistorial advocate Oldrado da Ponte, probably written in the first decade of the fourteenth century. Oldrado's *consilium* in fact includes passages in support of alchemy. The opening contains a quotation from the ninth- or tenth-century *Canon Episcopi,* a document intended to prohibit belief in witches, who, according to certain old pagan beliefs, could assume monstrous shapes.[54] First Oldrado quotes the part of the *Canon Episcopi* he intends to refute: "[It seems] that the art of alchemy should be prohibited, because the *Canon Episcopi,* question 26, 1, says that 'whoever believes that anything created [*creaturam*] can be either mutated or transferred into another species or into another similitude, except by the creator Himself, is an infidel, and worse than a pagan.' "[55] It is peculiar that Oldrado should have taken a document that originally had no concern with alchemy to apply directly to that art. Is it possible that others, seeing a similarity of language between the *Canon Episcopi* and the *De congelatione,* had preempted him in this regard? If one inspects the passage above without respect to its original context, it could indeed seem to be a sort of official decree of the message propounded by the *Sciant artifices*. The passage explicitly states that only God himself can transmute species, and that anyone who believes otherwise is not a Christian.

Oldrado's response to the *Canon Episcopi* is also revealing. Instead of replying that this edict has nothing to do with alchemy, he answers in the following manner: "[Alchemists] do not say that one species is mutated into another (as is imputed to them), because this is not possible. But they say that one species of metal (such as gold) can be produced from another species of metal (such as tin)."[56] Oldrado's rebuttal does not maintain that specific transmutation is possi-

[53] Halleux, *Textes alchimiques* (cit. n. 8), pp. 124, 126–127, 126 n. 30; C. Narbey, "Le moine Roger Bacon, et le mouvement scientifique au XIIIe siècle," *Revue des Questions Historiques,* 1884, *35*:157; and Migliorino, "Alchimia" (cit. n. 45), p. 16, n. 32 (on Eymeric).

[54] Norman Cohn, *Europe's Inner Demons: An Enquiry Inspired by the Great Witch Hunt* (London: Chatto, Heinemann, for Sussex Univ. Press, 1975), pp. 210–211.

[55] Oldrado da Ponte, *Consilium 74, de sortilegia, num. I,* in Fanianus, *De iure artis alchemiae* (cit. n. 3), p. 211.

[56] *Ibid.*

ble; he points out instead that the species of the metal is not transmuted, but only the metal itself. The origin of this strange-sounding claim was probably a late thirteenth-century alchemical work ascribed spuriously to Roger Bacon, the *Breve breviarium*. It is quite likely that the author of the *Breve breviarium* originated this defense himself, as it seems to be developed at greater length here than in any other medieval alchemical text. By claiming that the species of the metals are not transmuted, but only the metals themselves, the *Breve breviarium* means that the group of characteristics that make silver silver (its argenteity) and gold gold (its aureity) do not change if an individual piece of silver is transmuted into an individual piece of gold.[57] Gold will still be defined, for example, as a "yellow, soft, malleable, fusible, heavy, body," and silver as a "white, soft, malleable, fusible body, of moderate weight." Nonetheless, an individual piece of silver can be physically transmuted so that its matter will conform to the definition of gold. Hence, the physical characteristics of the individual piece of silver will have been changed to the degree that they now belong to the species of gold.

Oldrado's *consilium,* although taking the same approach as the *Breve breviarium,* differs from that text in its motivation. While the *Breve breviarium*'s argument seems to be directed solely against the *De congelatione,* Oldrado is responding to the *Canon Episcopi,* which explicitly said that only God could transmute species. Oldrado's response is therefore intended to bear the onus of doctrinal correctness, whereas the *Breve breviarium*'s—at least overtly—is not. We have already seen Thomas Aquinas and Giles of Rome treat alchemy in a theological context, where Thomas even mentioned alchemists in the same breath as demons. Like Thomas and Giles, Oldrado sees alchemy in a theological light, while the defenders of alchemy had focused merely on its naturalistic implications. This growing tendency to theologize the issue of alchemy, I propose, provides the main reason for the increased number of condemnations tendered against it during the late thirteenth and fourteenth centuries. We should not forget that Innocent III and Gregory IX had already established the papal Inquisition in the first half of the thirteenth century, and that by the second half that dreaded institution was "fully organized."[58] Oldrado's need to answer to the *Canon Episcopi* was not necessarily an anomaly: it may well have reflected the obsession with heterodoxy that had begun with the Albigensian Crusade and eventually resulted in the witch hunts of the sixteenth and seventeenth centuries.

Despite the efforts of Giles, John XXII, and later Nicholas Eymeric, Latin alchemy could not be wiped out by proclamation or other official means. The

[57] Pseudo–Roger Bacon, *Breve breviarium,* in *Sanioris medicinae magistri D. Rogeri Baconis Angli de arte chymiae scripta* (Frankfurt, 1603), pp. 123–126, on pp. 125–126: "Sic revera species non mutantur, sed individua: et sic illud intelligitur . . . species ergo argenti, quae est argenteitas non permutatur in speciem auri, quae est aureitas; quoniam species vere permutari non possunt, quia non sunt subiectae per se accretionibus [*sic codex;* MS Oxford, BL, Digby 119, fol. 66r, *leg.* actionibus *ut vid.*] sensibilibus, nec in se compositionem partitam habent, vel contrariam, quae sit causa permutationis vel subiectum. . . . Ex hoc argentum vel aurum factum est subiectum alterius speciei, quam alia complevit et induxit materiae purgatio atque digestio." The text of the *Breve breviarium* exists in fragmentary form in a manuscript that, according to oral communication from M.-Th. d'Alverny, derives from the late 13th century (MS Paris, BN, Lat. 6514, fols. 126–129). The inauthenticity of the ascription seems assured by the text's dependence on Albertus Magnus's *De mineralibus* for the theory that sulfur and other reagents contain a three-fold humidity (pp. 110, 165, etc.). For a description of this theory, and Albert's source, see Newman, "The *Summa perfectionis*" (cit. n. 14), Vol. III, pp. vii–xi.

[58] Cohn, *Europe's Demons* (cit. n. 54), p. 24.

vision of human power in the realm of technology raised by the *Book of Hermes,* Roger Bacon, and Paul of Taranto was too seductive to be repressed for long. The role of Hermetic literature in the message of such propagandists for the "dignity of man" as Pico della Mirandola is now well documented. Alchemy's place in shaping the reformatory vision of Paracelsus (d. 1541) cannot be overstated. The same may be said for the "archimage" of the sixteenth century, Cornelius Agrippa von Nettesheim, while the technological apologist John Dee, whose *Mathematicall Preface* of 1570 demonstrated the practical application of Euclid's *Elements,* was heavily indebted to the corpus of Roger Bacon, including a number of spuriously attributed alchemical works.[59]

[59] Frances Yates, *Giordano Bruno and the Hermetic Tradition* (London: Routledge & Kegan Paul, 1964); Walter Pagel, *Paracelsus: An Introduction to Philosophical Medicine in the Era of the Renaissance* (Basel: Karger, 1958); William Newman, "Thomas Vaughan as an Interpreter of Agrippa von Nettesheim," *Ambix,* 1982, *29*:125–140; Nicholas Clulee, "John Dee's Mathematics and the Grading of Compound Qualities," *Ambix,* 1971, *18*:178–211; and Clulee, "Astrology, Magic, and Optics: Facets of John Dee's Early Natural Philosophy," *Renaissance Quarterly,* 1977, *30*:632–680. Both articles by Clulee give ample witness to the heavy influence of Roger Bacon. Dee also owned a

APPENDIX: TEXT OF PAUL OF TARANTO'S DISCUSSION OF THE TYPES OF HUMAN TECHNOLOGY

Que naturalia etiam dicuntur materia intellectus vel artis et que artis instrumenta dicuntur Primum Capitulum [2r]

Cum igitur sub arte sumantur naturalia entia ut nove forme per artificium imprimantur in eis, tunc ipsa naturalia / que transmutantur ac formas ipsas recipiunt dicuntur materia intellectus vel artis. Entia vero quibus mediantibus aut aliquo modo agentibus imprimuntur ipse forme in id quod eas suscipit artis instrumenta dicuntur. Terminatur autem sic opus artis ad formam duplicis generis, nam aliquando ad / formam extrinsecam ut in arte pingendi, sculpendi, domificandique et similibus, et hec forma dicitur proprie forma artis, et aliquando terminatur opus artis ad formam substantialem intrinsecam, ut agricultura et medicina, et hec forma dicitur forma nature. Hec autem differentia surgit a / differenti modo sumendi naturam ut instrumentum, cum enim instrumentum sit de genere activorum, activum autem in natura sit aliqua virtus eius, necesse est ut natura dicatur sumi per aliquam suam virtutem, cum sit instrumentum operationis sub arte. Et quoniam virtus nature que sit sumi // potest se habere duobis modis, ideo contingit [2v] per eam dari formam duobus modis dictis alterius nature subiecte. Est enim omnis virtus in natura quedam qualitas. Qualitates autem naturales quedam dicuntur prime, que sunt differentie / naturales in quatuor corporibus simplicibus primis scilicet elementis. Et hec forme sunt ille quatuor qualitates principales ad omnia nature opera, videlicet calidum, frigidum, siccum, et humidum, que sunt in secunda specie qualitatis, que dicitur potencia naturalis. Alie vero naturales /

Foliation and lineation of the Latin text are derived from MS Paris, Bibliothèque Nationale, Latin 7159, fols. 1–55r, and replicated in William Newman, "The *Summa perfectionis* and Late Medieval Alchemy" (Ph.D. diss., Harvard Univ., 1986). For descriptions of the manuscripts and their families see *ibid.,* Vol. III, pp. 1–7. The manuscript sigla are as follows (this section does not appear in *Z* = Paris, Bibliothèque Nationale, Latin 14005):

P = Paris, Bibliothèque Nationale, Latin 7159.
M = Manchester, John Rylands Library, Latin 65.
A = London, British Museum, Additional 10764.

10–12 Que . . . capitulum *om. A* 10 vel] tractatus vel *P* 12 Primum *ego;* Primo

My purpose in this essay has not been to prove the continued influence of alchemy on the development of applied science and technology throughout the Scientific Revolution, but merely to show that here, in these obscure treatises of the thirteenth century, a propagandistic literature of technological development was born. During this innovative period, alchemical writers and their allies produced a literary corpus that was among the earliest in Latin to promote actively the doctrine that art can equal or outdo the products of nature, even if human art is learned by imitating natural processes. Similarly, these alchemical propagandists—or at least the bolder among them—did not shy away from the conclusion that man can even change the order of the natural world by altering the species of those products. This technological dream, however premature, was to have a lasting effect on the direction taken by Western culture.

number of important alchemical manuscripts, such as Oxford, BL, Digby 119, and Glasgow, Hunterian 253. Of the manuscripts listed in Dorothea Waley Singer, *Catalogue of Latin and Alchemical Manuscripts in Great Britain and Ireland* (Brussels: Maurice Lamertin, 1928), at least one contains works ascribed to Bacon with notes by Dee: London, BM, Sloane 2327, fols. 30r–v and 36r–38r.

APPENDIX: TRANSLATION

Which natural things are called the matter of intellect or art, and which are called the instruments of art

Therefore, since natural things are appropriated by art so that new forms be impressed in them through artifice, the natural things that are transmuted and receive the forms are themselves called the matter of intellect or art. But the entities by whose mediation or action these forms are impressed into that which receives them are called the instruments of art. Yet the work of art is restricted to a form of double genus as follows: sometimes [it is restricted] to an accidental, extrinsic form, as in the art of painting, sculpting, house building, and the like, and this form is properly called a "form of art"; and sometimes the work of art is restricted to a substantial, intrinsic form, as in agriculture and medicine, and this form is called a "form of nature." But this distinction arises from the different way of taking nature as instrument, for since an instrument is of the genus of "active things," but an active thing with regard to nature is some virtue of it [i.e. nature], it is necessary that nature be said to be taken through some virtue of itself, since it is the instrument of operation under art. And since the virtue of nature that is to be taken [2v] can exist in two manners, it therefore happens on account of it [the virtue] that the form of the other subjected nature [i.e., nature as matter] exists in the two said manners. For every virtue in nature is a certain quality. But certain natural qualities—which are the natural differences in the four first simple bodies, namely, the elements—are called "primary." And these forms are those four principle qualities [sufficing] for all the works of nature, namely, hot, cold, dry, and wet, which are in the second species of quality, which is called "natural capacity."[a] The other natural qualities are called secondary qualities, namely, those

[a] The second "species of quality" (*potencia naturalis*) belongs to a fourfold division of qualities that is derived ultimately from Aristotle, *Categories* 8b26–10a16.

PM 13 entia] etiam *P* 14 ipsa *om. M* 17 id] his *A* 18 suscipit] recipit *A* 23 agricultura] in agricultura *MA* 25 sumendi] resumendi *M* 28 suam] sui *M* 29 que sit] sicut *M* que sic *A*

2 subjecte] subjectis *M* 4 prime] proprie que sunt] proprie quedam *M* 6 hec *om. M* 7 videlicet] scilicet *M* 9 potencia passio *P*

qualitates dicuntur qualitates secunde, a primis scilicet causate, ut album, nigrum, dulce, amarum, durum, molle, acutum, obtusum, que sunt in tertia et quarta specie qualitatis, per passibilem qualitatem et per figuram. Cum igitur ars sumit pro instrumento nature virtutem que de genere / est secundarum qualitatum, ut se habet color in picturis vel figura anguli vel scabelli durities vel dolabre in sculpturis et dolacionibus sive similibus, tunc necesse est extrinsecus formam accidentalem induci, cuius ratio ista est: ars et artifex ex extra se habent ad patientem naturam in quam agunt. / Secunde autem qualitates predicte de se proprie active non sunt in naturam aliquam nisi per accidens, proprie enim de se sunt active in sensum per suas species secundum esse spirituale et intentionale quod habent, et non secundum eorum esse naturale, nisi per accidens. Color enim movet visum / secundum esse intentionale quod habet in perspicuo et non secundum esse naturale quod habet in re nature, et sapor ut sapor movet sensum et non naturam de se ex simili ratione, neque et sapor nutrit sed cibus et potus, scilicet aliqua substantia cuius // est sapor, unde sapor in naturam non agit nisi per aliud *[3r]* —scilicet calidum, frigidum, siccum, et humidum—que sunt in re saporosa. Neque aliqua secundarum qualitatum agere potest intra naturam et essentiam alicuius, nisi per qualitates / primas. Igitur quoniam ars et artifex sunt extra res patientes, et natura que sumitur ut instrumentum de se non est transmutativa substantie, nec ista accidentia sunt per se sed per accidens, nunquam ars opus terminare poterit nisi ad formam accidentalem ex extra. Cum vero ars sumit pro / instrumento nature virtutem que est de dictis qualitatibus primis, necesse est opus ad substantiam vel ad substantialia terminari, quoniam calidum, frigidum, siccum, et humidum sunt quasi manus nature et principales eius virtutes, per quas natura cuncta generabilia transmutat et facit. Et ars tunc / est proprie solum in ratione moventis et dirigentis, adminiculantis atque regentis, non autem in ratione facientis. Ipsa vero natura tunc se habebit ratione moventis, facientis, sive causantis, secundum quod Aristoteles in libro de generatione innuit distinctionem moventis, cuius est influere / formam motus, et facientis, cuius est influere formam rei per formam motus. Hinc est igitur quod in medicina terminatur ars ad formam nature que est dispositio complexionis intra, et in agricultura similiter ut est germen et fructus, qui non habentur nisi per cultum, sicut in hortis et in insertionibus / arborum et in agriculture similibus. Quoniam autem hi sumunt ipsa naturalia agentia et non solum nature materiam et passiva accidentia pro instrumento—sumit enim agricultor pro instrumento terram, aquam, aerem, calorem, et semina, et medicus similiter specierum virtutes—ideo essentialia // faciunt tales et *[3v]* non solum accidentalia, per virtutem et operationem nature. Et in omnibus quoniam natura facit, et ars solum ministrat coaptatque regitque, debet quidem effectus tribui potius nature quam arti, vel nature sub arte.

10–11 scilicet causate] quatuor causari *M* 12 specie] proprie *A* 16 scabelli] scalpelli *MA* 17 sive] vel *M* est *om. P* 19 ex *om. M* 20 proprie] prime *M* 20–24 proprie . . . accidens *om. A* 24 *post* naturale *add.* naturaleque *PM* 28 neque et] neque enim *M* 29 aliqua substantia] a quo substantiatus *M*

2 calidum] per calidum *MA* 3 Neque] atque *P* 7 per se sed] nisi *M* 8 nunquam] non quam *A* 9 ex *om. A* 10 de] ex *M* 13 et principales] principalesque *M* 15 solum] sola *A* 15–17 et . . . moventis *om. A* 16 adminiculantis] adminiculi *M* 18 sive causantis] causatis *P* 19 distinctionem] distinctionis *P* n 20 facientis] faciens *A* 24 et in insertionibus] et insertionibus *MA* 25 agriculture] agricultura *M* similibus] et similibus *M* 27 enim] ergo *M* 29 virtutes] virtutes pro primo esse *M* ideo essentialia] naturalia *M*

4 nature] melius nature *MA*

caused by the first, such as white, black, sweet, bitter, hard, soft, sharp, dull, which are in the third and fourth species of quality, through their ability to produce an affection,[b] and through [their having] figure [and shape].[c] Therefore, when art takes as instrument a virtue of nature that is of the genus of secondary qualities, such as color is held [to be] in pictures, or the figure of an angle, or the hardness of a knife, or pickax, in sculptures and carvings or the like, then it is necessary that an accidental form is extrinsically induced. The reason is as follows: art and artificer [in this case] are extrinsically related to the passive thing, nature, on which they act. But the foresaid secondary qualities do not of themselves properly act on any nature except accidentally, for of themselves they properly act on sense alone, through their own species, according to the spiritual and intentional being that they have, and not according to their natural being, except accidentally. For color moves the sight according to the intentional being that it has in the *perspicuum,*[d] and not according to the natural being that it has in the matter of nature; and flavor qua flavor moves the sense and not nature of itself, from a similar cause, nor does flavor nourish, but food and drink, namely, some substance that [3r] has flavor. Whence flavor does not act on nature except through something else—namely, hot, cold, dry, and wet—which are in the flavorful thing. And none of the secondary qualities can act on the nature and essence of anything else, except through the primary qualities. Therefore, since art and artificer are [in this case] external to the passive things and the nature that is taken as instrument is not of itself transmutative of substance, nor are these accidents [transmutative] *per se* but *per accidens,* art will only be able to lead its work as far as an external accidental form. But when art takes as an instrument the virtue of nature which is of the said first qualities, the work must extend as far as substance or substantial things, since hot, cold, dry, and wet are as it were the hands of nature and her principal virtues, through which nature transmutes and makes all things that come into being. Art, then, properly acts only by moving, directing, aiding, and ruling, but not by making. But nature herself behaves by moving, making, [and] causing, about which Aristotle intimates, in *De generatione et corruptione,* the distinction of the moving, whose [role] is to infuse the form of motion, and of the making, whose [role] is to infuse the form of the thing through the form of motion. Hence it is, therefore, that in medicine art is limited to the form of nature that is the disposition of a complexion within, and similarly in agriculture, as it is [in the case of] the seed and fruit that are not had except by cultivation, as in gardens, and in graftings of trees, and in similar agricultural matters. But since these men take up natural agents themselves as instrument, and not only the material and passive accidents of nature—for the agriculturist takes as instrument earth, water, air, heat, and seeds, and the doctor likewise the virtues of drugs —therefore [3v] such men make essential [changes] and not only accidental ones, [although] through the virtue and operation of nature. And since nature in all things makes, and art only administers, joins, and rules, the effect must surely be attributed to nature rather than to art, or to nature under art.

[b] I have translated *per passibilem qualitatem* as "through their ability to produce an affection." See Roy J. Deferrari et al., *A Lexicon of Thomas Aquinas* (Baltimore: John D. Lucas, 1948), under "Passibilis" and "Qualitas."

[c] See Aristotle, *Categories,* 10a11–16.

[d] The *perspicuum* is the transparent medium postulated by Aristotle in *De sensu et sensilibus* (Venice, 1572), p. 8.

Science as Handmaiden

Roger Bacon and the Patristic Tradition

By David C. Lindberg

EVERY HISTORIAN OF MEDIEVAL SCIENCE must sooner or later confront the enigma of Roger Bacon, the thirteenth-century Franciscan renowned for his attack on authority and his defense of mathematics and experimental science. Bacon has become the symbol, not exactly of medieval science, but of modern science struggling to emerge during the medieval period—prevented from doing so by ecclesiastical repression. Symbolism, of course, can be a vehicle for misunderstanding as well as understanding, and we must inquire, in Bacon's case, whether the symbol is consistent with the reality. In the process we stand to learn vital lessons about the nature of medieval science and its relationship to church and society.

The nineteenth-century image of Bacon, created by William Whewell, Emile Charles, and others, portrayed him as a bitter opponent of authority and scholastic method, a proponent of free inquiry and experimentation. Whewell argued that "Roger Bacon's works are not only so far beyond his age in the knowledge which they contain, but so different from the temper of the times, in his assertion of the supremacy of experiment, and in his contemplation of the future progress of knowledge, that it is difficult to conceive how such a character could then exist." Charles considered Bacon "the only man before modern times to claim the right of thought for its own sake, . . . as a method without which all progress becomes impossible."[1] A few years later Robert Adamson wrote of Bacon:

> It may seem but a small thing for a writer to reject authority, but one must reflect on what that meant in Bacon's time. It meant absolute revolt against the whole spirit of scholasticism; it was the assertion of freedom of thought, of the claim of science to push forward to its conclusions, regardless of fancied consequences, with implicit trust in the grand law that all truth is ultimately harmonious.[2]

Baconian hagiography reached its zenith, perhaps, in the work of Andrew Dickson White, who characterized Bacon as a lonely discoverer, risking his life in the battle for freedom of thought:

I acknowledge the very helpful criticism of William Eamon, Jeremiah Hackett, George Marcil, and two anonymous referees. Research support was supplied by the National Endowment for the Humanities and the Graduate School of the University of Wisconsin.

[1] William Whewell, *History of the Inductive Sciences*, 2 vols. (3rd ed., New York, 1858), Vol. I, p. 245; and Emile Charles, *Roger Bacon: Sa vie, ses ouvrages, ses doctrines* (Paris, 1861), p. 102.

[2] Robert Adamson, *Roger Bacon: The Philosophy of Science in the Middle Ages* (Manchester, 1876), p. 22.

> By him, more than by any other man of the Middle Ages, was the world brought into the more fruitful paths of scientific thought—the paths which have led to the most precious inventions; and among these are clocks, lenses, and burning specula, which were given by him to the world, directly or indirectly. . . . But it should be borne in mind that his *method* of investigation was even greater than its *results*. In an age when theological subtilizing was alone thought to give the title of scholar, he insisted on *real* reasoning and the aid of natural science by mathematics; in an age when experimenting was sure to cost a man his reputation, and was likely to cost him his life, he insisted on experimenting, and braved all its risks. Few greater men have lived.[3]

Enthusiasm for Bacon's achievements was somewhat tempered in the twentieth century by the debunking efforts of Lynn Thorndike, who showed that Bacon's "experimentalism" reflected well-established medieval traditions, and Henry Osborn Taylor, who credited Bacon with noble aims while criticizing him for failure to achieve them owing to confusion and inconsistency.[4] Sir William Dampier echoed Taylor's opinion, noting that Bacon was hampered by preconceptions, such as

> his agreement with the scholastic view, which in other ways he assailed vehemently, that the end of all science and philosophy was to elucidate and adorn their queen, theology. Hence came some of the confusion and the inconsistencies which at every turn are seen in his writings, mixed with originality and insight far beyond his time and even in advance of the next three centuries. Struggle as he might, he never cast off the shackles of the mediaeval habit of mind; it mars his work, as the persecution of his Order marred his life.[5]

More recently D. E. Sharp, Theodore Crowley, and Etienne Gilson have carefully studied aspects of Bacon's philosophy. Stewart Easton has published an intellectual biography, heavy with psychological analysis. And Raoul Carton, sensitive to the handmaiden status of Bacon's science, has left us three volumes of detailed exposition of Baconian doctrines.[6] These works, which attempt to view Bacon in historical context, go a long way toward restoring balance to his

[3] Andrew Dickson White, *A History of the Warfare of Science with Theology in Christendom,* 2 vols. (New York, 1896), Vol. I, p. 387.

[4] Lynn Thorndike, "Roger Bacon and Experimental Method in the Middle Ages," *Philosophical Review,* 1914, *23*:283–292; and Henry Osborn Taylor, *The Medieval Mind,* 2 vols. (4th ed., Cambridge, Mass.: Harvard Univ. Press, 1962), Vol. II, pp. 515–516.

[5] Sir William Dampier, *A History of Science and Its Relations with Philosophy and Religion* (Cambridge: Cambridge Univ. Press, 1940), p. 100.

[6] D. E. Sharp, *Franciscan Philosophy at Oxford in the Thirteenth Century* (Oxford: Clarendon, 1930), pp. 115–171; Theodore Crowley, O.F.M., *Roger Bacon: The Problem of the Soul in His Philosophical Commentaries* (Dublin: James Duffy, 1950); Etienne Gilson, *History of Christian Philosophy in the Middle Ages* (New York: Random House, 1955), pp. 294–312; Stewart C. Easton, *Roger Bacon and His Search for a Universal Science* (Oxford: Basil Blackwell, 1952); Raoul Carton, *L'expérience physique chez Roger Bacon* (Paris: J. Vrin, 1924); Carton, *La synthèse doctrinale de Roger Bacon* (Paris: J. Vrin, 1924); and Carton, *L'expérience mystique de l'illumination intérieure de Roger Bacon* (Paris: J. Vrin, 1924). Specialized studies of great value have also appeared; see, e.g., George Molland, "Roger Bacon as Magician," *Traditio,* 1974, *30*:455–460; N. W. Fisher and Sabetai Unguru, "Experimental Science and Mathematics in Roger Bacon's Thought," *Traditio,* 1971, 27:353–378; Thomas S. Maloney, "The Semiotics of Roger Bacon," *Medieval Studies,* 1983, *45*:120–154; Jeremiah M. G. Hackett, "The Attitude of Roger Bacon to the *Scientia* of Albertus Magnus," in *Albertus Magnus and the Sciences: Commemorative Essays, 1980,* ed. James A. Weisheipl (Toronto: Pontifical Institute of Medieval Studies, 1980), pp. 53–72; and Hackett, "The Meaning of Experimental Science *(Scientia Experimentalis)* in the Philosophy of Roger Bacon" (Ph.D. diss., Univ. Toronto, 1983).

reputation. But they represent a beginning, rather than the final word. It remains, I believe, to be seen just how mathematics, natural philosophy, natural history, medicine, the bitter attack on authority, and the enthusiastic promotion of experimental science fit into the larger context of Bacon's thought and how Bacon fits into the scientific movement of the Middle Ages.

Moreover, the nineteenth-century stereotype refuses to die, even among scholars. A. C. Crombie's *Robert Grosseteste and the Origins of Experimental Science,* published in 1953, touched off a new round of debate and commentary regarding Bacon's contributions to the tradition of experimental science. In Crombie's opinion Bacon did not invent the experimental method, but promoted and practiced the experimentalism of his elder contemporary, Robert Grosseteste. The net effect of this debate has been to restore the question more or less to its original form: was Bacon the founder (or one of the founders) of modern experimental science? The answers have become a great deal more critical and informed, but the categories remain largely those of the nineteenth century.[7]

This article, then, is an attempt to rescue Bacon from the mythologizers and the debunkers. It does not seek to examine his scientific achievements in detail, but to understand their motivation and to see what they meant in thirteenth-century terms. Its thesis is that Bacon was not a modern, out of step with his age, or a harbinger of things to come, but a brilliant, combative, and somewhat eccentric schoolman of the thirteenth century, endeavoring to take advantage of the new learning just then becoming available while remaining true to traditional notions, patristic in origin, of the importance to be attached to philosophical knowledge. With the patristic period we must therefore begin.

THE CHURCH FATHERS AND THE LIBERAL ARTS

As educated pagans were converted to Christianity in the second and third centuries, they inevitably sought to accommodate their newfound faith to the philosophical principles and modes of discourse with which their education had imbued them. Christian teachers appeared, claiming to offer a learned exposition of Christian doctrine that would compare favorably with the level of teaching in the pagan schools. Pagan philosophy was even put to use in the development of an intellectually respectable Christian theology and in defense of the faith against heretics and skeptics.

Justin Martyr (d. between 162 and 168) was a philosopher before he was a

[7] A. C. Crombie, *Robert Grosseteste and the Origins of Experimental Science* (Oxford: Clarendon, 1953). See also Małgorzata Frankowska, *Scientia as Interpreted by Roger Bacon,* trans. Ziemsław Zienkiewicz (Warsaw: Scientific Publications Foreign Cooperation Center of the Central Institute for Scientific, Technical, and Economic Information, 1971); and William A. Wallace, *Causality and Scientific Explanation,* Vol. I: *Medieval and Early Classical Science* (Ann Arbor: Univ. Michigan Press, 1972), pp. 47–50. Frankowska's position resists easy characterization; she cannot be accused of urging a return to the nineteenth-century opinion, but neither is she willing to give up the notion that Bacon was profoundly progressive. In what can perhaps be regarded as a summary statement, she claims that Bacon "was the first to consider in such a large way the theoretical problems connected with science, he was also the first who had the vision of the unity of sciences based on the unity of method and purpose, which would have ensured to sciences the absolute certitude and verity of their conclusions. Moreover, he was the first to originate theoretical reflections concerning the nature of science and its aims—reflections that were to find mature expression much later in the time of Francis Bacon and Descartes, but which have always had and have to this day their share in the development of the human thought" (p. 134).

Christian. Learned in the major Greek philosophical traditions, he ultimately found satisfaction only in Christianity, to the defense of which he devoted the remainder of his life. Justin saw no fundamental incompatibility between philosophical activity and Christian belief; rather, he viewed reason as a divine gift and philosophy as a valuable apologetic weapon. He found support for Christian teaching in Platonic metaphysics and Stoic ethics; he explained the striking continuities between these philosophies and Christianity by proposing that the philosophers had obtained access to the Old Testament writings and that the divine *logos* revealed truth even to pagan authors who lived uprightly.[8]

Justin's outlook was developed by Athenagoras (fl. 177) and the Alexandrians Clement (d. between 211 and 215) and Origen (d. ca. 254). All found Platonism congenial, particularly for its monotheism and its teaching on the immortality of the soul, and all put it to apologetic use. In his *Apologia* addressed to the Emperor Marcus Aurelius, Athenagoras defended Christianity against the charge of atheism; in the process he went so far as to maintain that the existence of one God is susceptible of rational demonstration. Clement proclaimed that truth, wherever one finds it, is from God: "Barbarian and Greek philosophy have torn off a piece of the eternal truth . . . from the theology of the Logos who eternally is. And he who brings together again the divided parts and makes them one, mark well, shall without danger of error look upon the perfect Logos, the truth." Origen argued that all men, insofar as they proceed rationally, are illuminated by the divine *logos*; rational activity is to be esteemed, and the Christian is to appropriate whatever in pagan culture proves useful.[9]

Apologetics led naturally to hermeneutics. The tendency in the early church was to read the Old Testament typologically and prophetically; that is, the text was treated not simply as a historical narrative, but as an allegory to be deciphered, on the principle that the Old Testament possesses at every point a deeper meaning in which the ministry of Christ is prophesied or prefigured.[10] This allegorical method was developed and systematized by Clement and Origen. Borrowing from Philo and the Platonic tradition, Clement and Origen believed that the bodily meaning of the words conceals eternal mysteries, which must be ferreted out by the exegete; indeed, Origen was prepared to acknowledge that not all passages possessed a literal sense, though all had a spiritual meaning. One of the great benefits of the allegorical method in the eyes of Clement and Origen was its ability to harmonize Scripture with the best of Greek philosophy: if a literal reading of the biblical text made no sense or conflicted with the philosophers, particularly Plato, one could dismiss the difficulty by allegorizing the passage. Allegorical method thus became an indispensable apologetic weapon.[11]

[8] On Justin see A. H. Armstrong, ed., *The Cambridge History of Later Greek and Early Medieval Philosophy* (Cambridge: Cambridge Univ. Press, 1970), Ch. 9; and Karl Baus, *From the Apostolic Community to Constantine,* Vol. I of *History of the Church,* ed. Hubert Jedin and John Dolan (New York: Crossroad, 1965), pp. 174–177.

[9] On Athenagoras see Baus, *Apostolic Community to Constantine,* p. 178. Clement is quoted by E. F. Osborn, *The Philosophy of Clement of Alexandria* (Cambridge: Cambridge Univ. Press, 1957), p. 124. On Origen see Joseph Wilson Trigg, *Origen: The Bible and Philosophy in the Third-Century Church* (Atlanta: John Knox Press, 1983); and Armstrong, ed., *Cambridge History* (cit. n. 8), Ch. 11.

[10] *The Cambridge History of the Bible,* Vol. II, ed. G. W. H. Lampe (Cambridge: Cambridge Univ. Press, 1969), pp. 155–165; and Robert M. Grant and David Tracy, *A Short History of the Interpretation of the Bible* (2nd ed., Philadelphia: Fortress, 1984), Chs. 2–4.

[11] Grant and Tracy, *Short History* (cit. n. 10), pp. 55–62; and Trigg, *Origen* (cit. n. 9).

However, allegorization could also be carried to excess; if everything signifies something else, by what criterion does one distinguish a valid from an invalid interpretation? Reaction to what they judged allegorical excess led members of the school of Antioch to anchor their interpretations in the literal sense. Theodore of Mopsuestia and John of Chrysostom, for example, were prepared to acknowledge the reality of higher senses but did not believe they could be reached except through the letter.[12]

Augustine was heir to these exegetical and apologetic traditions. In his enormously influential *De doctrina Christiana,* which can be conceived as an exegetical manual (among other things), Augustine cautions against the evils of both excessive literalism and excessive allegorization in biblical exegesis: the literalist, mistaking signs for things, is trapped in "miserable servitude" to the letter, while the allegorizer runs the risk of taking as figurative anything that does not conform to his own customs or opinions. To escape these extremes one must first demand a thorough grasp of the literal meaning, in order ultimately to achieve spiritual understanding; a literal reading of the text becomes the foundation of sound biblical exegesis.[13] Augustine thus confirms the attempts of the Antiochenes to elevate the dignity of the literal sense.

But successful scriptural exegesis requires more than hermeneutic principles; it also calls for a variety of scholarly tools. To begin with, linguistic and philological training is needed. "Against unknown literal signs," Augustine writes, "the sovereign remedy is knowledge of languages"— specifically, Hebrew and Greek. Beyond languages, one must know the arts of disputation and the rules of inference. Rhetorical skills also are useful for preaching.[14] Mathematics must be grasped because "numbers and patterns of numbers are placed by way of similitudes in the sacred books as secrets which are often closed to readers because of [their] ignorance of numbers." Music, too, is "given an honorable position . . . in the Sacred Scriptures." And natural history must be known because Scripture often employs the characteristics of animals, plants, and stones as similitudes: the reader ignorant of the behavior of serpents will not comprehend the Biblical injunction to "be wise as serpents and innocent as doves."[15]

Shall we conclude, then, that Augustine esteemed pagan philosophy and advocated its study? On this question we must proceed cautiously—both because of the subtlety of Augustine's position and because it has been clouded by a century of polemic. There was no doubt in Augustine's mind about the utility of pagan philosophy, particularly the liberal arts, for the Christian. We are not, he insists, to avoid literature because Mercury is thought to have been its inventor or music because some have associated it with superstition.[16] He chastises those who have opened Christianity to ridicule by refusing to take knowledge about nature from those who possess it:

[12] Grant and Tracy, *Short History,* Ch. 7; and Beryl Smalley, *The Study of the Bible in the Middle Ages* (Notre Dame, Ind.: Univ. Notre Dame Press, 1964), pp. 20–24.

[13] Augustine, *On Christian Doctrine,* trans. D. W. Robertson, Jr. (Indianapolis: Bobbs-Merrill, 1958), pp. 84, 88. Augustine proceeds to develop exegetical rules, which we need not delve into here.

[14] *Ibid.,* pp. 43 (quotation), 67, 118.

[15] *Ibid.,* pp. 52, 53, 50 (Matthew 10:16).

[16] *Ibid.,* p. 54. On the polemic see White, *Warfare* (cit. n. 3), Vol. I, p. 325; and David C. Lindberg, "Science and the Early Church," in *God and Nature: Historical Essays on the Encounter between Christianity and Science,* ed. Lindberg and Ronald L. Numbers (Berkeley/Los Angeles: Univ. California Press, 1986), pp. 26–28.

> Even a non-Christian knows something about the earth, the heavens, and the other elements of this world, about the motion and orbit of the stars and even their size and relative positions, about the predictable eclipses of the sun and moon, the cycles of the years and the seasons, about the kinds of animals, shrubs, stones, and so forth, and this knowledge he holds to as being certain from reason and experience. Now it is a disgraceful and dangerous thing for an infidel to hear a Christian, presumably giving the meaning of Holy Scripture, talking nonsense on these topics; and we should take all means to prevent such an embarrassing situation, in which people show up vast ignorance in a Christian and laugh it to scorn.[17]

Repeatedly Augustine admonishes Christians "to steal," without hestitation, "the vessels of the Egyptians":

> If those who are called philosophers, especially the Platonists, have said things which are indeed true and are well accommodated to our faith, they should not be feared; rather, what they have said should be taken from them as from unjust possessors and converted to our use. Just as the Egyptians had . . . idols and grave burdens which the people of Israel detested and avoided, so also they had vases and ornaments of gold and silver and clothing which the Israelites took with them secretly when they fled, as if to put them to a better use. . . . In the same way all the teachings of the pagans contain not only . . . superstitious imaginings . . . , but also liberal disciplines more suited to the uses of truth, and some most useful precepts concerning morals. Even some truths concerning the worship of God are discovered among them.[18]

Pagan philosophers obtained truth originally from the prophets and patriarchs, and the Christian is merely reclaiming what is rightfully his. "Every good and true Christian should understand that wherever he may find truth, it is his Lord's."[19]

But all such remarks must be viewed in context. Insofar as pagan learning is of utility, Augustine wants Christians to employ it; but he does not believe that such utility extends far. His writings are laced with cautions against overvaluing pagan philosophy. In *De doctrina Christiana* he argues that "the knowledge collected from the books of the pagans, although some of it is useful, is . . . little as compared with that derived from the Holy Scriptures."[20] In the *Enchiridion,* a brief manual of basic Christian doctrine, he maintains that we should not

> be dismayed if Christians are ignorant about the properties and the number of the basic elements of nature, or about the motion, order, and deviations of the stars, the map of the heavens, the kinds and nature of animals, plants, stones, springs, rivers, and mountains; about the divisions of space and time, about the signs of impending storms, and the myriad other things which these "physicists" have come to understand, or think they have. . . . For the Christian, it is enough to believe that the cause of all created things, whether in heaven or on earth, whether visible or invisible, is nothing other than the goodness of the Creator, who is the one and the true God.[21]

[17] Augustine, *The Literal Meaning of Genesis,* 2 vols., trans. John Hammond Taylor, S.J. (Ancient Christian Writers: The Works of the Fathers in Translation, 41–42) (New York: Newman, 1982), Vol. I, pp. 42–43.

[18] Augustine, *On Christian Doctrine* (cit. n. 13), p. 75.

[19] *Ibid.,* pp. 64, 54 (quotation).

[20] *Ibid.,* p. 78.

[21] Augustine, *Enchiridion* 3.9, in Augustine, *Confessions and Enchiridion,* trans. Albert C. Outler (Library of Christian Classics, 7) (Philadelphia: Westminster, 1955), pp. 341–342.

This is no attack on scientific knowledge, but an assertion of its lack of fundamental religious importance. Augustine frequently points out that the visible world is to be used, not loved: "We should use this world and not enjoy it, so that the 'invisible things' of God 'being understood by the things that are made' may be seen, that is, so that by means of corporal and temporal things we may comprehend the eternal and spiritual."[22]

Augustine's evaluation of the liberal arts is revealed with particular clarity in *De doctrina Christiana,* where he outlines a program of Christian education. He argues that within pagan learning, "aside from the history of things both past and present, teachings which concern the corporal senses, including the experience and theory of the useful mechanical arts, and the sciences of disputation and of numbers, I consider nothing to be useful." Moreover, moderation is to be observed even among useful disciplines. He proposes that handbooks be compiled to provide all that the Christian requires to know of each discipline, with nothing superfluous:

> I think it might be possible, if any capable person could be persuaded to undertake the task for the sake of his brethren, to collect in order and write down singly explanations of whatever unfamiliar geographical locations, animals, herbs and trees, stones, and metals are mentioned in the Scripture. The same thing could be done with numbers so that the rationale only of those numbers which are mentioned in the Holy Scripture is explained.[23]

If there are contexts within which Augustine puts his own liberal education to good use and reveals his mastery of pagan philosophical sources—as in his *Literal Commentary on Genesis*—we nonetheless see here how narrowly he draws the limits of utility. The Christian can benefit from a modicum of pagan learning, but Augustine would draw the line at that which is required for correct belief and virtuous conduct. Knowledge worth having is restricted to the eternal verities and such temporal studies as are necessary for the attainment of the eternal verities.[24] Scholarship, for practical purposes, reduces to biblical scholarship.

BIBLICAL SCHOLARSHIP IN THE EARLY MIDDLE AGES

The religious culture of medieval monasticism was continuous with that of the patristic period. The principles of Christian education set forth in Augustine's *De doctrina Christiana* were embodied in monastic *lectio divina,* the reading and contemplation of Holy Scripture. The level of sophistication and precise character of these studies would, of course, vary with the particular monastic community, but all emphasized the Bible, while encouraging supplemental attention to patristic and profane sources as needed to promote sound exegesis and proper understanding of the biblical text. The goal of these studies tended to be spiritual and contemplative, rather than scholarly; consequently, stress was placed on allegorical interpretation, which opened the door to the spiritual sense.[25]

[22] Augustine, *On Christian Doctrine* (cit. n. 13), p. 10.

[23] *Ibid.,* p. 74.

[24] In this respect Augustine was very close to his pagan contemporaries; there is no warrant, therefore, for supposing that he had a detrimental influence on science: see Lindberg, "Science and the Early Church" (cit. n. 16), pp. 30–33.

[25] Jean Leclercq, O.S.B., *The Love of Learning and the Desire for God: A Study of Monastic Culture* (New York: Fordham Univ. Press, 1961), pp. 28–30, 118–124; and *Cambridge History of the*

The most ambitious attempt to bring scholarship into the monastery was that of Cassiodorus (ca. 480–ca. 575), who left public life in Rome to found a monastic community at Scylacium in Calabria. Cassiodorus established a scriptorium in his monastery and made study an essential monastic activity. In his handbook of Christian scholarship, *Institutiones divinarum et humanarum litterarum* (undoubtedly inspired by Augustine's *De doctrina Christiana*), Cassiodorus offers his monks a syllabus of studies that includes an impressive number of secular authors. It must be made perfectly clear, however, that Cassiodorus recommends them only insofar as they serve biblical studies. He neither denigrates nor values them for their own sake; he simply reiterates the Augustinian position on the relationship of secular to sacred learning.[26]

Despite the efforts of Cassiodorus, the Middle Ages did not see a great deal of serious biblical scholarship until the eleventh and twelfth centuries, when intellectual life shifted from the monasteries to the cathedral schools. Fresh results quickly followed. A series of masters, including Fulbert of Chartres (d. 1028), Anselm of Laon (d. 1117), and Peter Lombard (d. 1160), not only made original contributions to biblical exegesis but also produced a series of textbooks and aids to study. During their lifetimes we see the rapid emergence of glosses, summaries, introductions, extracts, concordances, questions, commentaries (of a continuous sort), and collections of opinions or *sententiae*. Anselm and his followers compiled the *Glossa ordinaria,* the standard collection of glosses on the Bible.[27] And Peter Lombard, of course, produced his *Sentences,* which was to become the standard theological textbook in the medieval universities. While experiencing reinvigoration, biblical scholarship was also changing character. There was a marked tendency to move beyond simple biblical exegesis toward what would become speculative theology; it became customary to raise difficulties and opposing views, to apply dialectical techniques, and to explore rational foundations; the spirit was one of contention and controversy. The exegesis of the cloister was beginning to yield to the disputation of the schools.[28]

It is against this background that we must view the program of instruction advocated by Hugh of St. Victor. Hugh, a member of the canons regular (living under the rule of St. Augustine), taught in the school attached to the Abbey of St. Victor in Paris from about 1125 to his death in 1141. The central purpose of Hugh's teaching was to bring recalcitrant learning back under the authority of religious studies. Hugh saw the new tradition of theological disputation, and the incipient struggle of the arts and sciences for autonomy in the schools, as a threat to the Augustinian ideal of biblical studies supported by the liberal arts as their well-disciplined handmaiden. The challenge he faced was to give new meaning to Augustine's vision in an age of growing knowledge and specialization. He met this challenge in various writings, including the *Didascalicon,* a guide to reading

Bible (cit. n. 10), Vol. II, pp. 183–184, 195, 196. On the precise meaning of *lectio divina* within monasticism, see Leclercq, *Love of Learning,* pp. 88–89.

[26] James J. O'Donnell, *Cassiodorus* (Berkeley/Los Angeles: Univ. California Press, 1979), pp. 202–214; Leclercq, *Love of Learning,* pp. 25–28; and Smalley, *Study of the Bible* (cit. n. 12), pp. 30–32.

[27] See Smalley, *Study of the Bible,* Ch. 2, on the revival of biblical scholarship, and pp. 56–66, on the *Glossa ordinaria.* On the new genres of literature see *Cambridge History of the Bible* (cit n. 10), Vol. II, pp. 197–207.

[28] On these developments see G. R. Evans, *Old Arts and New Theology: The Beginnings of Theology as an Academic Discipline* (Oxford, Clarendon, 1980).

and study in the tradition of Augustine's *De doctrina Christiana* and Cassiodorus's *Institutiones*.[29]

The purpose of study, in Hugh's view, is to recover the divine likeness lost in the fall—to return mankind to prelapsarian perfection: "This, then, is what the arts are concerned with, this is what they intend, namely, to restore within us the divine likeness, a likeness which to us is a form but to God is his nature." Philosophy, or the pursuit of wisdom, is meant to restore souls to "the proper force and purity of their nature."[30] Hugh places particular stress on the traditional seven liberal arts, which reveal "pure and simple truth"; thereafter, "if time affords," the student may read poetry and other literature, "for sometimes we are better pleased when entertaining reading is mixed with serious, and rarity makes what is good seem precious. Thus we sometimes more eagerly take up a thought we come upon in the midst of a story."[31] Hugh's approach to learning seems liberal. He admonishes students to "hold no learning in contempt, for all learning is good. Do not scorn at least to read a book, if you have the time. If you gain nothing from it, neither do you lose anything." On the other hand, there is no time to read everything, and one must not allow the inferior to displace the superior; moreover, "perverse doctrine" must be shunned like poison.[32]

How, then, can we summarize Hugh's view of the relationship of philosophy to biblical studies? First, Hugh is convinced that proper biblical exegesis requires an exceedingly broad base of secular studies. Hugh accorded the literal sense of Scripture a higher status than (in general) had his immediate predecessors, viewing it as the essential foundation for any sound exploration of the spiritual senses; and with emphasis on the literal sense came the necessity of grammatical, historical, and geographical studies. Hugh recalls, fondly, that as a boy he had learned many things of no obvious utility (the names of things, for example) that had subsequently proved to be of worth. Yet his often-quoted admonition to "learn everything; you will see afterwards that nothing is superfluous" must not be torn out of context; it was offered in connection with the grammatical and historical studies required for biblical exegesis.[33] Hugh thus shares Augustine's reverence for the letter of Scripture and advocacy of whatever secular studies are required to unfold the literal meaning, but he is considerably more liberal than Augustine in his understanding of the range of such studies. Augustine thought that a little secular learning would suffice; Hugh believes that almost anything can prove useful.

In the second place, Hugh gives philosophy certain prerogatives beyond its service to biblical exegesis, and here he departs more fundamentally from Augustine. A reading of the first three books of the *Didascalicon,* on the divisions of philosophy, is instructive in this regard: Hugh does not constantly remind us (as Augustine does in *De doctrina Christiana*) of the subordination of all such en-

[29] *The Didascalicon of Hugh of St. Victor: A Medieval Guide to the Arts,* ed. and trans. Jerome Taylor (New York: Columbia Univ. Press, 1961), pp. 7, 11–18; see also Smalley, *Study of the Bible* (cit n. 12), pp. 83–97. On differences between the *Didascalicon, De doctrina Christiana,* and the *Institutiones* see Taylor in *Didascalicon,* pp. 28–34.

[30] *Didascalicon,* ed. and trans. Taylor, pp. 61, 48 (see also p. 51).

[31] *Ibid.,* pp. 88–89; for Hugh's definition of "art" see pp. 59–62; on Hugh's classification of the arts see Taylor's introduction, pp. 31–34.

[32] *Ibid.,* pp. 96–97.

[33] *Ibid.,* pp. 136–137.

deavor to biblical studies. On the contrary, he presents philosophy, of itself, as the pursuit of divine Wisdom, the path to Christian perfection and virtue: "We are restored through instruction, so that we may recognize our nature and learn not to seek outside ourselves what we can find within." The study that Hugh advocates is not, in any narrow sense, confined to the Bible. As Jerome Taylor has put it, "the pursuit of the arts becomes, in effect, convertible with religion."[34]

This is not, of course, the same as secularization. Hugh may allow philosophy to escape the grip of biblical studies narrowly defined, but surely not to free itself from religion. Indeed, Hugh goes to considerable lengths to reclaim philosophy for the faith, against the secularizing tendencies of such contemporaries as William of Conches.[35] Although Hugh advocates a wider range of secular studies than had Augustine or Cassiodorus, his purpose, no less than theirs, was to bring rebellious learning back into captivity to Christ.

ROGER BACON AND THE NEW LEARNING

In the middle of the thirteenth century Roger Bacon found himself in a situation roughly analogous to that of Hugh of St. Victor in the first half of the twelfth, but considerably graver. Each perceived the threat of scholarship divorced from biblical studies and theology. But if, early in the twelfth century, the handmaiden named "philosophy" appeared ready to dally with the pleasures of insubordination, by the middle of the thirteenth a full-scale rebellion did not seem out of the question. The translations of the twelfth and early thirteenth centuries had brought to Christendom an enormous body of Greek and Islamic learning, impressive in scope and depth, but much of it claiming no religious function or allegiance. A first reading of the Aristotelian corpus, for example, was not religiously reassuring: the eternity of the world was proclaimed; divine providence and personal immortality were denied; and authority in matters of belief was claimed for reason rather than revelation. And if the danger of an autonomous, secularized philosophy was not enough to worry about, there was the equally appalling opposite risk—namely, that the new learning would be judged impious and condemned without an adequate hearing, its potential benefits thus forever lost. That the latter was a genuine risk had become apparent with the banning of Aristotelian natural philosophy at Paris early in the century. This, then, is the dilemma that dominated the intellectual life of the thirteenth century and determined the shape of Bacon's career.

Bacon had no intention of giving up the new knowledge. As a master of arts, knowledge was his business, and surely his natural impulse was to enlarge, rather than to diminish, the body of teachable arts and sciences. Besides, to Bacon the utility of the new sciences of mathematics, astronomy, optics, and medicine was too obvious to permit even the thought of giving them up. Bacon's goal, then, had to be to reclaim secular learning for the faith—to demonstrate that, purged of a few errors, it was benign and useful. This was not an altogether novel goal, of

[34] *Ibid.*, pp. 47 (Hugh), 15 (Taylor). Taylor's view, to which I am much indebted, is spelled out more fully on pp. 28–32.

[35] See Taylor's introduction, *ibid.*, pp. 19–28; and M.-D. Chenu, O.P., *Nature, Man, and Society in the Twelfth Century: Essays on New Theological Perspectives in the Latin West*, trans. Jerome Taylor and Lester K. Little (Chicago: Univ. Chicago Press, 1968), Ch. 1.

course; it was a vigorous restatement, or perhaps recasting, of the traditional (Augustinian) ideal of knowledge in the service of religion, at a point in history when there was a great deal more purported knowledge, some of it quite unruly, to be pressed into servitude.

The chronology of Bacon's career is still open to dispute.[36] It seems that he lectured in the faculty of arts at Paris during the 1240s, producing a series of Aristotelian commentaries. Late in that decade or early in the next he returned to Oxford, began the investigations of mathematical and experimental science that would ultimately distinguish him from the great mass of arts masters, and eventually joined the Franciscan Order. We do not know when or how his plan for a reform of learning originated, but during the 1250s he seems to have been at work on pieces that would be fitted together in the 1260s to form the great *opera* for Pope Clement IV. It will be enlightening to examine the program of scholarship set forth in these works.

Bacon opens the *Opus maius,* the largest and most systematic of the works for the pope, with a discussion of four obstacles to the apprehension of truth—sources of error that have suppressed knowledge of the utmost beauty and value. "Out of these deadly pestilences arise all of the afflictions of the human race; for [on account of them] the greatest, most useful, and most beautiful teachings of wisdom, as well as the secrets of all arts and sciences, remain unknown. Worse yet, men blinded by the darkness of these four [errors] do not perceive their own blindness." The first cause of error is "adherence to flawed and unworthy authority," by which the ignorant are doomed. The second is the persistence of custom, which consistently favors the false over the true. The third is popular prejudice, which produces obstinacy and confirms men in their error. Bacon hastens to add that he has no objection to the "solid and true" authority of the church, the saints, and other experts in the pursuit of wisdom—a claim that he proceeds to demonstrate by appealing to such (no doubt) "solid and true" authorities as Seneca, Aristotle, Cicero, and Jerome, all of whom scorn reliance on custom and vulgar opinion.[37]

The fourth, and most serious, cause of error is the tendency to cloak ignorance in a show of wisdom; this fault, universal in mankind, is a "wild beast, which devours and destroys all reason." This source of error is not only dangerous in itself, but also contributes to the other three, for it leads us to rely on weak authority, to convert our own achievements into custom (if we can), and to disseminate our madness as popular prejudice. It is owing to such unfortunate tendencies that Aristotle is rejected on many points in favor of Plato, despite Aristotle's manifest superiority as a philosopher; and that the philosophical works of Avicenna and Averroes have been banned at the University of Paris. Recognizing this fourth cause of error, we should not allow the ignorant to suppress

[36] Every Baconian scholar eventually addresses this question. I have done so in David Lindberg, *Roger Bacon's Philosophy of Nature: A Critical Edition with English Translation, Introduction, and Notes of* De multiplicatione specierum *and* De speculis comburentibus (Oxford: Clarendon, 1983), pp. xv–xxvi.

[37] *The Opus Majus of Roger Bacon,* ed. John Henry Bridges, 3 vols. (London: Williams & Norgate, 1900), Vol. III (containing the revised text of *Opus maius* 1–3, which supersedes the version originally published in Vol I), 1.1, pp. 2–3 (major quotation); 1.2–3, pp. 4–8 (henceforth cited as **Bacon, *Opus maius***). I have avoided using Robert B. Burke's unreliable translation (Philadelphia: Univ. Pennsylvania Press, 1928); all translations are my own.

books; rather we must liberally permit works to be read, trusting that ultimately, with divine help, truth will prevail.[38]

To buttress his case, Bacon sets forth a series of historical examples of opposition to the truth. The law was given to Moses, but was scarcely received with enthusiasm by the pharaoh and the Egyptians, or even by the children of Israel. Christ and the apostles met opposition because of the novelty of their message and the ignorance of their audience. Again, many of those now acknowledged to be doctors of the church, such as Jerome, were first considered heretics: "Similarly, resistance to the truth has confronted all the doctors of sacred Scripture; for in their restoration of learning, they always encountered contradiction and hindrances. Nonetheless, truth grows and will continue to grow until the Antichrist comes."[39]

Bacon's campaign in defense of the new learning unfolds in Part 2 of the *Opus maius,* where he assembles arguments to demonstrate the legitimacy and value of philosophy. Bacon was not a particularly systematic thinker, and the arguments tumble helter-skelter from his pen. Looming over them, however, is the Augustinian conception of one perfect wisdom, contained (in kernel, at least) in Scripture and expounded by philosophy (or, as Bacon puts it, by canon law and philosophy). Just as writers on canon law are contributing to the development of the leaves, flowers, and fruit of a truth "whose roots and trunk are to be found in sacred Scripture," Bacon argues, so "philosophy is nothing except the unfolding of divine wisdom by teaching and writing."[40]

Philosophy and biblical exegesis enjoy an intimate and complex linkage. On the one hand, exegesis of the literal sense requires a knowledge of the nature of things, acquired from philosophy. On the other (and this is where Bacon places the emphasis), the true natures of creatures are revealed in Scripture, and there the philosopher must look for them if he is to succeed at the business of philosophy. Bacon uses the rainbow as an example:

> Scripture apprehends [the natures of] creatures far more certainly and more truly and better than philosophical labor could search them out. Instead of an infinity of examples, let the rainbow serve for the moment. The philosopher Aristotle confuses us with his obscurities, nor do we obtain any useful knowledge [of the rainbow] from him. And this should not surprise us, since his principal imitator, Avicenna, . . . acknowledges that he himself does not clearly understand the nature of the rainbow. And the reason for this is that [these] philosophers were ignorant of the final cause of the rainbow and, being ignorant of the purpose of the rainbow, they were ignorant of the things that serve this purpose. . . . But the final cause of the rainbow is the dissipation of the aqueous vapor, as is shown in the book of Genesis.[41]

Nor is the ability of Scripture to supply vital philosophical perspective or information an isolated phenomenon. Bacon argues that every creature, "in itself or in its similitude, as a universal or an individual, from the highest of the heavens to their furthest limits, appears in Scripture," placed there to assist our understanding; philosophy, which has no purpose other than to unfold the natures and

[38] Bacon, *Opus maius* 1.9, Vol. III, pp. 18–19, 21; 1.13, p. 28.

[39] *Ibid.,* 1.9, p. 20.

[40] *Ibid.,* 2.1, 2.2, 2.14, pp. 36, 39, 68.

[41] *Ibid.,* 2.8, p. 52.

properties of created things, is thus founded at every point on biblical exegesis.[42]

To support these contentions Bacon appeals to the authority of the church fathers: Augustine, Jerome, Cassiodorus, and others. Augustine's *De doctrina Christiana* is several times quoted or paraphrased: "wherever the Christian finds truth, it is his Lord's"; truth is to be reclaimed from the philosophers, "as from unjust possessors." Bacon quotes Cassiodorus in defense of the quadrivium, and Jerome in defense of eloquence. And he introduces Bede's admonition to Christians that they seize all that is useful in the liberal arts and apply it to sacred purposes.[43]

Philosophy is also justified by its pedigree. Bacon argues that it was revealed in its fullness to the patriarchs and prophets, "the only true philosophers, who knew all things." God also gave them increased longevity to provide time for the completion of that which had been revealed to them. This knowledge was passed down through various ancient sages to Plato and Aristotle, the latter "aspiring to that perfection of philosophy possessed by the ancient patriarchs." Bacon acknowledges that philosophy, as it has come down to his own age, is incomplete; but far from justifying its rejection, this merely obligates us to "supply what the ancients lacked."[44]

The divine origin of philosophy is proved by the presence of the principal articles of the faith in the writings of pagan philosophers—doctrines such as the existence of God, the resurrection of the dead, and the immortality of the soul. Moreover, virtue is taught by pagan authors, whose teachings can be "reduced to contempt for riches in place of covetousness, disregard for honors in place of pride, scorn for pleasures in place of lust and gluttony, and activity and receptivity of the soul in place of anger, envy, and sloth." Philosophy is justified, finally, by its utility in proving the faith and persuading the unbeliever.[45]

Philosophy, then, is a divine gift, an instrument of biblical exegesis and theology, and a weapon against the infidel; it is integral to the pursuit of wisdom, and its cultivation is obligatory. But Bacon cannot simply argue that all knowledge, in principle, serves the faith and leave it there; he must also demonstrate that, in fact, the new learning will submit to faith's command. To this task he devotes the remainder of the *Opus maius*.

Part 3 of the *Opus maius* is an appeal for the study of grammar and foreign languages. The need is obvious, because the original languages of the Bible were Hebrew and Greek, and philosophy has been acquired from Hebrew, Greek, and Arabic sources. Since "it is impossible for the peculiarities of one language to be preserved in another," our understanding will necessarily remain incomplete until we have mastered the original languages.[46] Bacon launches into a discussion of faulty translations and the errors that have arisen through linguistic ignorance;

[42] *Ibid.*, pp. 51–53.

[43] *Ibid.*, 2.1, 2.3–4, pp. 37, 39, 41–44. The first passage quoted from Augustine is from *De doctrina Christiana* 2.18, *Opera,* Vol. IV, Pt. 1 (Corpus Christianorum, Series Latina, 32) (Turnhout: Brepols, 1932), p. 54; the second is paraphrased from 2.40, *ibid.*, p. 73. See also the passages as quoted at nn. 18 and 20 above.

[44] Bacon, *Opus maius,* 2.9, 2.14, 2.13, 2.15, Vol. III, pp. 53, 67, 66, 70.

[45] *Ibid.*, 2.15, pp. 70–71; 7.3–4, Vol. II, p. 266 (the philosophers thus oppose the seven cardinal sins); and 2.18, Vol. III, pp. 75–76.

[46] *Ibid.*, 3.1, Vol. III, p. 80.

he complains about the Paris edition of the Bible and takes his readers on a brief excursion through the Hebrew and Greek alphabets.

Bacon concludes this part with a systematic presentation of reasons for the study of languages. They are required, first, for the government of the church: so that foreign words employed in the divine office and the administration of the sacraments will be properly pronounced and comprehended; so that the church will be able to rule successfully over foreign peoples; and so that numero-alphabetical studies of the Hebrew, Greek, and Latin Scriptures, which can predict the events of the end of the age, will be encouraged. Linguistic studies are also required for the government of the Christian commonwealth, in its commercial and diplomatic dealings with various nations. Finally, languages are necessary for missionary activity and for the repression of those who refuse conversion.[47]

If Bacon had no difficulty defending grammatical and linguistic studies, mathematics would seem to present a somewhat stiffer challenge. Bacon believes that mathematics is one of four sciences so basic as to constitute the foundation for all others. (The remaining three are language, perspective, and experimental science.) However,

> the gate and key of these [four] sciences is mathematics, which holy men discovered at the beginning of the world . . . and which has always been used by holy and wise men more than any other science. Its neglect for the past thirty or forty years has ruined the whole educational system of the Latins, since he who is ignorant of it cannot know the other sciences or the things of this world. . . . And what is worse, men ignorant of it do not perceive their own ignorance, and therefore they do not seek a remedy. And, on the contrary, knowledge of this science prepares the mind and elevates it to sure knowledge of all things, so that if one grasps the basics of wisdom concerning this science and applies them correctly to knowledge of other sciences and things, one will be able to know all things that follow, without error or doubt, easily and powerfully.[48]

Bacon devotes hundreds of pages to the elaboration of this claim, and we will not attempt to follow the argument in detail. One of his most serious points is that causal demonstration, the most powerful form of demonstration in the realm of natural things, requires a mathematical cause; therefore, the mathematician has significant advantages over the physicist or natural philosopher, who knows only the physical effects.[49]

Many other points follow. Bacon resorts again to authority (citing Boethius, Ptolemy, and Aristotle) to show the ubiquity of mathematics in all branches of philosophy. All sciences, he argues, employ mathematical examples. Mathematics is the most innate of all the sciences, and the most easily learned. Mathematics was the first of the sciences to be discovered. Mathematics alone yields certainty, and only as other sciences make use of it can they find secure foundations: "A science that is full of doubts and sprinkled with opinions and obscurities cannot be certified or made manifest or verified except through

[47] *Ibid.*, 3.11–14, pp. 115–125. By "numero-alphabetical studies" I mean schemes that assign numerical values to each letter of the alphabet for the sake of extracting hidden meanings from the text.

[48] *Ibid.*, 4.1.1, Vol. I, pp. 97–98.

[49] *Ibid.*, 4.4.15, pp. 168–169. For a detailed examination of the argument see David Lindberg, "On the Applicability of Mathematics to Nature: Roger Bacon and His Predecessors," *British Journal for the History of Science*, 1982, *15*:3–25.

another science, known, verified, clear, and certain to us, as in a conclusion reached through premises. But only mathematics . . . remains certain and verified to the limits of certitude and verification. Therefore, all other sciences must be known and certified through mathematics."[50]

Bacon proceeds to examples that demonstrate the necessity of mathematics for astronomy, astrology, meteorology, and optics, and the utility of these sciences for the church. Astronomy teaches us how to correct the calendar. Optics furnishes mirrors and other optical devices, through which we can inspire terror in unbelievers and defend Christendom against invasion. Astrology enables us to predict the future, as well as understand the complexion of the human body and the course of disease.[51] And all of these disciplines assist in biblical exegesis:

> We see that the literal sense rests on an understanding of the natures and properties of creatures, so that through suitable adaptations and similitudes the spiritual senses might be elicited. For thus it is explained by the saints and all of the ancient sages. . . . Wherefore, the theologian requires an excellent knowledge of creatures. But it has been shown that without mathematics nothing can be known; therefore, mathematics is altogether necessary for sacred knowledge.[52]

The third special science to be treated is *perspectiva*, or the science of vision. *Perspectiva,* Bacon points out, is the most beautiful of the sciences because vision is the noblest of the senses, best revealing the differences of things and offering "sure experiences of all that is in the heavens and on earth." He maintains at one point that *perspectiva* is especially useful for the pursuit of wisdom and that without it nothing can be accomplished in the other sciences. He admits at another that although other sciences may have greater utility, none has greater charm and beauty. In either case it is an admirable science, neglected by the Latins, and therefore deserving of special attention.[53] We might add that it was also the science Bacon knew best, and he could not pass up the opportunity to display his knowledge.[54]

As Bacon develops the science of *perspectiva* in Part 5 of the *Opus maius,* he mainly sticks to technical detail with little attention to practical application. However, vision leads inevitably to an analysis of sense perception, with its psychological and epistemological implications. Vision also leads to the laws of radiation, including the radiation of force, thus presenting us with principles of causation (including astrological causation).[55] Further practical applications are noted in the *Opus tertium,* where Bacon argues that nothing is more frequent in Scripture than references to light, color, vision, and mirrors; the biblical exegete

[50] Bacon, *Opus maius* 4.1.3, Vol. I, p. 107.

[51] On the power of optics see *ibid.,* 5.3.3.3, Vol. II, pp. 164–165; for the other examples see *ibid.,* 4, Vol. I, pp. 109ff.

[52] *Ibid.,* 4, Vol. I, p. 175.

[53] *Ibid.,* 5.1.1.1, Vol. II, p. 2 (quoted); and *Part of the* Opus tertium *of Roger Bacon, Including a Fragment Now Printed for the First Time,* ed. A. G. Little (Aberdeen: Univ. Aberdeen Press, 1912), p. 20.

[54] For a fuller analysis of Bacon's contribution to the discipline of *perspectiva* see David Lindberg, "Roger Bacon and the Origins of *Perspectiva* in the West," in *Mathematics and Its Applications to Science and Natural Philosophy in the Middle Ages,* ed. Edward Grant and John E. Murdoch (Cambridge: Cambridge Univ. Press, 1987), pp. 249–268.

[55] The laws of radiation are dealt with fully in Bacon's *De multiplicatione specierum.* Note also that Bacon places the laws of radiation under mathematics as well as optics.

therefore requires *perspectiva*. For example, the apostle Paul's remark about "now" seeing "through a glass darkly, but then face to face" (I Cor. 13:12), reveals the apostle's own ignorance of the principles of reflection. Moreover, optical instruments can reveal things distant and hidden; small things can be made to appear large, and large things small; and a man, through multiple reflections, can be made to appear as a multitude. Such wonders as these are useful in converting the infidel, who is more apt to be persuaded by visible demonstrations than by preaching; and the same wonders can be employed to terrorize those who refuse conversion.[56]

Bacon turns finally to what he calls "experimental science," an ill-defined category that embraces both a methodology applicable to all sciences and a body (or several bodies) of practical knowledge and lore, presumably gained by experimental or experiential means. As a methodology it certifies the conclusions of rational argument, providing the assurance that can only come from a direct confrontation with reality. For example, no rational argument can ever certify the injurious nature of fire as putting your hand in a fire can. Experimental methodology also dispels myth and refutes faulty authority: "Although argument does not suffice for the certification of truth, authority suffices far less. . . . Therefore, this science [experimental science] wishes to teach that nothing is to be examined by argument or by authority unless there is some [confirming] experience."[57]

In order to guard against misunderstanding, I must note that "experimental science," as a methodology, denotes an extremely broad range of activities. Although it certainly includes the use of instruments to poke nature and probe her secrets, its more general sense is simply that of "experience." And even experience is defined in the broadest possible way, to include reports of other people's experiences and the internal experience of divine illumination.[58]

Experimental science also discovers truths within the boundaries of other sciences not attainable by those sciences alone—that is, truths that cannot be reached demonstratively, by deduction from the first principles of the discipline.[59] For example, although the principles of the spherical astrolabe are supplied by the mathematician, the actual production of one requires the special skills of the experimental scientist. Again, there are secrets for the prolongation of life, unknown to the art of medicine, which experimenters have discovered; they "have lain in wait for beasts, in order to discover the powers of herbs, stones, metals, and other things by which they restored their bodies in many, shall we say, miraculous ways."[60]

Finally, experimental science discovers natural powers wholly outside the other sciences. Bacon has in mind the marvelous powers of plants and stones, the wonders of fireworks, the phenomena of magnets, and other mysterious virtues. Experimental science also offers astrological predictions without astronomical intricacies; it produces useful inventions, which will save the church from the infidel and the Antichrist; it investigates magical practices and distinguishes

[56] Bacon, *Opus tertium* (cit n. 53), pp. 40–42.

[57] Roger Bacon, *Tractatus in quo fit sermo de experientia in communi*, ed. Hackett, in "Experimental Science in the Philosophy of Bacon" (cit. n. 6), p. 276.

[58] Bacon, *Opus maius* 6.7.1, Vol. II, pp. 169–172.

[59] *Ibid.*, 6, p. 202; and Bacon, *Opus tertium* (cit n. 53), p. 44.

[60] Bacon, *Opus maius* 6, Vol. II, p. 208.

philosophical truth from superstitious belief; and, like every other science Bacon has examined, it contributes to biblical exegesis and conversion of the unbeliever.[61]

CONCLUSIONS

Viewed from the perspective of traditional Christian discussions of the uses of secular learning, going back to Augustine and before, Bacon's much-trumpeted modernity disappears without a trace. He was not the champion of autonomous, secularized natural science against a repressive church. Quite the contrary: the growing autonomy and secularization of science were among the things he feared. But he also feared suppression of the new learning. He was endeavoring, therefore, to steer a middle course between two equally dangerous extremes. In this he reflects, in microcosm, the dilemma of the thirteenth century.

Against those who feared the new learning and would condemn it if they could, Bacon presents traditional arguments about the usefulness of the arts and sciences for determining the literal sense of Scripture. He argues that philosophy in its fullness was given to the patriarchs and prophets and that we must aspire to recover their perfect wisdom. He notes the presence of certain articles of the faith in pagan writings. He quotes Augustine and Jerome and Cassiodorus, who called upon Christians to reclaim philosophy from its pagan possessors and put it to its proper use. Bacon proceeds to identify sources of error that help to explain the disrepute in which the new learning is held by the ignorant; unworthy authority, custom, popular prejudice, and pretended wisdom, not to speak of bad translations, have led to attacks on Aristotle and his commentators by those who do not understand what they are attacking. And finally, Bacon displays considerable originality in the discovery of new uses, beyond biblical exegesis, to which the church can put the new learning. Calendar reform, astrological predictions, the prolongation of human life, the discovery of nature's secret powers, and the invention of optical devices—all are of vital importance for the government of Christendom, conversion of the infidel, and defeat of the Antichrist. Nobody saw the religious possibilities of the new learning more vividly (which is not to say more accurately) than did Bacon.

But the possibilities that he saw *were* religious ones. Bacon may have endeavored to enlarge the range of useful studies beyond that contemplated by Hugh of St. Victor, and certainly beyond anything Augustine would have condoned; but he shared their determination to define utility in religious terms. In none of his writings is there so much as the hint of a preference for an autonomous philosophical enterprise. There is "one perfect wisdom," he argued in the *Opus maius,*

> and this is contained in sacred writ, in which all truth is rooted. I say, therefore, that one science is mistress of the others—namely, theology, for which the others are integral necessities and which cannot achieve its ends without them. And it lays claim to their virtues and subordinates them to its nod and command. Or better, there is only one perfect wisdom, which is totally contained in sacred Scripture and expounded by canon law and philosophy.

[61] *Ibid.*, pp. 215–222; and Bacon, *Opus tertium* (cit. n. 53), pp. 48–53.

Wisdom (the "way of salvation") is vitally in need of philosophy and the sciences, both for its achievement and for its exposition. Bacon emphatically rejects the opinion that knowledge is valuable for its own sake: "For every investigation of man that is not directed toward salvation is totally blind and leads finally to the darkness of hell."[62]

But how would one reply to the proposal that Bacon's claims about the religious utility of philosophy were rhetoric or subterfuge, offered only to placate the authorities and purchase enough freedom for the philosophical enterprise to proceed without interference? I would counter by suggesting that such a proposal could spring only from an iron determination to view Bacon as a modern, since it is not supported by a shred of evidence. Besides, there are persuasive arguments to the contrary. In the first place, Bacon was not known for tact or political acumen; he was capable of referring to one of the leading theologians of his own order as "the worst and most stupid author of those errors, . . . the most infamous among the stupid multitude." He was also prepared to violate the rules of his Order when he found it in his interest to do so; and he had the courage (or foolhardiness) to align himself with the radical "spiritual" wing of the Franciscan Order despite the significant dangers of such a stance.[63] Furthermore, if Bacon's purpose were simply to create breathing room for philosophy, he need only have argued that philosophy serves the faith. But he claimed much more: he developed the Augustinian notion that no rational activity can hope to succeed unless founded on Christian faith and set to work on the data of revelation; the pathway even to nature's secrets, he insisted, leads through biblical studies.[64]

The refusal to take Roger Bacon at face value is part and parcel of the refusal to accept the Middle Ages on its own terms; the opinion that forces heroism and martyrdom onto Bacon is the same opinion that demands incompatibility, if not confrontation, between medieval religion and medieval natural philosophy. Bacon's entire career was motivated by quite a different view of the matter. Bacon was certainly an early and enthusiastic defender of the new learning, but not as a body of knowledge distinct from, or antagonistic to, the faith. His defense, though ingenious in its details and impressive in its fervor, was the obvious one: the new learning could, despite various perils, be made to serve the faith and submit to its command. This was the Augustinian formula stretched to fit the burgeoning intellectual life of the thirteenth century.

To put it somewhat crudely, then, Bacon's campaign did not look forward to modern science, but backward to the patristic tradition. It does not follow, however, that Bacon was a reactionary. Although traditional in his conviction that knowledge is to be pursued only insofar as it serves religious functions, Bacon evinces an unusual command of the sciences and an unprecedented determination to bring them into the religious fold. Whereas Augustine had been fundamentally pessimistic about the utility of secular disciplines, Bacon was inclined to suppose that everything could be of religious use. If Augustine was prepared to accept a little secular learning, Bacon wanted it all. Bacon therefore stretched

[62] Bacon, *Opus maius* 2.1, Vol. III, p. 36.

[63] Roger Bacon, *Compendium studii theologiae*, ed. Hastings Rashdall (Aberdeen: Univ. Aberdeen Press, 1911), p. 52, referring to Richard of Cornwall; and Lindberg, *Bacon's Philosophy of Nature* (cit n. 36), pp. xxiii–xxiv.

[64] See the passage quoted at note 41 above.

the Augustinian formula to cover a vast array of sciences having, on the surface, no obvious theological or religious utility—an array that would certainly have dismayed Augustine. And it was Bacon's success in thus wielding the Augustinian formula that helped to overcome resistance to the new learning, to establish it in the universities and the religious orders of the thirteenth century, and ultimately to introduce it into the intellectual life of Europe. Bacon's traditionalist strategy thus proved a vehicle for fundamental change.

Nonetheless, we must be careful not to exaggerate Bacon's importance. The campaign that we have been examining was certainly not his alone; the integration of newly acquired Greek and Arabic learning into Christian thought was the central intellectual problem of the century, and it attracted Europe's finest minds. Moreover, new strategies soon appeared alongside Bacon's. The purpose of this paper, in any case, has not been to identify heroes and recount heroic deeds, but to take a man who has become the symbol of a medieval attitude that never existed and to see him for what he was. In the process, we stand to deepen our understanding of his age.

Getting the Big Picture in Perspectivist Optics

By A. Mark Smith

UNTIL RECENTLY the medieval science of *perspectiva* has been regarded almost exclusively as a physical venture and, more to the point, as a straight offshoot of geometrical optics—the reason being that it is firmly grounded in ray theory. So characteristic is this grounding that it has been used to gather certain representative works, from Alhazen's *Kitāb al-manāẓir* to Witelo's *Perspectiva,* into a special "perspectivist" tradition.[1] As a result that tradition has been approached with two basic issues in mind: how its model of light-radiation was applied to the analysis of reflection and refraction, and what methodological commitments that analysis entailed. This in turn has inspired a spate of studies devoted to such technical aspects of perspectivist optics as its account of the rainbow or its foundations in mathematical experimentalism.[2]

The trouble with this approach is not so much that it is inherently misguided as that it is accidentally misleading. By focusing so narrowly on its physical applications, these studies have tended to obscure the real intent of perspectivist optics. Like its classical forebear—the visual-ray theory—perspectivist optics was "optical" because its chief reference point was the eye.[3] In short, its fundamental aim was to explain sight, not light. Hence the perspectivist account of light-radiation cannot be properly construed unless its express purpose to provide an explanatory basis for visual theory is taken into full consideration.

This point has gained increasing recognition of late and has been turned to

This paper was written during my sojourn as a research associate at the Pontifical Institute of Mediaeval Studies, Toronto, in 1980–1981. I wish to thank Beryl Wells of the Institute for her help with the diagrams.

[1] Alhazen's *Kitāb al-manāẓir* (*The Book of Optics;* early 11th century) was translated into Latin in the mid-13th century as *De aspectibus.* In the 1260s Roger Bacon composed his *Perspectiva* (*Opus majus,* Pt. V) on the basis of this work, but along lines suggested by Robert Grosseteste. Soon after appeared Witelo's *Perspectiva* (c. 1275) and John Pecham's *Perspectiva communis* (c. 1278/1279). These four tracts form the core of what is commonly called the perspectivist tradition—Grosseteste here being excepted because, lacking access to Alhazen's work, he was not within the mainstream stemming from it. Alhazen's and Witelo's treatises are both published in *Opticae thesaurus,* ed. Friedrich Risner (Basel, 1572; facs. ed. [with intro. by David C. Lindberg] New York: Johnson Reprint, 1972). For Bacon's *Perspectiva* see *Opus majus,* ed. John H. Bridges (Oxford: Clarendon Press, 1897); for Pecham's *Perspectiva communis* see David C. Lindberg, *John Pecham and the Science of Optics* (Madison: Univ. Wisconsin Press, 1970).

[2] See, e.g., Carl B. Boyer, *The Rainbow: From Myth to Mathematics* (New York: Yoseloff, 1959); Alistair C. Crombie, *Robert Grosseteste and the Origins of Experimental Science, 1100–1700* (Oxford: Clarendon Press, 1953) (the classic on perspectivist methodology); William A. Wallace, *The Scientific Methodology of Theodoric of Freiberg* (Fribourg: Fribourg Univ. Press, 1959).

[3] The Greek term ὀπτικά, from ὤψ, meaning "eye," literally signifies "things to do with sight." So too the medieval Latin neologism *perspectiva,* from the root *spec,* "sight," connotes visual concerns. Thus, although light, as visible, could be a topic of study for classical and medieval optics, it was by no means, as it is today, the primary focus of that science.

especially good account in three recent works.[4] In this paper, however, I go even further to suggest that the ulterior concern of the perspectivists was epistemology and, therefore, that *perspectiva* should be understood as the science not of visual perception alone, but of visual *cognition.* To support this thesis, I place the perspectivist theory within the broader context of the medieval theory of "cognition by abstraction" and show how, within that context, vision was effectively reduced to a coherent succession of stages through which we forge a mental passage to objective reality. It is only by construing it in this way that we can make real sense of perspectivist optics, both as a scientific and as a historical construct.

In the first section I outline the theory of abstraction, discussing first its conceptual basis, then its psychological-epistemological basis, and last its causal basis. My purpose throughout is to show how these bases, and thus the theory itself, were not only paramountly Aristotelian, but also eminently sensible. In the second section I draw the perspectivist account of vision within the bounds of the theory of abstraction and show stage by stage how that account unfolds coherently within those bounds.[5] This procedure should make clear that what the perspectivists actually offered was less a scientific theory of light or vision than a scientifically justified world view.

THE THEORY OF ABSTRACTION

The Conceptual Basis. Every physical object in the Aristotelian universe is an embodiment of defining characteristics.[6] More than that, each is a particular manifestation or exemplification of supervening type. Every triangular object exemplifies Triangularity in general, and by the same token every human object somehow exemplifies Humanity. To know is therefore to grasp not mere particular manifestations, but rather *what* they manifest. And those "whats," or "quiddities" to use scholastic jargon (from *quid est,* the Latin for Aristotle's term τι εστιν), are Universals. The ontological status of Universals has been hotly debated since at least the time of Plato, the central issue being whether they really exist independent of their particular manifestations. Fortunately we can sidestep this debate, for within the continual round of arguments over ontology there is one untroubled spot: a general agreement that we are all privy to the *notion* of Universals, since whether they be taken as real or imaginary, Universals are undeniably the stuff of mental discourse, whereby we at least *think* we have a conceptual grasp of objective reality. So, problems of ontology aside, let us see how, in Aristotelian terms, we come to a conceptual grasp of supervening type from physical particulars.

"The natural way of doing this," Aristotle assures us, "is to start from the things which are more knowable and obvious to us and proceed towards those which are clearer and more knowable by nature." There is nothing cryptic about

[4]David C. Lindberg, *Theories of Vision from al-Kindi to Kepler* (Chicago: Univ. Chicago Press, 1976); Gary Hatfield and William Epstein, "The Sensory Core and the Medieval Foundations of Early Modern Perceptual Theory," *Isis,* 1979, *70*:363–384; A. I. Sabra, "Sensation and Inference in Alhazen's Theory of Visual Perception," in *Studies in Perception,* ed. Peter Machamer and Robert Turnbull (Columbus: Ohio State Univ. Press, 1978), pp. 160–185.

[5]A. C. Crombie hinted at this approach in "Early Concepts of the Senses and the Mind," *Scientific American,* May 1964, *210*(5):108–116.

[6]I.e., physical objects are categorically defined by their attributes, which fall under ten generic heads: substance, quantity, quality, relation, place, time, position, possession, action, and passion or affection. Cf. *Categories,* esp. 4, 1B25–2A4.

this. What Aristotle means is that we proceed inductively, passing from the realm of physical particulars, which is most immediate to us as physical, sentient beings, to the realm of conceptual representations or ideas, which is most immediate to us as percipient, reasoning beings.[7] The crucial point, however, is that sense particulars are immediately accessible to us only through their physical, sensible qualifications, and those qualifications form a complex.[8] Hence our first grasp of things is confused and multifaceted, consisting, say, in an impression of "short, laughing, snub-nosed, balding, two-legged thing." But our eventual grasp of this thing's supervening type, Humanity, is utterly simple or incomplex. It is the thing shaved to the nub, knowable, as Aristotle puts it, "without qualification."[9]

The Aristotelian world of concepts, then, is not just an abstract projection of objective reality through the senses into the mind; it is that reality made wholly intelligible and, perhaps more to the point, absolutely simple. Thus, cognition achieved through the senses unfolds in an inductive passage from the vague awareness of individuals as defined by their jumble of sensible attributes to a clear comprehension of the singular intelligible quiddities that make them simply what they are. That is how we make sense of the external world. Faced initially with a confusing welter of brute sense impressions, we extrapolate or isolate certain features from it to differentiate among the physical objects giving rise to these impressions. In this way we not only separate individual objects from each other; we also define each according to its special attributes to gain an impression of "this thing." From such impressions we note similarities among certain individuals and isolate those features according to which they are similar. Eventually, granted enough experiences of this sort, we are able to extract a notion of universal type comprehending all these similar things. This grasp of universal type is what finally allows us to order objective reality through recognition of "this or that *kind* of thing."[10]

Take the concept of Triangularity. Assume that, innocent of previous sense experience, we are confronted with a batch of variously shaped objects, some roughly triangular. At first we would get the "feel" of this confrontation, taking impressions of colors, smell, touches, and so forth. Yet on the basis of this rough sensory play we would begin to distinguish among objects, grouping them vaguely by their immediate complexes of sensible attributes. From here we would proceed to a more sophisticated differentiation according to less immediate but simpler attributes, such as size or, more important, shape. Next, we would single out the triangular objects according to the even simpler criterion of three-sidedness and then would isolate the criterion itself to form abstract impressions of "this or that triangularity." From this, finally, we would be led to grasp the general, absolutely simple impression of Triangularity by virtue of which these particular triangularities are alike. Once properly grasped, this conceptual impres-

[7]Quoting *Physics* I 1, 184A17–19, trans. R. P. Hardie and R. K. Gaye in *Aristotle's Works,* Vol. II, ed. W. D. Ross (Oxford: Clarendon Press, 1930). See also *Posterior Analytics* I 2, 71B33–72A4; II 19, 100B3–5.

[8]*Physics* I 1, 184A21–184B9: "Now what is to us plain and obvious at first is rather confused masses, the elements and principles of which become known to us later by analysis. Thus we must advance from generalities to particulars; for it is a whole that is best-known to sense-perception, and a generality is a kind of whole, comprehending many things within it, like parts" (trans. Hardie and Gaye). These "many things . . ., like parts" are the sum total of categorical attributes the thing has.

[9]Cf. *Posterior Analytics* I 2, 71B8–15. The τι ἔστιν alone, being absolutely simple and general, is knowable without qualification.

[10]*Posterior Analytics* II 19, 100A14–100B3; and *Metaphysics* I 1.

sion of Triangularity would permit us to recognize all further instances of triangularity almost by instinct. Thus recognition is only a memorative phase of cognition.

From this summary it should be evident that as far as Aristotle is concerned, we elicit deeper meanings from brute sense impressions in much the same way we seem to draw deeper meanings from inkblots in a Rorschach test. We literally pull them out. But the abstracted meanings in sense experience are veridical—not merely read *into* the phenomena, but actually drawn from them. Thus, while concepts give true, deep significance to sensible particulars, sensible particulars truly imply their deeper significance. Concepts are immanent in objective reality for Aristotle, shrouded, yet somehow implied, by the nexus of sensible and physical characteristics defining each individual within that reality. So cognition by abstraction is simply the act of unveiling those implications by, as it were, extricating them from their physical, sensible shells.

The Psychological-Epistemological Basis. The act of cognition, then, involves three basic levels or stages, the first and lowest of which—brute sensation—is actually twofold: establishment of physical contact between the sense organ and an external object, either immediately (as in touch) or mediately (as in sight); and a subsequent grasp of raw *sensibilia,* which Aristotle dubs "proper sensibles."[11] Although very primitive, the ensuing sensory impression is nonetheless abstract insofar as it stands apart from its objective, physical source. Likewise, the impression that is drawn out during the second stage—perception—stands abstracted from the sensory embodiment giving rise to it. Comprising both the so-called common sensibles (such attributes as size and shape not immediately accessible to sense)[12] and the proper sensibles in perceptible form, this impression yields a fully abstract representation of the external object giving rise to it. Yet despite its truly abstract nature, this representation is not universal; on the contrary, it is as particular as the physical object it represents.[13] Still, it forms the immediate source of intellectual abstraction, accomplished at the third and final stage of induction—which I shall call "apperception." It is at this, the highest, level that we attain a full conceptual grasp of external particulars by abstracting the *intelligibilia* (e.g., pure Triangularity or Humanity) from their perceptible representations, and that furnishes us with the wherewithal or, if you prefer, the premises of syllogistic reasoning.[14]

These three stages of induction were understood by Aristotle as material functions: functions, that is, of the physical sense organ and the material soul (the "mind," loosely speaking). Hence the act of sense cognition is determined doubly, by the nature of what is abstracted and by the nature of what abstracts. In explicating the latter, Aristotle fell back upon the notion of "faculties" or capacities:

[11] According to Aristotle, each sense has a sole "proper sensible" or "proper object" it is designed by nature to grasp. Thus, odor is the proper object of smell, just as sound is of hearing or savor of taste (*De anima* II 6, 418A10–13). Of the five senses, two (touch and taste) are immediate, so the remaining three occur over distance through a medium (*De sensu* 1, 436B13–20). That sight *must* occur through a medium is empirically verified by the fact that, when placed directly upon the eye, an object cannot be seen (*De anima* II 7, 419A12–14).

[12] Having no specific organ and so being proper to none, the common sensibles—which Aristotle enumerates as movement, rest, figure, magnitude, number, and unity—are perceived incidentally through the five special senses and thus inferred from the primary sense data (*De anima* II 6, 418B18–26 and III 1, 425A14–19).

[13] Cf. *De anima* III 3.

[14] *Posterior Analytics* II 19, 100B5–17.

each sense organ has a specific capacity to elicit its proper object from a near chaos of physical impingements, and by extension the material soul, as a sort of organ, is endowed with faculties that allow it to perform its inductive functions properly.[15] These faculties—common sense, imagination, reason, and memory in the medieval scheme (although Aristotle himself lodged all memorative functions in the faculty of the imagination)—work together in the following way.

Once each sense organ has abstracted its proper sensible, it passes it to the common sense, which acts as a conduit, distinguishing among the raw *sensibilia* and thereby rendering them discrete members of a sensory set in which they are all grasped as equivalent in kind.[16] The common sense also seems to serve as a conduit for the common sensibles, which it draws from the *sensibilia.* In addition, it transmutes the *sensibilia* themselves into perceptible form. The overall function of the common sense thus seems to be to initiate perception by abstracting perceptible impressions from sensory input and channeling them into the imagination. The imagination gathers these impressions into a single representation or "image" of the object initially sensed. Impressed there, these images are retained as specific short-term memories. The imagination also accounts for our ability to bring to mind "images" of past impressions, for these images are themselves impressions, whose duration depends on how forcefully they are made.[17] The faculty of reason in turn acts on these images, juggling them discursively to infer general conclusions through various combinations or "associations."[18] These general conclusions form simple mnemonic tokens or concepts of both the images juggled and the juggling process itself. Each token represents a specific process of discursive reasoning in all its component elements, but grasped as a whole and all at once. As such, it forms the content of memory insofar as it is subject to recollection, and recollection consists of recapitulating the contents of these mnemonic tokens in the imagination, where they are recombined in image form and thus quite literally re-membered.[19]

Although he ascribed these faculties and functions somewhat vaguely to the material soul, Aristotle never provided a physiological model of sensation or psychology within which to place them. Galen, however, did, lodging perceptual and mental functions in the brain and basing them on its charge of animal spirits.

[15] *De anima* II 5, 418A4–6. "Mind" and "material soul" are not strictly identical. Taken as what reasons discursively, mind is attributable to the material soul; yet as what actually knows or understands, it transcends material constraints (*De anima* III 5, 430A20–25 and III 7, 431B12–19).

[16] Aristotle's common sense is not a separate organ like the special sense organs, but a general sensibility subsequent to their special sensibility, one that enables us to accept the separate sensibles of a given object as a unity. Thus it is through the common sense that we form an internal impression of "this sweet, red, hard thing," even though "sweet," "red," and "hard" are mutually exclusive as special sensibles (*De anima* III 1, 425A14–425B4).

[17] *De memoria* 1, 449B30–450A31; esp. 450A25–31.

[18] Several times Aristotle insists that we cannot think without images (see, e.g., *De anima* III 7, 431A14–17 and III 8, 432A3–9, and *De memoria* 1, 449B30–450A6). W. D. Ross points out that with this conception of "thinking in images," Aristotle offers a prototypic notion of "association of ideas" (*Aristotle: Parva naturalia* [Oxford: Clarendon Press, 1955], pp. 36, 245).

[19] E.g., when we reach an unqualified scientific conclusion—say a demonstrable mathematical fact—through a set of logically associative (i.e., syllogistic) steps, at the instant we reach the final step we get a sudden flash of insight during which we truly understand the conclusion. And that flash of understanding includes both the conclusion and the steps leading up to it as if they were one. However, if we wish to remember that conclusion as truly understood, we must recapitulate the process of discovery, and if we do so correctly, we consider the conclusion as both known and memorized (cf. *De memoria* 2). This recollective process entails re-forming images along the line of previously made rational associations and pertains only to man (*De memoria* 2, 453A4–8).

More important, he also provided a virtual physiological model of the brain in terms of its ventricular structure, within which each of the four chambers he had isolated was presumed to be a locus of certain brain functions, ranging from sensation and thinking to bodily motion. Further, Galen made vision a paradigm of sensation by delineating a complete anatomical and physiological pathway through the eye, locating the point of first sentiency in the crystalline lens and tracing the passage of that sentiency through the optic nerves to the brain.[20] In other words, by reducing the eye to a physical and physiological extension of the brain, he reduced vision specifically, and sensation generally, to the same sort of act as conceiving.

What Galen did was to build an eminently suitable physiological basis upon which to graft Aristotle's psychology of faculties. That point did not long escape notice: there was an increasing tendency from late antiquity to "Galenize" Aristotelian psychology by transposing it into the brain and explaining it in terms of the physiology of spirits.[21] Peaking in the Middle Ages, this tendency yielded a physical theory of abstraction based on the model of a three-chambered brain within which the old faculties, transformed into "internal senses," were canonically located. Thus the common sense and imagination were assumed to lie in the anterior chamber, reason in the central chamber, and memory in the posterior chamber.[22] Within these locations, the internal senses served basically the same function as their faculty counterparts. The common sense accepted the various sensory forms presented to it by the five external senses and passed them, along with the common sensibles, to the imagination in coalesced perceptible form. From there, such perceptible forms—technically called "sensible forms" or

[20]See e.g., *De placitis* I–V, in Galen, *On the Doctrines of Hippocrates and Plato,* ed. and trans. Phillip de Lacy (Corpus Medicorum Graecorum, 4, 1, 2) (Berlin: Akademie Verlag, 1978); *De locis affectis* III 9, IV 3, in Galen, *On the Affected Parts,* trans. Rudolph Siegel (Basel: Karger, 1976) pp. 86–88; 109–111; *De usu partium,* in Galen, *On the Usefulness of the Parts of the Body,* trans. Margaret Tallmadge May (Ithaca: Cornell Univ. Press, 1968), Vol. II, pp. 463–503. Galen isolated four ventricles in the brain and, adducing various symptoms of dysfunction, hinted that they regulated certain mental and animal functions with their load of *pneuma psychikon.* But it was only with the medieval Galenists that the notion of each ventricle having exclusive control over a specific set of operations became fixed; see Rudolph Siegel, *Galen on Psychology, Psychopathology, and Functions and Diseases of the Nervous System* (Basel: Karger, 1975), pp. 238–239. The vehicle of optical sentiency was later called "visual spirits," although, as Siegel claims, Galen himself assumed that vision was a generic function of *pneuma psychikon* acting locally according to the nature of the eye; Siegel, *Galen on Sense Perception* (Basel: Karger, 1970), p. 74.

[21]See H. A. Wolfson, "The Internal Senses in Latin, Arabic, and Hebrew Philosophical Texts," *Harvard Theological Review,* April 1935, *25*(2):69–133, and E. Ruth Harvey, *The Inward Wits: Psychological Theory in the Middle Ages and the Renaissance* (Warburg Institute Surveys, 6) (London: Univ. London Press, 1975). See also Oswei Temkin, *Galenism: Rise and Decline of a Medical Philosophy* (Ithaca: Cornell Univ. Press, 1973).

[22]This is basically the model of Avicenna, al-Ghazzālī, and Averroes, although these authors disagreed about details, such as the actual number and nomenclature of the internal senses. One major difference between this scheme and Aristotle's is the inclusion of the *memorativa,* which provides an intellectual complement to the imagination. The standard medieval justification for lodging this internal sense in the occipital chamber of the brain is that a blow to the back of the head can easily cause amnesia. For mature Arabic thought on the internal senses see, e.g., Avicenna, *De anima* III 8, IV 1, in Avicenna Latinus: *Liber de anima seu sextus de naturalibus I–II–III,* ed. Simone Van Riet (Louvain: E. Peeters, 1972) and *Liber de anima . . . IV–V,* ed. Van Riet (Louvain: Editions Orientalistes, 1968); Avicenna, *Le livre de science,* trans. M. Achena and H. Massé (Paris: Belles Lettres, 1958), Vol. II, pp. 62–68; al-Ghazzālī, *Logica et Philosophia Algazelis Arabis* II 4, ed. Petrus Liechtensteyn Coloniensis (Venice, 1506); and Averroes, *Destructio destructionum philosophiae Algazelis in the Latin Version of Calo Calonymos,* ed. Beatrice Zedler (Milwaukee: Marquette Univ. Press, 1961), pp. 426–427.

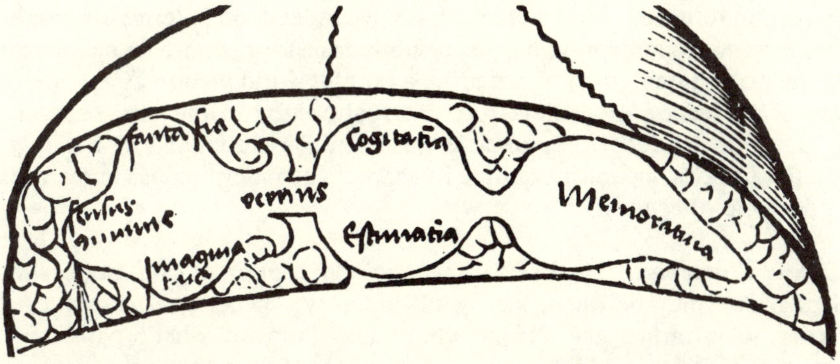

Figure 1. Detail from front cover, the three-chambered brain from Gregor Reisch, Margarita philosophica *(Basel, 1508), X, ii, 22. The anterior cell contains the* sensus communis *and the* imaginativa, *to which Reisch adds the* fantasia, *whose job it is to combine or "associate" the images retained in the* imaginativa *for use by the* cogitativa. *The* vermis *(the "wormlike" passageway) leads to the middle cell, divided into the* cogitativa *and the* estimativa. *The* memorativa *is located in the posterior chamber.*

"sensible *species*"—were presented to the reason, whose job it was to abstract "intelligible *species*" from them. These forms, finally, were passed to the memory to be stored as universal impressions or concepts.[23]

As clear as Aristotle's legacy is in this model, equally clear is the way it has been altered by specification. The very stress on the physiological nature of psychology is an obvious departure, as is the notion that psychological functions depend on both material ground (animal spirits) *and* physical location. Also, unlike the old faculties, the internal senses act in ways that suggest analogies with their external complements; each has its proper object, that of the common sense being the sensory form, that of the imagination being the sensible form, and that of the intellect (i.e., reason and memory) being the intelligible form. Moreover, the visual imagery of Aristotle's account—the notion of "imagination" or "thinking in images"—is reinforced in the language of *species* so natural to the medieval discourse of perception and apperception.[24]

The upshot of these changes is that the theory of cognition by abstraction, once

[23]See front cover for a popular early printed diagram of this model of the brain. See Edwin Clarke and Kenneth Dewhurst, *An Illustrated History of Brain Function* (Oxford: Sandforth, 1972), Ch. 3, for an extensive set of medieval and Renaissance diagrams (see esp. Robert Fludd's, Fig. 57, p. 38, for an illustration of how the internal senses are related to the *species* they abstract). It was primarily through Avicenna that the internal-senses scheme was adopted in the Latin West during the 13th century, although again there was not complete agreement over details. See, e.g., Bartholomaeus Anglicus, *De proprietatibus rerum* III 16 in *Bartholomaeus Anglicus On the Properties of Soul and Body,* ed. R. James Long (Toronto: Pontifical Institute, 1979); Albertus Magnus, *De anima* II iv, 7, ed. Clemens Stroick, Vol. VII, Pt. II of *Opera omnia* (Münster: Aschendorff, 1966); Thomas Aquinas, *Summa theologiae* I 78, Art. 4; and Roger Bacon, *Perspectiva* I i, 2–5, pp. 4–12. As an indication of how long this scheme persisted, see Francisco Suárez, *De anima* III 30 in *Opera omnia,* Vol. III, ed. D. M. André (Paris, 1856), pp. 702–709. Published posthumously in 1621, this tract gives an excellent survey of the internal-senses model (which Suárez himself repudiated in favor of a unified soul).

[24]*Species,* from *spec* (cf. *perspectiva* and such terms as "introspect" and "speculate"; see n. 3), translates the Greek εἶδος, derived from εἴδω "to see," perfect form οἶδα, "to know" ("idea" is from the same root). Hence both terms literally denote "what a thing looks like," yet both underlie terms denoting intellectual acts. In view of its equivocal nature, it is no wonder that *species,* even when intended as purely intelligible, was taken as a sort of pictorial representation in the mind.

understood in terms of visual metaphor, is now understood in terms essentially of a visual *model*. Apperception has been subtly transformed into an act almost of *seeing* the world intelligibly, of acquiring a faithful world picture by a continuous process of projection extending from external objects to the very back of the head. But in order fully to appreciate this cognitive model and its bearings on perspectivist optics, we must examine its theoretical underpinnings in the Aristotelian doctrine of causality.

The Causal Framework. The key to Aristotle's conception of causality lies in the primordial interplay between "act" and "potency," generic terms covering all contraries. Contraries, the "from what" and "toward what" governing all change, differ from contradictories in that they are of the same genus, so that, for instance, blackness can change to its contrary whiteness, but not its contradictory hotness. Act and potency are therefore not mutually exclusive principles but interconvertible expressions of one another—act of potency realized, and potency of act realizable. Everything in this world of change is subject to the two at once. Acorns are "acorns" by virtue not only of actually expressing the potential to be acorns, but also of having the potential to be oak trees and, equally significant, lacking the potential to be maples or dogs (maple-ness and dog-ness are contradictory to acorn-ness, whereas oak-ness, being contrary, is a legitimate "toward what" for acorns). In short, things are what they are because of their realizable potentials; they are what they are because they are pregnant with certain possibilities and yet constrained by the denial of others. Another way of putting this is to say that what things are consists in what they can and cannot express. Their very nature is therefore determined by their capacities for self-expression. A clear example of this is the propensity of animals to express their "nature" through self-reproduction.[25]

At first glance this may seem an odd way to understand the nature of things, but from an epistemological perspective it makes good sense. If, as Aristotle would have it, things are embodiments of certain characteristics, then how do we grasp them mentally? Surely not by taking them whole into the soul and storing them there bodily. But if not that, we must apprehend them through their embodied, formal characteristics. This means that physical objects must be able to express themselves "characteristically" and that we can accept such expressions as they are meant. Thus the act of apprehension or sense cognition is bilateral, contingent not only on the potential of external objects ultimately to express themselves intelligibly, but also on our power, as appercipient beings, to accept those expressions intelligibly, or literally, to *realize* them.

Recourse to the example of vision may help clarify this. According to Aristotle, what makes objects per se visible is their inherent color.[26] As visible, then, every object is defined and restricted by its capacity to express its nature through color. However, this same expression is equally restricted by what it redounds on: the visible expressions of things are realizable only in eyes, not in tongues or ears. Actual realization in turn is an expression of the eye's own sentient nature, so the act of seeing entails bivalent potencies. The object for its part has the power through its color to be seen, while the eye has the complementary power of realizing that capacity through its sentiency. When the two converge, seeing ensues.

[25] See, e.g., *Physics* I 6; *Metaphysics* X 4; *Physics* II 1, 193A30–193B19, II 8; *Metaphysics* V 4.
[26] *De anima* II 7, 418A26–418B3.

To complete the causal account of this act requires an understanding of one further interplay of potencies, between matter and form. Unformed matter is pure characterization without characteristics, whereas matterless form is pure characteristic in search of characterization. Alone, each is potential in the fullest sense, matter having a limitless capacity to be defined, form having an equally limitless capacity to be definite.[27] In actuality they coexist in a peculiar symbiosis that limits or specifies these potentials, form giving needed definition to matter, and matter giving needed definiteness to form. Physical objects manifest this symbiosis insofar as they are definite embodiments of defining form or characteristics—in fact, it is precisely because of this symbiosis that each such object actually exists. That is why we can speak of form and matter as "causes," because in symbiosis they truly determine not only what a thing is, but also that it is and, in a sense, why it is.

Returning to the visual act with this relation between form and matter in mind, we see that the formal cause of vision has already been stipulated in color. We have also stipulated two material causes in the external object and the eye. But these are not adequate, for there must always be a spatial separation between object and eye; and color, being formal, cannot bridge that gap unembodied. To fill the intervening space, Aristotle adduces a continuous medium, such as air, which in addition to providing a material substrate has a potential for transparency. Made actually transparent by light, the medium acquires the capacity to be informed by color, a capacity which, when actualized, allows this "information" free passage to the eye.[28]

Though couched in terms of formal and material causation, this account nonetheless points the way toward understanding efficient and final causality. If, for instance, color is the formal expression of visibility, then the efficient cause will lie in what does the expressing, be it the color-informed object or the color-informed medium. In each case, efficiency depends upon precisely how the color is embodied. In symbiosis with the object, it can effectively express itself to the transparent medium but not to the eye (which is another way of saying that vision is a mediate act). In symbiosis with the transparent medium, however, it *can* express itself effectively to the eye where its expression will be effectively realized. This act of realization crowns the entire process and so constitutes the final cause. Every stage is rigidly constrained by a particular set of potencies. The object itself cannot possibly express its visible nature without the physical intervention of a transparent medium, which, lacking transparency, could not take on the color that causes visibility. Moreover, without this intervention the eye could never grasp the visible nature of that color. Consequently, for all its *power* to express its visibility, no object could actuate that power without this chain of intervening causes, nor for that matter could the eye realize its own power of visual expression.

Within such a network of potencies and causes there is obviously no place for

[27] *Physics* IV 2, 209B9–10.

[28] Aristotle himself characterizes transparency as the "color" of light (*De anima* II 7, 418B12). Light is therefore not considered visible per se, but as a catalyst for transparency, so it is only inferentially sensible. This peculiar ancillary status of light is also evident in the visual-ray theory of Euclid and Ptolemy. Both acknowledge the existence of light, yet both locate the primary cause of vision in the ray emitted by the eye; thus both treat light as an accidental cause of vision. The "passage" to the eye is instantaneous and so is no real passage at all (*De anima* II 7, 418B20–27). Despite this, Aristotle's theory is generally characterized as intromittive, which is legitimate if we keep in mind that the direction of propagation is causally, not temporally, determined.

"action" or "passion" pure and simple. Action is invariably "interaction," and agents are only as untrammeled as their corresponding patients will let them be. To undergo action is itself a kind of action, a positive expression of power.[29] Hence in the analysis of vision it is not enough to say that the external object or its inherent color causes sight. The medium and the eye are causes too. Limited and specified by their inherent capacities, they not only provide the means for the object's self-expression, but in essence they determine it by realizing it. So cognition by abstraction is an expression of our innate capacity, as agents, to act reciprocally upon external reality by specifying its objective self-expressions as sensible, perceptible, or intelligible. Even the air has a necessary share in this activity. Although fundamentally passive, it nevertheless has the intrinsic power, through its transparency, to act directly on agent-objects by specifying their self-expressions as "visible" and thus to abstract such information.[30] So too the eye can act immediately upon the medium to specify its visible self-expression and, in so doing, abstract visual information. By the same token, finally, but at ever greater removes from the objective source, the mind can abstract perceptible and intelligible information—all in a chain of acts linked tightly by the four causes. At each reprise the object's potential self-expression is more fully realized, because the information is subtly transformed through symbiosis with a different material cause possessing different inherent capacities.[31]

Earlier we likened sense experience to what seems to happen in a Rorschach test, the inkblot representing objective reality and the elicitation of meanings from it representing abstraction. Now if instead of "potencies" we use the terms "meanings" or "intentions" suggested by that analogy, our causal analysis of abstraction falls outright into order. Objects, like inkblots, are what we make of them, and that depends upon what we bring to bear upon them—our natural aptitude for self-expression, if you will. In essence, then, objects imply or intend certain things to us, but we also intend toward objects by specifying what they will mean to us. We intentionally realize things which, conversely, if not intended will not be realized.

Abstraction therefore involves a reciprocal play of volitions. Objective reality

[29] *Metaphysics* V 12, IX 1.

[30] *De anima* II 7, 418B4–32. Bacon relies on this notion of positive specification in claiming that the propagation of *species* is "caused . . . by a true alteration and drawing-out of active power from the patient's matter"; *De multiplicatione specierum* I 3, in *Opus majus,* Vol. II, ed. J. H. Bridges (Oxford: Clarendon Press, 1897), p. 433.

[31] Thus we can understand the material cause in terms of the media, both aerial and "spiritual," through which the form or *species* is propagated. As it passes from the crass external medium through the visual and animal spirits, it becomes increasingly refined or "spiritualized." But it is the physical nature of the organs of sense, perception, and intellection that helps determine their inherent capacities. See *De sensu* 2, 438B16–439A2, where Aristotle discusses the nature of the various sense organs by recourse to the four elements—the eye, e.g., being primarily water rather than air, because water is transparent, as is air, but denser and more easily contained within the eye (2, 438A15). In *De memoria* he likens memory in the imagination to the impression of seal upon wax and reduces the explanation to the basic contrarieties of moistness and slipperiness versus dryness and hardness, claiming that poor memory is due either to taking impressions too quickly and superficially, as running water would take a stamp, or to resisting them too much, as old stuccoed walls resist clear impressions (1, 450B1–12). In *De memoria* 2, 453A15–32 he attributes poor recollection to, among other things, such temperamental disorders as melancholy, which disturb the requisite balance between moistness and dryness. These scattered notions are taken up in earnest by Bacon, who applies the moist/dry dichotomy to the brain as well as to the eye and its humors (see *Perspectiva* I iv, 1–2, pp. 20–23, and I i, 2–3, pp. 4–7). These passages imply that the propagation of forms through the eye into the mind, like the radiation of light through the transparent medium, is a matter of "physics."

cannot be understood if it will not, nor can we understand it if we will not. To use a homely example, when we act unintentionally, we operate in a sort of oblivion, stumbling into things we do not really see or missing clear implications. Just so cognition through the senses requires an attentive expression of intention, whose fulfillment lies in realizing the intelligible expression of objective intentions. This notion of intentionality is the linchpin of perspectivist visual theory, the means by which that theory explains how we abstract an intelligible world picture from a continuous inward flow of formal characterizations that starts with the physical radiation of light and color.

THE PERSPECTIVIST ACCOUNT OF VISION

The Physics of Light-Radiation. The perspectivist account of radiation was framed essentially in response to two flaws in Aristotle's theory. First, by remanding light to the distinctly secondary role of actualizing transparency, Aristotle flew in the face of the apparently self-evident fact that light is inherently visible. Second, his analysis of the propagation of forms is holistic in every way. The color form at the object's surface, as well as the color form reaching the eye, is taken as a totality (one consequent difficulty is that there is no coherent way to explain how such a holistic form "shrinks" to fit the pupil), and so indeed is the process by which the form arrives at the eye, since it happens instantaneously through a perfectly continuous medium. As a consequence, Aristotle's theory of propagation does not lend itself meaningfully to geometrical analysis, because it does not entail radiation in the true sense of the term.

For this and other reasons—not the least of which were recognition of the explanatory value of the classical visual-ray model and general acceptance of Neoplatonist light metaphor in ontology[32]—the perspectivists saw fit to import three crucial changes into Aristotle's theory. To start with, they elevated light to primacy, granting it the power to cause sight immediately by striking the eye, and mediately by illuminating colors to render them effectively visible. In the second place, they supposed that light or illuminated color radiates promiscuously from points and is therefore diffused spherically.[33] That assumption spelled a virtual end to the holism of Aristotle's visible forms. For if light and color were assumed to emanate from discrete points, the object's surface had to be understood as a composite of individual point-sources rather than as a whole. Therefore, its visible effect had to be understood as a composite of point-forms as well. Finally, by adopting a model of spherical propagation, the perspectivists opened their account explicitly to radial analysis. The sphere of propagation could now be readily interpreted as a locus of point-forms lying equidistant from a central point-source and therefore linked to it radially.

In order to adapt this model of radiation to Aristotle's theory, the perspectivists first assumed that light is an inherent formal characteristic of external

[32]The Neoplatonist account of the procession of Being from the One through Mind and Soul relies upon a metaphor from light: Existence diffuses downward from the point-source of creation (the One) and "creates" by illuminating or defining what it touches. This ontological model was adapted to the Aristotelian causal scheme in various ways, first by Byzantine commentators and later by such Arabic commentators as al-Fārābī and Avicenna.

[33]Primacy: Alhazen, *De aspectibus* I 1, Prop. 1, p. 1; Bacon, *Perspectiva* I viii, 1, p. 54; Witelo, *Perspectiva* III, Sup. 2, p. 84; and Pecham, *Perspectiva communis* I, Props. 1–2, p. 62 (henceforth, exc. Bacon, cited by author only). Radiation: Alhazen I 5, Prop. 1, pp. 7–8; Bacon, *De multiplicatione specierum* II 1, p. 458; Witelo II, Def. 10, p. 61; and Pecham I, Prop. 6, p. 82.

objects, giving them the "color" of luminosity. Called *lux* in that state, it has the power to express itself formally as *lumen* in all directions through a continuous transparent medium.[34] Furthermore, when the exercise of that power—illumination—is realized in colors, they acquire the power to express themselves visibly in the same way.[35] Finally, since light is assumed to be visible per se, as well as the agent of color's visibility, the medium can now be considered as inherently transparent. *Lux* and color are therefore formal causes in external objects, whereas *lumen* and illumined color, their physical effects, are formal causes in the medium—from which it follows that object and medium, each in its own way, are material causes. The medium, meanwhile, is inherently apt to be informed by *lux* and illuminated color because of its transparency, and that aptitude is physically expressed in the omnidirectional radiation of *lumen* and illumined color.[36]

Thus for the perspectivists, as for Aristotle, the physical propagation of visible forms (now in the guise of color *and* light) entails a convergence of potencies or intentions. Whatever its visible intentions, the luminous or illuminated object can only express them physically by means of light and color through a continuous transparent medium. Conversely, no medium that is not transparent can intentionally realize that physical expression, which means that the radiation of color or light is an act of abstraction. *Lumen* and illumined color, as physical representations of *lux* and illuminated color, are intentional forms elicited by the transparent medium from external objects. Their peculiar nature at this level of intentionality is a function of both what they represent (efficient cause) and *in* what they represent it (material cause).

So far the perspectivists have effectively incorporated light, as a primary visual agent, into the Aristotelian scheme. But there are other problems. Aristotle's account pivots on the supposition that the physical world and all its actions are absolutely continuous and holistic. The reason for this supposition was to avoid action at a distance. Yet by accepting the idea of radiation from points, the perspectivists seem to have repudiated that pivotal supposition and therefore to have accepted action at a distance. Worse, they have raised the apparently insoluble problem of how we can possibly see something that emanates from discrete points and strikes us along "breadthless lengths." Why do we see objective reality as continuous if its form of visible expression is not only discontinuous but dimensionless?[37]

The perspectivists responded by conceding that physical radiation is only virtually punctiform, that it actually occurs from indefinitely tiny spots rather than points and, consequently, that "rays" are really thin shafts instead of mathematical lines.[38] This concession allowed them to preserve the principle of

[34]The distinction between *lux* and *lumen* is crystal-clear in perspectivist writings. See, e.g., Bacon, *De multiplicatione specierum* I 1, p. 409: "We say that solar *lumen* in the air is the *species* of solar *lux*, which is in [the sun's] body."

[35]Alhazen I 2, Prop. 3, pp. 2–3; Bacon, *Perspectiva* II iii, 1, p. 103; Witelo III, Sup. 8, p. 84; and Pecham I, Prop. 2, p. 62.

[36]Alhazen I 5, Prop. 28, p. 17; Witelo II, Prop. 4, p. 64; and Pecham I, Prop. 51, p. 132. I purposely differentiate "illuminated color" and "illumined color" to clarify the distinction between activated (i.e., *lux*-informed) color in the object and its formal effect or activity in the medium.

[37]According to Euclid, *Elements* I, Def. 1, a point has no dimension at all, while, according to I, Def. 2, a line has only length. So if light-radiation actually occurred from true points along true mathematical lines, we could not see its effect, since it would lack lateral dimension.

[38]Alhazen IV 3, Prop. 16, p. 112; Bacon, *De multiplicatione specierum* II 2, pp. 459–460; Witelo II, Prop. 3, pp. 63–64; and Pecham II, Prop. 18, p. 168. The perspectivists, then, agreed that the mathematical ray is fictive, a mere representational device.

absolute continuity while nonetheless analyzing the visible world as if it were discontinuous. Accordingly, neither the visible form nor the form of radiation itself needed to be treated holistically any longer. Thus for Roger Bacon the propagation of forms becomes a function of "multiplications"—spot-forms (or *species*) passing from spot to contiguous spot of the medium in a continuous succession of agent-patient interchanges, each constituting a microscopic act of abstraction, and the whole succession amounting to a "ray."[39]

Basically, then, the perspectivists had it both ways. They could exploit Aristotle's causal analysis to justify their physical account of radiation. Yet by reducing that analysis to a microscopic scale, they were able to exploit the ray model. The key to success lay in drawing an implicit but clear distinction between "actual" and "virtual" in their explanation. Though actually continuous, the sphere of radiation could still be treated as virtually discontinuous, as a lateral succession of spot-forms propagated outward, spot by spot through the medium in a rectilinear succession of multiplications or abstractions.

Visual Sensation. Given their model of physical radiation, the perspectivists had to face the issue of how the eye, specifically the crystalline lens, transforms the medium's physical information into visual information. For partial answer, they assumed that what happens in the lens is an amalgam of straightforward physical contact and sensitive reaction. The contact is invariably explained through a seal-and-wax analogy—light and color, like seals, being formally impressed on the surface of the lens.[40] Unlike wax, however, the lens does not take these impressions in actual depth, so their intensity is contingent on the brightness rather than the mechanical force of impressions. The sensory obverse of this stamping-process, variously called *aspectus* or *aspectus simplex,* represents that curious fusion of agency and passivity so typical of "action" in the Aristotelian universe. Undeniably an act, since it constitutes the lens's sensitive response to physical contact, *aspectus simplex* is also a passion in that it takes form as a low-level pain that tells us we are seeing.[41] In fact, if carried too far beyond the normal threshold by an overly intense impression, this pain can become literally blinding, as when we stare at the sun.

[39] In *De multiplicatione specierum* II 1, p. 457, Bacon describes the multiplication of *species* as a succession of alterations through the medium such that "the first part of the patient (i.e., the medium), having been transmuted and possessing the *species* in act, transmutes the second, and the second the third, and so on." In I 1, p. 414, he describes this process as one of "assimilation," in which the patient takes on the *species* of the agent (and becomes like it) without undergoing substantial change. In all, Bacon offers a model of radiation in which the medium is understood as virtually discontinuous but actually continuous, yet by emphasizing the successivity of the multiplication of *species*, Bacon logically concludes that the process must take real time (*Perspectiva* I ix, 4, pp. 71–73).

[40] The perspectivists agreed with Galen that the crystalline lens (or "glacial humor") is the seat of visual sensation (Alhazen I 5, Prop. 16, pp. 8–9; Bacon, *Perspectiva* I iv, 2, pp. 27–28; Witelo III, Prop. 4, pp. 85–87; and Pecham I, Prop. 36, p. 120). The seal-and-wax analogy was suggested by our retention of an afterimage after looking at a bright light or color and averting or closing our eyes. A physical "explanation" is deduced from the fact that the anterior shell of the lens is harder than the succeeding humors and so ought to be more suited to hold impressions for a time. See Alhazen I 1, Prop. 1, p. 1; Bacon, *Perspectiva* I iv, 2, pp. 27–28; Witelo III, Prop. 6, pp. 87–88; and Pecham I, Props. 1–2, p. 62.

[41] This pain alerts us to pay attention (Alhazen I 5, Prop. 26, pp. 15–16; Bacon, *Perspectiva* I iv, 2, pp. 27–28; Witelo III, Prop. 16, p. 91; and Pecham I, Prop. 43, p. 126). *Aspectus simplex* is Witelo's phrase, derived from Alhazen's simpler *aspectus* (Witelo III, Prop. 51, pp. 108–109, and Alhazen II 3, Prop. 64, p. 67). Bacon and Pecham designate this phase as "vision by sense alone" (visio/cognitio solo sensu); see Bacon, *Perspectiva* I x, 3, III iii, 2, pp. 79, 162; and Pecham I, Prop. 56, p. 136.

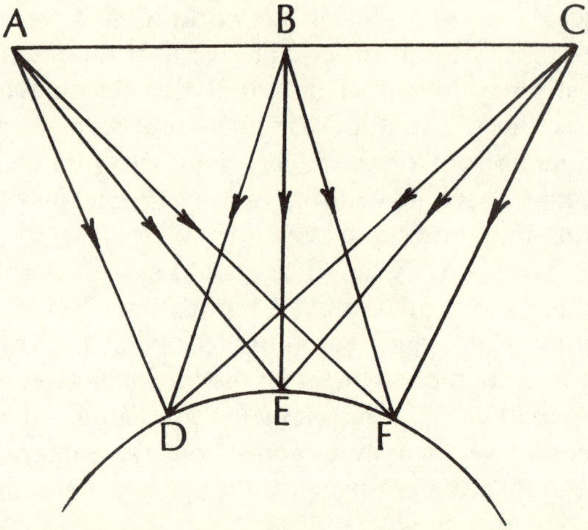

Figure 2. AC *represents a visible object with spots* A, B, *and* C *arbitrarily chosen;* D, E, *and* F *lie on the surface of the lens. Since the form of* A *strikes* D, E, *and* F*—as do the forms of* B *and* C*—these forms intermingle at all three spots on the lens.*

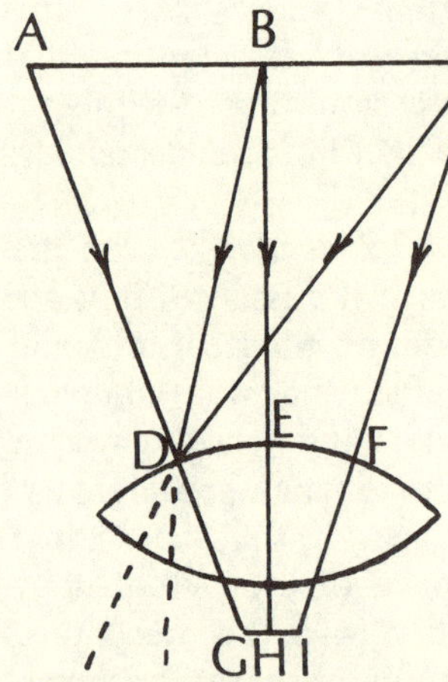

Figure 3. A, B, *and* C *are points on the visible object;* D, E, *and* F *are points on the lens. Rays* AD, BE, *and* CF, *being perpendicular to surface* DEF, *will pass unbent to project image* GHI, *a point-by-point representation of object* ABC. *Oblique rays* BD *and* CD *will be refracted out and will thus not interfere with the projected image.*

Such a primitive play of physical contact and brute sensation does not tell the whole story, though; still unanswered is the question of how the lens abstracts a coherent visible form from the multitude of physical impressions striking it. The crux of the problem is that, since each spot on any visible object radiates its form omnidirectionally, this same form must reach every spot on the facing surface of the lens. At the same time, every spot on the object must send its form to the same spot on the lens (see Fig. 2).[42] Thus what actually strikes the lens is a chaos of overlapping spot-forms amounting to sheer visible "noise." How can the lens abstract a coherent, point-by-point representation from such a chaos?

At first the perspectivist solution seems to involve two disparate factors: the lens's physical nature as a refractive instrument, and its sentient nature as a visual instrument. Now by its physical nature, the lens, like any other refractive body, will let only perpendicular rays pass unbent, so that it might seem plausible to conclude that the lens simply filters out a coherent image by letting only the perpendicular rays pass through and refracting out all oblique, incidental radiation (see Fig. 3).[43] But that is not how it is understood by the perspectivists. The "optical" selectivity of the lens is not a matter of its brute physical nature at all, but of its *virtus visiva,* or "visual power," which lets it choose a coherent image by accepting only those forms that strike it orthogonally and, in effect, refusing

[42] Alhazen I 5, Prop. 15, p. 8; Bacon, *Perspectiva* I vi, 2, pp. 37–39; Witelo III, Prop. 17, pp. 92–93; and Pecham I, Prop. 28, pp. 108–110.

[43] See, e.g., Alhazen I 5, Prop. 17–18, pp. 9–10; Witelo III, Prop. 17, p. 92; and Bacon, *Perspectiva* I vii, 1, pp. 47–49.

those that do not. Consequently, the crystalline lens *intentionally* realizes a coherent visual impression by dint of its innate sensitivity.[44]

Why use an extra agency to yield what the inherent refractivity of the lens would yield anyway? Because besides abstracting coherent images from the formal confusion striking it physically, the crystalline lens must also make visible sense of that confusion. The visual power enables it to do so, to transmute mere physical impressions into visible impressions. Therefore, the lens's ability to abstract coherent images and to render them visible is actuated by the same power, which is in turn animated by the constant charge of visual spirits flowing from the brain through the optic nerve. As an optical instrument, then, the crystalline lens is not just a refractive body; it is a refractive body informed by visual spirits and thereby brought to "life." Hence, its capacity for self-expression, its intentionality, is greater than, and different from, that of mere refractive media—and likewise its realization of this capacity.[45]

The visible form or *species* abstracted by the lens is a mosaic of visible light and color, each of whose spots conforms with a corresponding spot on the object that is its source. In addition, since the form lies within a cone whose base is on the object and whose vertex is in the eye, it has "shrunk" to fit the pupil.[46] This mosaic is passed radially through the lens to the vitreous humor, which accepts it integrally and passes it onward to the optic nerve, its integrity guaranteed by the vitreous humor's "custodial" virtue (see Fig. 4). So the vitreous humor, like the crystalline lens, has a sensitive life of its own that is expressed by its ability to preserve the coherence of visible *species* as they pass on to the optic nerve. Similarly, the optic nerve has its own custodial power, animated by the visual spirits that fill its hollow as it twists and turns its way from the eye to the brain. That is why visible *species* are multiplied in perfect integral order through those twistings and turnings to reach the forefront of the brain intact.[47]

Aspectus simplex therefore unfolds in a tight succession of abstractions from the lens through the vitreous humor and the optic nerve to the brain. Each stage in this succession is governed by a particular intentionality. That of the lens determines its ability sensibly to elicit coherent visible *species* from the confusion of *lumen* and illumined color sent to it from external objects through the medium.

[44]To emphasize the intentionality of the visual act, Bacon (*Perspectiva* I vii, 3, pp. 50–52) contends that, just as objects send *species* to the eye, the eye radiates *species* to objects to complete the visual act. In this he contravenes Alhazen, Witelo, and Pecham, all of whom stoutly deny any emission from the eye (cf. Alhazen I 5, Prop. 23, pp. 14–15; Witelo III, Prop. 5, p. 87; and Pecham I, Prop. 44, pp. 126–127). For the notion of a distinct *virtus visiva,* see Alhazen I 4, Prop. 4, pp. 3–4; Bacon, *Perspectiva* I iv, 2, pp. 27–28; Witelo III, Prop. 4, pp. 85–87; and Pecham I, Prop. 31, pp. 112–114. The actual selection is a matter of sensing, because the lens will only feel or distinguish forms striking it perpendicularly (Alhazen I 5, Props. 18–19, pp. 9–12; Witelo III, Prop. 17, pp. 92–93). However, as Alhazen warns in II 1, Prop. 4, p. 26, "the reception of forms in the sentient member (the lens) is not like the reception of forms in [merely] transparent bodies, for the sentient member receives and feels these forms; and they pass through it according to both its transparency and its sensitive virtue."

[45]See Alhazen I 4, Prop. 4, p. 4; Bacon, *Perspectiva* I v, 3, pp. 33–34; Witelo III, Prop. 7, p. 87; and Pecham I, Prop. 31, p. 112. As Bacon puts it, "according to the necessity and nobility of the soul's operations, *species* in an animate medium . . . disregards the common laws of natural multiplication, enjoying the special privilege of the soul" *Perspectiva* I vii, 1, p. 49.

[46]Alhazen I 5, Prop. 19, pp. 10–12; Bacon, *Perspectiva* I vii, 1, pp. 35–37; Witelo III, Prop. 18, p. 93; and Pecham I, Prop. 37, p. 120).

[47]Alhazen II 1, Props. 4–6, pp. 25–27; Bacon, *Perspectiva* I vii, 1, pp. 35–37; Witelo III, Props. 20–21, pp. 94–95; and Pecham I, Props. 33, 40, pp. 118, 122–124. Pecham refers to the passage to the brain as occurring "according to the laws of spirits." Since we have two eyes and thus two optic nerves joining toward the front of the brain, visual spirits are also responsible for merging the separate species conveyed by both into a coherent single form at that juncture.

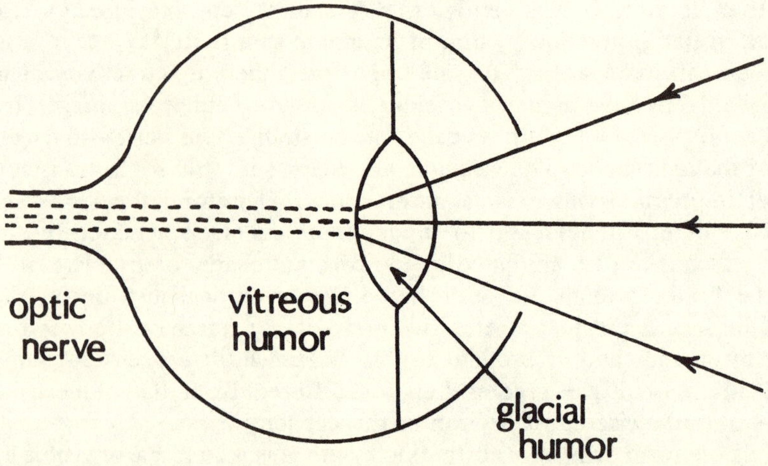

Figure 4. The rays of a visible form are "refracted" at the interface of the glacial and vitreous humors so that they continue to the optic nerve in a bundle, whose thinness is commensurate to the nerve's diameter.

By means of that intentionality, moreover, it realizes the primary visible intentions (i.e., light and color) of those objects and thus extrapolates visible *species* as intentional representations. The vitreous humor for its part expresses its intentionality by accepting the visible *species* from the lens and preserving it intact as it passes through. And in much the same way the optic nerve intentionally accepts and maintains the visible *species* as it wends its way to the brain. The multiplication of forms throughout this visual stage is subject to the same causal dictates as the multiplication of physical forms leading up to it.

Visual Perception. Like a pointillist painting, the intentional form that reaches the brain from the crystalline lens is only a two-dimensional array of luminous and colored spots. Its objective reality is therefore limited to the spots themselves and the contrasts among them. Yet, as in a painting, so in the visible form, these contrasts are all-important, for they carry us beyond the image's objective limitations to its underlying meaning by furnishing clues about what it is intended to represent.[48] Visible forms, like paintings, convey things not really in them by hinting at objective characteristics not actually there. Lacking actual depth, motion, or proper size, they still convey these things and thereby give us the impression of physical reality in all its aspects.

According to the perspectivists, such impressions come to us perceptually by means of the sensory form's twenty-two "visible intentions," which comprise, besides light and color, such attributes as shape, size, place, roughness, density, and even beauty.[49] Evidently derived from Aristotle's notion of common sensibles, these intentions are perceptible tokens of the defining physical qualities of external objects. As such, they are virtually but not actually contained in the

[48] Aristotle uses more or less the same analogy in *De memoria* 1, 450B21–26, where, discussing a depiction on a panel, he distinguishes between its meanings as "picture" and "likeness."

[49] Alhazen II 2, Prop. 15, p. 34; Bacon, *Perspectiva* I i, 3, p. 6; Witelo III, Sup. 2, p. 84; and Pecham I, Prop. 55, pp. 134–136. Although they are visible per se, light and color also have perceptible forms, so they are included among the visible intentions.

visible form and, taken as a whole, constitute its full potential for sensible self-expression. When realized they not only give us a complete abstract image of physical reality; they also allow us actually to specify what we see.[50] It is thus through the visible intentions that we mentally apprehend visible objects as physical particulars, and it is through this apprehension that we finally define them physically and recognize them.

Called *intuitio,* this act of perceptual abstraction takes place at the forefront of the brain in the *ultimum sentiens,* which, for all intents and purposes, corresponds to the common sense.[51] What empowers the *ultimum sentiens* is its *virtus distinctiva,* its capacity to discriminate, by which it elicits visible intentions from visual impressions, scrutinizes them, evaluates them by comparison, and then combines them into a definite perceptible representation. *Intuitio* is therefore the process according to which we realize the perceptible expressions of physical particulars. The discrimination carried out during *intuitio* is complex, involving both a straightforward grasp of visible intentions and a quasi-rational operation in which each intentional form is compared to similar types: through it we can both specify any given sensible impression and get a definite "idea" of what it represents.[52] And *intuitio* is the first consciously intentional act in the visual process, since it entails both judgment and attentiveness. For without judgment or evaluation we could never get a specific notion of what we see, and without attentiveness we can, and often do, misapprehend what we see.

This intuitive scrutiny, animated by the cerebral spirits contained within the *ultimum sentiens,* yields a faithful mental representation of the object if conducted with due care. The mental representation is intentionally verified or "certified," a process depending on our ability to scan the object with the visual axis (along which the clearest impression is obtained); it is then passed to the imagination, where it is impressed as a memory.[53] The first such impression, being somewhat indefinite, will not strike us very forcefully and so will not be deeply etched in the imagination, but repeated impressions of the same kind reinforce the memorative stamp until we do get a deep, lasting, and definite general impression of the given object. This final stamp, which Alhazen calls the *forma universalis,* gives us a standard against which to gauge all further impressions of the same specific kind. Consequently, with the *forma universalis* our initial superficial apprehension deepens into comprehension of a sort that lets us grasp and certify what we see *per intuitionem cum scientia precedente.*[54] The goal

[50] I.e., the more comprehensive the abstract image is, the more definite it is (Alhazen II 3, Prop. 63, p. 67; Bacon, *Perspectiva* I x, 3, p. 80; Witelo IV, Prop. 2, pp. 118–119; and Pecham I, Prop. 55, pp. 134–136).

[51] It is not clear whether the *ultimum sentiens* is meant as the common sense or as a special visual sense that feeds into the common sense. Bacon chooses the second alternative, placing the *ultimum sentiens* at the juncture of the optic nerves (*Perspectiva* I v, 3, pp. 33–34). Alhazen and Witelo, on the other hand, seem to opt for the first alternative (cf. Alhazen I 5, Prop. 26; II 2, Prop. 16, pp. 15–16, 34–35; and Witelo III, Prop. 27, pp. 94–95; see also Lindberg, *Theories of Vision,* p. 245, n. 110).

[52] Alhazen II 1, Prop. 11, p. 31 and II 3, Prop. 64, p. 67; Bacon, *Perspectiva* II iii, 2, pp. 104–106; Witelo III, Prop. 63, pp. 112–113; and Pecham I, Prop. 53, pp. 136–138. It is because judgment enters into this operation that we make errors in visual perception.

[53] Alhazen II 3, Prop. 65, pp. 67–69; Bacon, *Perspectiva* II ii, 1, pp. 92–93; Witelo III, Prop. 52, p. 109; and Pecham I, Prop. 33, p. 134. Passage into the imagination is discussed in Alhazen II 3, Prop. 66, p. 69; Bacon, *Perspectiva* I x, 3, pp. 79–81; and Witelo III, Prop. 58, p. 111.

[54] Alhazen II 3, Props. 67, 69, pp. 69–70, 72; Bacon, *Perspectiva* I x, 3, pp. 80–81; and Witelo III, Prop. 61, p. 112. Alhazen II 1, Prop. 11, p. 31 offers the example of recognizing written words. If we

of *intuitio,* then, is a specific recognition of what we have seen, which entails comparison of the particular sensible to a generalized impression retained in the imagination.[55]

From a causal perspective, the correspondence between *intuitio* and *aspectus simplex* is extremely close. The certified or universal form is intentionally conveyed by the visible form just as the visible form itself is intentionally conveyed by the physical form. Further, just as the crystalline lens uses its *virtus visiva* to abstract visible forms from physical forms in the medium, so the *ultimum sentiens* exercises its *virtus distinctiva* to abstract sensible forms from visible forms in the eye. Similarly, just as the vitreous humor and the optic nerve exercise a custodial power in maintaining the coherence of visible forms en route to the brain, the imagination exercises a custodial virtue in memoratively conserving sensible forms sent to it by the *ultimum sentiens.* Last, just as *aspectus simplex* provides a final cause for the multiplication of physical forms through the medium, *intuitio* provides a final cause for the multiplication of visible forms through the optic medium. The fundamental difference between the two lies in the level at which they express their intentionality; while *aspectus simplex,* being based on visual spirits, is wholly sensitive, *intuitio*, being based on animal spirits, is conscious and quasi-rational. Because it mediates between pure sensation and pure reason, and because it occupies a mediate position between the sense organ and the cogitative cell of the brain, *intuitio* shares the nature of both. The greater perfection and subtlety of the information abstracted by means of *intuitio* is due to the more perfect animate or "spiritual" nature of its material cause.

Apperception. From this account it might seem that, by yielding "universal forms," the process of *intuitio* is not just perceptual but also apperceptual. And in a sense it is. The *forma universalis* does in fact stand as an exemplar of specific sensible forms and is therefore the gauge by which those forms—and their object-sources—are deemed to be physically alike. Still, in spite of the generality that supervenes in its construction, the *forma universalis* is not truly universal, because it is only a general compendium of *sensible* attributes and is thus not properly intelligible or quidditative.[56] Hence to comprehend the *forma universalis* in the imagination is not, strictly speaking, to know or understand what it represents. Nevertheless, it is through these general forms in the imagination that we finally do arrive at such knowledge, for sensible forms, like their sensory counterparts, have specific intentions. And it is by means of those intentions—which are intelligible—that they can express the transcendent quidditative nature of what they represent.[57] The realization of that intelligible expression by reason

---

have never seen a given word before, the process of making it out letter by letter is time-consuming and tedious; but after repeated impressions we learn to read it instantly as a whole, because we learn to recognize it by signal features. Likewise, we come to recognize particular things without going through the laborious tasks of distinction and certification needed in first impressions.

[55] That is why Bacon calls this phase of vision "visio/cognitio per scientiam" (*Perspectiva* III iii, 2, p. 162); see also Pecham I, Prop. 61, p. 138, where he talks of grasping visible intentions "per distinctionem, argumentationem, et scientiam." For both Alhazen and Witelo the *forma universalis* occupies the imagination rather than a higher memorative faculty, just as in Aristotle.

[56] Bacon seems to imply this when he talks in *Perspectiva* I x, 3, p. 80 of the "vague (or indefinite) particular [which] is just as general as its universal and is interchangeable with it." Thus, the vague particular stands for, but is not, a true Universal.

[57] In contradistinction to the "sensibles *per se*" (i.e., the 22 visible intentions), Bacon calls the intentions of the sensible forms "sensible *per accidens.*" They are, moreover, proper to the

is cognition, and the abstract form of that expression is the true Universal or concept.

For example, repeated sensible impressions of individual people will eventually give us an abstract notion of Man as, say, "a featherless biped that laughs and talks." Now this notion or definition is certainly general, but it does not constitute true knowledge, because it is not comprehended without qualification. Quite the contrary, it is qualified (i.e., restricted) in several ways, among which we can count its exclusion of amputees and deaf-mutes. However, the fact that most men are bipedal and can therefore walk of their own accord, as well as the fact that most men can talk and laugh, betokens the deeper, simpler, and more comprehensive idea that men are animate and rational. Once we realize this by inference, we acquire a full conceptual grasp of Man as "rational animal."[58] Therefore, the general sensible impression that we have of Man in the imagination conveys or intends the more perfect intelligible form of "rational animality," which, when abstracted by reason and impressed in the memory, is completely understood.[59] Moreover, once actually understood, that form becomes a valid premise for scientific reasoning. Thus, if we know that man is a rational animal and that Socrates is a man, we can infer syllogistically that Socrates is a rational animal. Or to conclude with an optical example, given a full conceptual grasp of "sun," "moon," and "earth," we can associate these concepts syllogistically to reach the scientific conclusion that lunar eclipses are caused by the earth's interposition at times between the sun and moon.[60]

With cognition we have reached the end, in all respects, of the science of *perspectiva.* At rest in the memory, the concept or intelligible *species* is the last, subtlest, and most perfect of a stream of forms passing inward through a continuum of increasingly subtle and perfect media. It is the closest thing to pure reality that we, as physical entities, can aspire to intellectually, and our cognitive grasp of this reality is the fullest possible expression of that aspiration. Most important, as the culmination of the entire visual process, cognition is the final

---

"*cogitativa* and *memorativa* [which] are not ordinarily called 'senses,' although they belong to the sensitive soul; so the sensibles *per accidens* are called 'sensible' with respect to these [faculties]" (*Perspectiva* I x, 1, p. 75). "Hence," Bacon continues, "when I see a man, I see a substance and an animate thing; and so in a certain sense my sight falls upon his substantial nature and also upon his soul, which is a spiritual thing; but this is surely *per accidens*" (p. 76). Bacon expressly adds the *memorativa,* having realized that the imagination will not suffice for the retention of *intelligibilia* (cf. *Perspectiva* I i, 5, p. 11). Bacon is also quite definite in distinguishing the third phase of vision, which he calls "cognitio per syllogismum," from the previous two (*cognitio solo sensu* and *cognitio per scientiam*). He is equally definite in styling it the perfection of the visual act, "for by sense alone [*solo sensu*] we comprehend few things—for instance, light and color—and we do so imperfectly, grasping weakly whether they exist or that they do; but by 'science' [*per scientiam*] we comprehend in what way they exist and how [they differ] in kind, [for instance] whether the light [seen] belongs to the sun or moon, [or] whether the color is white or black. However, by syllogism [*per syllogismum*] we comprehend everything that pertains to light and color according to all twenty common sensibles. And so the first kind of comprehension is weak, the second is more complete, and the third is most perfect" (*Perspectiva* III iii, 2, p. 162). This threefold distinction is not made clearly, if at all, by Alhazen, although II 3, Prop. 68, p. 70 might seem to imply it (cf. also Witelo III, Prop. 66, p. 114). A good example of this triple process is the case of the written word (see n. 54). *Solo sensu* we would grasp it as black on white; by *intuitio* (i.e., *per scientiam*) we would parse it according to letter-shapes and juxtapositions to come finally to a recognition of the word itself as a whole; but by reason (*per syllogismum*) we would grasp its real meaning or intelligible intentions.

[58]This is the stock medieval definition of "man," usually with the added predicate "mortal."

[59]On the role of reason and memory, see Bacon's discussion in n. 57.

[60]For a similar example see Bacon, *Perspectiva* II iii, 2, pp.104–105.

cause par excellence, the ultimate "that for the sake of" everything leading to it, from physical light-radiation to perception.

THE BIG PICTURE

Beneath a prepossessing tangle of virtues, intentions, and causes, the theoretical grounds of the perspectivist theory are almost intuitive in their simplicity. The fundamental purport of that theory is to explain how external objects affect us mentally through sight, and its axiomatic core comprises four cardinal assumptions: that our mental apprehension of objective reality is formal or qualitative, that cause and effect are congruent, that physical objects embody formal characteristics, and that the physical world is a plenum (i.e., action at a distance is impossible).

From these assumptions it follows that, since the final effects of vision—concepts—are formal, the initial causes must be too. That is, if our visual grasp of physical objects is formal, then it must not be of the objects themselves, but of their formal characteristics. Further, since objects themselves cannot enter the mind through the eye, neither can most of their characteristics, such as size, shape, and the like. By default we are left with the inherent luminosity (*lux*) and color of objects; these alone seem to be per se visible. Yet even so, they cannot strike the eye directly, but must instead pass over the space between object and eye. Thus, because our first visual contact with objective reality is mediate, the formal effect of that contact—the visible *species*—must also be mediate. There must, in short, be a mediating formal cause—the so-called physical *species*, consisting of *lumen* and illumined color—to link the object and the visible *species* in the eye.

Comprising only visible light and color, the visible *species* has a meager store of actual information, but a wealth of potential or implicit information. Conveyed through the visible intentions, this information is passed to the *ultimum sentiens* at the anterior cell of the brain, where it gives rise to the sensible *species*. Impressed in the imagination, this provides a complete perceptible representation or mental image of the physical object, an image which has its own store of intentions. These intentions, when passed to the intellect at the central cell of the brain, yield the intelligible *species,* which is then comprehended as a true concept and stored memoratively in the brain's posterior cell.

Therefore, according to the perspectivists, visual cognition unfolds through a multiplication of four distinct formal causes, the first of which, the physical *species* in the medium, is only a superficial expression of its objective source's real nature. Yet like a Chinese box, it contains within itself all possible forms of that object's self-expression. Stripped of its outer physical shell by the crystalline lens, it reveals its first intentional layer of meaning in the visible *species*. That form, stripped away in turn by the *ultimum sentiens,* yields the second intentional layer in the sensible *species*. And the sensible *species*, stripped away by the intellect, lays bare the ultimate significance of the object in the intelligible *species*. Because each form in the sequence is an immediate "cause" of its successor, which is at the same time the immediate "effect" of its antecedent, the dictate of cause-and-effect congruency is strictly enforced to ensure that the conceptual picture we get of reality truly corresponds to what is out there.

Meantime, in conjunction with this series of formal causes, there is an equivalent series of material causes. The first of these, the transparent medium, is apt by its relatively crass nature to abstract only the physical species of *lumen* and illumined color. On account of its inanimacy, moreover, it has no control over how those *species* are multiplied, letting them radiate willy-nilly. The second material cause or medium, consisting of the crystalline lens, the subsequent humors, and the optic nerve, is animated by visual spirits; so the information it abstracts (the visible *species*) is more refined, and its control over the multiplication of that information is greater. The *ultimum sentiens* and imagination, partaking of both sensitivity and rationality because of their charge of animal spirits and their intermediate position between sense organ and *cogitativa,* elicit even more refined information (the sensible *species*) and exert more or less conscious control over it. Finally, being wholly inspirited and animate, the *cogitativa* and *memorativa* can abstract and control the most perfect information of all—the intelligible *species*.

In addition to these chains of formal and material causes, there is a parallel chain of final causes ordering the entire process of abstraction. Seeing, for instance, is a proximate final cause of visible radiation. As such, it provides a *raison d'être,* and therefore an informing pattern, for the physical account of light-radiation. Similarly, perceiving is a proximate final cause of seeing, as apperceiving is of perceiving. Each directly or indirectly governing its predecessors and governed by its successors, these causes are totally interdependent, thereby giving the perspectivist account final definite shape and coherence. For this reason, no single element in that account is meant to stand on its own "scientific" merits. The consistency of the perspectivist explanation of the physics of light and color, as well as its explanation of sight, perception, and apperception, hinges on the consistency of the whole.[61]

So what we have in the end is a comprehensive model of visual induction in which image-forms are continuously multiplied from objects through the external medium and the eye into the mind to yield a general picture of reality. As a process, this multiplication occurs in four intimately connected stages during which the forms induced become increasingly abstract as the media through which they are multiplied verge ever closer to pure spirituality, until both form and medium all but lose their physical moorings. The rest—the radial analysis of light and the account of sensation, perceptual discrimination, certification, memory, even syllogistic reasoning—is only an elaboration upon that basis.

The approach we make as historians of science to perspectivist optics should therefore be guided by the following considerations. First, neither as a whole nor according to its parts is the perspectivist theory gratuitously complex or obscurantist; it is in fact remarkably economical, reasonable, and coherent. Second, the theory itself pivots on the symbiotic relationship among formal, material, and

[61]The scientific canons by which perspectivist optics should be judged are not necessarily those governing optics since the 17th century. According to the latter, the medieval failure to derive the correct law of refraction constitutes a fatal flaw in the perspectivist analysis of light. However, within the overall account of induction and in view of the overwhelming success the perspectivists enjoyed in their mathematical analysis of *optics* proper and *catoptrics,* the same "failure" looms fairly small—which may well explain why the perspectivist theory persevered into the 17th century. Moreover, as I intend to show in a future paper, it was the 17th-century "discovery" of the sine-law and its justification through the least-time principle, far more than a clear-sighted wrapping up of loose ends, that destroyed the perspectivist model of induction, extending to its very metaphysical roots.

final causes. Without that relationship it falls apart completely. Third, the perspectivist account of physical light-radiation is only a part, but an integral part, of that theory. For this very reason it is shaped according to the overall structure of that theory, which is to say that there is no generic difference between the subjective and objective manifestations of light-propagation. The multiplication of light through the external medium is basically equivalent to the multiplication of its intentional forms through the internal media of visual and animal spirits. Fourth, the radial nature of light-propagation is, in actuality, an inference *per syllogismum* from the visible effects of light.[62] Hence the massive body of mathematical demonstrations that forms the bulk of (and is usually taken as the substance of) the science of *perspectiva* must be understood as descriptively corroborating one part of the whole cognitive model.[63] In short, to accept "geometrical optics" as the sole focus of the science of *perspectiva* is to mistake a single, albeit important, qualification for its real essence.

[62] The ray concept follows from the observation that light diffuses in all directions from any luminous source and is itself contingent on the a priori supposition that nature acts efficiently—i.e., along the shortest line (see, e.g., Witelo V, Prop. 5, p. 192; Pecham I, Props. 3, 27, pp. 63, 108).

[63] This is especially true since the perspectivists held that the "ray," as mathematical, is representational: the mathematical analysis of light is merely a way of clarifying the understanding of physical light and thereby making that understanding surer (cf. Bacon, *Opus majus* IV i, 3, pp. 103–108). For a discussion of the contrasting "realist" attitude among the Greek visual-ray theorists toward the mathematical analysis of optics, see A. Mark Smith, "Saving the Appearances of the Appearances: The Foundations of Classical Geometrical Optics," *Archive for History of Exact Sciences*, 1981, *21*.

The Impact of Archimedes on Medieval Science

By Marshall Clagett

THE importance of the role played by Archimedes in the history of science can scarcely be exaggerated. He was emulated and admired in his own day and at successive periods in later times. His name appears on the pages of the works of the great figures who fashioned the beginnings of modern mechanics. For example, Galileo owed a not inconsiderable debt to Archimedes —both direct and indirect. Galileo mentions Archimedes by actual count over one hundred times and in almost Homeric hyperbole, using such expressions as *suprahumanus Archimedes, inimitabilis Archimedes, divinissimus Archimedes,* and so on. Archimedes' significance for these founders of early modern science lay in the use of mathematics in the treatment of physical problems, as well as in the originality and fertility of his mathematical techniques. But all of this is well known.

Perhaps less well known is the role played by Archimedes in the middle ages. The problem of the medieval Archimedes is a classic one in the transmission of Greek learning. In some ways it has an advantage for study over other similar problems of transmission, since one is not overwhelmed by the abundance of materials as in the case of Euclid, or even more so in the case of Aristotle. That is to say, one has hopes both of defining the limits of the investigation and of presenting a substantially complete solution of the problem. The basic problem of the impact of Archimedes on medieval science can be subdivided into three questions:

(1) How much of the corpus of Archimedes' writings in their actual form — or some closely modified state — came directly into Latin?

(2) How much of Archimedes' methods and results was made available by translations from the Arabic or Greek of works which were influenced (but not so directly) by Archimedes?

(3) What use was made by medieval authors of the works and ideas available in Latin translation?

While I am preparing in detailed fashion the answers to these questions in a work, *Archimedes in the Middle Ages,* I think it possible in short compass to give the highlights of the answers as I conceive them.

Before answering the questions, two short observations ought to be made. The first concerns the temporal limits of the problem — the twelfth to fifteenth centuries. The latter terminus is arbitrary, the former is set by the state of

This paper was read at Oxford University in May 1956. A somewhat different form— much truncated—was presented to the History of Science Society, 29 December 1956, at the annual meeting.

the materials. It must be realized that there is virtually no Archimedes before the twelfth century in the Latin tongue. This has to be qualified by reference to some half a dozen brief indications of Archimedean ideas in the early medieval period. They mostly concern the density problem and I shall omit them here.[1]

The second preliminary observation concerns itself with the state of the Greek and Arabic texts of Archimedes available at the time of the translations. The Greek text extant in the Byzantine period consisted of at least three manuscripts, whose fate we can trace.[2] MS A, the principal manuscript, contained all of the works now known except the *Floating Bodies*, the *Method*, the fragmentary *Stomachion*, and the *Bovine Problem*. This was one of the two manuscripts available to Moerbeke. It is the source of all Renaissance copies of Archimedes — but it has been lost. MS B included the mechanical works — *Equilibrium of Planes, Floating Bodies* — the *Spiral Lines*, and the *Quadrature of the Parabola*. It too was available to Moerbeke. But it drops out of history after a reference to it in the early fourteenth century. Finally we can mention manuscript C, not available to the Latin West in the middle ages, or in modern times until its discovery by Heiberg in 1906 in Constantinople. It contains fragments of the *Sphere and Cylinder, Spiral Lines, Measurement of the Circle, Equilibrium of Planes, Stomachion*, most of the Greek text of *Floating Bodies*, and the brilliant work *Method* not present in the other manuscripts.

The apparent paucity of Greek manuscripts during the Byzantine period is probably the cause of the relative scarcity of Archimedean texts available to the Arabs. It now seems quite unlikely that they had access to any manuscript as complete as Greek manuscript A — the basis of most of our modern text of Archimedes. The Arabic Archimedes contains the following works:[3] (1) the *Sphere and Cylinder* and a portion of Eutocius' commentary on it; (2) the *Measurement of the Circle* — with perhaps Eutocius' commentary; (3) a fragment of the *Floating Bodies* (consisting of the enunciations without proofs of seven of the nine propositions of the first book and the first proposition of the second book — apparently the Arabs were not interested in the abstruse problems of the stability of conoidal segments found in Book II); (4) some indirect material from the *Equilibrium of Planes* found in other mechanical works translated into Arabic (such as Hero's *Mechanics*, the so-called Euclid

[1] I have discussed the various passages concerning hydrostatics available to early medieval schoolmen in my *Science of Mechanics in the Middle Ages* (Madison, 1959), chapter II. I do not know to which work of Archimedes Cassiodorus refers when, in Theodoric's name, he praises Boethius for having restored in Latin the mechanician Archimedes to the Sicilians (Cassiodorus, *Liber variorum*, Bk. I, 45, "Mechanicum etiam Archimedem Latialem siculis reddisti."). Certainly no Archimedean work has been found in any early medieval Latin codex.

[2] The standard account of the Greek manuscripts of Archimedes is still that of J. L. Heiberg, *Archimedis opera omnia*, Vol. *3* (Leipzig, 1915), Prolegomena. Cf. J. L. Heiberg, "Le Rôle d'Archimède dans le développement des sciences exactes," *Scientia*, 1916, *20:* 81-89, and E. J. Dijksterhuis, *Archimedes* (Copenhagen, 1956), chapter II. The manuscript designations A, B, and C are those followed by Heiberg, except that I have altered Heiberg's German Gothic letter to B.

[3] For the works of Archimedes translated by the Arabs, see H. Suter, "Die Mathematiker und Astronomen der Araber," *Abhandlungen zur Geschichte der mathematischen Wissenschaften*, 1900, Heft *10:* 36-37, 52; E. Wiedemann, "Beiträge zur Geschichte der Naturwissenschaften III," *Sitzungsberichte d. Phys.-med. Soz. in Erlangen*, 1905, *37:* 234, 247-50, 257.

On the Balance; Liber karastonis, etc.); (5) works perhaps based on originals of Archimedes now lost[4]— such as the *Lemmata (Liber assumptorum)*, a *Book of Triangles, On the Seven-part Division of a Circle, On Touching Circles, On Parallel Lines, On Data, On Properties of a Right-Angled Triangle,* and *On a Water-Clock.* Distinguished by their absence are the *Spiral Lines,* the *Quadrature of the Parabola,* the *Conoids and Spheroids,* and the *Method.* One final observation must be made concerning the Arabic Archimedes. Even though a number of the best treatises are missing, the Arabs nevertheless mastered the techniques that mark Archimedes' work, and in such a way as to show that they had made them their own — in fact, they did so much more readily on the whole than did the Latin mathematicians in the thirteenth and fourteenth centuries.

Now for our three questions. The first concerned the translation of the actual texts (or texts very close to actual texts) of Archimedes. Here we first should observe the importance of the date 1269, when the Flemish Dominican William of Moerbeke rendered virtually the whole corpus of Archimedes into Latin from the Greek text. Before that time the texts of Archimedes known to the Latin schoolmen came from the Arabic and were limited in number, although not in circulation. The first work of Archimedes translated into Latin was the *Measurement of the Circle.* It was translated from the Arabic twice in the twelfth century.[5] The first translation, which I have argued (but not surely) was done by Plato of Tivoli, was most inferior; just three manuscripts are known, only one of which is medieval.[6] Apparently not long after this first translation the great translator Gerard of Cremona again used the Arabic text and rendered the *Measurement of the Circle* into Latin. This time the translation was quite accurate, and so before 1187 a faithful version of this short but important treatise became available. We are fortunate that this version was included in MS Bibliothèque Nationale, Fonds latin 9335, a handsome codex of Gerard translations. Incidentally, this manuscript is one of the best examples of intelligent copying of scientific works. It has marginal variant readings which cite alternate copies. The drawings are carefully made. Even more important the transcription of numbers — even of six places — is almost perfect.

Gerard's translation of the *Measurement of the Circle* seems to have some importance for the textual study of the original Greek text. The extant Greek version is clearly far removed from the text as composed by Archimedes. Now the Gerard translation is not much closer, but it does include a corollary (giving the area of a sector of a circle in terms of the length of the arc of the sector and its radius) that seems to go back to the original Archimedean text although it is missing in the extant Greek text. Hero, who can be dated much earlier

[4] This list of works is largely drawn from al-Qīfṭī's account of Archimedes, as noted in Wiedemann, *op. cit.* in note 3, p. 247.

[5] My discussion of the *Measurement of the Circle* is based largely on my earlier article, "Archimedes in the Middle Ages: The *De mensura circuli,*" *Osiris,* 1952, *10:* 587-618. This article (abbreviated henceforth as "Archimedes") includes the texts of both the translations.

[6] Paris, BN lat. 11246, 13c (?), fols. 37v-39r; BN lat. 7224, 16c, fols. 63v-65r; and Trinity College, Dublin, D.2.9, 1565, fols. 54r-55r. The second of these manuscripts is a direct copy of the first. The third manuscript was copied by Francesco Barozzi.

than our extant text, specifically states that Archimedes proved this corollary in his *Measurement of the Circle.*[7]

Of the Gerard version there are at least nine manuscripts and no doubt others which I have overlooked.[8] Furthermore, it was the point of departure for ten different paraphrases or reworkings of Proposition I.[9] Proposition I, you will recall, relates the area of the circle to a right triangle whose two sides including the right angle are equal respectively to the circumference and the radius. These reworkings centered on elaborating by the use of Euclid's *Elements* the geometric steps only implied in the text of Archimedes. For example, by reference to the tenth book of the *Elements* (Proposition I) they demonstrate that, as we continually double the sides of a regular polygon described in a circle, more than half of the unexhausted area of the circle is thereby exhausted at each step — or, to put it in modern terms, they prove that the area of the inscribed polygon converges toward that of the circle as we double the sides.[10] These various reworkings of the Gerard version of Proposition I ought to be classified as a part of our answer to the third principal question on the use of Archimedes. Obviously at least the first proposition of Archimedes' *Measurement of the Circle* was standard mathematical fare for the geometers of the thirteenth, fourteenth, and fifteenth centuries, since it appears in one form or another in so many of the mathematical codices that date from these centuries. One ought to remark that on occasion the reworkings of the Gerard version appear in more than one manuscript, as the paraphrases themselves became standard for a given school or area.

Proposition II of the Gerard translation was also paraphrased along with many of these reworkings of Proposition I. Proposition III, involving the calculation of *pi,* seems to have frightened the scribes, no doubt because of the extensive arithmetical calculations given there. At any rate, I know of only three versions of Proposition III other than the initial translation by

[7] See Clagett, "Archimedes," pp. 589-590.

[8] In addition to Paris, BN lat. 9335, 13 or 14c, fols. 28v-29v, the following manuscripts can be cited: Dresden, Stadtbibliothek, Db 86, fols. 175v-176v, 178r; Oxford, Bodleian Library, Auct. F.5.28, 13c, fols. 101v-102v; Digby 174, 13c, fols. 133v-134v; Arch. Seld. B 13, 14c, fols. 2r-v; British Museum, Addit. MS 17368, 13c, fol. 69v; Paris, BN lat. 7378A, 14c, fols. 19r-20r; Vienna, Nationalbibliothek cod. 5303, 15c, fols. 19r-21v. The propositions are stated without proof in Digby 168, 14c, fol. 121r. Proposition II and the enunciation of Proposition III in the Gerard translation are given in Naples, Bibl. Naz., VIII.C.22, 13c, fol. 66v, which also includes Proposition I in an emended version (see footnote 9).

[9] I have published the texts of four of these versions in "Archimedes," pp. 607-616. These are the so-called Naples Version (MS Naples, Bibl. Naz. VIII.C.22, 13c, fols. 65v-66r), the Cambridge Version (MSS Cambridge, Gonville & Caius 504/271, 13c, fols. 108v-109v; British Museum, Royal 12E XXV, *ca.* 1300, fols. 150v-151v; and Florence, Bibl. Naz., Con. Sopp. J.V.18, 14c, fol. 92r-92v), and the two Florence Versions, Bibl. Naz., Con. Sopp. J.V.30, 15c, fols. 12r-v. In my *Archimedes in the Middle Ages* I shall publish six other distinct emended versions of Gerard's translation of Proposition I: (1) the version of "Gordanus" in Vat. Pal. lat. 1389, 15c, fols. 108r-111v (cf. Vienna, Nationalbibliothek 5257, 14c, fols. 67r-69r), which dates from the late thirteenth century, (2) a version attached to Bradwardine's *Geometry* in Vat. lat. 3102, fols. 111v-112v, (3) a variant form of No. (2), present in Florence, Bibl. Naz., Con. Sopp. J.IX.26, fols. 49v-50v; (4) *Questio Alberti de Saxonia de quadratura circuli,* in MS Bern, Bürgerbibliothek, A.50, 15c, fols. 169r-172r; (5) a version of Munich, Bay. Staatsbibl., cod. 56, 15c, fols. 182r-186v; (6) a version of Corpus Christi College, Oxford, MS 234, fols. 170r-172v. Versions (2), (3), and (4) also show the influence of the treatment of the quadrature problem found in the *Verba filiorum* of the Banū Mūsā.

[10] Clagett, "Archimedes," p. 590, n. 7.

Gerard.[11] The last of these versions given in a Florence manuscript (Bibl. Naz., Con. Sopp. J.V.30, fols. 9v-11v) is the most interesting, as it attempts to work out the calculations beyond the bare figures given in the Archimedean text; it does this somewhat in the manner of Eutocius in his commentary to the *Measurement of the Circle.*

Now the second work of the pre-1269 period which has a more or less direct relationship with a text of Archimedes is the so-called *De curvis superficiebus Archimenidis.* This work is a paraphrase and reworking of some of the propositions of Book I of the *Sphere and Cylinder;* it was done by a mysterious figure known in Latin as Johannes de Tinemue.[12] I have recently shown that it probably was composed in late antique or early Byzantine times.[13] This tract has the same objective as that of the *Sphere and Cylinder,* namely, the finding of the surface area and volume of a sphere. It has a number of propositions closely akin to the propositions in the *Sphere and Cylinder,* with demonstrations that are completely Archimedean in character, i.e., they employ the exhaustion procedure. These propositions, like Proposition I of the *Measurement of the Circle,* were to serve as models to medieval mathematicians when they attempted to apply this exhaustion procedure to entirely different propositions. The *De curvis superficiebus* was a work that was quite popular, being cited almost as often as the *De mensura circuli.* And hardly a geometer of the high and late medieval ages failed to cite this treatise. Thus in the fourteenth century it was used by such men as Bradwardine, Albert of Saxony, Johannes de Muris, etc. In my text of this treatise I have cited some eleven manuscripts of it,[14] and in one of these four additional and original medieval propositions with Archimedean type proofs were added, constituting further evidence of the use of Archimedes by medieval authors.[15]

In addition to the *De mensura circuli* and the *De curvis superficiebus* there circulated also a short fragment of the *Sphere and Cylinder.* This fragment was probably translated by Gerard of Cremona; it consists of some propositions which are cited in the introductions to the first and second books of the *Sphere and Cylinder,*[16] but no proofs were added. I have as yet found no citation of this fragment by later authors.

One further work translated from the Arabic by Gerard is quite close to the spirit of Archimedes' works. This is the so-called *Verba filiorum* or *Liber de geometria* of the eminent Arabic mathematicians, the Banū Mūsā, i.e., the sons

[11] (1) The version given in the *Verba filiorum* of the Banū Mūsā (see my article cited in note 5, p. 592); this version is of course not a reworking of the Gerard Cremona translation. (2) A version appearing as Chapter VIII of the *De ratione mensurandi* of Johannes de Muris (MSS Paris, BN lat. 7380, 7381, 7216, and 14736). This version may well be based on Moerbeke's translation from the Greek. (3) An anonymous version of Florence, Bibl. Naz., Con. Sopp. J.V.30, fols. 9v-11v.

[12] Vienna, Nationalbibliothek, cod. 5303, fol. 21v, has a fragment entitled *Demonstratio Joannis de Chinemue de quadratura,* which however does not follow the procedure of Archimedes.

[13] For the text and discussion of authorship, see M. Clagett, "The *De curvis superficiebus Archimenidis,*" *Osiris,* 1954, *11:* 295-358, and particularly, pp. 297-299. See also my notes in *Isis,* 1955, *46:* 281-282; 1957, *48:* 182-183.

[14] Clagett, "De curvis superficiebus," p. 300.

[15] *Ibid.,* pp. 347-358.

[16] M. Clagett, "A Medieval Fragment of the *De Sphaera et Cylindro* of Archimedes," *Isis,* 1952, *43:* 36-38.

of Mūsā ibn Shakir.[17] This tract contains the first and third propositions of the *Measurement of the Circle* and several propositions deriving from the *Sphere and Cylinder,* all with proofs in the Archimedean manner. It likewise contained solutions of two of the most fruitful problems of Greek geometry, the finding of two continually proportional means between two given quantities, and the trisection of any angle by a mechanical method. It further has the first known *demonstration* in Latin of Hero's formula for the area of a triangle in terms of its sides. Incidentally, this formula is attributed to Archimedes by an Arab mathematician.[18] When one appreciates the maturity of the geometry involved in the *Verba filiorum,* he must realize the great gulf that separates the elementary geometry of the early middle ages and the geometry of the period following the translations. The *Verba filiorum* was used by the famous mathematician Leonardo of Pisa, by Roger Bacon, and by others, as I shall show in my edition of this work.

In the foregoing remarks I have touched upon the highlights of direct translation before 1269. In that year Moerbeke made his translation from the Greek,[19] as I have already said. It included all of the texts known today with the exception of the *Sandreckoner,* the *Method,* the fragmentary *Stomachion,* and the brief *Bovine Problem.* It also included two of the three known commentaries of Eutocius, with only the commentary on the *Measurement of the Circle* missing. The complete Moerbeke translation exists in only one codex, a Vatican manuscript, Ottob. lat. 1850.[20] While the name of the translator is nowhere mentioned in this codex, consideration of historical circumstances and translating techniques convinces us that it was done by Moerbeke. In the first place, the translation is dated 1269 in all but two of the items (and these items are between works that are themselves dated 1269), and the whole manuscript is unquestionably in the hand of the original translator as the marginalia and corrections clearly reveal. Now it is well known that Moerbeke was at the Papal court of Viterbo at this time and engaged in other translations. He no doubt used the Greek Archimedean manuscripts A and B, which it appears were included in a collection of Greek manuscripts given to the Vatican by Charles of Anjou.[21]

17 A defective text of the *Verba filiorum* based on a single manuscript was published by M. Curtze in the *Nova Acta der Ksl. Leop.-Carol. Deutschen Akademie der Naturforscher* (Halle), 1885, *49:* 109-167. I am publishing a new text of this work based on six known manuscripts.

18 Al-Bīrūnī in a treatise *On Finding Sines in Circles* labels this theorem as being one by Archimedes. See H. Suter, "Das Buch der Auffindung der Sehen in Kreise etc.," *Bibliotheca Mathematica,* 1910-1911, 3. Folge, Bd. 11: 39.

19 For the Moerbeke translation, see J. L. Heiberg, "Neue Studien zu Archimedes," *Zeit. f. Math. u. Phys.,* 1890, *34,* Suppl., 1-84. Cf. Heiberg's edition of Archimedes' *Opera omnia* (Leipzig, 1915), pp. xliii-xlvii, and my article, "The Use of the Moerbeke Translations of Archimedes in the Works of Johannes de Muris," *Isis,* 1952, *43:* 236-242. See also E. J. Dijksterhuis, *Archimedes* (Copenhagen, 1956), pp. 37-40.

20 I have prepared a critical text of this whole corpus of Moerbeke translations and it will appear as the second part of my *Archimedes in the Middle Ages.*

21 See F. Ehrle, *Historia Bibliothecae Romanorum Pontificum tum Bonifatianae tum Avenionensis,* vol. 1 (Rome, 1890), pp. 96, 97, items 608 and 612; and A. Pelzer, *Addenda et Emendata etc.* (Vatican City, 1947), p. 93. Consult also J. L. Heiberg, *Archimedis opera omnia,* vol. 3, p. lviii. Both manuscripts A and B appear in the Vatican catalogue of 1311, but apparently through an oversight A was not listed in the earlier catalogue of 1295.

In the second place, comparison of this translation with others known to be by Moerbeke shows that its translator has precisely the same personal characteristics of translating as are apparent in the other Moerbeke translations. In brief, it shows all the translating preferences that Minio-Paluello has shown make up Moerbeke's style.[22] For example, just as in Moerbeke's known translations, we read *autem* as the translation of *de, enim* of *gar, itaque* of *dē, quidem* of *men, palam* of *dēlon, invicem* of *allala, igitur* of *oun, ergo* of *ara,* and so on.

While a number of mistakes in translation occur, for the most part the translation is careful and understandable, although quite liberal. The places that are particularly incoherent are usually where the Greek text was itself at fault. It would appear to me that Moerbeke went to quite some trouble to master the mathematics involved. This must have been particularly difficult, since there was so little Latin medieval precedence for most of the mathematics, say, of the *Conoids and the Spheroids,* or the *Spiral Lines,* or the *Quadrature of the Parabola.* His achievement is even more remarkable when we think of the variety of subjects involved in the various works of Aristotle and others that he translated. And Roger Bacon's comment that "William the Fleming knew nothing worthwhile in the sciences or in languages"[23] seems grossly unfair. I shall shortly speak of the use made of this translation in the answer to the third question.

The answer to the second question concerning works translated that showed less direct influence, I shall make brief. These works were particularly important in the fields of hydrostatics and statics. In hydrostatics, a pseudo-Archimedean *Floating Bodies,* reflecting *some* of the propositions of the original treatise, was composed in the thirteenth century and probably based on an Arabic translation. It circulated under the title of *De ponderibus Archimenidis* or *De insidentibus in humidum.*[24] We find in this treatise the first expression in Latin of the celebrated principle of Archimedes.[25] It contains the first use of the expression "specifically heavier," although, of course, not of the concept.[26]

In the field of statics several short works were translated, particularly an anonymous *De canonio* from Greek and a *Liber karastonis* of Thabit ibn Qurra, from the Arabic, by Gerard of Cremona.[27] These both are concerned with the problem of the steelyard or Roman balance (i.e. the balance of unequal arm lengths). These works show the influence of Archimedes in three respects. (1) They have adopted the Archimedean type of geometrical demonstration

[22] See L. Minio-Paluello, "Guglielmo di Moerbeke, traduttore della *Poetica* di Aristotele (1278)," *Rivista di filosofia neo-scolastica,* 1947, Anno *39,* pp. 1-17. I have compared the Archimedean translations of Ottob. lat. 1850 with a number of translations made by Moerbeke and will publish the results of that comparison in my *Archimedes in the Middle Ages.*

[23] R. Bacon, *Compendium studii philosophie,* Edition of J. S. Brewer (London, 1859), p. 472.

[24] I have discussed this tract and its antecedents in my *Science of Mechanics in the Middle Ages,* chapter II. The text was published in E. A. Moody and M. Clagett, *Medieval Science of Weights* (Madison, 1952), pp. 35-53.

[25] Moody and Clagett, *Medieval Science of Weights,* p. 42, "Omnis corporis pondus in aere quam in aqua maius est per pondus aque sibi equalis in magnitudine."

[26] *Ibid.,* p. 42, "Duorum equalium corporum, altero gravius esse in specie cuius pondus maiori calculorum numero adequatur."

[27] *Ibid.,* pp. 57-117, the texts and English translations of these works are given. A further discussion is found in Chapter II of my *Science of Mechanics in the Middle Ages.*

of statical theorems and the geometrical form implicit in weightless beams, weights that are really only geometrical magnitudes, etc. (2) They give reference in geometrical form to the law of the lever — and at least the reference in the *De canonio* is connected directly with Archimedes.[28] (3) They indirectly reflect the centers-of-gravity doctrine so important to Archimedes, in that in both treatises we find the practice employed of substituting for a segment of a *material* beam a weight equal in weight to this segment, but hung at the middle point of a weightless segment used to replace the material segment.[29] These treatises, so Archimedean in form (even if combined with Aristotelian ideas), were certainly to prove vastly stimulating to the development of the statics in the thirteenth century associated with the name of Jordanus. And if one grants the importance of the Jordanus treatises, as I believe one must certainly do, then once again we have evidence of the importance of Archimedes in the growth of science, in an area where it is not widely realized.

Now in suggesting the answers to the first and second questions posed at the beginning of this paper, I have partly answered my third question as to the use made by medieval mathematicians of these translations. An example of the original use of the *Measurement of the Circle* and the *Curved Surfaces* is found in a little-known treatise of the thirteenth-century geometer Gerard of Brussels, entitled *Liber de motu*.[30] It is devoted to kinematic problems arising from the rotation of geometrical figures. It has the objective of reducing the varying curvilinear velocities of such rotating figures to uniform movements of translation. Not only does Gerard cite the results of the two Archimedean treatises,[31] but, what is even more important, he borrows the basic kind of indirect proof found in these treatises, using it for the proof of wholly different theorems. Similarly, the mathematician Leonardo of Pisa makes original usage of the Archimedean type of proof, which he gained in part from reading the *Verba filiorum* of the Banū Mūsā.[32]

Less fruitful was the use of the hydrostatic ideas of the pseudo-Archimedean treatise *De insidentibus in humidum*.[33] The vague, non-mathematical, and non-experimental character of the Aristotelian treatment of the density problem dominated the scholastic discussion of this problem in the fourteenth century. However, certain schoolmen like Albert of Saxony and Blasius of Parma were not uninfluenced by the pseudo-Archimedean treatise. But at the

[28] The author of the *De canonio* assumes that the law of the lever has been demonstrated "ab Euclide et Archimede et aliis." See Moody and Clagett, *Medieval Science of Weights*, p. 56. An earlier Latin association of Archimedes with the balance is found in Gerard of Cremona's translation of the *Epistola Ameti filii Iosephi de proportione et proportionalitate*. See M. Curtze, *Anaritii in decem libros priores Elementorum Euclidis commentarii* (Supplement volume to *Euclidis opera omnia* of J. L. Heiberg and H. Menge [Leipzig, 1899], pp. xxvii-xxix).

[29] Moody and Clagett, *Medieval Science of Weights*, p. 64, lines 25-28; p. 104, lines 296-301.

[30] I have published the text of this work in my study, "The *Liber de motu* of Gerard of Brussels and the Origins of Kinematics in the West," *Osiris*, 1956, *12:* 73-175.

[31] *Ibid.*, p. 113, line 29 (where the *De curvis superficiebus* is called *De pyramidibus;* cf. p. 121, variant readings for lines 8-10). If one studies the analysis of contents given in the third part of the paper on Gerard, he will see the affinities in procedure between Gerard and Archimedes.

[32] See Clagett, "Archimedes," p. 597, n. 18.

[33] I have discussed the influence of the *De insidentibus in humidum* in Chapter II of my *Science of Mechanics in the Middle Ages*.

crucial point when one would expect them to use the principle of Archimedes, so succinctly expressed in the *De insidentibus,* to determine the specific gravities of bodies heavier than the fluid in which they are immersed, they fall back on the primitive method of comparing specific gravities by comparing the speeds of descent of the bodies in the liquid. However, a somewhat more faithful student of the pseudo-Archimedean hydrostatics was Johannes de Muris at Paris in the 1340's. He repeats much of the earlier treatise in his *Quadripartitum numerorum.*[34]

I have spoken earlier of the flowering of statics in the thirteenth century under Jordanus, and I have indicated this was in part indirectly influenced by the Archimedean statical tradition which joined with the more philosophical Aristotelian tradition to produce brilliant proofs of the law of the lever for straight and bent levers, the equilibrium of weights on oppositely inclined planes, etc. This combined Aristotelian-Archimedean statics produced in the middle ages was to circulate in the sixteenth century along with the more purely Archimedean statics of the *Equilibrium of Planes* and was to influence some of the early modern mechanicians in their approach to statics.

We have left one final problem on the use of Archimedes in the middle ages. To what extent did the new translation of the whole Archimedean corpus by Moerbeke influence the course of medieval mathematics. Heiberg felt there was no medieval influence whatsoever of the Moerbeke translation. Put so baldly, this is certainly not correct. Let us examine the evidence. The number of extant manuscripts of the Moerbeke translations at first glance seems to bear out Heiberg.[35] In addition to the single holograph manuscript of the whole corpus at the Vatican (Ottob. lat. 1850) there are only two other manuscripts with parts of the Moerbeke corpus. One is a Vatican manuscript of the early fourteenth century (Reg. lat. 1253, 14r-33r), which contains only the *Spiral Lines.* The other is a badly done manuscript of the early fifteenth century (Italian in origin but now at Madrid, BN 9119) which contains the *Measurement of the Circle,* the *Equilibrium of Planes,* the *Quadrature of the Parabola,* the *Floating Bodies,* and Eutocius' commentary on the *Equilibrium of Planes.* However, these three manuscripts do not tell the whole story, for in 1340 there was produced at Paris (or at least it circulated at Paris) a hybrid treatise entitled *De quadratura circuli,* which is compounded of a number of propositions of the *Spiral Lines* with the first proposition of the *Measurement of the Circle.*[36] This showed some real ingenuity on the part of the compositor, for he realized that in the first proposition of the *Measurement of the Circle* the real difficulty is not in the exhaustion proof showing that the circle is equal to the right triangle whose sides including the right angle are equal to the circumference and the radius of the circle. The proof is sound enough. But the real question is whether we can construct such a triangle, i.e., find straight lines equal to the curved lines. Now the first eighteen propositions of the *Spiral Lines* lead to the construction of such a triangle. So our compositor borrowed

[34] The main part of the hydrostatic section of the *Quadripartitum numerorum* has been given as *Document 2.4* of my *Science of Mechanics in the Middle Ages.*

[35] I have discussed the various manuscripts in "The Use of the Moerbeke Translations," pp. 236-242.

[36] *Ibid.,* pp. 238-239.

most of these propositions with their proofs from the *Spiral Lines* in the Moerbeke translation and, having constructed such a triangle, he then proceeded to give the first proposition of the *Measurement of the Circle* to find its area. The celebrated Nicole Oresme knew this hybrid treatise and perhaps more of the original Vatican autograph of Moerbeke.[37] For it is at least possible that Oresme was the continuator who finished off Johannes de Muris' *Ars mensurandi* after the latter's death. In the continuation of this treatise — whether by Oresme or some other fourteenth-century figure is for our present purposes immaterial — the continuator included many propositions from the *Sphere and the Cylinder* and the *Conoids and Spheroids,* both in the Moerbeke translation.[38] Further, there are a number of manuscripts including this continuation, which indicates that these Archimedean propositions were circulating.

So our conclusion, for fourteenth-century Paris at least, must be that the following works in the Moerbeke translation were known: *Measurement of the Circle, Spiral Lines, Sphere and Cylinder, Conoids and Spheroids.* These were coupled with all of the earlier works translated from the Arabic which can be shown to have been known not only at Paris but in other schools of the fourteenth century.

As we approach the Renaissance and the end of our trail of Archimedean texts and ideas, we ought to note in passing the considerable influence exercised by the Moerbeke translation in the Renaissance. Heiberg is rightly convinced that the fifteenth-century translation of the works of Archimedes done in *ca.* 1450 by Jacobus Cremonensis was made with an eye on the Moerbeke translation.[39] This Jacobus translation, after some correction by Regiomontanus, was published in Basel by Thomas Gechauff Venatorius with the first Greek edition in 1544.[40] The Moerbeke translation was also known and cited by Leonardo da Vinci.[41] Gaurico and Tartaglia pilfered it in the sixteenth century,[42] and actually the first translations of Archimedes to appear in print were the medieval translations of Moerbeke published at Venice by Gaurico in 1503 (the *Measurement of the Circle* and the *Quadrature of the Parabola*), and in 1543 when there was published directly from Gaurico's edition the same treatises and in addition, from Tartaglia's copy of the Madrid manuscript, the Moerbeke translation of the *Equilibrium of Planes* and Book I of the *Floating Bodies.* Both books of Moerbeke's translation of the *Floating Bodies* were published twice in 1565, the first time by Curtius Troianus in Venice, the second by Federigo Commandino in Bologna. Since the Greek text (MS B) of the *Floating Bodies* used by Moerbeke was by this time lost, Moerbeke's translation remained the only version of this important work until the discovery of Manuscript C in the twentieth century.

37 M. Clagett, "Three Notes," *Isis,* 1957, *48:* 182-83.

38 "The Use of the Moerbeke Translations," pp. 239-240, 241-242.

39 Heiberg, "Neue Studien" (cited above, note 19), p. 83.

40 The manuscript copy of this translation corrected by Regiomontanus was brought to Nuremberg and was published by Venetorius with the first Greek edition; it is, I take it, still at Nuremberg, Stadtbibliothek(?), with a number Cent. V. 15. Other manuscript copies are at Venice (Bibl. Naz. Mar. f. a. lat. 327), Paris (BN lat. 7220) and the Vatican (Urb. lat. 261).

41 W. Schmidt, "Zur Textgeschichte der 'Ochumena' des Archimedes," *Bibliotheca Mathematica,* 1902, 3. Folge, Bd. 3, 176-179.

42 Clagett, "The Use of the Moerbeke Translations," p. 236.

Of the sixteenth-century Latin versions of the works of Archimedes, Commandino's translations published at Venice in 1558 are among the most important. They perhaps demonstrate some acquaintance with the Moerbeke translations.[43] A second translation, or rather a paraphrase-translation, of Archimedes' works made by Francesco Maurolico over the period 1534-1549 was published in an almost completely destroyed edition of 1570 which was reprinted in 1685.[44] That Maurolico knew of the Gerard of Cremona translation of the *Measurement of the Circle* as well as that of Moerbeke can be shown.[45] Still one further translation of Archimedes' works was made in the sixteenth century, that of a mysterious figure Antonius de Albertis, who studied with Ficino's successor at the Florentine Academy, Franceso da Diacetto (1466-1522).[46] This translation is found in a Viennese codex;[47] it appears at first glance unrelated to any known translation, but further study of it is necessary.

As we glance back over the brief outline of the role of Archimedes in the middle ages which I have presented here, I think we must conclude that Archimedes played a modest, but nevertheless important, part in that thin tradition of Greek mathematics and physics that trickled down through the middle ages into modern times.

[43] On the whole, however, Commandino's translation is most original and mathematically perceptive. His *Archimedis opera nonnulla in latinum conversa et commentariis illustra* (Venice, 1558) contains translations of the *Measurement of the Circle, Spiral Lines, Quadrature of the Parabola, Conoids and Spheroids,* and the *Sandreckoner*. His correction (without a Greek text) of the Moerbeke translation of the *Floating Bodies* (Bologna, 1565) was brilliant, containing as it did considerable reconstruction of the text already much corrupt by the time of the writing of MS B which Moerbeke used.

[44] *Admirandi Archimedis Syracusani monumenta omnia mathematica, quae extant . . . ex traditione doctissimi viri D. Francisci Maurolici,* Palermo, 1685. The original dates of the various works are given in this edition. Since MS B was already lost, no translation of the *Floating Bodies* was made by Maurolico.

[45] Clagett, "Archimedes," p. 595, n. 14.

[46] O. Kristeller, *Renaissance Thought and Letters* (Rome, 1956), pp. 322n-323n; for Diacetto, see A. Della Torre, *Storia dell' Accademia Platonica di Firenze* (Florence, 1902), pp. 832-833.

[47] Vienna, Nationalbibliothek, cod, 10701. We read, fol. 1r, "Archimedis de quadratura paraboles liber e graeco tralatus Antonio de Albertis interprete." On the bottom of fol. 1r we find the scribe's name, "Mich. Creton (?)," with a date which I read as "1562" and a place "Bonon" (Bologna?). The table of contents gives the following works: *Quadrature of the Parabola, Sphere and Cylinder, Spiral Lines, Conoids and Spheroids, Measurement of the Circle, Equilibrium of Planes,* and *Sandreckoner*. But in the present manuscript, there is nothing after Proposition 24 of the *Conoids and Spheroids*.

Taddeo Alderotti and Bartolomeo da Varignana on the Nature of Medical Learning

By Nancy G. Siraisi

I

DURING THE SECOND HALF of the thirteenth century medicine emerged as a formally organized academic discipline in the *studium* of Bologna in close institutional association with liberal arts and philosophy.[1] In terms of institutional and professional history this constituted real and significant change. Public medical teaching leading to a license and doctorate in medicine and the formation of a student university and a doctoral college "of arts and medicine" were innovations. The appearance of learned physicians as a professional and academic group endowed with corporate privileges and prepared to compete with the prestige and privileges of the doctors and scholars of law was, indeed, a novelty to which the lawyers of Bologna found it hard to reconcile themselves.[2] But the significance of the change in terms of intellectual history or the history of science is less easy to determine. The creation of university faculties of arts and medicine in which logic, astronomy and astrology, and the *libri naturales* of Aristotle were studied as essential foundations of medical learning, and in which medical texts were subjected to scholastic analysis, did not introduce learned medicine, scholastic medicine, Aristotelian medicine, Galenic medicine, or Arab medicine to the West, for all had long since been present.[3] Nor, at any rate in Italy, did it divorce theory from practice, since

[1] On the origins and early years of the student University of Arts and Medicine and the College of Doctors of Medicine (and the arts) at Bologna, see A. Sorbelli, *Storia della Università di Bologna* (Bologna: Nicola Zanichelli Editore, 1944), Vol. I, pp. 106–115; V. Busacchi, "I primordi dell'insegnamento medico a Bologna," *Rivista di Storia delle Scienze Mediche e Naturali*, 1948, *39*:133–137; and V. L. Bullough, "Medieval Bologna and the Development of Medical Education," *Bulletin of the History of Medicine*, 1958, *32*:201–215.

[2] Busacchi, "I primordi," p. 136. Some discussion of the influence of the institutionalization of medical teaching in the *studia* upon the development of medicine as a profession is to be found in V. L. Bullough, *The Development of Medicine as a Profession: The Contribution of the Medieval University to Modern Medicine* (Basel/New York: S. Karger, 1966).

[3] P. O. Kristeller, "The School of Salerno: Its Development and Its Contribution to the History of Learning," *Studies in Renaissance Thought and Letters* (Rome: Edizioni di Storia e Letteratura, 1956), pp. 495–551, Pearl Kibre, "Hippocratic Writings in the Middle Ages," *Bull. Hist. Med.*, 1945, *18*:371–412, and M. Saffron, *Maurus of Salerno, Twelfth-Century "optimus physicus" with his Commentary on the Aphorisms of Hippocrates* (Philadelphia: American Philosophical Society, 1972) document the earlier background. See also C. H. Haskins, *Studies in the History of Mediaeval Science* (reprint ed., New York: Frederick Ungar, 1960), pp. 206–209, and M. T. d'Alverny, "Les traductions d'Avicenne (Moyen Age et Renaissance)," *Avicenna nella storia della cultura medioevale* (Rome:

practice, in the sense of advising the wealthy, was too lucrative to be abandoned by academic physicians.[4]

Yet institutionalization did have discernible effects in the intellectual sphere. At Bologna professors of medicine seem to have begun to re-examine alternative ways in which their own science might be conceived and be related to other branches of knowledge. In particular, noticeably different attitudes concerning these questions were expressed by two of the most respected masters to teach at Bologna in the early years of the faculty of arts and medicine: Taddeo Alderotti (d. 1295) and his sometime pupil and subsequent rival, Bartolomeo da Varignana (d. after 1321).[5] The views of these two masters merit examination, since, as will become apparent, their teaching undoubtedly helped to shape the approach to medical learning in the *studium* of Bologna. The influence of Bolognese medical teaching was in turn felt in most of the other Italian *studia* throughout the fourteenth and fifteenth centuries.

The problem to which these masters addressed themselves had in the broadest terms the following two contradictory aspects. On the one hand the institutional association of medicine with liberal arts and natural philosophy surely reinforced not only the venerable tradition linking the subjects[6] but also the concept of medicine as a fully learned discipline. Medical men could with renewed confidence reassert their long-standing claims to *scientia, ars* (of a kind comparable with the liberal arts in dignity), and *philosophia*—claims which had not always been respected by scholars in other fields.[7] On the other hand, medicine had a proud tradition as an autonomous science, with its own authorities and its own methods. Too much insistence on an association with philosophy and the arts might risk subordinating medicine to the Aristotelian natural philosophy that had become the mainstay of university teaching in arts. Discussion revolved around the issue of the appropriate balance to be maintained.

Accademia Nazionale dei Lincei, 1957), p. 76, for twelfth-century translators and translations of medical works from Greek and Arabic. No statutory evidence for the curriculum in arts and medicine at Bologna in the thirteenth and early fourteenth centuries appears to survive, but the extant writings of masters known to have taught there indicate that it was as described.

[4]Although municipal governments normally endowed one or two salaried chairs in medicine in the Italian *studia,* and although provisions were made for students to pay fees to their masters, there is little doubt that some of the leading professors of medicine in these schools made most of their living by medical practice of the type indicated. Their numerous extant *consilia* for individual cases alone serve to indicate this (see notes 16 and 17 below). Moreover, the strong Italian surgical tradition appears to have ensured that the contempt for surgery felt by learned physicians in the university centers of the north did not develop to anything like the same extent in Italy.

[5]For their careers, see M. Sarti and M. Fattorini, *De claris archigymnasii Bononiensis professoribus,* ed. C. Albicini and C. Malagola (Bologna, 1888–1896), Vol. I, pp. 554–564, 568–572, and further bibliography below.

[6]E.g., Isidore of Seville had declared that all the arts were necessary to medicine and that medicine was a second philosophy (*Etymologiae* 4.13.1–5), concepts which were certainly not original with him (see further, text and n. 39 below). On the association of medicine and philosophy in Galen's teaching, see Owsei Temkin, *Galenism: Rise and Decline of a Medical Philosophy* (Ithaca: Cornell University Press, 1973), pp. 1–94.

[7]See, e.g., Hugh of St. Victor, *Didascalicon de studio legendi,* ed. C. H. Buttimer (Washington, D.C.: Catholic University of America Press, 1939), pp. 37–39, 43–44, where medicine is categorized as one of the mechanical or adulterine arts which have no relation to theoretical or speculative philosophy and which provide suitable occupation for manually skilled members of the lower classes. Thomas Aquinas held the more moderate view that medicine, although wholly practical in purpose and excluded from the liberal arts because of its association with the human body, could be held to be subalternated to *physica,* that is, natural science [*Expositio super librum Boethii de Trinitate,* ed. Bruno Decker (Leiden: E. J. Brill, 1955), pp. 168–171].

II

The name of Taddeo Alderotti has long been associated with Aristotelianism in medicine, while his activities have been linked with the vigorous school of radical Aristotelian or Averroïst philosophy that was beginning to flourish among the masters and scholars of arts at Bologna in his lifetime.[8] "Medical Aristotelianism" is a conveniently vague term which can indicate one or more of the following: (1) the belief that (Aristotelian) logic should provide the methodology for medical discussion; (2) the belief that the study of the *libri naturales* of Aristotle constitutes an essential or highly desirable part of medical education; (3) a preference for Aristotle over Galen or other medical authorities on specific points concerning human physiology and anatomy. Taddeo qualified as a medical Aristotelian on all three counts, and it is therefore not surprising to find in him a warm advocate of the close association of medicine with *scientia naturalis,* that is, Aristotelian natural philosophy. His approach had the advantage of stressing the dignity of medical learning as an intellectual activity and of linking it to the most vital currents in the philosophy and science of the day, but it could also threaten the autonomy of medical science. Such a threat seems, at any rate, to have been perceived by Bartolomeo da Varignana, who can be categorized as a medical Aristotelian only in the first two senses of that term.

Taddeo Alderotti discussed the definition and nature of medical science together with its relation to other branches of knowledge at some length in the opening pages of his commentary on the medical *Isagoge* of Johannitius.[9] To preface a medical work with an exposition of this kind was of course entirely conventional. The same topics are treated at the beginning of the *al-Maleki* of Haly Abbas and the *Canon* of Avicenna, to name only two of the most authoritative medical works known to Taddeo, while discussions of these or related themes are a commonplace of fourteenth- and fifteenth-century medical literature. Taddeo's treatment of the subject is indeed largely a commentary upon Avicenna's definition of medicine as "the science by which the dispositions of the human body are known . . . so that health may be maintained or, if lost, recovered."[10] But whereas Avicenna, while admitting the necessity of some *sapientia physicalis* (for example, regarding the elements) on the part of the physician, had stressed the strictly limited medical value

[8] Bruno Nardi, "Note per una storia dell'averroismo latino, V: L'averroismo bolognese nel secolo XIII e Taddeo Alderotto," *Rivista di Storia della Filosofia,* 1949, *4*:11–22; M. Grabmann, "Das Aristotelesstudium in Italien zur Zeit Dantes," *Mittelalterliches Geistesleben* (Munich: M. Hueber Verlag, 1956), Vol. III, pp. 197–212. On Bolognese Averroïsm and Aristotelianism in the late thirteenth and early fourteenth centuries in general, see Charles Ermatinger, "Averroism in Early Fourteenth Century Bologna," *Medieval Studies,* 1954, *16*:35–56, the same author's "Some Unstudied Sources for the History of Philosophy in the Fourteenth Century," *Manuscripta,* 1970, *14*:67–87, and Z. Kuksewicz, *Averroïsme bolonais au XIVᵉ siècle* (Wroclaw/Warsaw/Cracow: Éditions de l'Académie Polonaise des Sciences, 1965), and works therein cited.

[9] *Thaddei Florentini Expositiones in arduum aphorismorum Ipocratis . . . In subtilissimum Joannitii Isagogarum libellum* (Venice, 1527; at the New York Academy of Medicine), fol. 343^{r-v}. This edition will be referred to hereafter as Taddeo (1527).

[10] "Dico quod medicina est scientia, qua humani corporis dispositiones noscuntur, ex parte, qua sanatur, vel ab ea removetur; ut habita sanitas conservetur et amissa recuperetur" (*Canon* 1.1.1, proemium). Elsewhere, in his commentary on the *Aphorisms* of Hippocrates [Taddeo (1527), fol. 2^{v}], Taddeo discussed the sense in which medicine could be said to be an art, admitting that the type of medicine practiced by Hippocrates ("medicina secundum sui inventionem") was more appropriately termed art than science.

of such knowledge and had criticized physicians who taught natural science,[11] Taddeo proceeded along different lines. The whole thrust of his discussion, it seems to me, was directed toward emphasizing the links, rather than the differences, between medicine and the liberal arts, especially between medicine and natural philosophy.

Starting from Avicenna's assertion that medicine is a science, Taddeo began by defining *scientia,* citing the *Ethics* and *Posterior Analytics. Scientia* is to be distinguished from the qualities of prudence and wisdom, from opinion (which can be either true or false, whereas *scientia* can only be true), and from the immediate apprehension of first principles without recourse to reason. *Scientia* is knowledge strictly demonstrable from premises. The subject of medical science is the human body; this distinguishes it from such sciences as astrology and music, which deal with the disposition of the heavenly bodies and with harmony. Medicine is also distinct from *scientia naturalis,* which does include the study of the human body as a mobile physical object but not its health or sickness.

Medicine, continued Taddeo, is divided into theory and practice, and the end of theory is practice. Yet although practice serves to show forth the antecedent theory and lead it to its ordained end, it cannot be said to prove theory, since that which comes after does not prove that which goes before. Theory is defined by Aristotle and Avicenna as knowledge for its own sake; by Constantinus Africanus as certain information about things which are grasped by the intellect alone. Although *practica* is not grasped exclusively by the intellect, *practica,* as Avicenna says, is itself a *scientia intellectualis.* Taddeo firmly denied that *practica* was to be identified with *operatio,* despite the assertions of Johannitius and Constantinus Africanus to the contrary. No science, according to Taddeo, consists of *operatio,* and *practica* is undoubtedly a science. It is properly defined as a *scientia operativa,* which teaches how and why theory should be put into operation manually and sensibly. For example, a physician knows from theory that fever is caused by heat; *practica* teaches him to cure by opposites, and he therefore prescribes that the patient be given cold water, which is a manual operation. This kind of medical practice, which is a science involving theoretical knowledge, is to be sharply distinguished from "the usual practice that old women carry on."[12] It is true, however, that medical practice is different from the practical parts of the other sciences, since the latter involve intellectual not manual *operatio.* The practice of grammar, for example, consists in teaching how to construe, and the practice of dialectic, how to reason by syllogisms.

Taddeo then went on to defend the inclusion of material relating to *physica* (in the general sense of natural science) in a medical work, despite the definition of the subject of medicine as the human body. He also took up the question of the position of medicine in regard to the three philosophies, rational, moral, and natural, which were commonly held to divide speculative science. Obviously, medicine could not be regarded as part of moral philosophy, and despite

[11]"Harum vero rerum, quaedam sunt, de quibus medico nihil aliud est agendum, nisi ut apprehendat: seu cognoscat: quid sit, vel earum essentiam apprehensione scientiali, et utrum sint, vel non sint, doctori sapientie physicalis credat. . . . Cum aliquis medicorum loqui inceperit, et affirmari elementa esse, et complexionem, et quae ista sequuntur, ex eis, quae scientia naturalis subiecta existunt, aggressus fuerit: errabit" (*Canon* 1.1.1).

[12]"Propter hoc seperatur practica que est scientia a practica usuali quam faciunt vetule" [Taddeo (1527), fol. 343v].

the use of logic in medical argumentation, its subject matter was not the same as that of rational philosophy or dialectic. "Saving the reverence due to a monk," Taddeo rejected the contention of Constantinus Africanus that because logic was used in medical discussions, medicine was partly subordinated to dialectic and partly to natural philosophy. Instead, he maintained strongly that it should be placed entirely under natural philosophy. Many of the received principles of medicine were drawn from natural philosophy, as Avicenna himself had admitted; as a result, medicine was subalternated to natural philosophy and therefore subject to it.

Taddeo's exposition appears to demonstrate certain very definite intentions: he has defined medicine as a science and has indicated also that it is comparable in scope to the quadrivial arts of astrology and music; he has shown that although medicine is distinct from *scientia naturalis,* it is properly regarded as subalternated to it, and, further, it is a subdivision of *philosophia naturalis* in the scheme of the three philosophies; and he has demonstrated that medical practice as well as medical theory is included in the foregoing categorization. Effectively this surely amounts to a justification of the alliance of medicine with the subjects normally studied in the arts schools, of the study of those subjects by physicians, and of the place of medicine—even practical medicine—in a university curriculum.

III

It seems reasonable to suppose that Taddeo's views on the nature of medical learning would be highly influential among his colleagues and contemporaries, given the extent of his contemporary and immediately posthumous celebrity as a medical teacher, author, and practitioner. At Bologna, where he was active from before 1268 until his death in 1295, his fame was such that the civic authorities extended to him and his pupils all the privileges enjoyed by professors and scholars of law.[13] This was an apparently unprecedented favor for a teacher of medicine and no doubt in itself marks an important stage in the institutionalization of medicine at Bologna. Moreover, his fairly numerous pupils became, almost without exception, well-known medical authors and university professors of arts and medicine.[14] Nor did Taddeo neglect practice. His privilege, unlike earlier agreements with medical men made by the Bolognese municipality, specifically exempted him from obligatory attendance upon the wounded, while reserving to him the right to engage in private practice *ad libitum.*[15] At his

[13] *Chartularium Studii Bononiensis,* Vol. VIII, p. 22. For his death date, sometimes given as 1303 but in reality 1295, see P. Pansier, "Les maîtres de la faculté de médicine de Montpellier au moyenage," *Janus,* 1904, 9:511. Taddeo's privilege is printed in A. Gaudenzi, "Gli antichi statuti del comune di Bologna intorno allo studio," *Bullettino dell' Istituto Storico Italiano,* 1888, 6:130. See also Sarti and Fattorini, *De claris,* Vol. I, p. 520; Sorbelli, *Storia,* Vol. I, p. 108; Busacchi, "I primordi," pp. 133–136.

[14] Physicians who are held, with varying degrees of certitude, to have been pupils of Taddeo include Dino del Garbo, William of Brescia, Mondino de' Liuzzi, Henri de Mondeville, Bartolomeo da Varignana, and Gentile da Foligno, all of whom were noted medical authors.

[15] One may compare Taddeo's privilege with the agreement made by the commune with the surgeon Ugo of Lucca in 1214. In exchange for Bolognese citizenship and a money fief (*feudum*), Ugo was retained to reside in Bologna and treat Bolognese citizens gratis in time of peace, and to accompany Bolognese military expeditions and treat the wounded in time of war. No mention was made of teaching (Sarti and Fattorini, *De claris,* Vol. II, pp. 214–215). Ugo's fief was subsequently

death he left a charitable bequest of 10,000 Bolognese lire, a sum unlikely, alas, to have been acquired by university teaching, however distinguished; he himself referred to his lucrative *operatio.*[16] Moreover, he was the author of a large number of *consilia* describing cases of particular interest and of a vernacular *regimen* for Corso Donati, subsequently leader of the Black Guelfs in Florence.[17] By the time Dante came to write the *Paradiso,* some twenty years after Taddeo's death, his name had become synonymous with medical learning and professional ambition.[18] For Filippo Villani, who included Taddeo's biography in the *Liber de civitatis Florentiae famosis civibus,* his works constituted a medical *glossa ordinaria,* while he himself was a new Hippocrates, as great an interpreter of medicine as was Accursius of law.[19] That this judgment represented the consensus in the fourteenth century is amply proven by the numerous respectful references to Taddeo and his teaching scattered through the writings of Italian physicians of that period.

The terms of Villani's eulogy indicate that Taddeo's reputation was based primarily on his teaching and his theoretical writings. Most of the latter seem to date from fairly early in his professional career: by 1283 he had already produced commentaries on the *Aphorisms* of Hippocrates, the *Tegni* of Galen, and the *Isagoge* of Johannitius "and upon many other books concerning the art of medicine as well." The composition of these works was interrupted, he told his readers, by civil war in Bologna (probably the disturbances of 1274) and by his medical practice.[20] For these or other reasons, several of

---

inherited by his son, who held it under the same conditions (G. Zaccanigni, "La vita dei maestri e degli scolari nello studio bolognese nei secoli XIII e XIV," *Archivum Romanicum,* Ser. I, 1926, 5:111).

[16]Will of Taddeo Alderotti, printed in Sarti and Fattorini, *De claris,* Vol. II, pp. 223–230; Taddeo (1527), fol. 1ʳ. The salaries paid to professors of medicine by the government of Bologna in the early fourteenth century amounted to no more than 100 Bolognese lire per year (Gaudenzi, "Gli antichi statuti," pp. 134–135). Taddeo's wealth and avarice became proverbial. Villani repeated of him the story also told of other learned physicians, most notably Peter of Abano, that he demanded 100 *aurei* per day to treat the pope. The unnamed pope, after rebuking his physician's avarice, presented him with 10,000 *aurei* on being cured. F. Villani, *Liber de civitatis Florentiae famosis civibus,* ed. G. C. Galletti (Florence, 1847), pp. 26–27.

[17]Taddeo Alderotti, *I "Consilia,"* ed. G. M. Nardi (Turin: Editore Minerva Medica, 1937); *Libello per conservare la sanità del corpo; fatto per maestro Taddeo da Firenze. Testo inedito del buon secolo della lingua toscana,* ed. F. Zambrini (Imola, 1852).

[18]
Non per lo mondo, per cui mo s'affanna
di retro ad Ostiense e a Taddeo
ma per amor della verace manna
in picciol tempo gran dottor si feo. . . .
Paradiso, XII, 82–85

For discussion supporting the identification of this Taddeo—unfavorably contrasted by Dante with St. Dominic—as Taddeo Alderotti, see C. Calcaterra, *Alma mater studiorum* (Bologna: Nicola Zanichelli Editore, 1948), pp. 52–54.

[19]Villani, *Liber de . . . famosis civibus,* pp. 26–27.

[20]Taddeo (1527), fols. 1ʳ, 194ᵛ. Taddeo's commentaries upon the *Aphorisms,* the *Prognostics* and the *De regimine acutorum* of Hippocrates, as well as that on the *Isagoge* of Johannitius, are printed in the edition just cited. His commentary on the *Tegni* of Galen was apparently printed at Naples in 1522, and his "digressiones" on the *De differentiis febrium* of Galen are to be found with *Libri de differentiis febrium a Galeno editi tres translationes . . .* (Pavia, 1519; at the New York Academy of Medicine). Several other commentaries and *questiones* remain in manuscript and are listed in Lynn Thorndike and Pearl Kibre, *A Catalogue of Incipits of Mediaeval Scientific Writings in Latin* (2nd ed., Cambridge: Mediaeval Academy of America, 1963), hereafter cited as TK.

his commentaries are unfinished and show signs of haste.[21] In them, however, Taddeo showed himself thoroughly familiar with the logical works, *libri naturales*, and *Ethics* of Aristotle (as is well known, he made an abbreviated Italian translation of the *Ethics*),[22] and with the *Canon* of Avicenna and some of the works of Averroës. He was also aware of current disputes at the University of Paris, since certainly before 1283 and probably a few years earlier he discussed the question of monopsychism and apparently alluded to the Parisian ecclesiastical condemnations of 1270 or 1277.[23] While Taddeo's information concerning contemporary trends in Parisian philosophy was fairly up to date, his natural philosophical learning does not strike one as particularly profound, nor was his range of reference especially wide. For example, he was certainly far less well read in Greek and Arabic natural philosophy and astronomy than his younger contemporary Peter of Abano (d. at Padua c. 1316).[24]

IV

Great as was Taddeo's fame, his views on the nature of medical science were not universally accepted in the faculties of arts and medicine. Physicians equally learned, equally academic, equally "scholastic" could and did insist upon the autonomy of medical authorities and restrict the role of Aristotelianism in medicine to the provision of essential logical methodology. The point may be illustrated by some statements regarding the place of medicine among the sciences and the nature of medical learning which are in the preface to a commentary on the *De interioribus* or *De locis affectis* of Galen contained in a Vatican manuscript.[25] This commentary is anonymous, but the body of the

[21] In the Venice edition of 1527, Taddeo's commentary on the *Aphorisms* breaks off at the end of the sixth book; his commentary on the *Prognostics* ends in the middle of book three; his commentary on *De regimine acutorum* also does not complete book three; while the commentary on the *Isagoge* ends with an apology for inadequate treatment. However, contrary to an earlier presumption, Taddeo's work does not appear to have been interrupted by impaired eyesight, for F. Puccinotti, *Storia della medicina* (Leghorn, 1855), Vol. II, Pt. 1, p. xxx, is in error in supposing that the Taddeus de Florentia who sent an emotional letter to the Augustinian evangelist Blessed Simone Fidati of Cascia bewailing the damage done to the writer's eyesight by watching an eclipse of the sun is to be identified with Taddeo Alderotti. Simone Fidati died in 1348 after a preaching career lasting twenty-seven years [Nicola Mattioli, *Il beato Simone Fidati da Cascia* (Rome, 1898), p. 27 and elsewhere], which apparently reached its peak in the 1330s; as noted, Taddeo Alderotti died in 1295. (The letter of Taddeo of Florence to Blessed Simone is printed by Mattioli, pp. 436–438, as well as by Puccinotti.)

[22] It was apprently incorporated by Bono Giamboni as Book VI of his Italian version of Brunetto Latini's *Trésor* and was characterized by Dante as "laido" (*Convivio* 1.10). See C. Marchesi, "Il compendio volgare dell'Etica Aristotelica e le fonti del VI libro del 'Trésor,'" *Giornale Storico della Letteratura Italiana*, 1903, *42*:1–74, and L. Frati, "Per due antichi volgarizzamenti: II, L' 'Etica' di Aristotile volgarizzato da Taddeo di Alderotto," *ibid.*, 1916, *68*:192–195.

[23] Taddeo (1527), fols. 360^{v}–361^{r}, "an intellectus omnium hominum sit unus vel non?" For a very full discussion of this passage, see Nardi, "L'Averroismo bolognese," *passim*. As noted, the commentary on the *Isagoge* of Johannitius in which the passage occurs was written before the commentary on the *Aphorisms*, begun in the ninth year of Taddeo's teaching and completed, after interruptions, in 1283. Taddeo was an established teacher in Bologna by 1268. F. van Steenberghen, *Aristotle in the West* (Louvain: Nauwelaerts, 1955), p. 198, dates the emergence of the controversy over the unicity of the intellect at Paris between 1260 and 1265.

[24] On Peter of Abano's natural scientific learning, see Lynn Thorndike, *A History of Magic and Experimental Science* (New York: Columbia University Press, 1923), Vol. II, pp. 875–946, and Nancy G. Siraisi, *Arts and Sciences at Padua* (Toronto: Pontifical Institute of Mediaeval Studies, 1973), Chs. III and IV, and bibliography there cited.

[25] MS Vat. Lat. 4452, fols. 83^{r}–102^{r} (TK, col. 857). The proemium occupies fols. 83^{r}–84^{r}.

text is the same as that of the commentary on *De interioribus,* without proemium, ascribed to Bartolomeo da Varignana in another Vatican manuscript.[26] It is to be presumed therefore that the remarks in the proemium attached to the commentary in the first manuscript referred to are also to be attributed to Bartolomeo.

Bartolomeo da Varignana, like his master Taddeo Alderotti, was both a highly successful medical practitioner and teacher and a fairly prolific writer on medical theory. The son of a physician, he appears to have studied under Taddeo at some time during the 1270s and 1280s.[27] Subsequently he began to attract students on his own account, with such success that in 1292 Taddeo formally accused him of luring away his own pupils.[28] The dispute between the two professors of medicine was submitted to arbitration, with friends or associates of each party offering security for the observation of the settlement. Conceivably Bartolomeo was also a political rival of Taddeo, since the latter may have received his privileges from the municipality during the popular regime of the 1280s,[29] while Bartolomeo secured the important position of Prior of the Anziani of Bologna in 1303 when the communal government had fallen into the hands of the White Guelf nobles.[30] By 1293, two years before Taddeo's death, Bartolomeo had added a lucrative and aristocratic professional practice to his academic activities.[31] He also served the commune of Bologna in a medico-legal capacity (his duties including at least one autopsy) until 1310, notwithstanding the expulsion from power in 1306 of the White Guelf party, of which he was a prominent member.[32]

By April 1311, however, Bartolomeo had joined the entourage of his best-known patient, the Emperor Henry VII, and as a result he earned a sentence of banishment from Bologna in October of the same year.[33] When Henry died suddenly in 1313, Bartolomeo defended the emperor's confessor from the charge of murdering him with a poisoned host. According to his physician, the emperor had ignored medical advice (perhaps that proferred in Bartolomeo's *consilium* for the emperor, still preserved in a Munich manu-

[26]MS Vat. Lat. 4454, fols. 55ʳ–75ᵛ (modern numbering of folios). At end: "expliciunt recollectiones libri de interioribus secundum M. Bartholomeum de Varignana." See TK, col. 531, where, however, a second set of folio numbers, also found in the manuscript, is used. The only differences between the text of this commentary and that contained in Vat. Lat. 4452 appear to be that the proemium and a final paragraph "de priapismo" present in 4452 are here omitted.

[27]Sarti and Fattorini, *De claris,* Vol. I, p. 569.

[28]*Ibid.,* Vol. II, p. 223.

[29]On the dating of Taddeo's privileges, see Gaudenzi, "Gli antichi statuti," pp. 118, 129, 132. Statutes bearing the dates 1272 and 1291 are found in the collection that includes the grant to Taddeo. On the political history of this period, see V. Vitale, *Il dominio della Parte Guelfa in Bologna (1280–1327)* (Bologna: Nicola Zanichelli Editore, 1902), pp. 1–105.

[30]Sarti and Fattorini, *De claris,* Vol. I, p. 569; Vitale, *Il dominio,* pp. 88–92.

[31]Sarti and Fattorini, *De claris,* Vol. I, p. 569; G. Fantuzzi, *Notizie degli scrittori Bolognesi* (Bologna, 1790), Vol. VIII, p. 153.

[32]L. Münster, "Alcuni episodi sconosciuti o poco noti sulla vita e sulla attività di Bartolomeo da Varignana," *Castalia: Rivista di Storia di Medicina,* 1954, *10*:213–214, and bibliography there cited; Vitale, *Il dominio,* pp. 198–103.

[33]In April 1311 Henry entrusted Bartolomeo with the delivery of the Iron Crown to the abbot of the monastery of St. Ambrose in Milan (*Monumenta Germaniae Historica, Legum sectio IV, Constitutiones et acta publica imperatorum et regum,* Vol. IV, Pt. 1, No. 609). For Bartolomeo's banishment, see Sarti and Fattorini, *De claris,* Vol. I, p. 570.

script)[34] and had consequently died of natural causes.[35] After Henry VII's death, Bartolomeo spent much of the remainder of his life at Genoa.[36]

V

Bartolomeo chose to introduce Galen's work on the external manifestations of injuries and diseases of internal organs with a critical exposition of the definition of medicine as *philosophia corporis*.[37] He drew the phrase from the proemium of a commentary on the *De sectis* of Galen—probably from that ascribed to Johannes Alexandrinus, which has been shown to be familiar to another member of the circle of Taddeo Alderotti, but perhaps from the closely related work attributed, in a ninth-century manuscript in the Biblioteca Ambrosiana, to Agnellus the Iatrosophist.[38] These commentaries, one of which became available in Latin as early as the sixth century A.D., appear to have been an important channel whereby the ideas of Alexandrian physicians regarding the close relationship of medicine to philosophy reached the Latin West. For example, in the commentary attributed to Johannes Alexandrinus it is stated that philosophy is the medicine of the soul and medicine the philosophy of the body, and, according to Aristotle, that philosophy and medicine are two sisters. In the commentary said to be by Angellus the remark "medicine is the philosophy of the body" is specifically attributed to Aristotle. These definitions are not, of course, in reality Aristotelian, as Owsei Temkin long ago pointed out.[39]

According to Bartolomeo, the definition of medicine as the philosophy of the body is of only limited validity.[40] Not only does such a definition ignore the purpose of medicine—that is, in Avicenna's words, to maintain health and recover it when lost—but it is not generally true even of medicine considered

[34]TK, col. 814.

[35]Sarti and Fattorini, *De claris*, Vol. I, p. 571.

[36]Münster, "Alcuni episodi," pp. 214–215, demonstrates that the death date usually given for Bartolomeo, 1318 or 1319, is incorrect, since he lived at least until 1321 and died before 1330. Münster also documents attempts to secure Bartolomeo's services by the governments of Florence, Perugia, and Venice during his last years.

[37]MS Vat. Lat. 4452, fol. 83r.

[38]". . . assumpsi diffinitionem huius scientie quam ponit commentator libri sectarum in prohemio" (*ibid.*). The commentary ascribed to Johannes Alexandrinus is printed with works of Galen in some early editions (see TK, col. 179); the commentary of Agnellus is preserved in MS Ambrosiana G 108 Inf., 9c, fols. 22r–48v (I am grateful to the Mediaeval Institute of the University of Notre Dame for supplying a photocopy of this commentary). The relationship between the two commentaries is discussed and their contents analyzed in Owsei Temkin, "Studies in Late Alexandrian Medicine, 1: Alexandrian Commentaries on Galen's *De sectis ad introducendos*," *Bull. Hist. Med.*, 1935, *3*:405–430, and in A. Beccaria, "Sulle tracce di un antico canone latino di Ippocrate e di Galeno. III," *Italia Medioevale e Umanistica*, 1971, *14*:1–23. In the former article it is pointed out that Mondino de' Liuzzi, the anatomist, who was a pupil of Taddeo Alderotti, made use of the commentary ascribed to Johannes Alexandrinus.

[39]Temkin, "Studies in Late Alexandrian Medicine," pp. 408–409, 418.

[40]". . . cum scientia medicinalis sit habitus quidam intellectualis erit etiam ipsius triplex diffinitio. una que est in comparatione ad subiectum ut pro universa diffinitio medicina est philosophia corporis. alia que est in comparatione ad finem ut illa diffinitio que asignatur a commentatore libri sectarum in prohemio, medicina nichil aliud est quam aditio et subtractio. Alia autem est diffinitio que est in comparatione ad subiectum et ad finem, et ista est completa, ut illa quam asignatur Avicenna prima canone: medicina est scientia qua humani corporis, etc. Ex hiis autem apparet . . . quod diffinitio medicine asignata est incompleta quoniam sola data per subiectum non autem per finem" (Vat. Lat. 4452, fol. 83r).

as a descriptive science. Knowledge about bodies, Bartolomeo pointed out, can be either mathematical or natural, and medicine is certainly in no way connected with mathematics. Nor can medicine be equated with natural philosophy of the body in general, because in that instance it would simply be a branch of *scientia naturalis,* which is not the case. It is true that the subject matter of medicine can be described as a certain type of *philosophia naturalis* specifically pertaining to the human body and its properties and passions,[41] yet medical knowledge, unlike that acquired by the study of *philosophia naturalis,* always serves an ultimately practical purpose. The whole of medicine is directed toward *opus,* and therefore medicine is to be categorized as a practical-intellectual skill.[42] Its object is thus quite different from that of *scientia naturalis,* which seeks the truth for its own sake. Because sciences with different ends cannot be subalternated one to another, medicine cannot properly be regarded as in any way subalternated to *scientia naturalis,*[43] even though the latter discipline does provide theoretical knowledge of such aspects of the human body as generation, health, and sickness, and even though the method of demonstration is the same in both medicine and *scientia naturalis.* Practical considerations concerning the human body are touched upon in Aristotle's *Politics* and in the study of military science, but here too the object is very different from that of medicine.[44] The passage concludes with some fairly conventional remarks about the nobility and usefulness of medicine and with a further emphatic statement about the practical purpose of medical knowledge. Thus Bartolomeo provided his readers with a firm defense of the autonomy of medical science, somewhat unusual in its insistence on the total divorce of even theoretical medicine from *philosophia* or *scientia naturalis.* The contrast with Taddeo's views could scarcely be more striking.

It should be made clear, however, that Bartolomeo certainly did not advocate the total omission of the study of Aristotelian logic or natural science from

[41] ". . . est intelligendum quod duplex est corpus, scilicet mathematicum et naturale, et corpus naturale duplex . . ., scilicet, in genere et in specie. modo si medicina sit philosophia corporis non est philosophia cuiuscumque quia non mathematici. quoniam si esset philosophia corporis mathematici aut esset philosophia corporis mathematici in genere aut in specie. non in genere quia tunc non differeret a scientia mathematica in communi. non in specie quoniam tunc esset subalternata scientie mathematice in communi, que ambo sunt falsa. Non etiam est philosophia corporis naturalis in genere quoniam tunc non differeret a scientia naturali in communi et esset ea subalternata, quod falsum. Sed est philosophia corporis naturalis in specie, non cuiuscumque, sed corporis humani quoniam est considerans eius proprietates et passiones" (*ibid.*).

[42] ". . . cum scientie maxime distinguantur per finem . . . et finis habitus intellectualis practici sit opus et etiam talis sit finis medicine, ideo medicina supponitur habitui intellectuali practico" (*ibid.*). This may be contrasted with Taddeo's more reserved statement, following Avicenna, that the end of medical theory was practice [Taddeo (1527), fol. 343^r; "nam practica est finis theorice"].

[43] ". . . inde tamen quod si [medicina] sit considerans tales passiones [corporis humani] non considerat illas propter verum sed propter opus. scientia autem naturalis illas considerat propter verum. et quia ille scientie que fine differunt una non subalternatur alteri, ideo ex dictis apparet quod medicina non subalternatur scientia naturali" (Vat. Lat. 4452. fol. 83^r).

[44] "Verum quod ex habitibus scientie naturalis quidam sunt theorici quidam practici considerantes de humano corpore: theorici in quibus determinatur de generatione hominis et de sanitate et egritudine et de aliis passionibus corporis humani; practici sicut liber politicorum et quidam specialis pars ipsius que dicitur scientia militaris. ab illis utriusque differt medicina. quoniam illi habitus theorici considerant passiones corporis humani propter verum. medicina autem propter opus. etiam differt a practicis, quia tales habitus illas passiones considerant propter corpus animatum, medicina autem propter corpus. . . . erit idem modus demonstrationis medicine et ipsius scientie naturalis . . . tamen quod modus demonstrationis scientie naturalis est certior" (*ibid.*).

the training of physicians. On the contrary, immediately before discussing the definition of medicine, he was at pains to demonstrate his own Aristotelian learning by means of an elaborate logical justification of the practice of beginning every work with an invocation of the deity: God is prior to everything and therefore should be at the beginning of everything. This rather commonplace assertion was buttressed by Bartolomeo with a formidable array of citations from the *Physics,* the *Metaphysics, De anima,* and the *Ethics,* and the pseudo-Aristotelian *De causis.* Later he was to insist on the necessity of a knowledge of logic for the study of medicine.[45]

In striking contrast, however, was his use of sources in the body of his commentary on Galen's work. Here citations of Aristotle are rare indeed among the innumerable references to the *Canon* of Avicenna. The main purpose of the commentary, moreover, seems to be to discuss and where possible to reconcile differences between Galen and Avicenna. But Bartolomeo, at any rate in the commentary on *De interioribus,* showed no interest in reconciling the differences as to human physiology between Aristotle and Galen. This topic, already discussed by Albertus Magnus in the mid-thirteenth century,[46] was still of absorbing interest to Bartolomeo's contemporary Peter of Abano.[47] *De interioribus* 3.8 contains a clear statement of Galen's views on one of the points on which he differed most sharply from Aristotle—that concerning the nervous system. Aristotle had believed that the nerves took their origin in the heart,[48] whereas Galen, of course, held them to be linked to the brain.[49] Bartolomeo's discussion of Book 3 of *De interioribus* contains no exposition of Aristotle's views on the subject.[50] By way of contrast, one may note that Taddeo Alderotti had not only discussed the differences between the statements about the nerves in the Aristotelian works on animals and the opinion of "all medical authorities," but had come down decisively on Aristotle's side, although adding a rather lame attempt to harmonize his sources.[51]

VI

In sum, then, Bartolomeo da Varignana endorsed the well-established tradition that the study of Aristotelian logic was necessary to the physician; he probably

[45]Vat. Lat. 4452, fol. 90ʳ; Vat. Lat. 4454, fol. 62ʳ. Bartolomeo may also be among the relatively few learned physicians who themselves commented on Aristotelian works; he is named as the author of a commentary on the *Oeconomica* (P. O. Kristeller, *Iter Italicum,* Vol. II, p. 291). See N. G. Siraisi, "The *Libri Morales* in the Faculty of Arts and Medicine at Bologna: Bartolomeo da Varignana and the Pseudo-Aristotelian Economics," *Manuscripta,* 1976, *20*:105–118.

[46]See George W. Corner, *Anatomical Texts of the Earlier Middle Ages* (Washington, D.C.: Carnegie Institute of Washington, 1927), p. 40, for discussion of early treatments of the subject by Latin writers. For Albertus' views, see his *Opera omnia,* ed. A. Borgnet (Paris, 1891), Vol. XI, pp. 201–233.

[47]*Conciliator,* diff. 41, "An nervi oriuntur a cerebro vel corde."

[48]*Historia animalium,* III, 5, 515a 27–28.

[49]See, e.g., Galen, *On the Usefulness of the Parts of the Parts of the Body,* trans. Margaret T. May (Ithaca: Cornell University Press, 1968), Vol. I, pp. 362–363. The full text of *De usu partium* was not translated into Latin until 1317, but Galen's views on this subject are also stated in *De placitis Hippocratis et Platonis,* a work attributed to Galen and apparently indirectly known to twelfth- and thirteenth-century Latin writers via allusions in treatises by Arab authors (see Ynez Violé O'Neill, "The Fünfbilderserie Reconsidered," *Bull. Hist. Med.,* 1969, *42*:236–245), and, of course, in *De interioribus* itself.

[50]Vat. Lat. 4454, fols. 65ʳ⁻ᵛ; Vat. Lat. 4452, fols. 93ʳ⁻ᵛ.

[51]". . . dico: quod nervi simpliciter: et secundum veritatem ortum habent ex corde sicut dicit Aris" [Taddeo (1527), fol. 353ʳ].

regarded the reading of other Aristotelian works as desirable although perhaps not essential in terms of the physician's general scientific, philosophical, and moral culture; and he simply ignored Aristotle as a source of information about physiology (at any rate in his commentary on *De interioribus*). But the foregoing should not be understood as implying that Bartolomeo was either less learned than Taddeo or that he advocated less reliance upon ancient authority in medical science. On the contrary, it is plain from their theoretical writings that the two men shared a common interest in presenting classical Greek medicine to their students.[52] By the late thirteenth and early fourteenth centuries, of course, this interest was in no way either novel or unusual, and Taddeo's choice of Hippocratic and Galenic works on which to comment was quite conventional.[53] But it is perhaps worth remarking that Bartolomeo turned his attention to certain works of Galen which up to that point had been relatively little studied by Latin scholars. In addition to his work on *De interioribus,* he also produced questions or commentaries on *De complexionibus*[54] and *De accidenti et morbo.*[55] *De interioribus* was translated into Latin by Burgundio of Pisa in the twelfth century and by at least one other translator, perhaps Constantinus Africanus,[56] before Bartolomeo wrote on it; *De complexionibus* was translated by both Gerard of Cremona and Burgundio of Pisa. But in all three instances Bartolomeo's appears to be among the earliest two or three known Latin commentaries. Bartolomeo's interest in this group of treatises may, however, have been stimulated by Taddeo Alderotti himself, since the latter seems to have prepared a corrected version of Burgundio of Pisa's translation of *De interioribus*[57] and may have written a work of his own on the subject of complexions.[58] All three of this group of Galenic works appear among the texts prescribed for the study of theoretical medicine by the statutes of the student University of Arts and Medicine of Bologna in the year 1405.[59] It is possible that they were introduced into the Bolognese curriculum by Bartolomeo before 1311, or even by Taddeo himself before 1295.

VII

Taddeo Alderotti and Bartolomeo da Varignana, master and pupil, professors in the same *studium* and heirs to a common tradition, shared many of the

[52] Bartolomeo, like Taddeo, was also the author of *consilia* and perhaps of a *regimen;* see TK for manuscript listings.

[53] The *Aphorisms, Prognostics,* and *De regimine acutorum* of Hippocrates and the *Tegni* of Galen formed the core of the widely used collection of medical tracts known as the *ars medicine* or *articella.* On the formation of this collection in the twelfth century, see Kristeller, "The School of Salerno," p. 515.

[54] Vat. Lat. 4451, fols. 57ra–87^{v} (TK, p. 1013); Vat. Lat. 4454, fols. 77^{r}–85^{v}; see also TK, col. 1202.

[55] Vat. Lat. 4452, fols. 67^{r}–82^{v}.

[56] The commentary in Vat. Lat. 4452 is based on Burgundio of Pisa's version but contains at fol. 90^{v} a reference to "alia translatio." Another translation beginning "Loca autem nominant particulas corporis non iuniores solum" is doubtfully attributed to Peter of Abano in TK, col. 831. Yet another incipit perhaps indicating the existence of a third version of the same work is listed in TK, col. 1273.

[57] Vat. Urb. Lat. MS 247, fol. 254^{r}: "Incipit liber de interioribus . . . Medicorum non solum moderni . . ."; fol. 281: "Explicit . . . correptus per magistrum Tadeum per alium translationem."

[58] Lynn Thorndike, "*De complexionibus,*" *Isis,* 1958, *40*:399–400.

[59] C. Malagola, ed., *Statuti delle università e dei collegi dello Studio Bolognese* (Bologna, 1888), pp. 274–277.

same attitudes and priorities in regard to medical learning. The differences between them, although real and significant, could easily be overstressed. In part, indeed, they may be explained by the nature of the works commented on. The *Isagoge* of Johannitius, broad and all-embracing, might seem to call for a more "philosophical" treatment than the more specialized *De interioribus.*

Both men were concerned to defend the validity of medicine as a science. Yet, in the context of the nascent faculty of arts and medicine at Bologna, it is easy to see Taddeo's stress on the traditional links between medicine, arts, and philosophy and his thoroughgoing Aristotelianism as being directed toward reinforcing the intellectual respectability and academic privilege of physicians who were practitioners as well as teachers—in a society in which a poet could express his disdain for anyone who "cheapens himself by the practice of shameful arts, performing mercenary operations for the cure of the sick body."[60] Bartolomeo's criticism of the association of medicine and *philosophia,* his refusal to subalternate medicine to *scientia naturalis* or identify it as a subdivision of *philosophia naturalis,* his Galenism in physiology, and his more emphatic statement of the ultimately practical purpose of all medical learning may represent a necessary reaction to Taddeo's views. Neither of these approaches to medical science was new, but in the early years of the institutionalization of faculties of arts and medicine in the Italian *studia* the old questions evidently acquired a new relevance.

60 "venales operas ad corporis aegri
disponens curas, turpes vilescet in artes"

Albertino Mussato, *Epistola ad magistrum Marsilium Physicum Paduanum,* in J. G. Graevius, *Thesaurus antiquitatum et historiarum Italiae* (Leiden, 1722), Vol. VI, Pt. 2, p. 49. Mussato (d. 1326) had a number of connections with the College of Doctors of Arts (and medicine) of Padua; the recipient of this verse epistle was the celebrated Marsiglio of Padua, author of the *Defensor pacis,* himself a physician who received part of his early training at Padua.

Arnald of Villanova and Bradwardine's Law

By Michael McVaugh

I

OF ALL THE contributions made to physical theory by the medieval scholars of Merton College, Oxford, among the most famous is the mathematical law formulated by Thomas Bradwardine to express the relationship between the dynamic principles regulating local motion. The view generally held by natural philosophers at the beginning of the fourteenth century derived (perhaps illegitimately) from Aristotle's authority and supposed that an object's speed was arithmetically related to the ratio between its motive force and resistance—supposed, that is, that $V \propto F/R$. In order to correct certain deficiencies of this rule, Bradwardine proposed a law of proportionality, arguing that "the proportion of velocities in motions follows the proportion of the power of the motor to the power of the thing moved"; his elaboration of this statement makes it plain that he is suggesting the relationship $F_2/R_2 = (F_1/R_1)^{V_2/V_1}$, or more simply $V = \log_a (F/R)$, when $a = F_1/R_1$. This so-called Bradwardine's Law, appearing in his *Tractatus de proportionibus* of 1328, exerted an immediate influence upon philosophers in England and on the Continent and was widely accepted in the later Middle Ages, introducing a number of fruitful ideas into medieval mechanics in the process.[1]

The search for the sources of Bradwardine's law has been neglected by modern critics, who prefer to study its importance for subsequent European science. It may well be that those who picture Thomas Bradwardine as a fourteenth-century Newton would suppose him to have independently worked out a formulation to satisfy the inadequacies of the Aristotelian rule. There is, however, good reason to believe that his law is not a piece of creative mathematics, but is instead the adaptation of an expression having considerable currency among another group of contemporary scientists—the medical theorists of Montpellier. It has already been remarked by Marshall Clagett that certain similarities exist between the law presented in Bradwardine's *Tractatus* and the system devised by the Arabian philosopher Alkindi in the ninth century to measure the qualitative intensity of compound medicines.[2] In fact, the resemblance is closer and the connection between the two still more probable than Clagett seems to have suspected.

1 See Marshall Clagett, *The Science of Mechanics in the Middle Ages* (Madison: Univ. Wisconsin Press, 1959), Ch. 7.

2 *Ibid.*, p. 439.

From the time of Galen to that of Bradwardine and beyond, medical theory recognized four intensive degrees in any medicinal quality, degrees corresponding to four different levels in the action of drugs on the human body; these ranged from the first degree, which could barely be sensed, to the fourth, which eventually destroyed the body to which it was given. But the Alkindian theory was the first which *also* tried to relate these varying degrees systematically to numerable changes in the qualitative powers producing them. The specific relationship proposed in Alkindi's one medical work, *Quia primos* (the incipit of the Latin translation made by Gerard of Cremona in the twelfth century), is one which links a geometric increase in the number of parts of a quality to an arithmetic increase in sensed effect. A temperate, qualitatively balanced medicine has, according to this theory, equal parts of hot and cold; one hot in the first degree has twice as much hot as cold; one hot in the second degree four times as much; and, ultimately, one in the fourth degree has sixteen times the hotness of a temperate medicine.[3] The arguments on which all this is founded are less logical than numerological. The real purpose of *Quia primos* is not so much to argue the metaphysical validity of the theory as to show the proper method of calculation with it: by adding up the number of hot parts and of cold contained in the simple constituents (each of known degree) of a compound medicine, and determining their ratio, the degree I of the compound can easily be found, since $H/C = 2^{\text{degree}}$ or $I = \log_2(H/C)$. Such a system is indeed a perfectly possible precursor of Bradwardine's law:

> Since this treatise of al-Kindi was translated by Gerard of Cremona, it was no doubt available to Bradwardine. Furthermore, Bradwardine was one of the first to make the analogy between velocity as intensity of motion and the intensity of permanent qualities. It is not then much of a step to go from (1) an exponential relationship between qualitative degrees of intensity and qualitative powers to (2) an exponential relationship between degrees of motion and the ratio of motive and resistive powers. It is certainly within the realm of probability that Bradwardine took precisely this step.[4]

It comes therefore as something of a surprise to find that *Quia primos* remained virtually unread in the Middle Ages. One indication of its unpopularity is the scarcity of manuscripts of the work—only three are known today.

[3] Alkindi nowhere states this principle succinctly. He does, however, repeatedly describe the numerical relationships that obtain between each of the first five stages in a doubling geometrical progression and the other four, e.g.: ". . . in gradu similiter tercio quod continetur caliditas duplum existit eius quod est in secundo ex caliditate, quemadmodum ostendimus ex incremento dupli. Sed quod in secundo gradu continetur ex caliditate quadruplum existit eius quod est in equali ex caliditate et quadruplum etiam eius quod in eodem continetur ex frigiditate. Oportet igitur ut sit quod in gradu tercio existit caliditatis octuplum caliditatis que est in equali et octuplum etiam frigiditatis eius, cum mensuracio ad frigiditatem facta fuerit. Et cum caliditas gradus tercii octupla fuerit frigiditatis que est in equali, erit quod est in ipso ex frigiditate octava pars eius quod est in eo ex caliditate; ergo erit caliditas partes quatuor et frigiditas pars media. In gradu quoque quarto quod est ex caliditate duplum existit. . . ." Alkindi, *Quia primos,* MS Paris Bibliothèque Nationale Lat. 9335, fol. 137v. (*Iacob Alchindi . . . de medicinarum compositarum gradibus investigandis libellum,* in *Ioannis Mesuae . . . Opera* [Venice, 1589], II, fol. 271rA). He details the entire sequence in three different contexts, and the result is to make the general principle clear through repetition.

[4] Clagett, *op. cit.*, p. 439, n. 35.

From another standpoint, of course, that unpopularity is perfectly understandable. The work would have been difficult to grasp even in perfect translation, and Gerard of Cremona's Latin version was by no means perfect. Moreover, it was true of thirteenth-century physicians that they were far more interested in the practice than in the theory of their art; on the whole, they were simply not ready for the idea of a mathematical medicine. Roger Bacon is one of the very few who seem to have had a knowledge of Alkindi's book, and even Bacon, who suggests in his writings the next century's attempt to quantify pharmacy, mentions it in terms which show that he has no close acquaintance with it.[5]

Yet after all it remains probable that, indirectly, the Alkindian mathematics provided the inspiration for Bradwardine's law. One man who did study *Quia primos* carefully was Arnald of Villanova (d. 1311), physician to the house of Aragon, papal counselor, theological visionary, and legendary alchemist. The only peaceful period in Arnald's turbulent career was his tenure as a master at Montpellier, from 1292 to 1299, during which he wrote the bulk of his medical works. It was apparently also during this period that he discovered Alkindi's book, since among the probable Montpellier writings is one (to which Arnald subsequently refers as the *Aphorismi de gradibus*) that was the first serious, detailed medieval appraisal of *Quia primos*.[6]

In form and substance the *Aphorismi* is Arnald's own creation: a commentary upon thirty-seven statements by an "unknown author" (evidently Arnald himself) on the nature and determination of qualitative intensity in compound medicines. First stating the definitions and postulates appropriate to the science, it proceeds to justify the general application of measure to qualities, and then appraises the merits of several specific mathematical pharmacies. It decides eventually for the Alkindian system with a few additions and modifications, presenting its calculations faithfully and far more clearly than does Gerard's translation.[7] What is peculiarly Arnald's in the

5 "In proportione est triplex difficultas. Una est in differentia radicis et eorum que sequuntur radicem, ut sciatur proportio horum certa in omni medicina; et hoc est valde difficile, et ignotum vulgo medicorum. Secundum est ut habeatur ex mixtione gradus quem volumus caliditatis et aliarum qualitatum sicut in simplici medicina, et illud determinari non potest nisi per viam quam Alkyndi de gradibus docet, que via est difficilis valde et penitus ignota hiis temporibus apud Latinos medicos, sicut quilibet novit. Nam qui vellet artem dicti philosophi perficere, oporteret eum bene scire communia mathematice propter species maioris inequalitatis et minoris, et propter species proportionum et propter leges fractionum difficiles valde, quibus utitur auctor dictus." *De erroribus medicorum*, in *Opera hactenus inedita Rogeri Baconi*, ed. A. G. Little and E. Withington (Oxford: Clarendon Press, 1928), Vol. 9, pp. 116–117.

6 This work appears in the 16th-century editions of Arnald's *Opera* as the *Liber amphorismorum de graduationibus medicinarum*; the shorter title appears in the text of his *Speculum* (*Opera Arnaldi de Villanova* [Venice, 1505], fol. 25r, v).

7 Arnald summarizes the Alkindian doctrine briefly before going on to argue its superiority over the Averroïstic one:

"[Alkindus docet] quod omnis eorum [*sc.* graduum] differencia continetur in terminis continue duplacionis que est ab unitate usque ad 16 inclusive, ubi quatuor sunt termini duplacionis continue. Dicitur autem continua quando ibi incipit ubi terminatur, sicut si dicamus unum et unum duo, et duo et duo quatuor, et quatuor et quatuor 8, et 8 et 8 16. Primus ergo terminus duplacionis hic sunt duo, et secundus quatuor, et tercius 8, et quartus 16. Intendit ergo quod primus gradus contineat duas porciones qualitatis aucte, secundus quatuor, tercius 8, quartus 16 . . . et hoc est illud

work is an attempt to give the system natural-philosophical justification, one historically so important as to warrant speaking of an "Arnaldian" pharmacy in the fourteenth century. For with the *Aphorismi* Arnald initiated at Montpellier a habit of intensive inquiry into the action of compounds, starting a fifty-year discussion that, although centered on the original Alkindian system, was derived entirely from Arnald's work. Some of his followers omitted most of his natural philosophy and perhaps in consequence gained a much greater popularity: of the *Tractatus de gradibus* (1303) by Arnald's contemporary Bernard Gordon, there exist today at least two dozen manuscripts, of the *Aphorismi* five. But all the Montpellier authors—Arnald, Bernard, Jordan de Turre, Gerard de Solo—were instrumental in giving the basic mathematical formulations a European renown, and it seems far more likely that Thomas Bradwardine would have drawn on their writings than directly on *Quia primos*.

For this reason it is of considerable interest to be able to point to the existence of a link between the Arnaldian pharmacy and Merton College. It may be only coincidence that the best manuscript of the *Aphorismi*, potentially the most interesting of the Montpellier treatises to natural philosophers, is to be found in the Merton College library, since the occasion and date of its accession are unknown.[8] But it is quite certain that the mathematical elements of the pharmacy were appreciated by one of Bradwardine's associates. The mathematician and astronomer Simon Bredon was a fellow of the college during the years 1330–1341 at least, and the list of his writings suggests that he shared the particular inclinations of his more famous colleagues. Bredon is unusual in that he also held a medical degree, but his one medical work, the *Trifolium*, while in many respects unremarkable, reflects his mathematical tastes in its treatment of pharmacy: it is essentially a studied repetition of the Arnaldian-Alkindian system. The text, which seems to owe a particular debt to Bernard Gordon's *Tractatus*, does not try to substantiate the theory it presents, but it preserves the fundamental numerical hypothesis perfectly, that of an arithmetically increasing scale of intensity corresponding to a geometrically increasing ratio between opposed qualitative powers.[9] Admittedly it is impossible to make any judgment as to the date of the *Trifolium*, and by itself,

quod aliis verbis consuevit exprimi cum dicitur gradus est sexdecupla proporcio vel aliqua submultiplicium parcium eius continua duplacione prolata." *Aphorismi de gradibus*, MS Merton College 230, fol. 88r (aphorism 21). The statement which Arnald attributes to Alkindi, at the end of this passage, cannot be found in *Quia primos*.

The modifications which Arnald subsequently makes in the Alkindian system do not affect the numerical relationships holding between opposite qualities within the several degrees; these remain as before. Arnald differs from Alkindi merely over the problem of how to measure the influence of weight variations in individual medicines upon the final degree of intensity of a compound medicine prepared from them.

8 This is MS Merton College 230, fols. 83r–99r, from which all my references are taken. F. M. Powicke, *The Medieval Books of Merton College* (Oxford: Clarendon Press, 1931), was not able to trace this manuscript; but Dr. R. W. Hunt, Keeper of Western Manuscripts at the Bodleian Library, judges it to be of the first half of the 14th century, with marginal notes in an English hand of the same period. Its availability to the Mertonian natural philosophers thus remains perfectly possible.

9 The only copy of the *Trifolium* of which I know is that in MS Bodleian Digby 160, fols. 102r–222r.

therefore, it proves nothing about the presence of the Montpellier pharmacy at Merton College. Nevertheless, to find that a fellow of the college was familiar with the medical anticipation of "Bradwardine's" function, and found it attractive, makes it the more probable that this was equally true of Thomas Bradwardine himself.

II

Modern scholarship tends to neglect one side of Arnald of Villanova's intellectual personality. His controversial theology has been quite closely investigated: his relevant books and letters have been published, his kinship to the spiritual-Franciscan movement made clear. In contrast, his medical writings, which helped win him a lasting scientific reputation, have received comparatively superficial attention. Paul Diepgen's studies,[10] now fifty years old, are virtually unique, and happen to deal largely with works today thought to be wrongly attributed to Arnald, notably the *Breviarium practice*—though this does not necessarily invalidate their conclusion that Arnald characteristically relied upon observation and experience in his medicine to an unusual degree. Unfortunately, Diepgen took very little interest in Arnald's Montpellier writings, perhaps because (although obviously genuine) they are wholly in the tradition of scholastic medical theory. While Diepgen does admit that "ecclesiastical authority, religious mysticism, scholastic natural philosophy, and a Neoplatonic view of nature form the basis of his [Arnald's] medicine," this rather vague concession does not recognize that in their own way Arnald's theoretical writings are important evidence for the originality of his scientific outlook.[11]

In particular, the arguments of the *Aphorismi de gradibus* reveal that Arnald's undoubted empiricism, guided by a neo-Pythagorean predisposition to discover numerical relationships in nature, led him not only to admit into medical theory the mathematical function later to be adapted by Bradwardine to mechanics, as we have seen, but to anticipate the formulation of Bradwardine's dynamic law itself. In this he exemplifies a new direction in medical thought. There is apparent in the writings of a few late-thirteenth-century medical writers an emerging belief that pharmacy should somehow be quantifiable, and Arnald himself seems to have shared this conviction before ever he mastered *Quia primos*. In his *De simplicibus*[12] he offers some simple

10 Paul Diepgen, "Studien zu Arnald von Villanova: I. Der Lebens- und Bildungsgang Arnalds von Villanova," *Archiv für Geschichte der Medizin*, 1909, *3*: 115–30; "II. Zur Echtheitsfrage des Breviarium," *ibid.*, 188–96; "III. Arnald und die Alchemie," *ibid.*, 369–96; "IV. Arnalds Stellung zur Magie, Astrologie und Oneiromantie," *ibid.*, 1911, *5*: 88–115; "V. Zur Charakteristik Arnalds von Villanova," *ibid.*, 115–20. "Zweite Folge: I. Zur Echtheitsfrage des Arnald von Villanova und anderen zugeschriebenen Traktates über die Unfruchtbarkeit (de sterilitate)," *ibid.*, 1913, *6*: 380–91.

11 Paul Diepgen, "Arnald von Villanova und die Medizin des Mittelalters," *Lychnos*, 1939, p. 239. Diepgen gives somewhat fuller consideration to the scholastic element in Arnald's thought in his article, "Die Weltanschauung Arnalds von Villanova und seine Medizin," *Scientia*, 1937, *61*: 38–47; this was subsequently reprinted in a collection of papers, *Medizin und Kultur* (Stuttgart: Enke, 1939), pp. 176–185.

12 The date of composition of the *De simplicibus* is unknown. There is no certain reference to it (so far as I know) in any of Arnald's other writings, but its authenticity has never been

numerical rules for the compounding of medicines (governing their weights only), explaining what was not yet common medical practice with the statement that "excellence of action in all things comes only with their proper and harmonious proportion."[13] The quasi-logical, numerological proofs of *Quia primos* were probably sufficient to cause him to adopt the Alkindian system. But harmony of proportion was not a property with much appeal for the majority of thirteenth-century physicians, who were still firmly convinced that medicine was above all an art, to be perfected with practice alone. The *Aphorismi* suggests that Arnald felt it important to answer the objections of any colleagues who might claim that the abstract mathematical approach of *Quia primos,* divorced from any appeal to experience, made it unfit to serve as the basis for the practice of a systematic quantitative pharmacy. For after citing Alkindi's original numerological arguments, Arnald adds evidence of his own to show that experience too proves the validity of this pharmaceutical theory:

> Everyone must come to accept it, impelled by reason and convinced by suitable experiment. For physicians find reason in those things which we discover in the objects of the other senses. But it is apparent in the case of the objects of other senses that if one of them cannot affect someone's sense, so that it be perceived by him to a given extent at a given distance, it is necessary that the force with which it has to affect that sense be at least doubled in it—as can be seen with all visible objects, whether luminous, like a flame, or a quantity, etc. For if a flame set in a given place cannot be seen by the eye, hindered by the complete unsuitability of the medium, in order that it be seen by the same at the same distance it must be at least doubled; likewise with other things visible. The same is true of things audible. If a sound cannot be heard at all by someone at a given distance, though medium and organ be well disposed, it must be at least twice as loud in order to be heard by the same. The same will also occur with things smelt and tasted.[14]

Arnald's argument is clearly that the Alkindian formulation relating medicinal qualitative intensity exponentially to causal power is only one expres-

---

questioned, and its stylistic and substantive affinities with other texts convince me that it is really his work. In that case, it must surely predate the *Aphorismi*: Arnald continued to maintain the doctrines of the latter work (as references in his *Speculum medicine*, for example, show), while the *De simplicibus* describes a different and much cruder analysis of medicinal interaction.

13 ". . . quia bonitas actionis omnium rerum non est nisi in sua proportione recta et armonia." *Opera*, fol. 276vA.

14 "Ad hoc autem ponendum necessario inducuntur omnes, racione cogente et congruo experimento confirmante; racionem vero accipiunt medici ex hiis que in obiectis aliorum sensuum experimur. Est autem in obiectis aliorum sensuum evidens, quod si aliquid illorum sensum alicuius movere non possit, ut aliqualiter ab eo percipiatur ex aliqua distancia, necesse est ut potencia qua movere habet sensum illum ad minus dupletur in eo, ut patet in quibuscumque visibilibus, sive sit lucidum, ut ignis, sive quantitas aliqua, et cetera. Si enim aliquis ignis in aliquo loco positus ab aliquo oculo non videatur omni ineptitudine medii circumscripta, necesse est ad hoc ut ab eodem videri possit ex eadem distancia, ut ad minus dupletur; et idem est in aliis visibilibus. Similiter autem in audibilibus veritatem habet. Si enim aliquis sonus ab aliquo nullatenus audiatur ex aliqua distancia, medio et organo bene dispositis, est necesse ad minus ut in duplo alcior sit, ad hoc ut ab eodem audiatur. Exinde et illud eciam contingit in odorabili et gustabili." *Aphorismi de gradibus*, MS Merton College 230, fol. 88vA.

sion of a more general law of experience governing *all* sense impressions, a medieval hint at psychophysics. This is his justification for then applying the mathematical concepts of Alkindian theoretical pharmacy to the interpretation and direction of the numerical techniques that physicians have used to measure the actual intensities of simple medicines since the time of Galen. The result is that at Arnald's hands pharmacy becomes the first and perhaps the only medieval science to link theory and practice in one quantified paradigm.[15]

Given the similarity between Arnaldian pharmacy and Mertonian mechanics, it now becomes of considerable interest to discover how general this law was meant to be: does the reference to "quantity" mean that Arnald believed a change in quantity of any measurable force to be geometrically related to the change in its intensity? Specifically, would he have applied this physical law to the forces involved in local motion? Apparently he would have; this, at least, is the impression given by a passage further on in the *Aphorismi*, defending the Arnaldian law against another set of critics—the medical Averroïsts. These men based their arguments upon Averroës' medical encyclopedia, the *Colliget*, whose fifth book sets out a new system for calculating the degree of a compound medicine.[16] This work does not really attempt to disprove Alkindian pharmacy; its criticism reduces to the simple assertion that a compound's degree is arithmetically proportional to the difference between the opposing qualitative powers involved, not geometrically proportional to their ratio. Averroïstic doctrine thus replaced $I = \log(H/C)$ with the much less involved $I = H - C$. Subsequently, Latin readers of the *Colliget* tried to show that the formulation was incompatible with experience, claiming that no force in nature is known to continue to exert an effect when confronted with a greater amount of its opposite.

Arnald of Villanova, however, found no trouble in exposing the fallacies

15 The failure of natural philosophers to develop a fully quantified physics is treated by A. C. Crombie, "Quantification in Medieval Physics," *Isis*, 1961, *52*: 143–160.

16 There has long been a dispute over the origin of the Latin translation of the *Colliget*: it has been attributed both to a Paduan Jew, Bonacosa, writing in 1255, and to Armengaud Blaise, who was producing scientific translations from the Arabic at Montpellier in the 1290's. There is very little evidence either way; Moritz Steinschneider (*Die Hebraeischen Uebersetzungen des Mittelalters* [Berlin, 1893], pp. 672, 973) decided for Bonacosa on the basis of a single Cesenan manuscript. But it may be remarked that since we now know Arnald of Villanova to have been the uncle of Armengaud Blaise (Miguel Batllori, "La documentacion de Marsella sobre Arnau de Vilanova y Joan Blasi," *Analecta Sacra Tarraconensia*, 1948, *21*: 75–120), accepting the latter attribution would nicely explain the genesis of Arnald's interest in compound medicines: he was teaching at Montpellier at the time in question, and had he been shown his nephew's new translation, he might easily have been led to *Quia primos* from Averroës' criticism of it.

This idea seems to be supported by the reference of Lynn Thorndike and Pearl Kibre (*A Catalogue of Incipits of Mediaeval Scientific Writings in Latin* [rev. ed., Cambridge, Mass.: Mediaeval Academy of America, 1963], col. 1171) to a Blaisian translation of the *Colliget* dated 1294. Unfortunately, this reference—to Durham Cathedral MS C. 1. 19—is mistaken. It derives from the catalogue of the cathedral manuscripts, which attributes an explicit to this effect to the *Colliget*; but what the catalogue's editor took for one work is in fact two. The *Colliget*, entire but without ascription to translator, is on fols. 234r–295v of the manuscript; it is followed on fols. 296r–301v by a fragment of Armengaud's translation of Avicenna's *Cantica*, and it is to this fragment that the attribution of translator and date properly belongs.

in the illustrations which the Averroïsts adduced against Alkindian pharmacy. These men were arguing

> that if one of two forces (*virtutes*) resisting each other equally is doubled, it will not just weaken the resisting one, but will overcome it entirely (as appears in the action of fire against the resisting logs, or in two weights resisting each other equally on a balance, or in two fighters of equal force); and therefore, if the *virtus* of some quality in the first degree is doubled, it consequently overcomes its opposite entirely. . . .

Arnald's reply to them incidentally makes clear some limitations of his general law:

> But that the instances which they adduce are fallacious is clear, in that they are improper and inconclusive–improper since they do not deal with conjoined forces nor with those of a mixture; inconclusive because one cannot measure how great the force of the fire is in respect to the logs, [cannot measure] either how great the force of resistance is in the logs or when it has doubled in the fire in respect to them–nor can he in the case of the fighters, who fight not with a natural force but partly with mental activity and partly with bodily strength, which affects the motive force. Thus anyone who uses such examples supposes to be known that which he himself does not know, simply following the drift of his imagination; he who has little evidence talks the more easily. As for the case of the weights hung from the balance, it deals with the forces of separated objects, whose motion depends not only on natural *gravitas* but on the position of the beam as well, since it has to lie over the point of the perpendicular.[17]

Arnald's rejection of these three very different Averroïstic counterinstances does not depend at all upon the qualitative nature of the forces involved. There is no evidence available to suggest that Arnald would not have applied his law to any dynamical confrontation in which force and resistance were measurable and were not separated.

What then of local motion? If Arnald had had to face this problem explicitly, he would probably have maintained that it too was described by

[17] "Si vero conetur probare istud sicut sequaces eius, dicens quod duarum virtutum equaliter sibi invicem resistencium, si altera dupletur non tantum frenabit resistentem sed vincet omnino, sicut apparet in accione ignis supra resistencia ligna, vel in duobus ponderibus equaliter sibi invicem resistentibus in statera, vel in duobus pugnantibus uniformis virtutis . . . dicemus quod istis accidit sicut melancholicis, qui pertinaciter ratiocinantur supra fundamentum quod corrumpta ymaginacio vel estimacio dedit eis. . . . Quod autem signa quibus utuntur sint fallacia patet per hoc quia impropria sunt et incerta. Impropria quidem quia non sunt de virtutibus mixtis sive mixtorum; incerta vero quia nemo potest mensurare quanta sit potencia ignis respectu talium lignorum, aut quanta sit in eis potencia resistendi, aut quando duplatur in igne respectu ipsorum; nec similiter in pugilibus qui pugnant non naturali virtute, sed partim industria racionis et partim robore corporali, quod pertinet motive virtuti. Unde qui talibus exemplis utitur subponit esse notum quod ipse ignorat, tantummodo sequens levitatem ymaginacionis, nam ad pauca respiciens de facili pronunciat. Exemplum vero quod est de gravibus apensis statere: patet quod est de virtutibus separatorum ad invicem quorum motus non solum a naturali gravitate dependet sed eciam a disposicione ferentis; nam supra punctum perpendiculi habet fieri. Discant ergo scienciam de ponderibus, et cognoscent exemplum ad propositum esse fallax." *Aphorismi de gradibus*, MS Merton College 230, fol. 89rB–89vA.

his law. Certainly the difficulties he has just mentioned do not apply to it. The criticism Arnald makes of the balance problem suggests that he thinks *gravitas*, and presumably local-motive forces generally, can be conclusively measured; and since local motion was universally recognized to derive from two conjoined opposing forces, the Arnaldian law should describe the dynamic relationship involved. It must be remembered, of course, that Arnald himself does not refer directly to mechanical situations other than the one above, relating to statics. Yet whatever the implications the Arnaldian law held for its author, a natural philosopher could certainly have seen in it (as we have) the suggestion that the pharmaceutical rule might be applicable to local motion; and had he attempted to pursue the analogy, he would have discovered that the physician and the natural philosopher are continually dealing with congruent problems: Arnald's attack on Averroës' $I = H - C$, for example, paralleled natural philosophy's consistent rejection of a dynamic $V \propto F - R$—actually the first of the possible alternatives to his law rejected by Bradwardine in the *Tractatus de proportionibus*. The passage from medicine to mechanics would have been extremely easy.

Still, there is no unmistakable proof that the Mertonian law did derive from Montpellier pharmacy, no continuity of language or method to give solidity to our speculations. Nor is this surprising, given the dissimilarity between the aims of the two sciences. To go from the pharmaceutical form of the Alkindian expression to the dynamical one requires only a slight modification of the common pattern, $I \propto \log(F/R)$: the generalization of the exponential base (always 2, for the pharmacists). But Arnald of Villanova and his successors apply the expression to a wide range of specific compound medicines and work out a number of definite problems with it, while Bradwardine formulates his law not in connection with actual measurements but in substantiation of certain statements in Aristotle's *Physics*. There is consequently no reason to expect any textual confirmation of a link between the two traditions, and none is to be found. Nevertheless, there remains a set of parallel circumstances so numerous as to provide strong indirect evidence of Montpellier's influence on medieval mechanics. In the end it is surely of equal importance to recognize that in anticipating Bradwardine's law by thirty years and discovering in nature an exponential law describing all dynamic processes, Arnald of Villanova was trying to provide contemporaries with the basis for a fully quantifiable natural science.

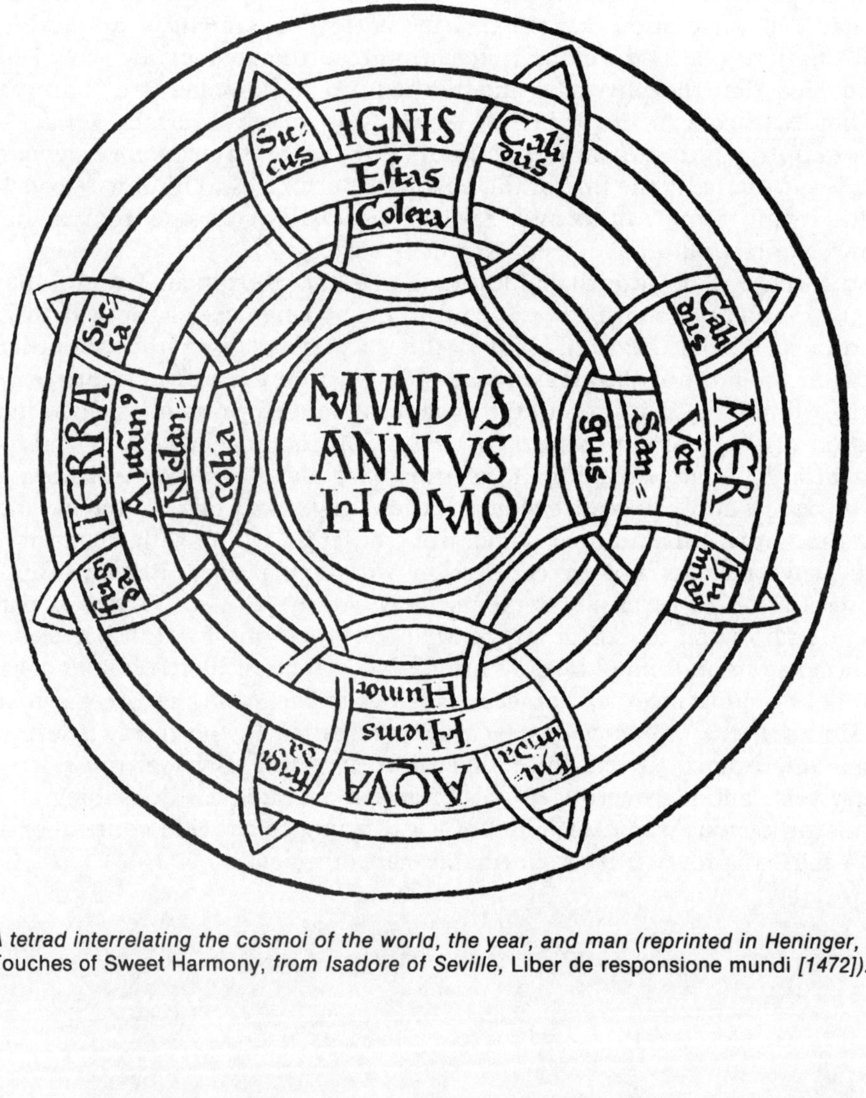

A tetrad interrelating the cosmoi of the world, the year, and man (reprinted in Heninger, Touches of Sweet Harmony, *from Isadore of Seville,* Liber de responsione mundi *[1472]).*

Did Science Have a Renaissance?

By Brian P. Copenhaver

WE COMMONLY USE THE WORDS *Renaissance* and *Scientific Revolution* to name large periods in Western intellectual history, eras that mark deep and partly concurrent transitions from premodern to modern times, yet the two terms imply conflicting notions of historical movement: one is retrospective, the other prospective. In order to decide what was revolutionary in the Scientific Revolution, historians have often read the pre-Newtonian evidence in a post-Newtonian light, though some have been less careless of whiggery and more inclined to hold the past responsible for its own past than to insist that it be prescient. In the starkest whig terms, postmedieval inquiry into nature counts as more or less important insofar as it resembles or promises modern science. But historians of Renaissance thought, in order to learn what was reborn after the Middle Ages, must sometimes look backward for canons of judgment. We admire the Renaissance as much for its mimetic as for its creative achievements. When the historian's topic is the study of nature during the Renaissance, questions of periodization and change become problematic because our instincts and experience tell us that good science must be progressive. For an important sector of early modern thought, some of the confusion disappears if we substitute the term *natural philosophy* for *science*. After all, Newton did not call his greatest work the *Mathematical Principles of* Science. The eventual demise of the category "natural philosophy" after the early modern period followed the disuse of mimesis and erudition as aims of inquiry into nature. After Descartes, these motives diminished among the authorities who came to be canonical, if not among all others who wrote about the natural world. G. W. Leibniz was the great anomaly in this respect—unlike Pierre Gassendi, who quickly became unreadable just because of his erudition. Gassendi, the beneficiary of humanism, knew that Epicurus and others had written natural philosophy in antiquity, so he looked in the ancient tradition for texts to be edited, ideas to be reborn, heroes to be emulated. For a time, at least, a Renaissance in natural philosophy seemed possible to Gassendi, Francesco Patrizi, Ulisse Aldrovandi, Konrad Gesner, and many others who searched for a new science in antiquity and through philology.[1]

A draft of this paper was read at the annual meeting of the Renaissance Society of America at Duke University in 1991, in a session organized by Nancy Siraisi; and again at the Center for Ideas and Society of the University of California, Riverside, in 1992. David Glidden, John Headley, and Alex Rosenberg made helpful comments on the paper, for which I am most grateful.

[1] The first edition of Newton's work was *Philosophiae naturalis principia mathematica* (London, 1687). For breaks and continuities between the culture of erudition and the culture of science see

But what actually happened to natural philosophy during the time called "the Renaissance"? If we take the year 1600 to be roughly the end of the period, we will find in it only a few figures or events important enough to make the textbooks: Leonardo, Copernicus, Andreas Vesalius, and Tycho Brahe stand out from a crowd in which Luca Pacioli, Erasmus Reinhold, Leonard Fuchs, Guillaume Rondelet, and many others remain dimmer figures. And if we take a closer look, much of what we see obscures any neat sense of progress through well-defined periods. By the standards of the day, some of the most interesting and influential natural philosophy was the work of die-hard Aristotelians teaching a scholastic curriculum. At the height of the Italian Renaissance, Pietro Pomponazzi was the prime advocate of a Peripatetic naturalism that seemed progressive to some, pernicious to others.[2] What could be bolder than a clever philosopher who doubted the rational grounds for belief in the soul's immortality? What could be more wrongheaded than a dogmatic Aristotelian who wrote about magic in

---

Henri Gouhier, *Les premières pensées de Descartes: Contribution à l'histoire de l'anti-renaissance* (Paris: Vrin, 1958); and Michel Foucault, *Les mots et les choses: Une archéologie des sciences humaines* (Paris: Gallimard, 1966), esp. pp. 32–59, and my discussion at the end of this paper. See also Anthony Grafton, *Defenders of the Text: The Traditions of Scholarship in an Age of Science, 1450–1800* (Cambridge, Mass.: Harvard Univ. Press, 1991), pp. 1–22; Grafton, "Humanism, Magic, and Science," in *The Impact of Humanism on Western Europe*, ed. Anthony Goodman and Angus McKay (London: Longman, 1990), pp. 99–117; and Lynn Joy, *Gassendi, the Atomist: Advocate of History in an Age of Science* (New York: Cambridge Univ. Press, 1987).

[2] See Charles B. Schmitt, *Aristotle and the Renaissance* (Cambridge, Mass.: Harvard Univ. Press, 1983), whose bibliography refers to Edward Cranz, Eugenio Garin, P. O. Kristeller, Edward Mahoney, Bruno Nardi, Antonino Poppi, and others who have clarified the role of Aristotelian thought in the early modern period. For Pomponazzi, the most recent comprehensive study is Martin L. Pine, *Pietro Pomponazzi: Radical Philosopher of the Renaissance* (Padua: Antenore, 1986); see also Giovanni Di Napoli, *L'immortalità dell'anima nel Rinascimento* (Torino: Societa Editrice Internazionale, 1963); F. Fiorentino, *Pietro Pomponazzi: Studi storici su la scuola bolognese e padovano del secolo XVI con molti documenti inediti* (Florence: Le Monnier, 1868); Eugenio Garin, "Phantasia e imaginatio fra Marsilio Ficino e Pietro Pomponazzi," *Giornale Critico della Filosofia Italiana*, 5th Ser., 1985, *5*:349–361; Etienne Gilson, *Humanisme et Renaissance* (Paris: Vrin, 1986), pp. 133–281; Franco Graiff, "I prodigi e l'astrologia nei commenti di Pietro Pomponazzi al *De caelo*, alla *Meteora* e al *De generatione*," *Medioevo*, 1976, *2*:331–361; Paul Oskar Kristeller, "Two Unpublished Questions on the Soul by Pietro Pomponazzi," *Medievalia et Humanistica*, 1955, *9*:76–101; Kristeller, *Eight Philosophers of the Italian Renaissance* (Stanford, Calif.: Stanford Univ. Press, 1964), pp. 72–90; Kristeller, "The Myth of Renaissance Atheism and the French Tradition of Free Thought," *Journal of the History of Philosophy*, 1968, *6*:233–243; Kristeller, *Aristotelismo e sincretismo nel pensiero di Pietro Pomponazzi* (Padua: Antenore, 1983); A. Maurer, "Between Reason and Faith: Siger of Brabant and Pomponazzi on the Magic Arts," *Medieval Studies*, 1956, *18*:1–18; Bruno Nardi, *Saggi sull'aristotelismo padovano dal secolo XIV al XVI* (Florence: Sansoni, 1958); Nardi, *Studi su Pietro Pomponazzi* (Florence: Le Monnier, 1965); Antonino Poppi, *Introduzione all'aristotelismo padovano* (Padua: Antenore, 1970); Poppi, *Saggi sul pensiero inedito di Pietro Pomponazzi* (Padua: Antenore, 1970); Lynn Thorndike, *A History of Magic and Experimental Science*, 8 vols., Vol. 5: *The Sixteenth Century* (New York: Columbia Univ. Press, 1941), pp. 94–110; Giancarlo Zanier, *Ricerche sulla diffusione e fortuna del "De incantationibus" di Pomponazzi* (Florence: La Nuova Italia, 1975); and Zanier, "Noterelle Pomponazziane," *G. Crit. Fil. Ital.*, 1979, *58*:211–225. For the primary sources see Pietro Pomponazzi, *Opera* (Basel, 1567); Pomponazzi, *Tractatus de immortalitate animae*, ed. G. Morra (Bologna: Nanni & Fiammenghi, 1954); Pomponazzi, *Libri quinque de fato, de libero arbitrio et de praedestinatione*, ed. Richard Lemay (Lugano: Thesaurus Mundi, 1957); Pomponazzi, *Corsi inediti dell'insegnamento padovano*, ed. Antonino Poppi (Padua: Antenore, 1966–1970); Pomponazzi, *De naturalium effectuum causis sive de incantationibus* (Basel, 1567; rpt. Hildesheim: Olms, 1970). *De immortalitate animae* is translated by William Hay and John Herman Randall in *The Renaissance Philosophy of Man*, ed. Ernst Cassirer, Paul Oskar Kristeller, and John Herman Randall (Chicago: Univ. Chicago Press, 1948), pp. 257–381; Henri Busson's translation, with some omissions, of *De incantationibus* is *Les causes des merveilles de la nature ou les enchantements* (Paris: Rieder, 1930).

clumsy Latin? In some ways Pomponazzi exemplifies the rebirth of learning characteristic of his age, yet his unrepentant Aristotelianism, his insistent occultism, and his considerable notoriety confound our understanding of the Renaissance. In order to test the idea of a Renaissance in natural philosophy, we may begin by imagining two types in the category "philosopher," one ideal, one not so ideal, who may help us in our inquiry.

Think of an Aristotelian philosopher, active in the early sixteenth century, who challenged Aristotle on several points. His innovations arose not only from newly recovered ancient sources, especially the Greek commentators on Aristotle and the Hellenistic schools that shaped their doctrine, but also from contemporary thinkers who constructed original philosophical systems, Marsilio Ficino most notably. This same person prized the autonomy of nature and the freedom of philosophy. He bravely drew the conclusions implied by his principles, even when they led toward materialism, determinism, and other views likely to displease the Church. This philosopher wrote several notorious books, two of them after having been called a heretic, and his later work argued that Aristotelian philosophy flatly contradicted Christian dogma on demons and miracles. More than once he suggested that Christian no less than pagan positions on these matters had more to do with schemes of thought control than with facts of nature. So thorough was his naturalism that he made religion itself, including the three faiths of Judaism, Islam, and Christianity, an effect of periodic natural causes. The age of Christian miracles was past, or nearly past, because natural forces that explain apparent miracles had reached the end of a cycle. The same natural forces cause the wonders that simple Christians attribute to angels and demons when they fail to understand philosophy, whose purpose is "to find the cause of natural effects."[3] All these things can be said plausibly of Pomponazzi, I believe. Readers who find his ideas progressive have assured his place in the traditional tableau of the Renaissance, which we like to furnish with such attitudes as naturalism, intellectual liberty, and hostility to superstition. Moreover, such habits of mind make it easy to see a Renaissance figure like Pomponazzi as a herald of the Scientific Revolution. Ernst Cassirer, who in the end denied that "Pomponazzi's path led to the modern view of nature," nonetheless admired his naturalism and determinism, claiming that he conferred "upon the naturalistic view exclusive power and dominion over the whole of intellectual life. . . . No region can possibly exist beyond the realm of complete determination." Cassirer credited Pomponazzi with having prepared natural philosophy to contain the physical laws proclaimed later by Kepler, Galileo, and Newton.[4]

Cassirer was writing about Pomponazzi's *De naturalium effectuum causis sive de incantationibus,* finished in 1520, widely circulated in manuscript but not printed until 1556.[5] This work, *On Spells or Causes of Natural Effects,* ends with

[3] Pomponazzi, *De incantationibus:* "philosophi sit caussam naturalium effectuum reddere" (p. 296) (unless otherwise indicated, all translations are mine). See Pine, *Pomponazzi,* pp. 16–30, 61–63, 75–77, 94, 99–103, 123–124, 139, 254–265, 277, 286, 297–301, 307, 338–343, 346, 350–352, 360–368.

[4] Ernst Cassirer, *The Individual and the Cosmos in Renaissance Philosophy,* trans. Mario Domandi (New York: Harper & Row, 1964), pp. 80–83, 103–109, 136–141, on pp. 106, 141; see also Cassirer, *Das Erkenntnisproblem in der Philosophie und Wissenschaft der neueren Zeit* (Darmstadt: Wissenschaftliche Buchgesellschaft, 1974), pp. 98–117.

[5] Pine, *Pomponazzi,* pp. 238–239; see also Ch. 3 (pp. 235–274), which is devoted to the work.

a submission to ecclesiastical authority so complete, so abject, so out of keeping with several hundred previous pages of close argument that some critics have dismissed it as merely expedient and flagrantly insincere. "Whatever has been affirmed by canonical scripture and universally decreed by the holy catholic church," wrote Pomponazzi, "must be held entirely, firmly, definitively, ineluctably, unshakably and without any trace of doubt." On one of the two main points in question, whether demons and angels exist to cause wondrous effects, he declared that

> although such effects may occur naturally and, as a matter of fact, do occur without their help, nonetheless, if the church has decided that something has been done by demons or by angels or by nature, one must hold exactly what the church decides. . . . But if any human reason or authority, whether of Aristotle or of Plato or of anyone else, stands opposed, it is inadmissible, . . . [and] if such things seem impossible to us, we do well to hold the mind captive.[6]

So much for the philosopher's liberty. The Latin in which Pomponazzi disputed lengthily and then surrendered theatrically reaches its stylistic peak in such moments as the leaden litany of allegiance just cited. "Quicquid affirmatum est," he chants, "illud totum, firmum, ratum, inevitabile, inconcussibile et prorsus sine aliqua dubitatione tenendum est." Rabelais could not have done much better, but Bruni would have. Even when he wrote books instead of *quaestiones*, Pomponazzi's powers of organization never broke the bonds of questions, replies, suppositions, and endless scholastic hesitations. The last chapter of *On Spells*, where Pomponazzi summarizes his own views, admits that "we seem to favor both sides in the foregoing" but promises to clear things up lest "those who look into our work, such as it is, think us indecisive." The same chapter makes it very clear that Pomponazzi's whole project, despite concessions to the Church and borrowings from newfound or newmade sources, had been conceived *"ad mentem Aristotelis"*—naturally enough for someone who spent his life professing Peripatetic philosophy. As an expert on Aristotelianism, he found no room in the

---

Zanier analyzes it and traces its *fortuna* through the early seventeenth century in *Ricerche* (cit. n. 2). For the reprint of the 1567 Basel edition cited here see note 2, which also lists the Busson translation and relevant secondary literature. In addition, see Brian P. Copenhaver, "Astrology and Magic," in *The Cambridge History of Renaissance Philosophy*, ed. Charles B. Schmitt *et al.* (Cambridge: Cambridge Univ. Press, 1988), pp. 264–300, esp. pp. 267–274; Daniel P. Walker, *Spiritual and Demonic Magic from Ficino to Campanella* (Studies of the Warburg Institute, 22) (London: Warburg Institute, 1958), pp. 107–111; Paola Zambelli, "I problemi metodologici del necromante Agostino Nifo," *Medioevo*, 1975, *1*:129–171; Zambelli, "*Aut diabolus aut Achillinus:* Fisionomia, astrologia e demonologia nel metodo di un aristotelico," *Rinascimento*, 1978, *18*:59–86; Zambelli, *Une réincarnation de Jean Pic à l'époque de Pomponazzi: Les thèses magiques et hérétiques d'un aristotélicien oublié, Tiberio Russiliano Sesto Calabrese* (*1519*) (Wiesbaden: Steiner, 1977); and Giancarlo Zanier, "Ricerche sull'occultismo a Padova nel secolo XV," in *Scienza e filosofia all'università di Padova nel Quattrocento*, ed. Antonino Poppi (Padua: Lint, 1983), pp. 345–372.

[6] Pomponazzi, *De incantationibus:* "Quicquid affirmatum est per scripturam canonicam et universaliter decretum est per sanctam ecclesiam catholicam, illud totum, firmum, ratum, inevitabile, inconcussibile et prorsus sine aliqua dubitatione tenendum est" (p. 318). "Et quamvis talia per naturam fieri possint et de facto fiant absque eorum praesidio, tamen si quod per ecclesiam determinatum est esse factum per daemones vel per angelos vel per naturam, sicuti ipsa ecclesia praecise determinat tenendum sic est. . . . Si autem ratio humana vel authoritas aliqua, sive sit Aristotelis sive Platonis seu cuiuscunque alterius, est in oppositum, ea non est admittenda, . . . [et] si nobis appareant impossibilia, bene oportet mentem captivare" (pp. 319–320).

Peripatetic system for angels, demons, or miracles. But he also accepted many reports of odd phenomena that made others believe in demonic magic and miracles, and, like other Aristotelian professors of his day, he accounted for them as effects of causes explained by the Peripatetic *scientiae* of natural magic and astrology.[7]

How do we fit this other Pomponazzi—the committed Aristotelian whose philosophy kept its distance from the new classicism while embracing natural magic—into our picture of the Renaissance, and how does a Renaissance that accommodates Pomponazzi's Peripatetic occultism connect with the age of scientific discovery from Copernicus and Vesalius through Galileo and Newton? An eminent student of these latter figures, Alexandre Koyré, wrote an essay in 1949, reprinted in 1966, on "the scientific contribution of the Renaissance" in which he weighed that contribution and found it wanting, confined to the influence of Ptolemy on Copernicus and of Archimedes on Kepler. Scholars still cite Koyré's essay in debating the periodization of late medieval and early modern thought; I mention it here because it summarizes opinions still current and worth correcting. Koyré finds that

> the ideal of civilization of the age correctly called the Renaissance *of letters and arts* is in no way an ideal of science but an ideal of rhetoric. . . . The great reform of logic that it attempted . . . tried to substitute a technique of persuasion for the technique of proof in classical logic. The character who embodies the context and spirit of the Renaissance is, obviously, the great artist . . . [and] especially the man of letters. . . . [But] the spirit of erudition . . . is not at all the spirit of science. . . . The great enemy of the Renaissance . . . was the Aristotelian synthesis, and . . . its great work was the destruction of that synthesis. But . . . credulity, belief in magic and so forth seem to me to be direct effects of this destruction. Once having destroyed Aristotelian physics, metaphysics and ontology, the Renaissance found itself without a physics or ontology, unable to decide what is possible or not possible. . . . Until a new ontology . . . was formed . . . in the seventeenth century, . . . there was no criterion . . . for deciding if a report received about one "fact" or another is true or not. . . . [Thus,] the Renaissance was . . . driven back to a magical ontology . . . [as in] Marsilio Ficino, Bernardino Telesio or even Campanella. . . . If one wishes to express the Renaissance mentality in one phrase, I would propose the formula: *everything is possible*.

On the side of progress, Koyré locates science, logic, and the critical spirit; on the other side he puts the aesthetic spirit, rhetoric, and erudition. He holds the Renaissance guilty of the last three sins and, worse, of the crime of magic, which not only delayed the findings of seventeenth-century science but also eroded the achievements of Aristotelian ontology. Unable to ignore so prominent a figure as Pomponazzi, Koyré puts him among "those who out of a certain duty had to defend Aristotelian ontology, the Paduan Averroists and Alexandrists, . . . contaminated by the spirit of the times; in Nifo and in Pomponazzi you will find the

[7] *Ibid.:* "Quoniam secundum praedicta utrique parti favere videmur, ne inspecturi has nostras qualescunque lucubrationes nos in hac materia indeterminatos existiment, ideo hoc ultimo capite quaedam per modum conclusionum apponere volvi, per quae quid determinate in hac materia sentiamus aperte intelligi possit" (pp. 297–298). See Pine, *Pomponazzi* (cit. n. 2), pp. 114, 236–243; for Agostino Nifo, Alessandro Achillini, and others who connected the Peripatetic tradition with occultism see Thorndike, *History of Magic* (cit. n. 2), Vol. 5, pp. 37–93; Schmitt, *Aristotle and the Renaissance* (cit. n. 2), pp. 98–103; and the articles by Copenhaver, Zambelli, and Zanier cited in note 5.

same magical ontology and the same belief in demonic powers."[8] In other words, it was duty rather than conviction that made Pomponazzi an Aristotelian. The real basis of his ontology was the same Neoplatonic virus that infected Ficino. The lesser point to make about Koyré's analysis is that Pomponazzi most emphatically did *not* believe in demons, at least as a matter of philosophical competence; Koyré misses the key distinction in the treatise *On Spells* between philosophically intelligible *natural* magic and *demonic* magic that must be taken on faith. Despite the enormous amount of work done on Renaissance magic since Koyré wrote in 1949, the force of this distinction may still elude students of the Renaissance whose post-Cartesian instincts make it hard to accept that an intelligent and well-informed person of any period can have taken magic seriously.[9] Another post-Cartesian attitude, however, which blames the Peripatetic system for everything retrograde in philosophy before Hegel, actually runs contrary to the larger point that I wish to question in Koyré. Unlike many other critics, Koyré aligns Aristotelian with post-Cartesian thought on the side of clarity and rigor, distinguishing both from the muddy "magical ontology" to which the Renaissance reverted.[10] If Koyré has his categories straight, the Pomponazzi who assented to magic must have been contaminated by something un-Aristotelian, something woolly, unrigorous, and Neoplatonic.

Pomponazzi's credentials as intellectual hero have tarnished in Koyré's estimation, a judgment that breaks with the party line established in the French tradition since the seventeenth century, when it was promulgated by Gabriel Naudé, a librarian, bibliophile, and polemicist associated with Gassendi, Marin Mersenne, and other "learned libertines." In 1852, in his epochal work *Averroes and Averroism,* Ernest Renan, by way of Naudé, Pierre Bayle, and Jean Pierre Nicéron, claimed that Pomponazzi's ideas were "the living thought of his century," expressed in a "materialism of great simplicity: that the immortality of the soul was invented . . . to keep the people in order; . . . that miraculous effects are only frauds or illusions; . . . [and] that religion was made only for the simpleminded." This Gallic apotheosis of Pomponazzi as enlightened libertine began with Naudé, gathered credit with Renan, and continued in the works of Jean-Roger Charbonnel, Henri Busson, René Pintard, and other twentieth-century scholars.[11] Naudé had good reason to be interested in Pomponazzi's treatment of

[8] Alexandre Koyré, "L'apport scientifique de la Renaissance," in *Études d'histoire de la pensée scientifique* (Paris: Presses Univ. France, 1966), pp. 38–47, on pp. 38–40; see also John M. Headley, "Tommaso Campanella and the End of the Renaissance," *Journal of Medieval and Renaissance Studies,* 1990, *20*:168–169.

[9] Walker, *Spiritual and Demonic Magic* (cit. n. 5), pp. 3–11, 36–59, 75–84, 107–111; Paola Zambelli, "Il problema della magia naturale nel Rinascimento," *Riv. Crit. Stor. Fil.*, 1973, *28*:271–296; Zambelli, "Platone, Ficino e la magia," in *Studia humanitatis: Ernesto Grassi zum 70. Geburtstag* (Munich: Fink, 1973), pp. 121–142; and Copenhaver, "Astrology and Magic" (cit. n. 5), pp. 264–300.

[10] Koyré, "L'apport scientifique" (cit. n. 8), pp. 39–42; for the connection between antioccultism and anti-Aristotelianism in post-Cartesian philosophy see Brian P. Copenhaver, "The Occultist Tradition in Seventeenth-Century Philosophy," in *Cambridge History of Seventeenth-Century Philosophy,* ed. M. Ayers and D. Garber (forthcoming).

[11] Ernest Renan, *Averroès et l'averroisme: Essai historique,* 2nd ed. (Paris: Michel Levy, 1861), pp. 354–358; Jean-Roger Charbonnel, *La pensée italienne au XVIe siècle et le courant libertin* (Paris: Champion, 1919); Henri Busson, *La pensée religieuse française de Charron à Pascal* (Paris: Vrin, 1933); Busson, introduction to translation of Pomponazzi, *Les causes des merveilles* (cit. n. 2), pp. 14–105; and René Pintard, *Le libertinage érudit dans la première moitié du XVIIe siècle* (Paris: Boivin, 1943). For other perspectives see Kristeller, "Myth of Renaissance Atheism" (cit. n. 2);

demons and miracles, for these topics were much in the air when he began his literary career in the early 1620s. For several decades after 1580, episodes of demonic possession from Marthe Brossier in 1598 in the Loire valley to the nuns of Loudun in 1634 had given scandal and caused bloodshed in France. A scare of another kind peaked in the summer of 1623, when posters notified Paris that the invisible Rosy Cross brothers had arrived; a year later chemical debates between Paracelsans and Peripatetics attracted crowds in Paris and worried the authorities who dispersed them. Meanwhile, within a few weeks of the posting of the Rosicrucian placards, Naudé had replied with his mocking *Instruction to France about the Truth of the Story of the Rosicrucian Brethren,* but in the face of hysteria few observers had his wit or the courage of the physician Michel Marescot, who had diagnosed Marthe Brossier as possessed by "much pretense, many natural ailments, none demonic."[12] For Pomponazzi's reputation, the pivotal event of these panicky years was the burning of Giulio Cesare Vanini in Toulouse in 1619. One of the two books that brought Vanini to the stake, his *Wondrous Mysteries of Nature, Queen and Goddess of Mortals,* had appeared in Paris in 1616, and Vanini had cribbed whole sections from Pomponazzi's *On Spells.* Vanini's execution naturally amplified Pomponazzi's ideas, making them all the more horrendous for the enemies that the book had accumulated since before its publication. In France, one of Pomponazzi's most important opponents was Pierre Le Loyer, author of a work entitled *On Ghosts or Apparitions* (1586) that went into a second, enlarged edition just after the turn of the century and became one of Naudé's main targets in his *Apology for All the Great Persons Falsely Suspected of Magic* (1625).[13]

Naudé is best remembered as an unphilosophical skeptic and a powerful critic of occultism. This reputation is accurate enough, as long as one recalls that his *Apology* clears Paracelsus, Agrippa, Peter of Abano, and older figures of *demonic* magic by explaining the wonders attributed to them as effects of *natural* magic, which, like Pomponazzi, Naudé regarded as a legitimate department of

---

Kristeller, "Between the Renaissance and the French Enlightenment: Gabriel Naudé as Editor," *Renaissance Quarterly,* 1979, *32*:41–47; and Richard H. Popkin, *The History of Scepticism from Erasmus to Spinoza* (Berkeley: Univ. California Press, 1979), pp. 87–109.

[12] Gabriel Naudé, *Instruction a la France sur la verité de l'histoire des freres de la roze-croix* (Paris, 1623); Naudé, *Iugement de tout ce qui a esté imprimé contre le Cardinal Mazarin depuis le sixiéme ianvier, iusques à la declaration du premier avril, mil six cens quaranteneuf* (n.p., n.d.), p. 311 (quotation); Busson, *Pensée religieuse,* p. 363; James V. Rice, *Gabriel Naudé, 1600–1653* (Johns Hopkins Studies in Romance Literatures and Languages, 35) (Baltimore: Johns Hopkins Press, 1939), pp. 52–62; Pintard, *Libertinage,* pp. 156–174, 251–270, 442–471; J. S. Spink, *French Free-Thought from Gassendi to Voltaire* (London: Athlone, 1960), pp. 3–26; Paul Arnold, *La Rose-Croix et ses rapports avec la franc-maçonnerie: Essai de synthèse historique* (Paris: Maisonneuve et Larose, 1970), pp. 43, 56, 63–68, 73–89, 115–140, 189–191; Robert Mandrou, *Magistrats et sorciers en France au XVIIe siècle: Une analyse de psychologie historique* (Paris: Seuil, 1980), pp. 153–312; Frances A. Yates, *The Rosicrucian Enlightenment* (Frogmore: Paladin, 1975), pp. 71–80, 91–98, 126–150, 279–280; and Daniel P. Walker, *Unclean Spirits: Possession and Exorcism in France and England in the Late Sixteenth and Early Seventeenth Centuries* (Philadelphia: Univ. Pennsylvania Press, 1981), pp. 1–42. On Naudé see also Brian P. Copenhaver, "The Power of Magic and the Poverty of Erudition" (forthcoming).

[13] G. Naudé, *Apologie pour tous les grands personnages qui ont esté faussement soupçonnez de magie* (Paris, 1625); Giulio Cesare Vanini, *De admirandis naturae reginae deaeque mortalium arcanis* (Paris, 1616); Pierre Le Loyer, *Discours et histoires des spectres* (Paris, 1605); Charbonnel, *Pensée italienne* (cit. n. 11), pp. 302–382; Busson, *Pensée religieuse,* pp. 316–335; and Busson, introduction to translation of Pomponazzi, *Les causes des merveilles* (cit. n. 2), pp. 79–81.

Peripatetic natural philosophy. In other words, the person who started Pomponazzi on his posthumous career as paraclete of French libertinism was an Aristotelian who believed in natural magic and who, less rigorously than Pomponazzi, also allowed that demonic possession might be possible in principle. A famous passage in Naudé's *Apology* describes Aristotle's natural philosophy as having been "rendered greatly suspect and dubious by a swarm of innovators . . . who truly have no other design than to shove aside this philosophy and ruin the great edifice that Aristotle and . . . his interpreters strove to build." Among the swarm of innovators, Naudé named Bernardino Telesio, Francesco Patrizi, Tommaso Campanella, Francis Bacon, Giordano Bruno, and Sebastien Basson, all of them anti-Aristotelians and few of them his particular favorites. He did not list Pomponazzi because he did not distrust him as an innovator, a rebel against Aristotle. He knew that the Peripatetic system was in trouble, but he thought of its decay as a movement of history, not an achievement of philosophy, roughly in the way that Pomponazzi described transformations of religion: Aristotelianism was dying not because it was unsound philosophically but because its time had come in the order of cyclical change.[14] Naudé had learned Peripatetic school philosophy at Paris and at Padua, and during his long stay in Italy he deepened his acquaintance with its last creative practitioners, the best known of whom were Cesare Cremonini (d. 1631) and Fortunio Liceti (d. 1657), both in Pomponazzi's tradition. Naudé's philosophical case against demons in the *Apology* has the same structure as the argument of the treatise *On Spells,* and his admiration for Pomponazzi is explicit in other published works, especially the *Advice on Arranging a Library,* where Pomponazzi appears several times, once with Cremonini, Liceti, Alessandro Achillini, Agostino Nifo, Jacopo Zabarella, and other Aristotelians in a list of "absolutely essential experts."[15]

For our purposes a more revealing and a more evasive text is the *Naudeana,* a collection of table talk published in the early eighteenth century that may or may not record Naudé's actual remarks but has certainly influenced later opinion, down to our own time. Of Italy Naudé was alleged to have said that it "abounds in the kind of people who dig as far as possible into nature and believe nothing more," but also that it was "a land of swindles and superstition, where some believe too little, others too much, and they talk of miracles all the time without reason or truth." Of Liceti the *Naudeana* claim that "he believes only in Aristotle and mocks all the religion of the Italians," of Cremonini that he was

> a great person, . . . one who had been wised up (*déniaisé*) and cured of stupidity, who knew the truth, but in Italy one dare not tell it. All the professors in that country,

[14] Naudé, *Apologie:* "toutes ces veritez naturelles qu'il dit luy avoir esté cognuës sont aujourd'huy renduës grandement suspectes et douteuses par un essain de novateurs qui se grossit de jour à autre sous la conduite de Telesius, Patrice, Campanella, Verulamio, Iordan Brun et Basson, qui n'ont veritablement autre dessein que de donner du coude à cette philosophie et ruiner ce grand bastiment qu'Aristote et plus de douze mil qui l'ont interpreté se sont efforcez de bastir par une si longue suite d'annees" (p. 331). See also Pomponazzi, *De incantationibus,* pp. 169, 293–294; Pine, *Pomponazzi* (cit. n. 2), pp. 252–263; Pintard, *Libertinage* (cit. n. 11), pp. 163–164, 445–446, 451–454, 466–468; and, for Naudé's critique of magic, Copenhaver, "Power of Magic" (cit. n. 12).

[15] Gabriel Naudé, *Avis pour dresser une bibliotheque* (Paris, 1627), pp. 54–56, 75, 80; see also Naudé, *Advice on Establishing a Library,* trans. Archer Taylor (Berkeley: Univ. California Press, 1950); and Pintard, *Libertinage,* pp. 158, 163–164, 168–172, 262.

> especially those in Padua, have wised up . . . and have been disabused of the vulgar errors of the ages, and they have a good understanding of Aristotle's thought, of whose spirit Cremonini is a . . . perfect epitome. . . . In Italy there are quite a few who believe no more than Cremonini: Machiavelli, . . . Cardano, . . . Pomponazzi, . . . Vanini.[16]

What the *Naudeana* say about Naudé's opinion of Pomponazzi is more complicated. First, that he was in "great danger of being burned . . . [and] made an *Apology* for his book that was worse than the book itself. I have never known a philosopher who did not praise Pomponazzi, even those who have written against him. . . . [He was] a good man. . . . No one . . . has been able to refute his arguments." Also that

> he wanted to give a natural reason for the miracle of the revived Lazarus in his book *On Spells*. A physician of Montpellier called La Porta [tried the same thing] around the year 1608, . . . but it's all just stories and empty phrases, pure impieties punishable by fire, by flame and iron. Pomponazzi was an atheist or at least a very dangerous libertine because he was smart. This La Porta was a Jew. . . .[17]

Loose, vicious talk of this sort made good reading for the intellectually mobile on the eve of the Enlightenment, and some critics find that the *Naudeana* reflect nothing but the prejudices of the early eighteenth century, nothing directly about Naudé, much less about Pomponazzi. This last point, certainly, is well taken and must be borne in mind when reading the modern literature from Renan onward, but I find it unlikely that the *Naudeana* tell us nothing about Naudé. Would a French controversialist writing after 1700 have spoken so warmly of late Italian Aristotelianism? I think not. I believe that the *Naudeana* leave us a plausible picture of Naudé as an Aristotelian (though not a very philosophical one) who admired Pomponazzi (with reservations, as the last quotation shows) as a source of late Peripatetic naturalism and as a critic of what he (Naudé) saw as superstition.[18] The critical point is that both Pomponazzi and Naudé were Aristotelians and that neither regarded natural magic as superstitious. For both, natural magic

[16] *Naudeana et Patiniana ou singularitez remarquables prises des conversations de messieurs Naudé et Patin* (Paris, 1701): "L'Italie est pleine et abonde en ces sortes de gens qui penetrent le plus autant qu'il leur est possible dans la nature et ne croyent rien plus" (pp. 6–7). "L'Italie est un païs de fourberie et de superstition, les uns n'y croyent pas assez, les autres y croyent trop, et à toute heure sans raison et sans verité on y suppose des miracles" (pp. 76–77). "Il ne croit qu'en Aristote et se moque de tout la religion des Italiens" (p. 99). "Ce Cremonin étoit grand personnage . . . , un homme deniaisé et gueri du sot, qui sçavoit bien la verité mais qu'on n'ose pas dire en Italie. Tous les professeurs de ce païslà, mais principalement ceux de Padoüe, sont gens déniaisez. . . . Ils doivent étre détrompez des erreurs vulgaires des siécles et bien connoître l'opinion d'Aristote, de l'esprit duquel ce Cremonin est un vrai Tiercelet et parfait abregé. . . . Il y en a bien en Italie qui n'en croyent pas plus que Cremonin: Machiavel, . . . Cardano, Castellanus, Pomponace, Bembe, . . . Vanini" (pp. 115–117). On the *Naudeana, Patiniana, Scaligerana,* and other such collections see Erich Haase, *Einführung in die Literatur des Refuge: Der Beitrag der franzosischen Protestanten zur Entwicklung analytischer Denkformen am Ende des 17. Jahrhunderts* (Berlin: Duncker & Humblot, 1959).

[17] *Naudeana:* "grand danger d'être brûlé . . . [et] fit une Apologie pour son livre qui étoit pire que le livre même. Je n'ai jamais vû philosophe qui n'ait loüé Pomponace, quoi qu'il eut écrit contre lui. . . . [Il était] un bon homme. . . . Personne . . . n'a pû renverser ses raisons" (p. 27). "Pomponace a voulu rendre une raison naturelle du miracle du Lazare ressuscité en son livre *De incantationibus*. Un medecin de Montpellier nommé la Porta, environ l'an 1608 [essayait la même chose], . . . mais tout cela sont des contes verbaque inania, ce sont de pures impietés punissables par le feu, flamma et ferro. Pomponace étoit un athée ou du moins un libertin tres-dangereux parcequ'il avait de l'esprit. Ce Laporta étoit un Juif . . ." (pp. 95–96).

[18] On these points see Kristeller's articles "Myth of Renaissance Atheism" (cit. n. 2) and "Gabriel

was a normal part of Peripatetic natural philosophy and a weapon against credulous supernaturalism.

Educated people in late medieval and early modern Europe believed in magic for reasons of three kinds: *doxographic, empirical,* and *theoretical.*[19] Doxographic reasons were arguments not just from authority but from authority conceived as tradition; the most important versions of the doxography of magic were variants on the ancient theology, wherein Zoroaster or Hermes Trismegistus or sometimes Moses stood at the head of a succession of ancient sages who made magic philosophically credible. Pomponazzi, though he mentions Zoroaster a few times in the treatise *On Spells,* was not much interested in doxography.[20] He was more taken by empirical evidence, but he seems not to have cared that the data he accepted were found in texts, not in direct experience, having been transmitted since antiquity in a literary tradition almost never tested by the people who referred to it again and again. Having barred demonic magic from Peripatetic explanations of puzzling phenomena, Pomponazzi had either to reject the phenomena or to account for them on better Peripatetic principles. "How shall we save experience," he asks, finding it "neither safe nor honest to make the usual claim of those who deny experience, that humans invented these things—like Aesop's fables—to edify the masses, or that they are tricks used by priests to snatch money. . . . With this view I cannot agree because people of the weightiest authority and greatest distinction in learning, both recent and ancient, Greek and Latin and barbarian, affirm that they are entirely accurate."[21] Pomponazzi, like other philosophers since classical times, believed that certain objects—the magnet, the remora, the electric ray, the basilisk, and dozens more—had properties repeatedly confirmed by experience yet unexplained by the elements and sensible qualities of Peripatetic matter-theory. Some, like the basilisk, we might call textual rather than natural objects; but others, like the magnet, are natural objects whose veridical properties fell outside the scope of Peripatetic physics.[22]

Naudé as Editor" (cit. n. 11); I differ only in thinking the *Naudeana* a possible source of evidence for Naudé's views, not for Pomponazzi's or those of any other Italian Aristotelians.

[19] Brian P. Copenhaver, "Natural Magic, Hermetism, and Occultism in Early Modern Science," in *Reappraisals of the Scientific Revolution,* ed. David C. Lindberg and Robert S. Westman (Cambridge: Cambridge Univ. Press, 1990), pp. 261–301; see also Copenhaver, "Scholastic Philosophy and Renaissance Magic in the *De vita* of Marsilio Ficino," *Renaiss. Quart.,* 1984, *37*:523–554; Copenhaver, "A Tale of Two Fishes: Magical Objects in Natural History from Antiquity through the Scientific Revolution," *Journal of the History of Ideas,* 1991, *52*:373–398; and Copenhaver, "Astrology and Magic" (cit. n. 5).

[20] Pomponazzi, *De incantationibus,* pp. 98, 106, 200. See also Daniel P. Walker, *The Ancient Theology: Studies in Christian Platonism from the Fifteenth to the Eighteenth Century* (London: Duckworth, 1972), pp. 1–21; Charles B. Schmitt, *Studies in Renaissance Philosophy and Science* (London: Variorum, 1981), Chs. 1, 2; and Copenhaver, "Natural Magic, Hermetism, and Occultism," pp. 261–270, 285–290.

[21] Pomponazzi, *De incantationibus:* "Verum cum Peripatetici hoc non admittant, quomodo igitur experimenta salvabuntur? . . . Mihi autem non videtur tutum neque sine verecundia dictum quod a plerisque dici solet, haec experimenta negantibus, haec scilicet esse ab hominibus conficta, velut Aesopi apologi, ad plebis instructionem, vel quod sunt sacerdotum aucupia ad subripiendas pecunias. . . . Ego inquam hanc sententiam non approbo quandoquidem viri gravissimi, doctrina eminentissimi, et novi et veteres, tam Graeci quam Latini ac barbari moribus, haec verissima esse affirmant" (pp. 113–114). See also Pine, *Pomponazzi* (cit. n. 2), pp. 104, 236–239; and Zanier, *Ricerche* (cit. n. 2), pp. 4–9.

[22] Copenhaver, "Natural Magic, Hermetism, and Occultism" (cit. n. 19), pp. 275–280; and Copenhaver, "Tale of Two Fishes" (cit. n. 19), pp. 373–381.

Two such objects with enigmatic properties—the remora or *echeneis* ("ship-holder," in Greek) and the torpedo or electric ray—especially caught Pomponazzi's attention. *On Spells* refers to one or both at least five times, and with good reason, for Aristotle, Pliny, Albertus, and many other authorities known and unknown to Pomponazzi had certified and recertified the wondrous faculties of these two fish over the course of two millennia, giving enormous credit to the belief that a ray could stun its victim without direct contact and that a tiny remora could slow or stop the motion of a great ship. Both fish are natural objects, though the remora's identity was more obscure than the torpedo's, and tradition had linked them ever since Pliny treated the two as a pair and as leading examples of magical power. By modern standards reports of the ray's stunning effect were true, though badly explained, while those about the remora as ship-holder were false. But the literary glue binding the two fish together and reinforcing their magic melted away only in the seventeenth century, while the torpedo resisted analysis and encouraged resort to occult qualities even later. Meanwhile, in discussing the production of occult phenomena by natural, nondemonic means, Pomponazzi reasoned that

> if a fish the size of your finger can hold back a loaded ship of two hundred feet or more, driven by wind and oar; if the torpedo-fish can stun at so great a distance with no sensible action; or the basilisk can infect so great an amount of air; and many such kinds of serpents do unbelievable things frequently described by philosophers and physicians; . . . then it seems to me that if demons do this by applying actives to passives, then humans can understand such things naturally as well, from which it also follows that many who were considered magicians and necromancers, like Peter of Abano and Cecco d'Ascoli, may have had no commerce with unclean spirits—indeed, with Aristotle they may have believed that demons do not exist.[23]

Since demonic magic could not help him with these odd cases, Pomponazzi used a theory of natural magic to account for them. The theory that he depended on had three components—which I shall call *psychological, astrological,* and *ontological*—all within the limits of natural action. The elements of this philosophical theory had been explicit since the thirteenth century in the Latin Peripatetic tradition, which Marsilio Ficino had recently enriched with Neoplatonic ingredients in his influential account of natural magic, *On Arranging Life in a Heavenly Manner,* the third of his *Three Books on Life* of 1489. In showing how heavenly

[23] Pomponazzi, *De incantationibus:* "Si enim piscis digitalis quantitatis navim onustam ducentorum pedum et amplius, undique ventis ac remis impulsam, retinere potest; et piscis torpedo ad tantam distantiam stupefacere insensibiliter; regulus sive basiliscus potest tantam aeris quantitatem inficere; et multa talia genera serpentum incredibilia facere, de quibus philosophi et medici multa dicunt; . . . mihi videtur quod si daemones faciunt hoc applicando activa passivis, et talia naturaliter ab hominibus cognosci possunt. Ex quibus ulterius sequitur quod fortassis multi habiti sunt magi et necromantici, ut Petrus Aponensis, Ciclus Esculanus, qui tamen nullum commertium habuerint cum spiritibus immundis—imo fortassis cum Aristotele crediderunt daemones non esse" (pp. 42–43).

On Peter of Abano and Cecco d'Ascoli see Thorndike, *History of Magic* (cit. n. 2), Vol. 2, pp. 874–968. The astrologer Cecco was burned in Florence in 1327, about ten years after the death of Peter of Abano, who, despite close relations with Pope John XXII and other Church authorities, was also attacked for his occultist teachings. Cecco taught at Bologna, Peter at Padua, and both professed an astrological naturalism roughly like Pomponazzi's, though neither seems to have gone as far in excluding demons. Naudé, *Apologie* (cit. n. 13), pp. 386–391, defends Peter.

influences, occult qualities, and powers of the soul could explain many puzzling phenomena, Ficino began with ideas long used by Christian thinkers to insulate magic from demonic evil. This same naturalism challenged orthodox demonology, however. Ficino feared the demons and did his best to keep them out of his magic. But his commitment to Neoplatonism required him—as philosopher and theologian—to admit their existence, and he knew that the ancient Neoplatonists saw them as eventually and inevitably involved in all magical action. If Pomponazzi ever shared Ficino's anxieties about demons, his fears left little trace in the treatise *On Spells,* whose theory of magic, unlike Ficino's, is entirely and aggressively naturalist, expressly intended to make demons redundant.[24]

The first or psychological part of Pomponazzi's natural magic appealed to faculties of the human soul, especially the imagination, which in Pomponazzi's opinion have the power to cause changes not just in the body of the imagining subject but even in other bodies. Physical vapors are the medium connecting subject and object, as in the case of persons thought to have the power to heal others by mere proximity. "I find it possible that a man might have such a [healing] disposition," he maintained,

> just as rhubarb has the power to purge choler. However, rhubarb does not purge choler unless vaporized when activated by natural heat. And so it is that there are people who have such powers in potency, and when the force of imagination and desire works on them actively, such a power becomes active and affects the blood and spirit, which seek release through evaporation and produce such effects. Reason and experience prove what I have said about this mode of action.

It was this psychological part of Pomponazzi's theory of magic that owed most to Ficino, but even here he had a Peripatetic basis, especially the pseudo-Aristotelian *Problemata.*[25] The other two parts of the theory depended on astrological

[24] Michael J. B. Allen, *The Platonism of Marsilio Ficino: A Study of His Phaedrus Commentary, Its Sources and Genesis* (Berkeley: Univ. California Press, 1984), pp. 8–31; Allen, *Icastes: Marsilio Ficino's Interpretation of Plato's Sophist* (*Five Studies and a Critical Edition with Translation*) (Berkeley: Univ. California Press, 1989), pp. 161–200; Carol Kaske and John Clark, introduction to Ficino, *Three Books on Life: A Critical Edition and Translation with Introduction and Notes,* ed. and trans. Kaske and Clark (Medieval and Renaissance Texts and Studies, 57) (Binghamton, N.Y.: MRTS, 1989), pp. 31–70; Copenhaver, "Astrology and Magic" (cit. n. 5), pp. 264–266, 274–288; Copenhaver, "Scholastic Philosophy" (cit. n. 19), pp. 531–546, 549–554; Copenhaver, "Renaissance Magic and Neoplatonic Philosophy: *Ennead* 4.3–5 in Ficino's *De vita coelitus comparanda,*" in *Marsilio Ficino e il ritorno di Platone: Studi e documenti,* ed. Gian Carlo Garfagnini (Florence: Olschki, 1986), Vol. 2, pp. 351–369; Copenhaver, "Iamblichus, Synesius, and the *Chaldaean Oracles* in Marsilio Ficino's *De vita libri tres:* Hermetic Magic or Neoplatonic Magic?" in *Supplementum Festivum: Studies in Honor of Paul Oskar Kristeller,* ed. James Hankins, John Monfasani, and Frederick Purnell (Medieval and Renaissance Texts and Studies, 49) (Binghamton, N.Y.: MRTS, 1987), pp. 441–455; and notes 30 and 33. See also Thorndike, *History of Magic,* Vol. 5, pp. 101, 104; Walker, *Spiritual and Demonic Magic* (cit. n. 5), pp. 107–110; and Pine, *Pomponazzi* (cit. n. 2), pp. 241, 247 (quotation), 252–253, who says of Pomponazzi's two possible lapses (*De incantationibus,* pp. 74, 104–105) on demonic power that they "cannot outweigh the entire argument which is directly opposed to it." For earlier Aristotelian treatments of magic see note 26; and for the less thorough naturalism of Nicole Oresme see Bert Hansen, *Nicole Oresme and the Marvels of Nature: A Study of His* De causis mirabilium *with Critical Edition, Translation, and Commentary* (Toronto: Pontifical Institute of Medieval Studies, 1985), pp. 50–85.

[25] Pomponazzi, *De incantationibus:* "Et possibile est apud me quod homo habeat talem dispositionem qualem diximus, sicut habet rhabarbarum virtutem expurgandi coleram. Rhabarbarum tamen

influences and occult qualities, forces well established in normative scholastic sources, including works by Albertus and Aquinas frequently cited in the treatise *On Spells*. Albertus, Aquinas, and many other Christian philosophers influenced by Aristotle and his Greek and Muslim commentators believed that generation-and-corruption, the fundamental process of change in terrestrial objects, was caused or at least mediated by the planets and stars, sources of the second or astrological set of magical powers in Pomponazzi's scheme.[26] As he saw it, all earthly change explicable by philosophy occurs because God's simple power becomes diversified through the celestial Intelligences, which move the stars and planets, whose influences in turn account for physical variety and change on earth below. "As far as Aristotle is concerned," he argued,

> even though God is the cause of all things, nothing can happen in this sublunary world unless the celestial bodies mediate as the instruments required, *per se* and necessarily, for the making and preservation of these sublunary things. . . . Aristotle believed . . . that a completely immaterial agent can have no effect of any kind on these lower things without the mediation of a body between the eternal things and those subject to generation and corruption, as is obvious from *Physics* 8, *On Generation* 2 and countless other places.[27]

Scholastic authorities also accepted the doctrine of occult qualities as part of the larger hylemorphic physics and metaphysics of form (*hulê*) and matter (*morphê*),

---

non expurgat coleram nisi fuerit activatum per calorem naturalem et in vaporem fuerit conversum. Sic contigit tales esse homines qui habeant huiusmodi vires in potentia, et per vim imaginativam et desiderativam cum actu operantur, talis virtus exit ad actum et afficit sanguinem et spiritum, quae per evaporationem petunt ad extra et producunt tales effectus. Quod autem dictum est de isto modo ratione et experimento probatur" (pp. 43–44). See also p. 140 and ps.-Aristotle, *Problems* 953a10–5a40; p. 31 and Marsilio Ficino, *Theologia platonica* 13.1, in *Opera omnia* (Basel, 1576), pp. 284–286; Walker, *Spiritual and Demonic Magic,* p. 107; and Pine, *Pomponazzi,* pp. 243–245.

[26] Pomponazzi, *De incantationibus,* pp. 23–24, 98, 129, 146, 194, 199–200, cites Albert's *On Minerals, On Animals,* and other works; most citations of Thomas, on pp. 24, 81, 104, 231–234, 304, 317, seem to refer to the questions (50–64, 105–117) of the first part of the *Summa Theologiae* on angels, demons, miracles, and related matters, but Pomponazzi also cited, contradicted, and questioned the authenticity of Thomas's genuine treatise *On the Occult Works of Nature,* for which see J. B. McCallister, *The Letter of St. Thomas Aquinas De occultis operibus naturae ad quemdam militem ultramontanum* (Washington, D.C.: Catholic Univ. Press, 1939). See also Thorndike, *History of Magic* (cit. n. 2), Vol. 2, pp. 548–592, 602–615, Vol. 5, p. 106; Walker, *Spiritual and Demonic Magic,* p. 107; Copenhaver, "Scholastic Philosophy" (cit. n. 19), pp. 531–546, 549–554; Copenhaver, "Astrology and Magic" (cit. n. 5), pp. 264–266, 274–288; Aloïs van Hove, *La doctrine du miracle chez Saint Thomas et son accord avec les principes de la recherche scientifique* (Paris: Gabalda, 1927); Thomas Litt, *Les corps célestes dans l'univers de Saint Thomas Aquin* (Louvain: Publications Universitaires, 1963); John M. Riddle and James A. Mulholland, "Albert on Stones and Minerals," in *Albertus Magnus and the Sciences: Commemorative Essays,* ed. James A. Weisheipl (Toronto: Pontifical Institute of Medieval Studies, 1980), pp. 203–234; Dorothy Wyckhoff, introduction to Albert, *Book of Minerals,* trans. Wyckhoff (Oxford: Clarendon Press, 1967), pp. xxvi–xli; and James J. Scanlan, introduction to Albert, *Man and the Beasts: De animalibus* (*books 22–26*), trans. Scanlan (Medieval and Renaissance Texts and Studies, 47) (Binghamton, N.Y.: MRTS, 1987), pp. 1–56.

[27] Pomponazzi, *De incantationibus:* "Quanquam Deus sic sit omnium causa, penes tamen Aristotelem nihil operari valet in hoc sublunari mundo nisi mediantibus corporibus coelestibus tanquam instrumentis per se et necessario requisitis ad factionem et conservationem horum sublunarium. Suppositio patet ex multis, tum quoniam Aristoteles existimavit hunc ordinem essentialiter esse et per se, et quod ab agente omnino immateriali non potest in haec inferiora aliquis effectus provenire nisi mediante corpore medio inter aeterna et generabilia et corruptibilia, ut patet 8 *Physicorum,* 2 *De generatione* et infinitis aliis locis" (p. 120). Among the "other places" see Aristotle, *Physics* 266a10–7b26, *On Generation* 336a14–7a15; see also Pine, *Pomponazzi,* pp. 235–237, 252–253.

substance and quality, the key ideas behind the third or ontological component of Pomponazzi's theory of natural magic. Aquinas devoted a special work to them, *On the Occult Works of Nature,* and Albertus discussed dozens of occult properties in his works *On Minerals, On Animals,* and elsewhere.[28]

Since his book *On Spells* replied to a physician who had asked if demons can cause magical cures, Pomponazzi began by showing that demons, as spiritual beings, cannot heal bodies directly. All they can do is "apply actives to passives," that is, they must bring active healing objects into contact with the sick bodies affected by them, according to patterns of natural action. Pomponazzi makes the idea absurd by imagining demon-medics flitting about with their little black bags, "like surgeons or druggists, carrying packets or satchels or tiny sacks full of plasters and potions."[29] Then he sorts curative substances into three classes: first, those that directly and perceptibly (*manifeste*) change the sensible qualities of bodies, making them more or less hot, cold, wet, or dry, as when fire heats water; second, those that do the same thing perceptibly but indirectly, by means of a vapor, as when rhubarb purges a choleric ailment; and third, those that change bodies by "an occult and invisible alteration" without effect on manifest qualities, as when a magnet attracts iron. The magnet, explains Pomponazzi, works without change either in its own manifest qualities, whether primary or secondary, or those of the iron; it acts

> through an unknown insensible quality, as daily experience teaches and as happens likewise in almost an infinity of cases. This hypothesis needs no proof because all universally recognize it. . . . There are occult powers almost without end. . . . Some who have knowledge of these actions . . . cause certain effects which ordinary people believe to come from God or angels or demons when they see them but can find no cause for them because a cause of this kind works imperceptibly. . . . Hence, if someone sees the fish *echeneis,* half-a-foot long, holding back a ship of two hundred feet with all its gear, would he not suppose it done by the power of demons or by a divine force? . . . If someone goes numb whose finger touches a spear in contact with an electric ray, who would think that this happens because of the fish's nature when no means of producing such an effect is apparent? And yet these claims are quite true and based on experience, as Albertus reports . . . from the opinion of Pliny and other authors of the greatest repute.[30]

[28] See note 26.

[29] Pomponazzi, *De incantationibus:* "daemones ipsi . . . sanitatem non inducunt immediate alterando, sed applicando activa passivis. . . . igitur daemones secum ferunt pyxides aut peras aut sacculos plenos emplastris et succis, veluti chirurgi et pharmacopolae" (pp. 18–20). "Daemones hoc facere non poterunt nisi applicando activa passivis quae naturaliter operantur" (p. 39).

[30] *Ibid.:* "Supponamus . . . omnia quae concurrunt ad artem medicam, imo ad omnes fere operationes humanas, corpora alterare aut immediate per qualitates manifestas manifeste alterando corpora in qua agunt, aut mediate alterare quoniam convertuntur in vaporem quae sic conversa postea alterant corpora, aut aliquando alterare per occultam alternationem et invisibilem. . . . Exemplum tertii, sicut magnes attrahit ferrum, neque per qualitatem unam de primis neque per unam de secundis neque quoniam magnes vertatur in vaporem, sed per qualitatem insensibilem quae ignoratur, sicut docet quotidiana experientia et consimiliter apparet fere in infinitis. Haec suppositio non indiget probatione quoniam ab omnibus universaliter conceditur. . . . Et sunt fere infinitae virtutes occultae. . . . Aliqui habentes cognitionem istorum sic operantium . . . inducunt aliquos effectus quos vulgares, videntes et nescientes reducere in causam (quoniam insensibiliter operetur huiusmodi causa), credunt talia fieri aut a Deo [a]ut ab angelis aut a daemonibus. . . . Unde si quis videret echeneidem de genere piscium semipedalis longitudinis detinere navim ducentorum et amplius pedum cum suis armamentis, nonne existimaret hoc fieri vi daemonum aut divina virtute? . . . Quis, si lanceam a pisce torpedine tactam digito attingeret ac in stuporem incideret, hoc ex natura piscis factum existimaret, cum ibi non ap-

Behind this brief account of *natural* magic properties, so certain in Pomponazzi's mind that it needed no proof, there was an elaborate physics and metaphysics of occult qualities and substantial forms that is not at issue in the treatise *On Spells*.[31] Pomponazzi took it for granted. On the other hand, he emphasized the *epistemic* consequences of magic's place in the larger arena of Peripatetic natural philosophy: that magic is "in itself a real practical science (*scientia*), dependent on natural philosophy and astrology, just like medicine and many other sciences; it is intrinsically good and a perfection of the intellect." If anyone could really learn magic from demons or use it for evil, the learning and the practice would be sinful; the magic would be illicit in origin and application. But the theory of magic is blameless in itself, though the principles of a licit theory of natural magic may be applied for good or bad results. In other words, while recognizing its practical dangers, Pomponazzi gave magic the same epistemic and moral status as medicine, another pragmatic science used for good or ill and one that motivated most of the curriculum at the universities where he taught.[32] He had to make natural magic a part of natural philosophy in order to conclude that "if anyone considers the occult and wondrous works of nature, the powers of the heavenly bodies, God, and the intelligences as they watch over all lower and human things, there is no need for demons." Celestial forces, imagination, and occult properties can explain any phenomenon that otherwise befuddles natural philosophers yet lies within the scope of their calling.[33]

Pomponazzi's occultism was a central feature of his Aristotelian naturalism, but Koyré felt obliged to explain it away as something impure in a Peripatetic ontology normally free of such pollution. Koyré knew, of course, that Aristotle had little to say about magic, demons, or astrology as such—a silence that Pomponazzi himself addressed—and he was right to think that Pomponazzi borrowed from Ficino and from Neoplatonic commentators when he corrected Aristotle. But Koyré seems not to have recognized that Aristotelian philosophy as Pomponazzi learned and taught it, without reference to any extra-Peripatetic influence, had long since invented its own theory of magic, whose elements were inferences from texts of the Aristotelian corpus, many of them cited in the treatise *On Spells*.[34] If I may suggest an analogy: Aristotle never heard of the Chris-

---

pareat modus talis effectus? Et tamen haec verissima sunt et experta, ut . . . refert Albertus ex sententia Plinii et aliorum gravissimorum authorum" (pp. 21–24). See Pine, *Pomponazzi* (cit. n. 2), pp. 241–245; and note 24.

[31] For a summary of the theory see Copenhaver, "Natural Magic, Hermetism, and Occultism" (cit. n. 19), pp. 270–275.

[32] Pomponazzi, *De incantationibus:* "Nemini dubium est ipsam in se esse veram scientiam factivam et subalternatam philosophiae naturali et astrologiae, sicut est medicina et multae aliae scientiae. Et in se est bona et intellectus perfectio" (pp. 73–74). See also pp. 63–64, 75–79, 199–200; and Pine, *Pomponazzi,* pp. 245–246.

[33] Pomponazzi, *De incantationibus:* "Si quis mirabilia et occulta naturae opera consideraverit, virtutes corporum coelestium, Deum et intelligentias humana et omnia inferiora curantes, nihil opus esse daemonibus" (p. 198). See also pp. 19–20, 98, 105–106, 134–136, 148, 298–299.

[34] Pomponazzi, *De incantationibus,* pp. 205–206. References to particular Aristotelian texts in *On Spells* are sometimes hard to specify, but the following tentative correspondences between Pomponazzi's citations and the relevant parts of the corpus may be useful. For physical and metaphysical principles see Pomponazzi, *De incantationibus,* pp. 107, 112, 115, 120, 133, 143, 298, 300, 309, 315; and Aristotle, *Physics* 266a10–7b26; *Metaphysics* 1071b3–6a4; *On the Heavens* 269b12–70b14, 286a3–b9, 292b25–3a14; *On Generation* 336a14–7a35; *On the Soul* 434a27–5b26. For methodological principles see pp. 106, 131, 325; and *Metaphysics* 982a4–10; *Topics* 119b16–30, 151b11–16. For psychology and

tian eucharist, but no one is astonished to learn that Peripatetic thinkers accounted for certain features of the eucharistic mystery in Aristotelian terms; likewise, a good case can be made that Aristotle never held the immortality of the individual human soul, but we are unsurprised to find that Christian philosophers found Aristotelian reasons to believe themselves immortal. The same is true of magic, whose theoretical basis centered on characteristic Peripatetic principles also used in philosophical defenses of the real presence and the personal immortal soul—the hylemorphic doctrine of substantial forms and accidental qualities.

The subtle Descartes, Naudé's contemporary, attacked the Peripatetic system on many fronts, but in dealing with the sacrament of the altar and the fate of the human soul he had either to contrive a peace with its physics and metaphysics or to leave the contest undecided. Retaining some vestige of hylemorphism to preserve the soul or the sacrament was common among seventeenth-century thinkers who for other purposes dispensed with Aristotle's forms and qualities. After Mersenne, Gassendi, and Descartes, however, no one wanted to save magic with these Peripatetic principles. On the contrary: seventeenth-century opponents of traditional natural philosophy belittled Aristotelian forms and qualities just because they were the main theoretical grounds for a belief in magic that was now despised. When the soul or the eucharist was at stake, pugnacious anti-Aristotelians retreated; otherwise, Peripatetic natural philosophy was fair game, and it was weakest in its attachment to natural magic. Post-Cartesian critics ended magic's career as an object of serious philosophical inquiry, but—unlike Koyré—they understood that Aristotelianism itself, not some miasma from Neoplatonism or elsewhere outside the system, was the real disease. The enemies of magic in early modern Europe who finally deprived it of philosophical interest—Mersenne, Gassendi, Descartes, Thomas Hobbes, Kenelm Digby, Robert Boyle, John Locke, and others—found Ficino and the Neoplatonists either irrelevant or else powerless. Peripatetic natural magic was the greater problem by far.[35]

The natural magic earnestly combated and ultimately stripped of its philosophical cover by these thinkers was precisely the natural magic that Pomponazzi so hotly defended at the beginning of the previous century by making it a department of Peripatetic natural philosophy. What does it do to our conception of Renaissance natural philosophy to learn that one of its great minds and famous heroes used traditional Aristotelian ideas to protect a belief now as discredited as

---

imagination see pp. 28, 30, 140–141; *Movement of Animals* 700b15–25; *Problems* 886a25–7b7, 953a10–5a40; *History of Animals* 631b5–19. For magical objects and phenomena see pp. 43, 179–180, 194; and *History of Animals* 505b18–23, 577a9–14, 596b9–10, 608b14–10a35, 612a2–5. For magical practices see pp. 108, 140–141; and *History of Animals* 522a13–19, 577a9–14, 605a2–7, 608b26–30, 616b19–24; *Rhetoric* 1407b1–5; *Problems* 953a10–5a40. On Aristotle and occultism see also Copenhaver, "Astrology and Magic" (cit. n. 5), p. 287; and notes 3, 8, 9, 10.

[35] Marin Mersenne, *L'impieté des deistes* (Paris, 1624), Vol. 2, pp. 450–455; Pierre Gassendi, *Opera omnia* (Lyon, 1658), Vol. 1, p. 273, Vol. 2, p. 114; René Descartes, *Oeuvres* (Paris: Vrin, 1964–1976), Vol. 3, pp. 420–421, 493, 505, 508; Robert Boyle, *The Sceptical Chymist* (London: Dent, 1964), p. 200. On Descartes and the eucharist see Jean-Robert Armogathe, *Theologia Cartesiana: L'explication physique de l'eucharistie chez Descartes et Dom Desgabets* (The Hague: Nijhoff, 1977); Richard A. Watson, "Transubstantiation among the Cartesians," in *Problems of Cartesianism*, ed. Thomas M. Lennon, John M. Nicholas, and John W. Davis (Montreal: McGill-Queens Univ. Press, 1982), pp. 127–148; and D. M. Clarke, *Occult Powers and Hypotheses: Cartesian Natural Philosophy under Louis XIV* (Oxford: Oxford Univ. Press, 1989). For the fate of the occult philosophy in the seventeenth century see the chapter by Copenhaver cited in note 10.

magic? If we picture the Renaissance as located on a vector of history whose extremes are well described as "primitive" and "modern," and if we accept the account of magic (popular since James Frazer's *Golden Bough*) as a distinctly primitive belief, we will find it hard to bring Pomponazzi into focus.[36] If we see Pomponazzi as having lived in a Renaissance whose historical mission was to destroy Peripatetic philosophy, our image becomes even more distorted. But if we think of the Renaissance as characterized by a new confidence in ancient sources that included a richer and more authoritative body of Aristotelian texts and commentary, and if we understand that Cartesian, Hegelian, Victorian, Foucauldian, and other historiographies, anthropologies, and archaeologies are as contingent as the visions of Joachim of Fiore, as apt to misconceive the past as Joachim was unfit to predict the future, then we may be able to look back at Pomponazzi's Renaissance a little more clearly and to accept what was true about it whether we like it or not: that there was a Renaissance of natural philosophy that embraced Aristotelianism and at the same time encouraged natural magic.

The Renaissance as a period of Western history is an old, even an old-fashioned conception, which owes much of its celebrity to Jacob Burckhardt's great work of 1860 and some of validity to older convictions of Renaissance people that theirs was an age of rebirth. Burckhardt, who wrote a chapter called "The Discovery of the World and of Man," certainly believed that changing views of nature contributed to this self-conscious revival of Greco-Roman antiquity as alternative to medieval culture, but, lacking a mature history of science, he had to apologize for the little he could actually say about medieval or postmedieval inquiry into nature.[37] By the time Michel Foucault wrote *Les mots et les choses* in 1966, the history of science had come a long way since the days of Pierre Duhem and George Sarton, not to speak of Burckhardt. Foucault's famous book bears on the present inquiry because its second chapter attempts to establish the epistemic character of the Renaissance (as distinct from what Foucault calls "the Classical Age") through an account of magic and natural history, by way of such late Renaissance authors as Ulisse Aldrovandi, Girolamo Cardano, Oswald Croll, Giambattista Della Porta, Paracelsus, and others credibly related within loose limits of chronology. Foucault puts Aldrovandi, his colleagues, and a learned magic on one side of an epistemic divide; on the other side are Descartes, his successors, and magic dislodged from the world of learning.[38]

Between Aldrovandi and Descartes, after the linguistic pathologies of Paracelsus but before the semiotic therapies of the *Port-Royal Logic,* Foucault traces the wanderings of Don Quixote across the endless plain of his own text, where the mournful hero "reads the world to prove his books" at a time when "writing had ceased to be the prose of the world," the epistemic condition that Foucault calls

[36] G. E. R. Lloyd, *Demystifying Mentalities* (Cambridge: Cambridge Univ. Press, 1990), pp. 39–72, presents an excellent analysis of the methodological problems involved in writing the history of magic and science.

[37] Jacob Burckhardt, *The Civilization of the Renaissance in Italy,* trans. S. G. C. Middlemore (London: Penguin, 1990), pp. 185–229, esp. p. 192; see also Wallace K. Ferguson, *The Renaissance in Historical Thought: Five Centuries of Interpretation* (Cambridge, Mass.: Riverside, 1948), pp. 1–67.

[38] Foucault, *Les mots* (cit. n. 1), pp. 32–59, 64–65. Gary Gutting, *Michel Foucault's Archaeology of Scientific Reason* (Cambridge: Cambridge Univ. Press, 1989), which connects Foucault with Gaston Bachelard and Georges Canguilhem, will be an excellent introduction for historians of science.

"the negative of the world of the Renaissance." The inverse state of affairs (also mandarin joy, in alluding to a title by Maurice Merleau-Ponty) is what Foucault means by calling the second chapter of his book "La prose du monde," though it is mainly about magic.[39] Between the Renaissance and the Classical Age he locates an epistemic shift from a ternary construct called *resemblance* to a binary order of *representation*. He treats magic as a belief tenable only within the former epistemic framework of resemblance.[40] What he means by resemblance is a beautiful, intricate, and orderly structure of ideas thinly based in the texts that he cites. I know of no Renaissance authority—not even Paracelsus, Croll, and others who propagated the doctrine of signatures—whose views on magic were guided by the delicate, expensive ship that sails so gracefully through Foucault's second chapter. In fact, I know of no early modern writer who was even aware *as such* of the elaborate construct that Foucault describes, though its ingredients were common knowledge. By contrast, there were other ideas and systems of ideas, equally lovely and just as involved, of which almost all Renaissance thinkers were cognizant and which they often invoked to explain magic. One such construct was the hylemorphic theory of substance and quality as understood by Pomponazzi and other late Latin Peripatetics. To be fair to Foucault, he seems not to have cared much about conscious declarations of this kind. At the end of his chapter on Classical theories of language, for example, he disclaims any intention of "doing a history of seventeenth and eighteenth century grammatical conceptions, [or] of . . . calculating the common denominator of opinions," hoping rather to define "the point of departure from which it was possible to have opinions, of one sort or another, about language."[41] The aims of Foucault's archaeological project, even in 1966, were not to excavate an intentional past. But, to be fair to the past, it must be said that what he wrote about magic was more like architecture than archaeology, more beautiful than true. His account of Renaissance magic is hard to falsify because it so elegantly transcends the texts whose epistemic basis he wants to uncover.

Foucault claims that the Renaissance expressed the episteme of resemblance through four figures: *conformity, imitation, analogy,* and *sympathy,* which I shall represent as **C, I, A,** and **S.** *Conformity* is a resemblance effected by place or proximity, connecting things as links in a chain are bound to one another. *Imitation* requires no contact or adjacency, permitting one thing to mimic another at a distance, like an object and its reflected image. *Analogy* depends on like relations of parts; two things are analogous because they have similar structures or assemblages. *Sympathy* is the most powerful of the four modes of resemblance because it is a force as well as a fact; it is the likeness that begets more likeness. Comparing early modern thought to a speaking person, Foucault maintains that each of the four resemblances acted as a *signature* whose purpose was to mark a *similitude.* The signature that marks and the similitude marked are both *resemblances;* in fact, both are the same four resemblances named above. To embody this con-

[39] Foucault, *Les mots,* pp. 35–36, 41, 47, 60–61, 72–75; and Maurice Merleau-Ponty, *The Prose of the World,* ed. C. Lefort, trans. J. O'Neill (Evanston, Ill.: Northwestern Univ. Press, 1973).

[40] Foucault, *Les mots,* pp. 48, 57–59.

[41] *Ibid.,* p. 135. For a current work on the doctrine of signatures see Massimo Luigi Bianchi, *Signatura rerum: Segni, magia e conoscenza da Paracelso a Leibniz* (Rome: Edizioni dell' Ateneo, 1987).

struct Foucault uses the walnut, which Renaissance people considered a good cure for ailments of the head. How did physicians know this secret? They read it in the prose of the world, in a three-term linguistic transaction. The third and ultimate term is a meaningful expression: that walnuts cure heads. The walnut, a natural object, is a signature, the first clue toward meaning. The hard shell covering its involuted meat resembles the brain enclosed by the cranium: this resemblance is a similitude, the second indicator of meaning. Walnuts and other such natural tokens resemble what they finally mean: in this case, the curative affinity of nut and head is first betokened by the cranial nut and then discovered by the homology between nut and head. The nut (a signature) marks the nut/head likeness (a similitude), which uncovers the nut/head affinity (a meaning). Nature is an open book. What you see is what you read, and both are the world. Knowledge, contrary to post-Cartesian understanding, comes neither from rational demonstration nor from empirical observation. It comes from reading the relation of one order of language, divinely written in signatures, to a higher order, likewise inscribed in similitudes, and leading finally to God-given meanings. But the meanings themselves, like the walnut's healing affinity for the head, are matters of likeness. The hermeneutic recapitulates the semiotic, but, according to Foucault, a gap to accommodate nature remains between signs and the meanings they (dis)-cover.[42]

Without trying to resolve the problems of semantic and syntactic hierarchy in this compressed account of a long, subtle argument, let me confess that I do not know what Foucault means even in grosser terms by this claim about nature, meaning, and signification in the Renaissance. In finer detail, he arranges the four modes of resemblance as similitudes in descending order of potency and as marked by the same modes taken as signatures, as follows, so that the similitude in the first column is marked by the signature appearing on the line below it in the second column, sympathy by analogy, analogy by imitation, and so forth:

<table>
<tr><th>Similitudes</th><th>Signatures</th></tr>
<tr><td>S</td><td></td></tr>
<tr><td>A</td><td>A</td></tr>
<tr><td>I</td><td>I</td></tr>
<tr><td>C</td><td>C</td></tr>
<tr><td></td><td>S</td></tr>
</table>

When Foucault said that the Renaissance imposed semiotics on hermeneutics, he meant that the finding and understanding of signs as signs (semiotics) overlapped the discovery of what signs mean (hermeneutics) because the objects of both enterprises (signatures for semiotics, similitudes and meanings for hermeneutics) are modes of resemblance. The table above represents this superimposition by showing similitudes lined up with signatures for modes **A**, **I**, and **C**; it represents displacement between the two orders of resemblance by showing **S** out of align-

[42] Foucault, *Les mots*, pp. 32–42 (on the four figures), 42–45, 54. Foucault gives two of the terms in Latin (*convenientia*, rendered as *convenance* in French, here *conformity;* and *aemulatio*, *émulation* in French, here *imitation*), two in French (*analogie* and *sympathie*); the hardest to express is *convenientia/convenance*, whose etymology in Latin or French supplies a sense of "coming together" that escapes the English *conformity*, but *convenience* in English seems even less appropriate.

ment. The misalignment is where Foucault puts nature, between semiotics as the search for signs and hermeneutics as the quest for meanings.[43] Again, I do not know what he means by putting nature between signification and interpretation.

Clearly, however, he believed that the intricate order expressed in this lopsided table of resemblances was the core of its epistemic force. Belief in magic, which he seems to have understood in the manner of nineteenth-century anthropology as necessarily involving some notion of likeness, was inevitable for Renaissance thinkers because thought itself was then a matter of relating various orders of resemblance. Magic ceased to be a cultural necessity in the seventeenth century, when binary representation of signifieds by signifiers displaced ternary resemblance among signatures, similitudes, and meanings as the dominant episteme. Erudition lost its grip on the high culture at the same time, for the same reason. The Renaissance, according to Foucault, had seen no difference between *marques visibles* and *mots lisibles,* between signs that God planted in nature at the creation and words that authors left in texts (biblical or classical) in antiquity. Both were taken as natural legible objects, sometimes as markers, sometimes as meanings, linked in a network of universal correspondence centered on the human microcosm. Ancient erudition recorded this magical world view and thereby sustained it; moreover, the texts exhumed by humanists actually embodied the same resemblances given in creation. Before pride confounded speech at Babel, language had been a transparent medium of resemblance, its words as easy to read as cracking a nut. If humanism came closer to this primordial clarity by resurrecting antiquity, its powers were indeed magical.[44]

Like Henri Gouhier, Foucault was right to connect the Renaissance culture of erudition with the postmedieval revival of learned interest in magic.[45] But his view of Renaissance culture as epiphenomenal to language is at best an incomplete explanation of the rise and decline of magic, at worst a misleading one. It is misleading if one takes Foucault's prose of the world as corresponding in any strong way to intentional or conscious views of early modern thinkers who worried philosophically about magic—misleading in a sense that cannot be charged, for example, against those large features of Peripatetic physics and metaphysics that Pomponazzi and others connected with magic. To put it another way, Foucault's analysis of Renaissance magic can tell us little about some of its leading critics and proponents. Naudé, for example, makes a poor fit between Foucault's Renaissance and his Classical Age, but he was surely no Don Quixote. His humanist education came at a time when the multiplication of ancient and contradictory authority led naturally to skepticism, not to more tilting at the windmills of resemblance. Yet skepticism could not liberate him from the occultist tradition as long as he clung to the crumbling edifice of Aristotelianism. It was also as an Aristotelian, not as any kind of humanist, that Pomponazzi had made his case for natural magic. Granted, some of the evidence that he took seriously, such as the old tale of two fishes, came to him by way of classical tradition, but

[43] *Ibid.*, pp. 43–45.

[44] *Ibid.*, pp. 47–54, esp. p. 48; see also note 40.

[45] Gouhier, *Les premières pensées de Descartes* (cit. n. 1). For a different point of view on magic and erudition in Foucault see George Huppert, "*Divinatio et eruditio:* Thoughts on Foucault," *History and Theory*, 1974, *13*:191–207.

the same cannot be said of the physical and metaphysical space from which he chased the demons while leaving plenty of room for natural magic and astrology. To call Pomponazzi's magic a relic of classicism, one must say the same of Aquinas, Albertus, and other medieval authorities, which leaves us reading a much different prose than Foucault attributed to the Renaissance. With apologies to General Bosquet and the heroes of the Light Brigade, I can only conclude by saying of Foucault's account of magic: *C'est magnifique, mais ce n'est pas l'histoire . . . ni l'archéologie*.

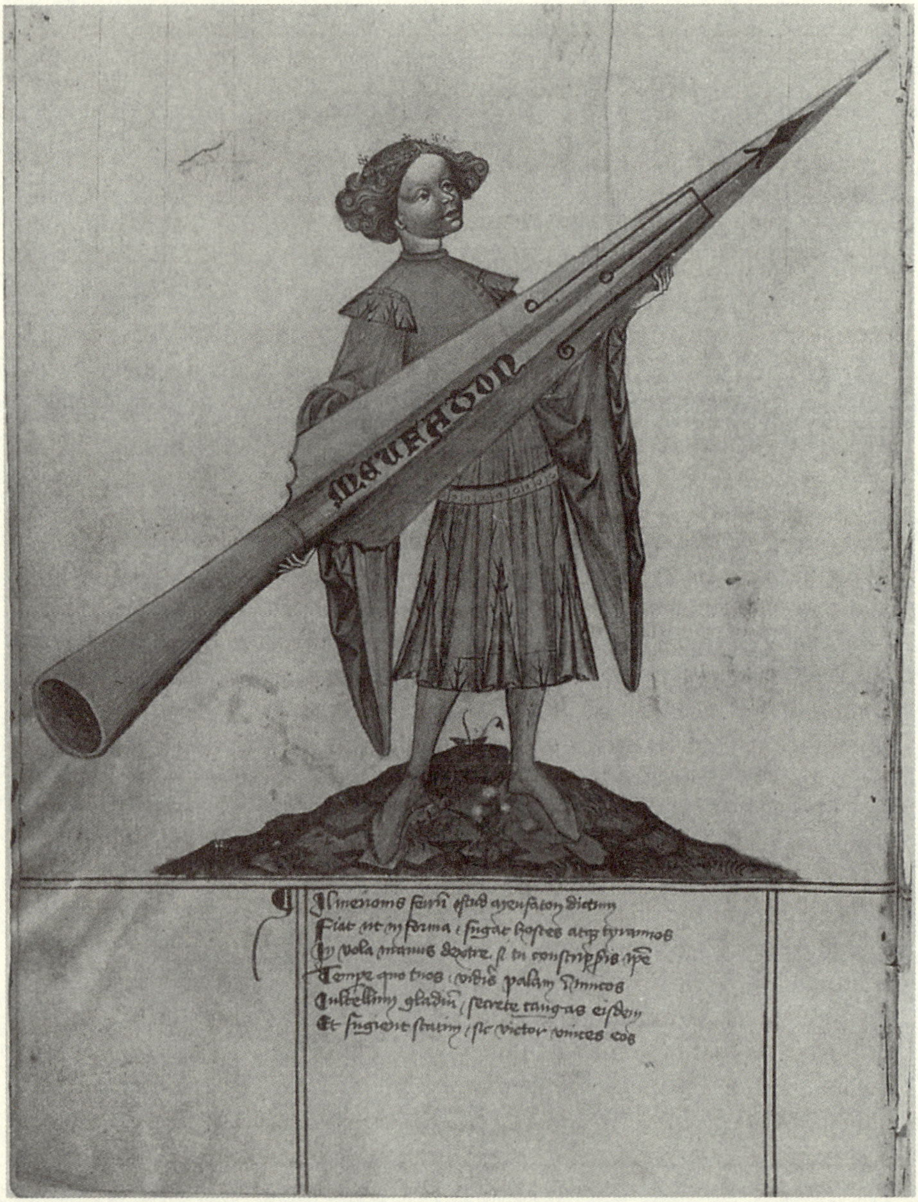

Alexander of Macedon holding a rocket. From Conrad Kyeser, Bellifortis, *Cod. phil. 63, fol. 11b. (Courtesy of the Niedersächsische Staats- und Universitätsbibliothek, Göttingen.) Kyeser venerated Alexander of Macedon as a commanding military leader from antiquity. Here he shows Alexander holding a powerful rocket inscribed with the mysterious, magically powerful letters* MEUFATON. *Whether or not* MEUFATON *stands for Methone (353 B.C.), a king conquered by the young Alexander, as one scholar has suggested, Kyeser depicts an Alexander endowed with technological and magical power. The representation of a powerful ruler handling military hardware brings together* praxis *and* techne *in a way that would have been contrary to ancient notions of* praxis.

Power, Patronage, and the Authorship of *Ars*

From Mechanical Know-how to Mechanical Knowledge in the Last Scribal Age

By Pamela O. Long

ABSTRACT

"Power, Patronage, and the Authorship of *Ars*" is a study of books on the mechanical arts that appeared in the fifteenth century from Italy and south Germany. At this time a significant proliferation of such books occurred, some profusely illustrated, others exclusively textual. They include Latin books on military apparatus and other kinds of machines, German-language codices on gunpowder artillery and machines, humanist treatises on painting, sculpture, architecture, and the military arts, and vernacular writings on such subjects by practitioners. The author argues that the patronage that encouraged the creation of such books developed because of a new close alliance between political and military *praxis* and the mechanical arts. She suggests that such authorship elevated the status of the mechanical arts by explicating them in writing, rationalizing them, and associating them with ancient traditions. They were thus prepared for appropriation by the new experimental philosophy of the next era.

MANUSCRIPT BOOKS ON THE MECHANICAL ARTS proliferated rapidly in fifteenth-century Europe, particularly in south Germany and north and central Italy. Despite some important ancient and medieval exemplars, there is simply no precedent for the great number and variety of such codices that appeared from the early years of the

The research for this article was supported by National Science Foundation grant DIR-9112729, by a fellowship from the Dibner Library of the National Museum of American History, Washington, D.C., and by a National Endowment for the Humanities Fellowship at the Folger Shakespeare Library. Aspects of this research were presented at the Folger Library, the Society for the History of Technology, London, the Departmental Colloquium at Johns Hopkins University, the New York Academy of Sciences, the Graduate Program in Science and Technology Studies at Virginia Polytechnic Institute and State University, the Renaissance Society of America in New York, and as the Guy F. Goodfellow History Lecture at Washington College, Chestertown, Maryland, for which I especially thank Nate Smith. I am grateful to members of the audiences, colleagues, and friends whose discussions and comments have greatly improved this study. For their careful and critical reading of the entire manuscript, I thank especially Priscilla Long, Gabrielle Spiegel, the anonymous referees, and the editor of *Isis*.

fifteenth century until its end. Both university-educated humanists and artisan-trained practitioners created treatises, some chiefly pictorial, others primarily textual, on constructive and other processes in the material world. Topics included gunpowder and artillery; machines, such as mills, pumps, fountains, and cranes; machine parts, particularly gears and pulleys; vehicles, including war carriages, carts, cars, and boats; and numerous other devices. Treatises also appeared concerning the ornamental and constructive arts, especially painting, sculpture, and architecture. These books usually have been studied within particular disciplines such as military history, the history of technology, and the history of art and architecture; they have been investigated primarily for their multifarious technical content. This essay focuses, rather, on the authors who created them, on the context of their authorship, and on its cultural implications.[1]

This study is further delimited in that it concerns authorship on the mechanical arts that occurred within the context of scribal rather than print culture. Some of the writings treated here (such as Leon Battista Alberti's *De re aedificatoria*) were later printed, but they were originally composed as handwritten and hand-copied works, either before the invention of the press or outside of its immediate influence. Despite the fact that the printing press was invented in Germany around the mid-fifteenth century, this investigation does not concern printing directly. Yet in its focus on the scribal proliferation of technical books in the era just before printing, it does call for a modification of Elizabeth L. Eisenstein's thesis that posits the press as the most important causal agent for the resurgence of technical literature in early modern Europe.[2]

This investigation concerns authorship on the mechanical arts within the political and cultural context of north Italy and south Germany, and it claims as well to have relevance to the historiography of early modern science. Among other issues, it elucidates the role of the artisan in the development of experimental philosophy. The thesis positing artisanal influence on early modern science originated most significantly in a series of important articles published by Edgar Zilsel in the early 1940s. Zilsel's views have been reiterated and elaborated by Paolo Rossi and, more recently, by J. A. Bennett, among others.[3] In this

[1] See esp. William Crossgrove, *Die deutsche Sachliteratur des Mittelalters* (Bern: Peter Lang, 1994), esp. pp. 103–144; Gerhard Eis, *Mittelalterliche Fachliteratur* (Stuttgart: J. B. Metzlersche, 1962); Bertrand Gille, *Engineers of the Renaissance*, trans. anon. (Cambridge, Mass.: MIT Press, 1966), esp. pp. 55–170; Bert S. Hall, "Der Meister sol auch kennen schreiben und lesen: Writings about Technology ca. 1400–ca. 1600 A.D. and Their Cultural Implications," in *Early Technologies*, Vol. 3, ed. Denise Schmandt-Besserat (Malibu, Calif.: Undena, 1979); Hall, *The Technological Illustrations of the So-Called "Anonymous of the Hussite Wars": Codex Latinus Monacensis 197, Part 1* (Wiesbaden: Ludwig Reichert, 1979), esp. pp. 121–133; and Max Jähns, *Geschichte der Kriegswissenschaften vornehmlich in Deutschland*, Vol. 1: *Altertum, Mittelalter, XV. und XVI. Jahrhundert* (Munich: Oldenbourg, 1889), pp. 241–443. More recently, the best introduction—especially to the Italian material—is Paolo Galluzzi, ed., *Prima di Leonardo: Cultura delle macchine a Siena nel Rinascimento* (Milan: Electa, 1991).

[2] Elizabeth L. Eisenstein, *The Printing Press as an Agent of Change: Communications and Cultural Transformations in Early-Modern Europe*, 2 vols. (Cambridge: Cambridge Univ. Press, 1979), esp. Vol. 2, pp. 520–574. Recent scholarship concerning particular kinds of early modern technical authorship includes William Eamon, *Science and the Secrets of Nature: Books of Secrets in Medieval and Early Modern Culture* (Princeton, N.J.: Princeton Univ. Press, 1994); Pamela O. Long, "The Openness of Knowledge: An Ideal and Its Context in Sixteenth-Century Writings on Mining and Metallurgy," *Technology and Culture*, 1991, *32*:318–355; and Mary J. Voss, "Between the Cannon and the Book: Mathematicians and Military Culture in Sixteenth-Century Italy" (Ph.D. diss., Johns Hopkins Univ., 1994).

[3] See esp. Edgar Zilsel, "The Sociological Roots of Science," *American Journal of Sociology*, 1942, *47*:544–562; and Zilsel, "The Origins of Gilbert's Scientific Method," in *Roots of Scientific Thought: A Cultural Perspective*, ed. Philip P. Wiener and Aaron Noland (New York: Basic, 1957), pp. 219–250. For Zilsel's life and work see Johann Dvořak, *Edgar Zilsel und die Einheit der Erkenntnis* (Vienna: Löcker, 1981). For a collection of Zilsel's essays that includes important biographical and bibliographical scholarship by the editors see Zilsel,

essay, I extend but also revise the Zilsel thesis in two ways. First, I stress authorship as an important and heretofore neglected modality of artisanal influence. Second, I question Zilsel's view of the artisan as separate in every way from learned and elite groups. Artisanal authorship came about through the patronage of oligarchs and rulers who supported the technological authorship of learned humanists as well. I argue that technical books attracted the patronage of elites because in the fifteenth century the practice and representation of rulership came to be closely associated in particular ways with technological power and the mechanical arts.

Artisan-trained practitioners authored a significant number of books concerning the mechanical arts, although by no means all of them. They represent a select group of artisans, most of whom had turned away from the support of craft guilds to the patronage of elite rulers; they wrote books in part to further that patronage. I suggest that not only artisanal practice per se, but artisanal authorship, influenced the development of thought by changing the status and character of certain mechanical arts.

Some historians focusing on seventeenth-century topics, such as Mario Biagioli and Steven Shapin, have emphasized the low status of the mechanical arts and the efforts of experimental philosophers to dissociate themselves from them. Others have elaborated a picture of more complex interaction.[4] In my view experimental philosophy, given its methodology of testing hypotheses by manipulating mechanical devices, must be said to have appropriated both values and specific knowledge from the mechanical arts. That appropriation was possible because those arts had increased in status and cultural significance in the preceding two centuries. Rather than contest the low status of the mechanical arts within particular local cultures in the late sixteenth and seventeenth centuries, I suggest that there is a complicated prior history concerning those arts that needs further study.

Another issue, underscored recently by Alan Gabbey, involves the transition from the mechanical arts to the science of mechanics. What, he has asked, is the conceptual history that transformed the mechanical arts from "artificial" human construction produced in opposition to nature (the traditional view) to the science of mechanics as a branch of physics devoted to the study of certain aspects of nature?[5] The present study, which is restricted to fifteenth-century manuscript books, obviously does not elucidate the entire

---

The Social Origins of Modern Science, ed. Diederick Raven, Wolfgang Krohn, and Robert S. Cohen (Boston Studies in the Philosophy of Science) (Dordrecht: Kluwer, forthcoming). Later scholarship that is indebted to Zilsel's work includes Paolo Rossi, *Philosophy, Technology, and the Arts in the Early Modern Era,* trans. Salvator Attanasio, ed. Benjamin Nelson (New York: Harper & Row, Harper Torchbooks, 1970); Pamela O. Long, "The Contribution of Architectural Writers to a 'Scientific' Outlook in the Fifteenth and Sixteenth Centuries," *Journal of Medieval and Renaissance Studies,* 1985, *15*:265–298; J. A. Bennett, "The Mechanics' Philosophy and the Mechanical Philosophy," *History of Science,* 1986, *24*:1–28; and Bennett, "The Challenge of Practical Mathematics," in *Science, Culture, and Popular Belief in Renaissance Europe,* ed. Stephen Pumfrey, Rossi, and Maurice Slawinski (Manchester: Manchester Univ. Press, 1991), pp. 176–190.

[4] Mario Biagioli, "The Social Status of Italian Mathematicians, 1450–1600," *Hist. Sci.,* 1989, *27*:41–95; Biagioli, *Galileo, Courtier: The Practice of Science in the Culture of Absolutism* (Chicago: Univ. Chicago Press, 1993); and Steven Shapin, *A Social History of Truth: Civility and Science in Seventeenth-Century England* (Chicago: Univ. Chicago Press, 1994), esp. pp. 355–407. For views that suggest greater interaction and parity see Alexander Keller, "Mathematics, Mechanics, and the Origins of the Culture of Mechanical Invention," *Minerva,* 1985, *23*:348–361; Bruce T. Moran, "German Prince-Practitioners: Aspects in the Development of Courtly Science, Technology, and Procedures in the Renaissance," *Technol. Cult.,* 1981, *22*:253–274; and Pamela H. Smith, *The Business of Alchemy: Science and Culture in the Holy Roman Empire* (Princeton, N.J.: Princeton Univ. Press, 1994).

[5] Alan Gabbey, "Between *Ars* and *Philosophia naturalis:* Reflections on the Historiography of Early Modern Mechanics," in *Renaissance and Revolution: Humanists, Scholars, Craftsmen, and Natural Philosophers in Early Modern Europe,* ed. J. V. Field and Frank A. J. L. James (Cambridge: Cambridge Univ. Press, 1993), pp. 133–145. See also W. R. Laird, "The Scope of Renaissance Mechanics," *Osiris,* 2nd Ser., 1986, *2*:43–68.

transformation to which Gabbey alludes. Yet an early part of that process, I suggest, involved the explication and rationalization of certain craft practices in books. To put it another way, the textual and pictorial elaboration of the mechanical arts in codices allowed their transformation from "know-how," available for constructing things in the world, to "knowledge" involving rational or mathematical principles—and taken to be true.

Books on the mechanical arts became useful in patronage relationships because of a newly perceived alliance between technology and military and political power. In the fifteenth century a close affiliation developed between the ancient, separate categories of *praxis,* meaning political and military action, and *techne,* referring to productive and constructive activities in the material world. Military strength and political legitimacy came to be closely associated with construction, technology, and technique. Rulers, princes, and military captains who wanted to consolidate their power achieved legitimation through the remodeling of urban space, the construction and decoration of great palaces, and the creation of painting, sculpture, and other material artifacts. As Christine Smith has put it, such projects were indicative of the "use of the built environment as evidence for the authority of the state."[6]

New political maps were carved by individuals and groups who did not enjoy the legitimacy provided by membership in the traditional noble classes. Urban patriciate elites, including bankers, merchants, and other beneficiaries of late medieval commercial capitalism, created new power configurations. Members of the nobility who did achieve power did so more through military skill than through kinship relationships. In oligarchic republics such as Florence, Venice, Nuremberg, and Augsberg, commercial activity and wealth did not automatically confer political legitimacy. Nevertheless, the German emperors relied on patriciate oligarchs in the imperial cities in their struggle against the traditional nobility. In Italy, men who started their careers as *condottiere,* military captains leading mercenary armies, often achieved autonomous power as princes of city-states. Initially they established control through military force—a consequence of their ability to lead soldiers. Thereafter, like the merchant oligarchs of the republics and the free imperial cities, they sought to augment their authority by other means. They created visible manifestations of power—by reordering urban space, building and decorating great palaces and churches, staging elaborate festivals, and creating courts that became centers of culture and learning.[7]

[6] Christine Smith, *Architecture in the Culture of Early Humanism: Ethics, Aesthetics, and Eloquence, 1400–1470* (New York: Oxford Univ. Press, 1992), p. 50. The distinctions between *episteme* (knowledge of unchanging things), *praxis* (action requiring *phronesis* or judgment in contingent situations), and *techne* (material construction) were emphasized in Aristotle, *Nicomachean Ethics* 6.4–6.5 (1140a–1141a). The analysis and recasting of such distinctions is at the heart of the work of political philosophers such as Hannah Arendt, *The Human Condition* (Chicago: Univ. Chicago Press, 1958); and Jürgen Habermas, *Theory and Practice,* trans. John Viertel (Boston: Beacon, 1973), esp. pp. 41–81, which treats the classical separation of *praxis* and *techne* and its subsequent fusion in the thought of Machievelli. See esp. Joseph Dunne, *Back to the Rough Ground: "Phronesis" and "Techne" in Modern Philosophy and in Aristotle* (Notre Dame, Ind.: Univ. Notre Dame Press, 1993), for an important study of Aristotelian and twentieth-century discussions; and Alasdair MacIntyre, *After Virtue: A Study in Moral Theory,* 2nd ed. (Notre Dame, Ind.: Univ. Notre Dame Press, 1984). Historians of science such as Owen Hannaway and Pamela Smith have much to say about the changing nature of these relationships: Hannaway, *The Chemists and the Word: The Didactic Origins of Chemistry* (Baltimore: Johns Hopkins Univ. Press, 1975); and Smith, *Business of Alchemy* (cit. n. 4), esp. pp. 56–92.

[7] Studies that elucidate relationships between political power and the decorative and constructive arts in the Renaissance include Richard A. Goldthwaite, *The Building of Renaissance Florence: An Economic and Social History* (Baltimore: Johns Hopkins Univ. Press, 1980); Goldthwaite, *Wealth and the Demand for Art in Italy, 1300–1600* (Baltimore: Johns Hopkins Univ. Press, 1993), esp. pp. 212–255; Mary Hollingsworth, *Patronage in Renaissance Italy: From 1400 to the Early Sixteenth Century* (Baltimore: Johns Hopkins Univ. Press, 1994); Thomas DaCosta Kaufmann, "Editor's Statement: Images of Rule: Issues of Interpretation," *Art Journal,* 1989,

No longer the result of mere craft know-how and skill, the built environment became the material manifestation of rational and mathematical principles that were elaborated in books. Construction projects were carried out with the help of machines that themselves came to be illustrated and explicated on the folios of codices. Similarly, the *praxis* of military leadership came to be closely associated with armaments and techniques—in contrast to most ancient models, in which generalship was perceived to rest on character and qualities of leadership rather than on technology.[8] In this environment technical books, whether lavishly illustrated and thus a visual manifestation of power, or predominantly textual, involving the explication of rational principles, came to be highly appropriate gifts for princes.

Such books necessitated the open display and explication of the technical, craft, and mechanical disciplines that constituted their subject matter. The degree and nature of such openness varied from one author to another. Some authors candidly explained and illustrated aspects of their subject while suggesting that further secrets might be revealed if employment or patronage were forthcoming. Others wrote in an unambiguously open fashion. In either case, the openness of fifteenth-century books on the mechanical arts constitutes a striking contrast to the well-documented secrecy of late medieval craft guilds.[9]

The argument I have outlined can be demonstrated only by careful attention to specific authors, the books that they created, and the contexts within which they undertook authorship. Fifteenth-century books on the mechanical arts can conveniently be separated into four groups: illustrated Latin treatises from the first half of the century; German treatises, composed throughout the century, that primarily although not exclusively treat gunpowder artillery; humanist Latin writings; and, finally, diverse treatises by practitioners, mostly Italian, mostly from the second half of the fifteenth century. It will be seen

---

48:119–122; Joanna Woods-Marsden, "Images of Castles in the Renaissance: Symbols of 'Signoria'/Symbols of Tyranny," *ibid.*, pp. 130–137; Kaufmann, *Court, Cloister, and City: The Art and Culture of Central Europe, 1450–1800* (Chicago: Univ. Chicago Press, 1995); Bram Kempers, *Painting, Power, and Patronage: The Rise of the Professional Artist in the Italian Renaissance*, trans. Beverley Jackson (London: Allen Lane, Penguin, 1993); F. W. Kent and Patricia Simons with J. C. Eade, eds., *Patronage, Art, and Society in Renaissance Italy* (Canberra: Humanities Research Centre, 1987); Lauro Martines, *Power and Imagination: City-States in Renaissance Italy* (New York: Random House, Vintage, 1980); Charles M. Rosenberg, ed., *Art and Politics in Late Medieval and Early Renaissance Italy: 1250–1500* (Notre Dame, Ind.: Univ. Notre Dame Press, 1990); Randolph Starn and Loren Partridge, *Arts of Power: Three Halls of State in Italy, 1300–1600* (Berkeley/Los Angeles: Univ. California Press, 1992); Roy Strong, *Art and Power: Renaissance Festivals, 1450–1650* (Berkeley/Los Angeles: Univ. California Press, 1984); and David Thomson, *Renaissance Architecture: Critics, Patrons, Luxury* (Manchester: Manchester Univ. Press, 1993).

[8] See Pamela O. Long and Alex Roland, "Military Secrecy in Antiquity and Early Medieval Europe: A Critical Reassessment," *History and Technology*, 1994, *11*:259–290, which suggests that the openness of much ancient military culture (with the exception of imminent battle situations) is based on the association of military success with the character of the military leader rather than with military technology.

[9] For secrecy and intellectual property attitudes within medieval craft contexts see Pamela O. Long, "Invention, Authorship, 'Intellectual Property,' and the Origin of Patents: Notes toward a Conceptual History," *Technol. Cult.*, 1991, *32*:846–884. Openness and secrecy within fifteenth-century technical authorship are complex issues that will be treated at length in my book in progress: Long, "Openness, Secrecy, Authorship, Ownership: Studies in Premodern and Early Modern Practical, Technical, and Knowledge Traditions." My point here is that the context of authorship within patronage relationships involved at least some requisite display and explication of technological material and led to the open explication of the mechanical arts in manuscript books. Here and elsewhere, issues of openness and secrecy are influenced closely by context. See William Eamon, "From the Secrets of Nature to Public Knowledge," in *Reappraisals of the Scientific Revolution*, ed. David C. Lindberg and Robert S. Westman (Cambridge: Cambridge Univ. Press, 1990), pp. 333–365; and Roger Hahn, "The Meaning of the Mechanistic Age," in *The Boundaries of Humanity: Humans, Animals, Machines*, ed. James J. Sheehan and Morton Sosna (Berkeley/Los Angeles: Univ. California Press, 1991), pp. 142–157, both of which associate openness more broadly with developments in science (such as "mechanization of the world picture") or with technological developments, such as printing.

that the books themselves often contain crucial evidence concerning patronage or hoped-for patronage. Moreover, the explication of machines and instruments is at times closely connected with assumptions about political and military power and the representations of that power. Finally, the mechanical arts are rationalized in a variety of ways and newly associated with learning, the liberal arts, and classical antiquity.

BETWEEN EMPIRE AND CITY-STATE: LATIN TECHNICAL AUTHORSHIP IN THE EARLY FIFTEENTH CENTURY

The first group of writings to be considered comprises Latin technical treatises from the early fifteenth century. The authors of these books, Conrad Kyeser (1366–1405), Giovanni Fontana (ca. 1393–ca. 1455), and Mariano Taccola (1382–1453), explicated numerous kinds of machines and weapons both in writing and pictorially.[10] The three men worked in the contiguous regions of north and central Italy and south Germany. The area was divided linguistically and encompassed a multitude of changing political and military territories, but it also contained common features and was characterized by porous boundaries. The German emperors had long-standing interests in Italian territories, certain components of which were formally if not effectively part of the empire. Italian princes received patronage from the emperor and fought for him as *condottiere.* Mercenary soldiers traveled back and forth between Italian and German regions; most gunners in north Italian armies in the early fifteenth century were German. Other facilitators of interchange between the two regions were the church, commerce, and banking. In important ways, the three authors shared a single political and military culture.[11]

Conrad Kyeser, the son of South German burghers, authored *Bellifortis* [*Strong in War*] between 1402 and 1405 while in exile in his home town of Eichstätt. A strikingly illustrated treatise written in Latin verse, *Bellifortis* is the earliest fifteenth-century technical book from the German territories. Kyeser was exiled in 1402 when Sigismund, king of Hungary and of Germany, defeated Prague, imprisoned his half-brother, the Bohemian king Wen-

[10] Kyeser, who had worked as a physician in the imperial army, wrote his military treatise, *Bellifortis,* between 1402 and 1405; see Conrad Kyeser aus Eichstätt, *Bellifortis,* ed. and trans. Götz Quarg (Düsseldorf: Verlag des Vereins Deutscher Ingenieure, 1967). Fontana's treatise, *Bellicorum instrumentorum liber,* treats military machines, fountains, and other mechanical devices and includes Latin text that is partially encrypted. See Eugenio Battisti and Giuseppa Saccaro Battisti, *Le macchine cifrate di Giovanni Fontana* (Turin: Arcadia, 1984). Mariano Taccola was a notary from Siena who authored *De ingeneis* in the 1420s and 1430s and *De machinis* in the 1440s. See Mariano Taccola, *De ingeneis: Liber primus leonis, Liber secundus draconis, addenda; Books I and II, On Engines, and Addenda (The Notebook),* ed. and trans. Gustina Scaglia, Frank D. Prager, and Ulrich Montag, 2 vols. (Wiesbaden: Ludwig Reichert, 1984); Prager and Scaglia, *Mariano Taccola and His Book "De ingeneis"* (Cambridge, Mass.: MIT Press, 1972); and Taccola, *Liber tertius de ingeneis ac edifitiis non usitatis,* ed. J. H. Beck (Milan: Il Polifilo, 1969). For *De machinis* see Taccola, *De machinis: The Engineering Treatise of 1449,* ed. and trans. Scaglia, 2 vols. (Wiesbaden: Ludwig Reichert, 1971); and Taccola, *De rebus militaribus (De machinis, 1449),* ed. Eberhard Knobloch (Baden-Baden: Valentin Koerner, 1984).

[11] For the older literature see Paolo Brezzi, "I comuni cittadini italiani e l'impero medioevale," in *Nuove questioni di storia medioevale* (Milan: Marzorati, 1969), pp. 177–207. See also Heinz Angermeier, "Die Sforza und das Reich," in *Gli Sforza a Milano e in Lombardia e i loro rapporti con gli stati italiani ed europei (1450–1535)* (Convegno internazionale, Milano, 18–21 maggio 1981) (Milan: Istituto Editoriale Cisalpino–La Goliardica, 1982), pp. 165–191; Hermann Kellenbenz, "Oberdeutschland und Mailand zur Zeit der Sforza," *ibid.,* pp. 193–227; and M. E. Mallett and J. R. Hale, *The Military Organization of a Renaissance State: Venice c. 1400 to 1617* (Cambridge: Cambridge Univ. Press, 1984), for much information about the relationship of one city-state, Venice, to the empire, including Venetian recruitment of German soldiers (pp. 21, 79). For German soldiers in Italy see Mallett, *Mercenaries and Their Masters: Warfare in Renaissance Italy* (London: Bodley Head, 1974), esp. pp. 27–36; Karl H. Schäfer, *Deutsche Ritter und Edelknechte in Italien während des 14. Jahrhunderts* (Paderborn: Ferdinand Schöningh, 1911); and Marino Viganò, ed., *Architetti e ingegneri militari italiani all'estero dal XV al XVIII secolo* (Livorno: Sillabe, 1994).

ceslas, and sent the court, presumably including Kyeser, into exile. Banished to his hometown of Eichstätt in Bavaria, Kyeser wrote his treatise in the last years of his life. The manuscript, of which there are at least fourteen exemplars, was illustrated by illuminators from the Prague scriptorium who were passing through Eichstätt after their own expulsion.[12]

Bellifortis is the response of a bitterly angry military physician to the imperial politics of his time. Kyeser had been present at Nicopolis on that "impious day" in 1396 when the Ottoman Turks had precipitated the flight "with unheard of audacity of Prince Sigismund, hermaphroditic Hungarian king," and his imperial army. He was a victim in the struggle between Sigismund and Wenceslas, both sons of the Luxembourg emperor Charles IV. Kyeser dedicated his treatise to the emperor Ruprecht (ruled 1400–1410) from the Palatinate (Pfalz), a weak emperor chosen by the electors after their unprecedented deposal of Wenceslas in 1400.[13]

In *Bellifortis* Kyeser offers technological strength to the emperor Ruprecht, advertising the efficacy of astrology, sorcery, mechanical technology, and weaponry. The lavish illustrations of the treatise depict cannon, rockets, chariots, a counterweight trebuchet, battering rams, mobile bridges, ships, mills, scaling ladders, incendiary devices, crossbows, and instruments of torture—weapons, machines, and devices both old and new. The large, expensive format, the striking illustrations, and the text in Latin verse all suggest that the emperor Ruprecht to whom the treatise was dedicated was also representative of the intended audience. Kyeser wrote a prince's book, not an engineer's.[14]

[12] Kyeser, *Bellifortis,* ed. and trans. Quarg (cit. n. 10), is an edition and German translation of the work that includes extensive notes and a facsimile reproduction of a manuscript at Göttingen dated 1405 (Niedersächsische Staats- und Universitätsbibliothek, MS philos. 63 cim.). However, see the critical review of Quarg's edition by Hermann Heimpel in *Göttingische Gelehrte Anzeigen,* 1971, *223*:115–148. The identity of the *Bellifortis* Kyeser with a "Master Conrad of Eichstätt" who authored medical tracts is uncertain. Lynn White, Jr., "Kyeser's 'Bellifortis': The First Technological Treatise of the Fifteenth Century," *Technol. Cult.,* 1969, *10*:436–441, on p. 438 n. 2, identified Kyeser as the "Master Conrad of Eichstätt" who composed a collection of medical recipes in an early fifteenth-century manuscript associated with the court of Wenceslas in Prague. But Günter Kallinich and Karin Figala, "Konrad von Eichstätt, eine Arztpersönlichkeit des deutschen Mittelalters," *Sudhoffs Archiv,* 1969, *52*:341–346, argued that the author of medical texts was a well-known physician from Eichstätt who lived in the early fourteenth century.

[13] Kyeser, *Bellifortis,* ed. and trans. Quarg, Vol. 1, p. 4 (fol. 3a): "diem impium," "inaudite audacie principis Sygismundi Regis ermifroditi ungarie." In a later passage (Vol. 1, p. 53, fol. 85a) Kyeser reiterates that the Turks had defeated "Sigismund, a deserter and madman, deceitful good-for-nothing [Sygmundum perfugum atque furibundum/Fallacem nequam]." The disadvantages of a weak emperor had become apparent after the electors had watched Ruprecht lead an embarrassingly incompetent imperial army in and out of Italy in 1402. The electors reverted back to the Luxembourg line in 1410, choosing the far more powerful, if less malleable, Sigismund. Thereafter, Sigismund became king of Bohemia in 1419 (after the death of Wenceslas) and king of the Lombards in 1431. He was crowned Holy Roman Emperor in 1433. For Ruprecht see Aug. Thorbecke, "Ruprecht III," in *Allgemeine deutsche Biographie* (Leipzig: Duncker & Humblot, 1889), Vol. 29, pp. 716–726; for his 1402 military adventure in Italy see Brian Pullan, *A History of Early Renaissance Italy: From the Mid-Thirteenth to the Mid-Fifteenth Century* (London: Allen Lane, 1973), pp. 234–235. For Sigismund's career see esp. János Bak, "The Late Medieval Period, 1382–1526," in *A History of Hungary,* ed. Peter F. Sugar *et al.* (Bloomington: Indiana Univ. Press, 1990), pp. 54–64; Wilhelm Baum, *Kaiser Sigismund: Hus, Konstanz und Türkenkriege* (Graz: Styria, 1993); Friedrich Bernward Fahlbusch, *Städte und Königtum im Frühen 15. Jahrhundert: Ein Beitrag zur Geschichte Sigmunds von Luxemburg* (Cologne: Böhlau, 1983); Heinrich Koller, "Sigismund, 1410–1437," in *Kaisergestalten des Mittelalters,* ed. Helmut Beumann (Munich: C. H. Beck, 1984), pp. 277–300, 375–376; Elemér Mályusz, *Kaiser Sigismund in Ungarn, 1387–1437* (Budapest: Akadémiai Kiadó, 1990); and Sabine Wefers, *Das politische System Kaiser Sigmunds* (Stuttgart: Franz Steiner, 1989). See also Bertalan Kéry, *Kaiser Sigismund: Ikonographie* (Vienna: Anton Schroll, 1972), a thorough study of the many portraits and other visual images of Sigismund. For the Battle of Nicopolis see Aziz Suryal Atiya, *The Crusade of Nicopolis* (London: Methuen, 1934).

[14] I refer to the illustrations in the facsimile edition—Kyeser, *Bellifortis,* ed. and trans. Quarg—of one particular manuscript (Göttingen, Niedersächsische Staats- und Universitätsbibliothek, MS philos. 63 cim.). The illustra-

As would befit a prince, the technology of *Bellifortis* includes a full panoply of magical and mechanical weapons. Kyeser enhances the prestige of that technology by tying it to both learning and classical antiquity. He insists upon the relevance of letters to military strength: "The ambition of Caesar [i.e., Sigismund] is foolish, for art is worth gold / And letters rule the shield and sword of the soldier / Letters alone govern the chasuble and the robe / All things are under letters, with which the One Almighty rules."[15] Kyeser praises the strong soldiers of Germany and simultaneously applauds her liberal and mechanical arts: "just as the sky shines with stars, Germany shines forth with liberal disciplines, is embellished with mechanics, and adorned with diverse arts."[16]

Kyeser's hero is Alexander of Macedon, conqueror of the east, whose triumph stands in stark contrast to Sigismund's eastern debacle at Nicopolis. Yet his portrait of Alexander as technologically efficacious in his own person is antithetical to ancient notions of military leadership. In *Bellifortis* Alexander stands with a rocket-like weapon inscribed with the mysterious letters *MEUFATON* (see Frontispiece). The accompanying verses explain the efficacy of the weapon against all enemies. Elsewhere, Kyeser reports that Alexander is the inventor of a huge war carriage (see Figure 1). He thus represents Alexander's power as rooted in technical competence that is both mechanical and magical.[17] Kyeser's Alexander brings together the categories of *praxis* and *techne.* In striking contrast to ancient mores, he is a military leader armed with magical, mechanical, and incendiary devices.

Kyeser presents himself as a dying man in 1405, the date inscribed at the end of his dedication to Ruprecht. His own portrait appears near the end of the treatise, accompanied by a self-composed funeral ode and epitaph, a list of rulers and princes with whom he had been associated (Wenceslas, Sigismund, and Francis of Carrara in Padua, among others), and a prayer for his about-to-be departed soul (see Figure 2). He includes verses for the arts that have informed his treatise—the seven liberal arts and also geomancy, theology, philosophy, law, canon law, physics, alchemy, the theurgical arts, and the military arts. He represents himself as famous and loved by all and as an expert on military matters whose *Bellifortis* will scatter armies. He will be widely mourned, by nobles, the wealthy, the poor, and even animals—quadrupeds, tripeds, birds, worms.[18]

Nothing in Giovanni Fontana's writings corresponds to the bizarre finale of Kyeser's treatise. Yet Fontana shared with his German counterpart both a delight in mechanical contrivances and the assumption that power is associated with technological efficacy. Fontana first took up technical authorship at the University of Padua, where he wrote three small treatises—two concerning clocks and one treating measurement from a distance.

---

tions in other copies (e.g., New York Public Library, Spencer MS 58) are not in every case identical to those of the Göttingen codex.

[15] Kyeser, *Bellifortis,* ed. and trans. Quarg, Vol. 1, p. 2 and n. 1 (fol. 2a): "Caesaris ambicio desipit, nam ars valet aurum / Et littera clippeum militis gerit atque mucronem. / Casulam cum stola gubernat littera sola / Litteris cuncta subsunt, agla quibus imperat unus." Quarg suggests that "*agla*" is an old Hebraic name for "Almighty," derived from the first four letters of an ancient Jewish prayer and used by Kyeser to display his linguistic knowledge. "*Stola,*" in line 3, can refer to a royal robe, in which case the line suggests that letters govern both priestly and royal realms.

[16] *Ibid.,* Vol. 1, p. 3 (fol. 2b): "sicud celum ornatur sideribus, sic alemania prefulget disciplinis liberalibus, honestatur mechaniis diversisque artibus adornatur."

[17] *Ibid.,* Vol. 1, p. 15, and Vol. 2 (fol. 11b), for Alexander holding the rocket; and Vol. 1, p. 17, and Vol. 2 (fol. 16a) for the war carriage. Quarg suggests that *MEUFATON* refers to Philip of Macedonia's conquest of Methone in 353 B.C., during which conflict the Macedonian king lost an eye. What Kyeser makes clear is the almost magical properties of the weapon, which will cause enemies to flee. Concerning the war carriage, Kyeser writes: "Hoc instrumentum ab allexandro repertum [this instrument was invented by Alexander]."

[18] *Ibid.,* Vol. 1, pp. 100–105 (fols. 135a–139a).

***Figure 1.** A giant tank invented by Alexander of Macedon. From Conrad Kyeser,* Bellifortis, *Cod. phil. 63, fol. 16a. (Courtesy of the Niedersächsische Staats- und Universitätsbibliothek, Göttingen.) Kyeser again emphasizes the technological skill of a powerful ruler with his depiction of this gigantic tank, which he claims was invented by Alexander of Macedon. To view the tank coherently, push the ram at the top down and flip the round wheels on the sides underneath. The soldiers enter by the door in front. The vehicle's defenses include a combination of scythes and cannon placed in alternating positions between the wheels, which are (we are told) hidden. Thus can the army safely approach the enemy and destroy it. The tradition extolling Alexander here enters a new phase, as the Macedonian king becomes the inventor of large-scale military hardware. The illustration beautifully evokes the combination of ancient warfare (scythes) with modern weaponry (cannon).*

After receiving his medical degree in 1421, evidence suggests, he worked as a military physician and was based, for a time at least, with the Venetian army in Brescia.[19]

[19] For a discussion of Fontana's life and writings see Marshall Clagett, "The Life and Works of Giovanni Fontana," *Annali dell'Istituto e Museo di Storia della Scienza di Firenze,* 1976, *1*:5–28; and Battisti and Saccaro Battisti, *Macchine cifrate di Fontana* (cit. n. 10), pp. 7–52. His three early treatises, bound together in the university library in Bologna (Bibliotheca universitaria, 2705), are *De horologio aqueo* (fols. 53r–84v), on water clocks; *Nova compositio horologii guod ex pulverum casu consistit* (fols. 1r–52v), on sand clocks; and *De pisce, cane et volucre* (from fol. 85r), on measurement from a distance. Fontana's illustrations were clearly influenced by Giovanni da Dondi, whose treatise on the astrarium he must have seen in Padua. See Jane Andrews Aiken,

Figure 2. *Portrait of Conrad Kyeser, author of the* Bellifortis. *From Conrad Kyeser,* Bellifortis, *Cod. phil. 63, fol. 139a. (Courtesy of the Niedersächsische Staats- und Universitätsbibliothek, Göttingen.) This remarkable portrait depicts Kyeser wearing a green shirt, the color of physicians and of those born under the sign of Jupiter (including Alexander of Macedon). The portrait appears in the codex accompanied by the author's self-composed epitaph and funeral ode. In contrast to the magical and technological powers emanating from the* Bellifortis, *Kyeser presents himself (with a deeply scarred face, presumably displaying a battle wound) as a dying man. The blank space that he left for his date of death was never filled in.*

The military context of such a career involves the early fifteenth-century expansion of Venetian rule to the *terra ferma* in the form of direct military control and the creation of a standing army to protect the newly acquired territory. Venetian conflicts included the successful war against the Carrara of Padua in 1404–1405, the defense against Hungarian invasions, prompted by Sigismund's expansionist policies, in 1411–1413 and 1418–1421, and the wars against Milan between 1426 and 1454. Venetian military expansion required the hiring of captains-general, a move undertaken with great trepidation by the Venetian senate. Fontana's presence in Brescia is presupposed by his documented role in transmitting a message from the senate to the captain-general of the Venetian army, the ill-fated Francesco Carmagnola. Although Fontana's whereabouts for a number of years are undocumented, he records in a treatise of the 1440s some of the details of his life as a physician in Udine, an outpost of the Venetian army.[20]

It seems probable that Fontana wrote his *Bellicorum instrumentorum liber* in the 1420s and intended it for Carmagnola. The Latin treatise, which concerns military machines, fountains, and other mechanical devices, is written partially in cryptographic code. It consists of seventy folio sheets and close to 140 drawings of numerous kinds of machines and devices: fountains and pumps, magic lanterns, machines for lifting and transporting heavy weights, clocks, defensive and offensive devices to protect forts and castles, battering rams, incendiary and other kinds of weapons, alchemical furnaces, magic lanterns, measuring instruments, a mechanical devil and witch, towers, locks, keys, and scaling ladders, to name a few. Although the illustrations are often self-explanatory, most sheets also contain textual explication that begins in alphabetic Latin and then shifts to encryption. Fontana uses identical encryption in a later work, the *Secretum de thesauro experimentorum,* that treats experiments and natural philosophy.[21]

The encryption of the *Bellicorum instrumentorum liber* can be seen as a display of writing technique involving codification rather than as a means of concealing technical information. In a cogent analysis, the editors of the facsimile edition conclude that the single existing manuscript of the work and the manuscript of Fontana's *Secretum* were both produced by scribes. They hypothesize that Fontana developed the code when he wrote the *Secretum* and that he then instructed scribes to use it to copy the earlier *Bellicorum instrumentorum liber.* In the latter, encrypted text usually appears directly under alphabetic Latin on each page, regardless of the content. On one sheet, for example, Fontana illustrates a fountain and in alphabetic Latin exclaims upon his love for fountains and

"Truth in Images: From the Technical Drawings of Ibn Al-Razzaz Al-Jazari, Campanus of Novara, and Giovanni de' Dondi to the Perspective Projection of Leon Battista Alberti," *Viator,* 1994, *25*:325–359; and Carlo Maccagni, "Le scienze nello studio di Padova e nel Veneto," in *Storia della cultura Veneta dal primo quattrocento al Concilio di Trento,* Vol. 3, ed. Girolamo Arnaldi and Manlio Pastore Stocchi (Vicenza: Neri Pozza, 1981), pp. 135–171, esp. pp. 142–147. Fontana's other writings include *Tractatus de trigono balistario* (Oxford, Bodl. libr. canon. misc. 47), a treatise on measurement using the triangle, composed in Udine in 1440; and a natural history that exists in a printed edition—*Liber Pompilii Azali Placentini De omnibus rebus naturalibus* . . . (Venice: Ottaviano Scoto, 1544). He also wrote a work, presumed lost, on perspective that he dedicated to the Venetian painter Jacopo Bellini. Battisti and Saccaro Battisti hypothesize that it may be the partially extant perspectival treatise attributed to Toscanelli in Alessandro Parronchi, "Le fonti di Paolo Uccello, II: I 'filosofi,' " *Paragone,* 1957, no. 95, pp. 3–33; see Battisti and Saccaro Battisti, *Macchine cifrate di Fontana,* pp. 18–24.

[20] For Venetian military matters in the early fifteenth century see Mallett and Hale, *Military Organization of a Renaissance State* (cit. n. 11), esp. pp. 20–43, upon which this account depends. Battisti and Saccaro Battisti, *Macchine cifrate di Fontana,* pp. 27–28, notes Fontana's mission to Brescia. Fontana's view of the captain-general Carmagnola's execution by order of the senate in 1432 can perhaps be inferred from his recording of the evil omens that he saw beforehand.

[21] See Battisti and Saccaro Battisti, *Macchine cifrate di Fontana,* pp. 35–37 (concerning the encryption), 141–158 (for the *Secretum de thesauro*). The manuscript is in Paris, Bibliothèque Nationale, Lat. nouv. acq. 635.

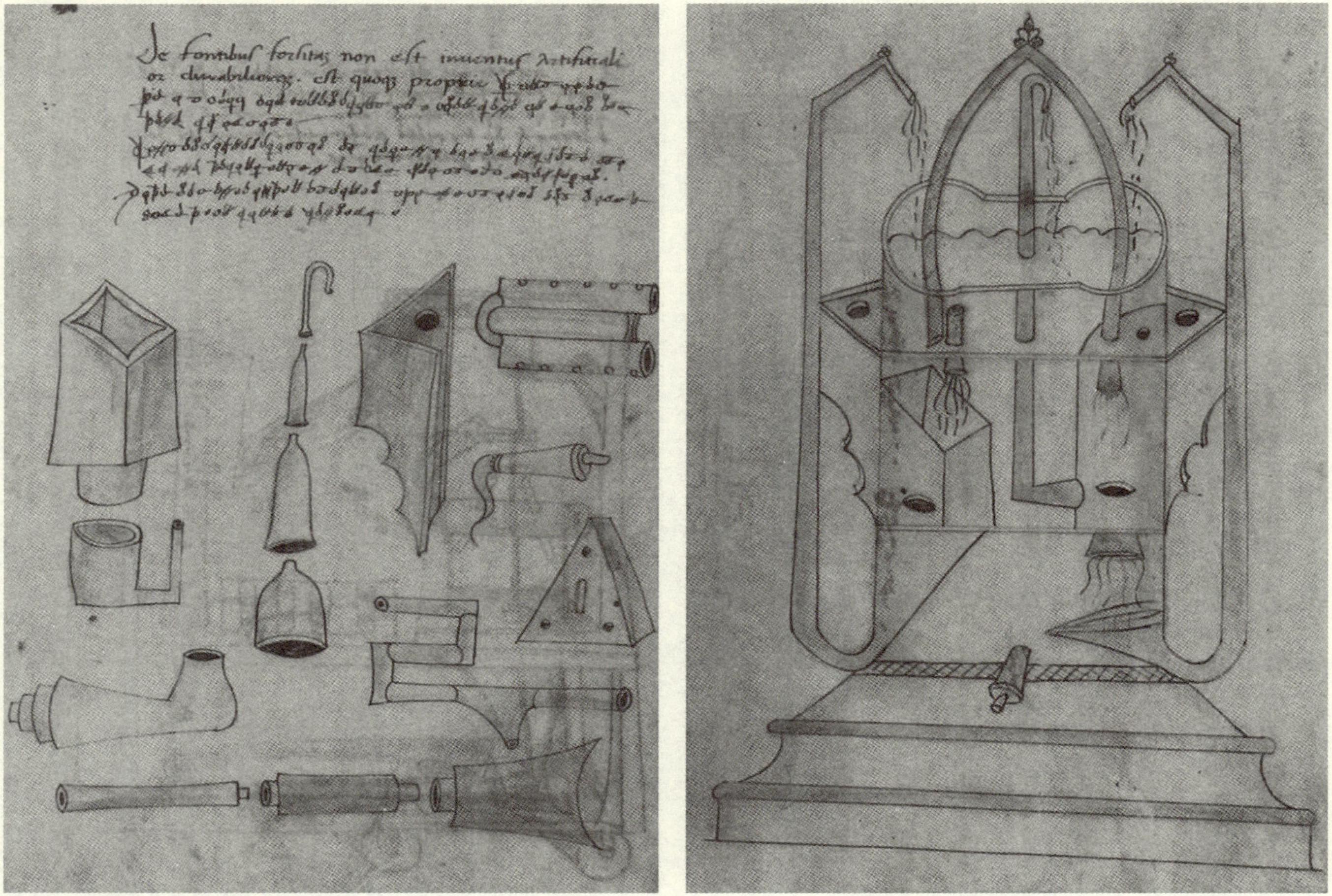

Figure 3. *Parts of a fountain* (left) *and a transparent view of the assembled fountain* (right). *From Giovanni Fontana,* Bellicorum instrumentorum liber, *Cod. icon. 242, fols. 62v and 63r. (Courtesy of the Bayerische Staatsbibliotheque, Munich.) Here Fontana shows the parts of a fountain and how they are assembled into a working fountain. He records in alphabetic Latin that he delights in the study of fountains, whereas in the encrypted Latin he assures the reader that everything has been explained clearly. This example suggests that Fontana did not use encryption to hide technical information.*

his great skill and ingenuity in inventing them (see Figure 3). The encrypted section includes the following: "All parts of the Fountain, and the complete figure of this thus are depicted clearly so that you can understand with facility."[22] The secrecy embodied in the encryption is unrelated to technological detail. As the editors of the work suggest, Fontana was interested in the idea of the code itself rather than in hiding technological secrets per se.

Fontana's contemporary Mariano Taccola, from Siena, authored two pictorial machine books: *De ingeneis,* created in the 1420s and early 1430s, and *De machinis,* composed in the Veneto in the 1440s. Although they are by no means identical, both works are filled with drawings of constructions and machines—cannon, trebuchets, incendiary weapons, winches, pumps, mills, gears, cranes, and ships, among many others. Most sheets also contain explanatory Latin text.[23]

Mariano came from a middle level of Sienese society—his father was a wine dealer, and his wife, Madonna Nanna, was the daughter of a leather merchant. Evidence suggests that his early education included the surprising combination of craft practice and notarial training, including Latin. In 1424 he became a *camarius,* or secretary, of the Domus Sapientae, a hospital and student lodging that served as a temporary residence for foreign visitors. Guests included some of the entourage of Sigismund during the emperor's nine-month Sienese stay in 1432, an interlude necessitated by negotiations with the papacy concerning the imperial coronation. Later, Mariano became a *stimatore* for Siena, one who tallied completed work at a building site, assigned wages, and perhaps estimated the amounts of required materials. In 1441 he was appointed *viaio,* a superintendent of streets, fountains, and bridges.[24]

On one level, Mariano's technical treatises speak to the concerns and problems of the

[22] Battisti and Saccaro Battisti, *Macchine cifrate di Fontana,* p. 96 (fol. 62v of manuscript): "Partes omnes fontis, et figura eius completa ita clare depingantur, ut cum levitate intelligas." See Frank D. Prager, "Fontana on Fountains: Venetian Hydraulics of 1418," *Physis,* 1971, *13*:341–360, for a detailed study of Fontana's fountains. Battisti and Saccaro Battisti's conclusion that the single extant copy was produced by scribes is made on the basis of analysis of mistakes, which are of the sort that would have been made by a copyist rather than by the author. They call Fontana's cryptographic system a simple, rational cipher based on signs without letters or numbers. Fontana had devised a "complete and organic system of signs" that was equivalent to and could substitute for the conventional alphabet. They conclude that Fontana's cryptographic system responded "not so much to the criterion of secrecy as to an awareness of a conscious symbolism of signs and their systems." He was interested in "a system of significance" rather than a system "of the occultization of significance." Battisti and Saccaro Battisti, *Macchine cifrate di Fontana,* pp. 36–37 (my translation of the Italian).

[23] For a facsimile of the first two books (Munich, Bayerische Staatsbibliothek, Codex latinus monacensis 197, Pt. II) with transcription, translation, and notes see Taccola, *De ingeneis,* ed. and trans. Scaglia *et al.* (cit. n. 10). See also Prager and Scaglia, *Taccola and His Book "De ingeneis"* (cit. n. 10), which contains reproductions of the sheets, transcription, and translation of books 3 and 4. For a facsimile edition of books 3 and 4 see Taccola, *Liber tertius de ingeneis,* ed. Beck (cit. n. 10), which includes transcriptions of the documents pertaining to Taccola as well. For *De machinis* see Taccola, *De machinis,* ed. and trans. Scaglia (cit. n. 10), which includes a facsimile edition of Munich, Bayerische Staatsbibliothek, Codex latinus monacensis 28 800; see also Taccola, *De rebus militaribus (De machinis, 1449),* ed. Knobloch (cit. n. 10), which includes a facsimile of Paris, Bibliothèque Nationale, Codex parisinus latinus 7239. An excellent introduction to Taccola and fifteenth-century Sienese culture is Paolo Galluzzi, "Le macchine senesi: Ricerca antiquaria, spirito di innovazione e cultura del territorio," in *Prima di Leonardo,* ed. Galluzzi (cit. n. 1), pp. 15–44.

[24] The evidence for his background and training includes his commissions to carve heads for the choir stalls in 1408 and to produce sculptures in 1441, both for the Sienese cathedral; and his nomination in 1417 as an applicant for admission to the Sienese Guild of Judges and Notaries. For a summary of Taccola's career see Taccola, *De ingeneis,* ed. and trans. Scaglia *et al.,* Vol. 1, pp. 11–15; and Prager and Scaglia, *Taccola and His Book "De ingeneis,"* pp. 3–21. Most of the relevant documents are published in Gaetano Milanesi, *Documenti per la storia dell' arte senese,* Vol. 2: *Secoli XV e XVI* (1854; rpt., Holland: Davaco, 1969), pp. 284–286. For the role of the *stimatore* see Nicholas Adams, "The Life and Times of Pietro dell'Abaco, a Renaissance Estimator from Siena (Active 1457–1486)," *Zeitschrift für Kunstgeschichte,* 1985, *48*:384–395.

Sienese state. They explicate numerous aspects of ships, harbors, and naval warfare, reflecting Siena's efforts to control Tuscan rivers and to develop a commercial port at Talamone on the Tuscan Sea. Taccola's many depictions of pumps, dams, siphons, aqueducts, and other water-controlling devices constitute a response to the perennial Sienese problem of a water supply insufficient for the needs of the population and local industries, especially leather and textile manufacture. His multifarious depictions of cannon, firearms, cannon lifting and transport devices, and weaponry of various sorts suggest a response to Sienese military problems that included the frequent ravaging of the *contada,* or countryside, by marauding mercenary companies and the constant pressure of Florentine expansionist efforts (see Figure 4).[25]

If Taccola's authorship developed out of Sienese concerns, he also used it in a quest for imperial patronage. In 1432 he presented Sigismund with a copy of books 3 and 4 of the *De ingeneis* that included two portraits of the emperor.[26] Taccola specifies that he desires a position in Sigismund's Hungarian court. Pictorially, he has Saint Dorothy recommend him as a master of machines for waterworks in the Hungarian regions (he has drawn her pointing to the relevant text), and he includes authorship (involving a compilation of the emperor's "places") as one of the services he can perform. Although he probably never made it to Hungary, there is indirect evidence that Taccola worked in the Veneto and northern Italy in the late 1430s and 1440s and that he wrote his later treatise *De machinis,* completed in 1449, in the Veneto as well. For example, a copy was made for the Venetian captain-general Bartolomeo Colleoni; his authorial efforts may also have been prompted by renewed hope for imperial patronage, this time from the early Hapsburg emperor Frederick III, whose coronation in Rome was being discussed in Italy in 1447.[27]

De machinis is a pictorial treatise, written as an aid to "fighting the infidels." It is divided into ten sections or books and contains more than two hundred illustrations, many of which

[25] For an introduction to Sienese history and culture in the fourteenth and fifteenth centuries see esp. Mario Ascheri, *Siena nel Rinascimento: Istituzioni e sistema politico* (Siena: Edizioni "Il Leccio," 1985); Duccio Balestracci and Gabriella Piccinni, *Siena nel trecento: Assetto urbano e strutture edilizie* (Florence: Cooperativa Libraria Univ. Studi Florentini, 1977); William M. Bowsky, *A Medieval Italian Commune: Siena under the Nine, 1287–1355* (Berkeley/Los Angeles: Univ. California Press, 1981); Bowsky, "Siena," in *Dictionary of the Middle Ages* (New York: Scribner's, 1987); David L. Hicks, "The Sienese State in the Renaissance," in *From the Renaissance to the Counter-Reformation: Essays in Honor of Garrett Mattingly,* ed. Charles H. Carter (New York: Random House, 1965), pp. 75–94; and Judith Hook, *Siena: A City and Its History* (London: Hamish Hamilton, 1979). The Sienese aqueducts are the subject of a dissertation-in-progress by Michael P. Kucher (Univ. Delaware). See also Balestracci *et al., I bottini medievali di Siena* (Siena: Alsaba, 1993). The aim of the harbor was to rival those at Pisa and Genoa and to assist in importing grain from Sicily in the face of chronic food shortages.

[26] The manuscripts of books 3 and 4 of *De ingeneis* are in Florence, Biblioteca Nazionale, Palat. 766, fols. 27–57 and 58–76, respectively. See Prager and Scaglia, *Taccola and His Book "De ingeneis"* (cit. n. 10), pp. 70–71, for the portraits and the accompanying text, which appears on the verso along with an illustration of the measurement of the height of a tower. For the portrait of Sigismund see James H. Beck, "The Historical 'Taccola' and Emperor Sigismund in Siena," *Art Bulletin,* 1968, *50*:309–320; Lionello G. Boccia, "L'armamento defensivo nei trattati di Mariano di Iacopo detto il Taccola," in *Prima di Leonardo,* ed. Galluzzi (cit. n. 1), pp. 47–56, esp. pp. 52–53; and Gustina Scaglia, "An Allegorical Portrait of Emperor Sigismund by Mariano Taccola of Siena," *Journal of the Warburg and Courtauld Institutes,* 1968, *31*:428–434.

[27] Prager and Scaglia, *Taccola and His Book "De ingeneis,"* pp. 15–16 (for the titles that he received as a result of his presentation), 144–145 (for his recommendation by St. Dorothy). Although Sigismund seems never to have granted Taccola's wish to join the court in Hungary, he did name him *familiares* in 1432, and in the next year in Mantua, on his return to German territories, he granted him and some other Sienese citizens the title *comites palatini.* For (inconclusive) evidence concerning Taccola's authorship of *De machinis* in the Veneto see Milanesi, *Documenti,* Vol. 2 (cit. n. 24), pp. 299–300; Prager and Scaglia, *Taccola and His Book "De ingeneis,"* pp. 16–20; and esp. Taccola, *De machinis,* ed. and trans. Scaglia (cit. n. 10), Vol. 1, pp. 19–23, 29–58 (for a detailed discussion of the manuscripts).

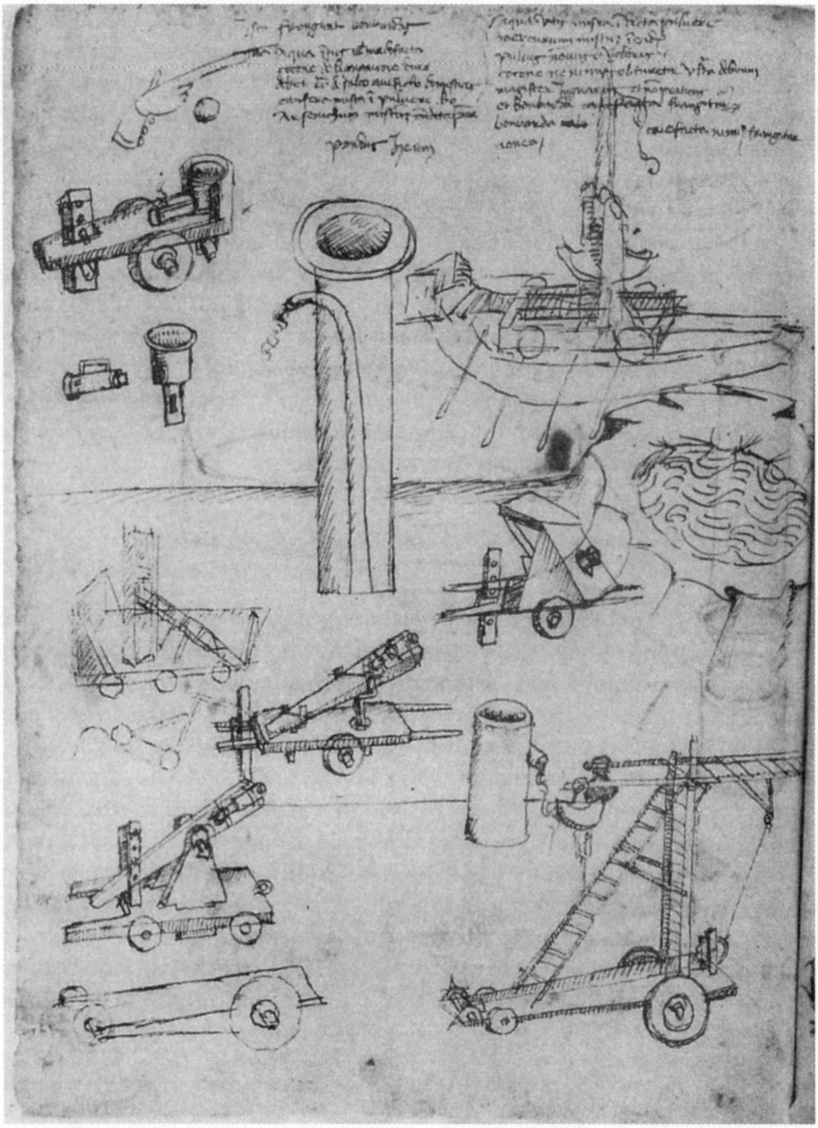

Figure 4. *A sheet of Taccola's* De ingeneis *depicting cannon, a pump, and other devices. From Mariano Taccola,* De ingeneis, *Codex Latinus Monacensis 197, Pt. II, fol. 52v. (Courtesy of the Bayerische Staatsbibliothek, Munich.) This sheet is one of several hundred in Taccola's notebook. It includes sketches of machines and devices central to both Taccola's own interests and those of Siena—military equipment, including mortars, cannon, cannon carriages, and scaling ladders, and hydraulic equipment such as a pump and a syphon.*

are explained textually. The table of contents represents an early attempt to rationalize types of machines into coherent groups, such as "Bridges and Structures on Water," "Bombards and Guns," and "Weights Raised Higher and Lowered Down."[28]

[28] Taccola, *De machinis,* ed. and trans. Scaglia, Vol. 1, p. 18 (for the list of subjects), 59–60 (for Taccola's detailed statement about the uses of his book for fighting the infidels and "bad Christians" who try to usurp the land and power of other Christians).

In this, his later treatise, Taccola calls himself the "Archimedes of Siena." Archimedes' fame as an inventor of military machines was widespread in the fifteenth century, especially through the account in Plutarch's *Life of Marcellus.* Yet Plutarch's Archimedes explicitly declined to write down his military and mechanical inventions because of the vulgarity of such a utilitarian activity.[29] In striking contrast, Taccola, the Sienese Archimedes, embraces technical authorship, presenting a great array of machines with both pictorial and textual explication; he leaves no doubt that they constitute both the proper subject of rational thought and the essential instruments of civic success and military victory.

Although the treatises of Kyeser, Fontana, and Taccola exhibit important differences, they also share a number of characteristics. They are written in Latin, are lavishly illustrated, and display a variety of mechanical contrivances and weapons in lucid detail. They are dedicated to rulers or, in some cases, seem to have been written for rulers or military captains. They directly associate military and political power with technological power in a new way. They enhance the status of mechanical disciplines by explicating them in Latin and by associating them with antiquity and with learning.

GERMAN BOOKS ON GUNPOWDER AND ARTILLERY

Alongside these Latin books, numerous German-language treatises, primarily concerning gunpowder artillery and machines, appeared beginning in the early decades of the fifteenth century and continuing until its end. Although the specific authors of these books are often unknown, clearly many of them were practitioners in the burgeoning field of gunpowder artillery. Their books contain a wealth of technical information, such as how to make gunpowder and how to load cannon. Yet they also attempt to rationalize their practice by treating the causes of particular phenomena, and they treat ethical issues such as the desirable attitudes and deportment of both practitioners and patrons. These books transmitted technical knowledge concerning gunpowder and artillery. Yet they also elevated such practices by making them rational, by explicating them in writing, and by holding practitioner and patron to a set of ethical norms.[30]

An example from the early fifteenth century, Cgm 600 in the Bayerische Staatsbibliothek in Munich, is a small book of twenty-two sheets that treats the manufacture of gunpowder

[29] *Ibid.*, pp. 59–60: "However, I, Ser Marianus Taccola, also called Archimedes of great and magnificent Siena, have not designed these engines, machines and weapons that they may operate against Christians, but have invented, composed and designed them that they may go against the infidel and barbarian people." For Plutarch's account see *Plutarch's Lives,* 11 vols., Vol. 5: *Agesilaus and Pompey, Pelopidas and Marcellus,* trans. Bernadotte Perrin (London: William Heinemann, 1917), pp. 479–481 ("Life of Marcellus" 17.4). For Archimedes in the fifteenth century see W. R. Laird, "Archimedes among the Humanists," *Isis,* 1991, *82*:629–638; and D. L. Simms, "Archimedes and the Invention of Artillery and Gunpowder," *Technol. Cult.,* 1987, *28*:67–79.

[30] Over a hundred individual examples of fifteenth-century manuscript military treatises of German provenance are extant, including multiple copies of particular texts. The most complete recent list is Hall, *Technological Illustrations* (cit. n. 1), pp. 121–133; see also Jähns, *Geschichte der Kriegswissenschaften,* Vol. 1 (cit. n. 1), pp. 241–443. For gunpowder artillery and other military technologies that form the technological context of this literature see esp. Kelly DeVries, *Medieval Military Technology* (Peterborough, Ont.: Broadview, 1992), pp. 143–168; Bert S. Hall, *Weapons and Warfare in Renaissance Europe: Gunpowder, Technology, and Tactics* (Baltimore: Johns Hopkins Univ. Press, 1997); Gerhard W. Kramer, *Berthold Schwarz: Chemie und Waffentechnik im 15. Jahrhundert* (Munich: Oldenbourg, 1995); and Volker Schmidtchen, *Bombarden, Befestigungen, Büchsenmeister: Von den ersten Mauerbrechern des Spätmittelalters zur Belagerungsartillerie der Renaissance: Eine Studie zur Entwicklung der Militärtechnik* (Düsseldorf: Droste, 1977). See also Erich Egg, *Der Tiroler Geschützguss: 1400–1600* (Innsbruck: Universitätsverlag Wagner, 1961); and Egg, *Das Handwerk der Uhr- und der Büchsenmacher in Tirol* (Innsbruck: Universitätsverlag Wagner, 1982), esp. pp. 184–201.

Figure 5. *Loading a cannon barrel and igniting the powder. From Cod. germ. 600. (Courtesy of the Bayerische Staatsbibliothek, Munich.) The illustration on the left shows two gunners, depicted in an early fifteenth-century book on gunpowder artillery, loading a cannon barrel. As the text explains, one is hammering in wooden wedges between the stone ball and the interior surface of the barrel. The other holds containers with other substances to be poured in—loam, oakum (part of flax), or hay. On the right, a gunner is igniting the cannon. The text warns of the danger of the barrel exploding. The gunner should stand back and ignite the powder with a long rod. The clamps that secure the barrel to the carriage are shown but not mentioned in the text. Notice the serious expressions, the industrious demeanor, and the simple clothing of the artisans, each of whom is depicted with his own distinctive clothing, suggesting his individuality.*

and loading and firing cannon. On most pages a pictorial image dominates. For example, on one page instructions are given concerning how to load the barrel of the cannon. Another illustration (see Figure 5) shows a gunner igniting a cannon, a highly dangerous task: sometimes the weapon itself exploded rather than discharging the ammunition. The text cautions the gunner to stand some distance away and to use a rod when lighting the powder.[31] Cgm 600 transmits technical information but also presents images of practitioners deeply absorbed in technical procedures. Each of the human figures depicted in the woodcuts is distinguishable from the others—each has a distinct physiognomy, wears his

[31] Munich, Bayerische Staatsbibliothek, Cgm 600. For a detailed description see Karin Schneider, ed., *Die deutschen Handschriften der Bayerischen Staatsbibliothek, München,* new ed., Vol. 5, Pt. 4 (Wiesbaden: Otto Harrassowitz, 1978), p. 227. See also Wilhelm Hassenstein, *Das Feuerwerkbuch von 1420* (Munich: Verlag der Deutschen Technik, 1941), pp. 48, 56.

own particular clothes, and sports his own hat or hairstyle. Some work in pairs, some alone. All are dressed plainly and clearly belong to the artisan class. All are busy at work, with serious demeanor, carrying out the procedures described in the text below. Although authored anonymously, Cgm 600 validates a new group of practitioners who are represented in the process of doing their work, their individuality exhibited along with their virtues of seriousness, industriousness, and care.

The practitioner is also a central figure in the *Feuerwerkbuch,* an anonymous treatise that was written around 1420 and widely distributed in manuscript copies before it was printed in 1529. The treatise describes itself as concerned with the guns, missiles, gunpowder, and pyrotechnics by which princes, counts, lords, knights, and cities protect themselves when besieged. These princes and cities should have "servants who are godly and steady people," who will voluntarily give their bodies, lives, and property in the fight against the enemy before surrendering. They should also be "wise people" who know how to bring up guns and missiles in an emergency, how to attack, how to build walls and bulwarks, and how to use arms effectively. Finally, they need to get along with each other, not quarrel, and settle their disputes according to the best advice.[32]

The author suggests one reason why he wrote his treatise—that technical details of gunpowder manufacture are too complex to remember without the help of writing. Yet concerns beyond the transmission of technical information are also evident. The treatise begins with twelve questions, some of which indicate an interest in causes. For example, the first question asks whether it is the fire or the force of the vapor produced by the fire that drives the stone from the cannon. The author argues that it is the vapor and proposes an experiment to prove it. Take a pound of good powder, put it in a thick wine cask, and close the cask well so that no vapor may come out. Then ignite the powder at the hole; the powder at hand is burned and the vapor destroys the cask. The second question asks whether the saltpeter or the sulfur pushes the stone. It is both. When the powder is kindled, the sulfur is hot and the saltpeter cold. Since heat won't admit cold and cold won't admit heat, they are two contradictory things that won't tolerate each other. Other questions concern technique: How much powder should be used? Should the wedges pushed against the stone ball in the barrel be made of soft wood like linden or hard wood like oak or beech? Is the gunpowder better corned or in the form of a fine powder?[33]

The *Feuerwerkbuch* also enumerates the qualities of a good gunner. "Above all he should honor and hold God before his eyes." In addition, he should exercise great care when handling guns and powder. He should "rest content with the world in which he

[32] For a reprint of the 1529 edition and a translation into High German see Hassenstein, *Feuerwerkbuch von 1420,* pp. 15–16; see pp. 41–43 for the modern German: "diener / die als from[m] unnd fest leüt seyen"; "sy durch eeren willen / ir leyb lebe[n] und güt / und was in Got ye verleyhen hat gegen iren feinden / darstrecken und wagen dürfften"; "weissleut." The author further admonishes princes and rulers that they need good smiths, bricklayers, carpenters, cobblers, and also good gunners; they must be good masters to their gunners, giving them all the supplies and equipment that they need. Kramer, *Berthold Schwarz* (cit. n. 30), pp. 52–231, includes a facsimile edition and transcription of one manuscript version of the *Feuerwerkbuch* (Freiburg, Universität Freiburg, HS. 362). See also Jähns, *Geschichte der Kriegswissenschaften,* Vol. 1 (cit. n. 1), pp. 392–408.

[33] Hassenstein, *Feuerwerkbuch von 1420,* p. 16: "And thereupon since the subjects belonging to it [gunnery] are so many, which every good gunner should know, and which a master without writing cannot remember in his mind," all the necessary details are provided in the manual. ["Und darumb wann der stuck sovil sind die darzügehöred / die ein yetlicher gütter püchsenmaister künden soll / und die ein mayster on die geschrift in seinem sinne nit gedencken kan."] The treatise provides instruction on how to treat the ingredients of gunpowder (sulfur, saltpeter, and charcoal), how to mix them together, how to load the barrel, and how to discharge the cannon in diverse situations and with various kinds of ammunition such as flaming bolts and stone balls. See also *ibid.,* pp. 16–17 and 43–45 (for modern German).

travels" and should be a "strong, undaunted man." In war, he should "hold himself pleasantly," for such people are greatly trusted. "The master should also know how to write and read"—all the elements of the art cannot be held in the head, but they can be read in the book at hand. He should know something about purifying and distilling and about defensive operations, and he should understand weights and measures. He should teach with honesty and friendliness in both his words and his works, should be thoughtful and avoid drunkenness at all times.[34]

The author proffers advice to rulers as well. Many have found themselves besieged and haven't taken proper care for defense, nor have they brought people to them "who through their technical wisdom [*Kunst Weissheit*] advise and aid in fighting their enemies." As a result, many God-fearing princes and lords of the Holy Roman Empire have come to destruction. Thus, "the advice of a true advisor of all princes" is that they, as well as other nobles and cities, provide themselves with sober, skillful gunners to whom they give adequate provisions and supplies.[35] The *Feuerwerkbuch* addresses, then, a double readership—gunners and the princes and nobles who employ them. It provides a written text that recommends and enhances an image of learning, character, and technical competence for gunner and patron alike.

Much more study is needed of the many further manuscript treatises on weaponry and other kinds of machines and devices that proliferated in the fifteenth-century German empire. They are diverse in both their specific subject matter and their form; some are illustrated, some written in prose, others in verse. Examples include HS. 25,801, in the Germanisches Nationalmuseum in Nuremberg, which depicts various ways in which gun carriages and their aiming mechanisms (such as the elevating arch and the screw) might be constructed; the manuscript known as the "Anonymous of the Hussite Wars," which illustrates and describes machines such as mills, cranes, and aquatic devices, as well as military apparatus; and the *Mittelalterlichen Hausbuch,* from about 1480, which illustrates, among other things, cannon, weight-lifting machines, mills, a spinning wheel, a wagon train, and mining activities and machines.[36]

In the last third of the fifteenth century, several prominent gunners in the region of southwest Germany supplemented their military practice with authorship. One was Johann Formschneider, active in Nuremburg between 1460 and 1470, whose name is attached to a particular group of writings and drawings of cannon.[37] Martin Merz (d. 1501) served the

[34] *Ibid.,* pp. 18 and 47 (for modern German): "sol er got eren und vor allen dingen vor auge[n] haben"; "sich auch bescheidenlich mit der welt halten / mitt der er dann wandelt"; "ain endlicher unverzagter man sein"; "sich tröstlich halten"; "d[er] maister sol auch künden schreiben un[d] lesen."

[35] *Ibid.,* pp. 31 and 69 (for modern German): "durch der kunst weissheit / radt und hülff / sy iren feinden widerstan . . ."; "radt der getrew radtgeb allen fürsten."

[36] Nuremberg, Germanisches Nationalmuseum, HS. 25,801, "Kriegsmachinen," is an early fifteenth-century manuscript that provides explanatory text in verse. Hall, *Technological Illustrations* (cit. n. 1), includes a facsimile edition of the "Anonymous of the Hussite Wars," with a transcription and English translation of the text and a technological commentary. For the last-named text see Helmuth Th. Bossert and Willy F. Storck, eds., *Das mittelalterliche Hausbuch nach dem originale in Besitze des Fürsten von Walburg-Wolfegg-Waldsee im Auftrag des deutschen veneins für Kunstwissenschaft* (Leipzig: E. A. Seemann, 1912). For descriptive inventories of fifteenth-century German manuscripts of this kind see Hall, *Technological Illustrations,* pp. 11–25, 118–133; and Jähns, *Geschichte der Kriegswissenschaften,* Vol. 1 (cit. n. 1), pp. 243–443.

[37] Formschneider's works include Munich, Bayerische Staatsbibliothek, Cod. germ. 734, fols. 60v–71r, of which fol. 60v is signed "Johann Formschneider, büchsen meister und gutter aben teurer"; a group of similar drawings is found in fragments in Munich, Bibliothek des Deutschen Museums, MS 1949/258, and Nuremberg, Germanisches Nationalmuseum, HS. 719 (formerly Kr. 300), among others. See esp. Hall, *Technological Illustrations,* pp. 21–22, 127–129; Jähns, *Geschichte der Kriegswissenchaften,* Vol. 1, pp. 392, 411; Hassenstein, *Feuerwerkbuch von 1420* (cit. n. 31), p. 86; and Eduard A. Gessler, "Hans Formschneider von Nürnberg," in *Die deutsche Literatur des Mittelalters: Verfasserlexikon,* 4 vols., ed. Wolfgang Stammler and Karl Langosch (Berlin: Walter de Gruyter, 1933–1953), Vol. 1, col. 637.

energetic duke Friedrich I of Siegreich (ruled 1451–1476), who reorganized his state administratively while expanding his territorial power. In 1471, apparently while in Friedrich's service, Merz wrote a treatise on the mathematics of aiming cannon. His tomb is in the parish church of Amberg. The oldest memorial of a European gunner, it is covered by a full-length relief sculpture of Merz dressed in a richly ornamented cloak. Inscriptions declare that he was "famous before others in the mathematical art of gunning" and that the dukes of Pfalz always received his heart and work and were truly served by him.[38] Another Pfalz gunner, Philipp Mönch, following Merz's example, wrote a military treatise near the end of the century and included his own portrait on the title page.[39]

A very different kind of military codex was created when the emperor Maximilian, as part of a reorganization and augmentation of the imperial army, ordered the arsenals throughout his far-flung territories to be inventoried by means of large, spectacularly illustrated books—*Zeugbücher.* For each arsenal location, cannon and other weapons were listed on one or more folios and illustrated in beautiful, hand-painted images. Other German princes created similar books by collecting images from the many technical books of the preceding century in large, hand-painted weapons books.[40] Large codices containing exquisitely rendered drawings of cannon and other machines served as new visual representations of princely power.

The many German manuscript codices within this tradition of authorship need much more study with regard to both their substantive content and their relationships to one another. The variety and richness of these books suggest the probability of diverse uses. Certainly the transmission of technical information from one practitioner to another can be included among these uses. Also evident are the value of the artillery codex in patronage exchanges and its cultural utility as a representation of the power of the prince or emperor.

[38] One of his eyes is covered with a patch, his hands hold rosary beads in prayer, and his feet rest on a cannon barrel. The sculpture displays two coats of arms, one depicting a basilisk and the other a heavy gun built into a wheeled gun carriage. The treatise is in Munich, Bayerische Staatsbibliothek, Cod. germ. 599, fols. 66r–101v. For a description see Schneider, ed., *Deutschen Handschriften,* Vol. 5, Pt. 4 (cit. n. 31), pp. 225–226. The location of the copy mentioned in earlier sources—Vienna, Bibliothek des Fürsten Liechtenstein, Cod. MS 3, 1471—has to my knowledge been unknown since the end of World War II. For Merz see Eduard A. Gessler, "Martin Merz (Mercz)," in *Deutsche Literatur des Mittelalters,* ed. Stammler and Langosch, Vol. 3, cols. 368–370; George Sarton, *Introduction to the History of Science,* Vol. 3, Pt. 2: *Science and Learning in the Fourteenth Century* (Washington, D.C.: Carnegie Institution, 1948), p. 1553; and Schmidtchen, *Bombarden, Befestigungen, Büchsenmeister* (cit. n. 30), pp. 153, 182–183. For Friedrich of Siegreich see Henny Grüneisen, "Friedrich I. der Siegreiche," in *Neue deutsche Biographie* (Berlin: Duncker & Humblot, 1961), Vol. 5, pp. 526–528.

[39] Mönch's treatise is in Heidelberg, Universitätsbibliothek, Cod. palat. germ. 126. See Eduard A. Gessler, "Philipp Mönch," in *Deutsche Literatur des Mittelalters,* ed. Stammler and Langosch, Vol. 3, col. 427; Hall, *Technological Illustrations* (cit. n. 1), pp. 23, 125; Jähns, *Geschichte der Kriegswissenschaften,* Vol. 1 (cit. n. 1), p. 271; and Hans Wegener, *Beschreibendes Verzeichnis der deutschen Bilder-Handschriften des späten Mittelalters in der Heidelberger Universitäts-Bibliothek* (Leipzig: J. J. Weber, 1927), pp. 99–101.

[40] One example is Munich, Bayerische Staatsbibliothek, Cod. icon. 222, a large vellum codex of 296 folios of hand-painted illustrations of cannon with their stone or iron balls, culverins (light field guns), and other armaments from specific sites in the empire, by Bartholomeus Freysleben, a locksmith by trade who became master gunner at Innsbruck for the emperor Maximilian I between ca. 1493 and 1509. The compilation of *Zeugbücher* was an aspect of the reorganization and strengthening of the imperial army that included the creation of a center of cannon founding and arms production and a central arsenal at Innsbruck. See especially Erich Egg, "From the Beginning to the Battle of Marignano—1515," in *Guns: An Illustrated History of Artillery,* ed. Joseph Jobé (Greenwich, Conn.: New York Graphic Society, 1971), pp. 26–31; Egg, *Tiroler Geschützguss* (cit. n. 30), pp. 49–94; Hall, *Technological Illustrations,* pp. 23, 126; and Schmidtchen, *Bombarden, Befestigungen, Büchsenmeister* (cit. n. 30), esp. pp. 172–174. Another example is the large *Kriegsbuch*—Erlangen, Universitätsbibliothek, Ms. 1390 (B26)—an illustrated encyclopedia of military drawings taken from fifteenth-century books, done by Ludwig von Eybe zum Hartenstein (the younger) (1450–1521) for Pfalzgraff Philipp bei Rhein, ca. 1500. See Hall, *Technological Illustrations,* p. 124; and Otto Pültz, *Die deutschen Handschriften der Universitätsbibliothek Erlangen,* ed. Armin Dietzel and Günther Bauer (Wiesbaden: Otto Harrassowitz, 1973), pp. 39–40.

A careful look at particular codices suggests another point. Take, for example, a codex in the New York Public Library—Spencer MS 104. The codex retains its fifteenth-century binding, made of beautiful hand-tooled calf leather stretched over wooden boards, and its iron clasp, lock, and key. Inside are careful copies of two treatises—the *Feuerwerkbuch* and a German translation of Kyeser's *Bellifortis.* The illustrations for the latter are only partially completed, and the specific ownership history of the codex in the early modern period is unknown.[41] Yet the two texts—very different treatises, which I have treated in different sections here—were put together for readers or collectors who (the beautiful binding tells us) were far from lowborn. My point is that the tradition of authorship discussed here also had the effect of creating a readership made up of individuals who valued, collected for their libraries, and perhaps even read books on the mechanical arts.

THE HUMANIST SYNTHESIS OF *TECHNE* AND *PRAXIS:* ALBERTI, CUSANUS, VALTURIO

Unlike the books of German gunners, humanist writings on the mechanical arts have been the focus of extensive scholarship. Humanism was a broadly based cultural and educational movement that favored rhetoric and ethics over logic and philosophy. In general, humanists advocated a life of active civic participation over one of contemplation. They attempted to reinstate classical, Ciceronian Latin in the place of the "barbarisms" of medieval Latinity. Finally, they wrote histories, both ancient and contemporary, the latter often undertaken as part of their service in the courts and republics of patrons.[42] To put it another way, the humanists embraced *praxis,* that is, political and military action, and the linguistic strategies that shaped it. Some humanists joined this interest in *praxis* with a concern for *techne* or *ars,* construction in the material world, and they undertook to write about that construction.

Three key figures in the humanist synthesis of *praxis* and the constructive arts were contemporaries and probably acquaintances—Leon Battista Alberti, Nicholas Cusánus, and Roberto Valturio. Alberti (1404–1472), the illegitimate son of an exiled Florentine banker, was one of the most important humanists of the fifteenth century. Supported by ecclesiastical positions and benefices, as well as the patronage of princes, he wrote on a wide range of subjects—literary, moral, and technical, including painting, sculpture, and architecture—and was involved as well in urban reconstruction and building design.[43] The German cardinal Nicholas of Cusa (1401–1464) attended the University of Heidelberg

[41] New York Public Library, Spencer MS 104.

[42] For an introduction to humanism, including extensive further bibliography, see esp. Albert Rabil, Jr., ed., *Renaissance Humanism: Foundations, Forms, and Legacy,* 3 vols. (Philadelphia: Univ. Pennsylvania Press, 1988); Pamela O. Long, "Humanism and Science," *ibid.,* Vol. 3, pp. 486–512; and Anthony Grafton, *Defenders of the Text: The Traditions of Scholarship in an Age of Science, 1450–1800* (Cambridge, Mass.: Harvard Univ. Press, 1991).

[43] The foundational study of Alberti's life and work was Girolamo Mancini, *Vita di Leon Battista Alberti,* 2nd ed. (Florence: G. Carnesecchi, 1911). See also esp. Franco Borsi, *Leon Battista Alberti* (Milan: Electa, 1975); Cecil Grayson and Giulio C. Argan, "Leon Battista Alberti," in *Dizionario biografico degli Italiani* (Rome: Istituto della Enciclopedia Italiana, 1960), Vol. 1, pp. 702–713; and Joan [Kelly] Gadol, *Leon Battista Alberti: Universal Man of the Early Renaissance* (Chicago: Univ. Chicago Press, 1969). Recent studies include Mark Jarzombek, *On Leon Baptista Alberti: His Literary and Aesthetic Theories* (Cambridge, Mass.: MIT Press, 1989); Paolo Marolda, *Crisi e conflitto in Leon Battista Alberti* (Rome: Bonacci, 1988); and Giovanni Ponte, *Leon Battista Alberti: Umanista e scrittore* (Genoa: Tilgher-Genova, 1981). See also Susannah Foster Baxendale, "Exile in Practice: The Alberti Family In and Out of Florence, 1401–1428," *Renaissance Quarterly,* 1991, *44*:720–756; and Susannah Kerr Foster, "The Ties That Bind: Kinship Association and Marriage in the Alberti Family, 1378–1428" (Ph.D. diss., Cornell Univ., 1985). A brief, useful introduction to Alberti's life is Martin Kemp, "Introduction," in Leon Battista Alberti, *On Painting,* trans. Grayson (London: Penguin, 1991).

before coming to Italy, where he absorbed humanist learning. He pursued an illustrious career in the papal court, often served as a diplomat between Italian and German regions, and wrote brilliantly original treatises devoted primarily to theological subjects. Roberto Valturio (1405–1475) was born in Rimini, where he received a humanist education, then worked in Rome as secretary in the papal court, and finally returned to his native city to serve the lord, Sigismund Malatesta. Unlike Alberti and Cusanus, Valturio wrote only one book, *Elenchus et index rerum militarum.* It became the most widely disseminated military treatise of the fifteenth century.[44]

Battista Alberti was a preeminent man of letters who was also deeply involved in the constructive arts. Although his complex literary production cannot be recapitulated here, two points deserve emphasis. First, in addition to a base of ecclesiastical service and benefices, Alberti's authorship was fundamentally grounded in the patronage of oligarchs and princes such as the Este of Ferrara, the Gonzaga of Mantua, the Rucellai of Florence, Sigismund Malatesta of Rimini, and the Florentine pope Tommasso Parantucello—Nicholas V. Second, Alberti himself became active in the constructive arts as a consultant and designer of building projects, such as churches and the redesign of cities.[45]

Arriving in Florence sometime after the ban against his family was lifted in 1428, Alberti was deeply influenced by Florentine developments in the visual and constructive arts. He came in time to observe the spectacular achievements of Brunelleschi: his engineering feat of raising the immense dome of the Florentine cathedral, his classically proportionate building design, and his discovery of painter's perspective. Alberti also observed the perspective paintings of Masaccio and others, the gilded bronze relief sculptures by Ghiberti on the doors of the Florentine baptistery, and Donatello's newly classical sculpture.[46] He

[44] Among his other achievements, Nicholas of Cusa contributed to the resolution of the "great schism" by arguing for the authority of councils (*De concordantia catholica,* 1433) and wrote a proposal for negotiation with the Hussites. For an introduction see J. E. Hofmann, "Nicholas Cusa," in *Dictionary of Scientific Biography,* ed. Charles C. Gillispie, 18 vols. (New York: Scribner's, 1970–1990), Vol. 3, pp. 512–516; Jasper Hopkins, *A Concise Introduction to the Philosophy of Nicholas of Cusa* (Minneapolis: Banning, 1978); and Pauline Moffitt Watts, *Nicolaus Cusanus: A Fifteenth-Century Vision of Man* (Leiden: Brill, 1982). For Valturio's life see Friedrich Klemm, "Roberto Valturio," in *Dictionary of Scientific Biography,* ed. Gillispie, Vol. 13, pp. 567–568; and Pier Luigi Bassignana, ed., *Le macchine di Valturio nei documenti dell'Archivio Storico Amma* (Turin: Umberto Allemandi, 1988), pp. 13–16.

[45] An example of Alberti's use of authorship to acquire patronage can be seen in his treatise on painting. He dedicated the Latin version of *De pictura* to Giovan Francesco Gonzaga (1407–1444), "illustrious prince of Mantua." Alberti praises the prince's kindness, glory in arms, and skill in letters. The aim of the dedication is explicit: "You could know my character and learning, and all my qualities best, if you arranged for me to join you, as I indeed desire. And I shall believe my work had not displeased you, if you decide to enrol me as a devoted member among your servants and to regard me as not one of the least." Alberti's dedication brought the desired results. Although Giovan Francesco died in 1444, his son Lodovico became an important patron of Alberti's architecture, including two major churches in Mantua—S. Sebastiano and S. Andrea. See Alberti, *On Painting,* trans. Grayson (cit. n. 43), p. 36; and Leon Battista Alberti, "De pictura," in *Opere volgari,* Vol. 3, ed. Cecil Grayson (Bari: Laterza, 1973), pp. 7–107 (for the Latin text and Alberti's Italian translation). For Alberti's architecture, in addition to works cited above, see esp. Smith, *Architecture in the Culture of Early Humanism* (cit. n. 6), pp. 19–39 (for his relationship to Brunelleschi), 98–129 (for Pienza). For Alberti's involvement in Nicholas V's Rome see Carroll William Westfall, *In This Most Perfect Paradise: Alberti, Nicholas V, and the Invention of Conscious Urban Planning in Rome, 1447–1455* (University Park: Pennsylvania State Univ. Press, 1974); and Manfredo Tafuri, " 'Cives esse non licere': The Rome of Nicholas V and Leon Battista Alberti: Elements toward a Historical Revision," *Harvard Architecture Review,* 1987, *6*:60–75.

[46] For an introduction to these developments see Frederick Hartt, *History of Italian Renaissance Art: Painting, Sculpture, Architecture,* 4th ed., revised by David G. Wilkins (New York: Abrams, 1994). On artist's perspective see Samuel Y. Edgerton, Jr., *The Renaissance Rediscovery of Linear Perspective* (New York: Basic, 1975); and John White, *The Birth and Rebirth of Pictorial Space,* 3rd ed. (Cambridge, Mass.: Harvard Univ. Press, Belknap, 1987).

made his own significant contributions to this visual revolution, in part through technical authorship.[47]

Traditional scholarship portrayed Alberti as one who changed the status of painting and architecture from mechanical to liberal arts by making those disciplines mathematical. It emphasized those aspects of Alberti's work that concerned design and mathematical proportion and tended to disregard his interests in engineering and material construction. More recently, the traditional erasure of Alberti's grounding in material manipulation and construction has been corrected. Christine Smith has cogently argued that Alberti's admiration for Brunelleschi was based not so much on his design of proportionate buildings as on the engineering triumph represented by the dome of the Florentine cathedral and the many ingenious weight-lifting machines that he invented to build it. Similarly, John Oppel suggests that although it is usually said that Alberti "raised architecture to the status of one of the liberal arts," it is "at least as true and historically more appropriate to argue that he did just the reverse, that he brought the liberal arts down to the level of the mechanical ones." For Alberti, indeed, "the whole of architecture is based on mechanics."[48] I would say, rather, that for Alberti both painting and architecture remain firmly grounded in material construction but that they are mathematical as well and are placed emphatically within the ethical world of human society. Instead of "mere" mechanical arts, they now exist as a focal point of ethical and civic life, of material construction, and of mathematics.

Alberti produced two versions of his treatise on painting—one in Latin, dedicated to a potential patron, Giovan Francesco Gonzaga, and the other in Italian, dedicated to Brunelleschi.[49] He begins *De pictura* by insisting that he is writing not as a mathematician

[47] In addition to the treatises on painting and architecture that I discuss in greater detail, Alberti's technical writings include the following:

Elementi di pittura, probably written in Florence in the early 1430s, on geometric figures and their transposition onto foreshortened planes. See Leon Battista Alberti, "Elementi di pittura," in *Opere volgari,* ed. Grayson, Vol. 3 (cit. n. 45), pp. 109–129, 341–351; and Alessandro Gambuti, "Nuove ricerche sugli 'Elementa picturae,' " *Studi e Documenti di Architettura,* 1972, *1*:133–172.

The *De statua,* a treatise on sculpture that deals with human proportions and with a device for reproducing such dimensions in stone or clay. See Alberti, *"On Painting" and "On Sculpture": The Latin Texts of "De pictura" and "De statua,"* ed. and trans. Cecil Grayson (London: Phaidon, 1972), pp. 117–144; and Jane Andrews Aiken, "Leon Battista Alberti's System of Human Proportions," *J. Warburg Courtauld Inst.,* 1980, *43*:68–96.

The *Ludi rerum mathematicarum* (ca. 1450), on measuring distances, dimensions, and weights. See Alberti, "Ludi rerum mathematicarum," in *Opere volgari,* ed. Grayson, Vol. 3, pp. 131–173; and Luigi Vagnetti, "Considerazioni sui 'Ludi matematici,' " *Stud. Doc. Architettura,* 1972, *1*:173–259.

A treatise on the use of an instrument to survey the city of Rome—*Discriptio urbis Romae.* See Vagnetti, "La 'Discriptio urbis Romae,' uno scritto poco noto di Leon Battisti Alberti," *Quaderno,* 1968, *1*:25–78.

A now-lost work that explains how he attempted to raise a sunken Roman galley from Lake Nemi—*De navis;* and another lost book on lifting weights. See Gadol, *Leon Battista Alberti* (cit. n. 43), p. 204; Mancini, *Vita di Alberti* (cit. n. 43), pp. 281 (on *De navis*), 286–288 (on *De motibus ponderis*). On *De motibus ponderis* see esp. Alberti, *Opere volgari,* ed. Grayson, Vol. 3, p. 431.

A tract on cryptography, *De compondendis cifris,* that explains his invention of a code wheel. See Gadol, *Leon Battista Alberti,* pp. 207–211.

[48] Smith, *Architecture in the Culture of Early Humanism* (cit. n. 6), pp. 19–39; and John Oppel, "The Priority of the Architect: Alberti on Architects and Patrons," in *Patronage, Art, and Society,* ed. Kent and Simons with Eade (cit. n. 7), pp. 251–267, on pp. 251, 262. The traditional view is elaborated most importantly in the classic work: Rudolf Wittkower, *Architectural Principles in the Age of Humanism,* 4th ed. (London: Academy, 1973), esp. pp. 1–56.

[49] Alberti, *On Painting,* trans. Grayson (cit. n. 43), p. 34, where Alberti praises the great achievements in the arts carried out by Brunelleschi, as well as by "our great friend the sculptor Donatello" and by Ghiberti, Luca della Robbia, and Masaccio. Alberti describes the achievements of these men as "in no way inferior to" those of the ancients and asks that Brunelleschi correct any aspect of his treatise that seems to need amendment. For the dedication to Francesco Gonzaga see note 45, above.

but as a painter. What he means is that "mathematicians measure the shapes and forms of things in the mind alone and divorced entirely from matter." He, on the other hand, wants to speak in terms of Minerva, about visible things. Speaking as a painter—one who does not separate mathematical and theoretical concerns from the visible world—Alberti also frequently alludes to ancient texts and the ancient art of painting. He speaks, that is, not as an ordinary but as a learned painter. *De pictura* treats the process of vision, the creation of the illusion of three-dimensional space on a flat surface by one-point perspective, and procedures of composition, light, and color. It also concerns the painter's creation of *istoria,* that is, a dramatic effect in the painting that moves the soul of the viewer. Finally, it enumerates desirable moral attributes of the painter, of which the first is that he should "be a good man, well-versed in the liberal arts."[50]

Alberti presented his architectural treatise, *De re aedificatoria,* to Pope Nicholas V around 1452, during the time when both were involved in the urban renovation of Rome.[51] In this, the most important architectural treatise of the Renaissance, Alberti separates architecture from craft practice by means of mathematics and design. Yet he also grounds it in engineering for the improvement of civic life. The architect is not a carpenter but, rather, one "who by sure and wonderful reason and method, knows both how to devise through his own mind and energy, and to realize by construction, whatever can be most beautifully fitted out for the noble needs of man, by the movement of weights and the joining and massing of bodies." Architecture provides houses, walks, swimming pools, and baths to maintain human health. It supplies vehicles, mills, timepieces, and other smaller inventions that play a vital role in everyday life. It has devised methods of drawing up vast quantities of water for many purposes, and it has produced buildings for divine worship. Further, by cutting through rock, tunneling through mountains, filling in valleys, restraining the waters of the sea and lakes, draining marshes, building ships, altering the course of rivers, dredging the mouths of rivers, and constructing harbors and bridges, the architect has not only "met the temporary needs of man, but also opened up new gateways to all the provinces of the world," allowing nations to exchange food and goods as well as experience and knowledge. Architecture moreover has provided ballistic engines and machines of war, fortresses, and other things that protect liberty.[52]

Alberti repeatedly emphasizes that architecture, civic life, and human nature are closely linked. "When you erect a wall or portico of great elegance . . . good citizens approve and express joy for their own sake, as well as for yours, because they realize that you have used your wealth to increase greatly not only your own honor and glory, but also that of your family, your descendants, and the whole city." Types of buildings should be commensurate, Alberti insists, with types of people. The patron should take account of his

[50] Kemp, "Introduction," in Alberti, *On Painting,* trans. Grayson, pp. 12, 37. The allusion to Minerva is to Cicero's *De amicitia,* where the ancient author refers to a more popular kind of knowledge; for Alberti's use of *De amicitia* 5.19 see Leon Battista Alberti, *On Painting,* rev. ed., trans. John R. Spencer (New Haven, Conn.: Yale Univ. Press, 1966), pp. 18–20. For the desirable moral attributes of the painter see Alberti, *On Painting,* trans. Grayson, p. 87. On *istoria* see esp. Jack M. Greenstein, "Alberti on Historia: A Renaissance View of the Structure of the Significance of Narrative Painting," *Viator,* 1990, *21*:273–299.

[51] Leon Battista Alberti, *L'architettura (De re aedificatoria),* ed. and trans. Giovanni Orlando, 2 vols. (Milan: Il Polifilo, 1966); and Alberti, *On the Art of Building in Ten Books,* trans. Joseph Rykwert, Neil Leach, and Robert Tavernor (Cambridge, Mass.: MIT Press, 1988).

[52] Alberti mentions several of his own inventions that have to do with rigging gangplanks of ships to prevent boarding by an enemy and other devices for sinking and burning enemy ships. He promises that he will deal with war machines at greater length elsewhere, implying, perhaps, a plan to write a treatise on the subject: Alberti, *Art of Building,* trans. Rykwert *et al.,* pp. 3–4 (quotation), 135–136 (war machines and his own military inventions); and Alberti, *L'architettura (De re aedificatoria),* ed. and trans. Orlando, Vol. 1, pp. 6–13, 386–393.

social standing when he commissions a building because "it is a sign of a well-informed and judicious mind to plan the whole undertaking in accordance with one's position in society and the requirements of use." The variety of buildings is the result of the differences in human nature. Just as various human types, such as the wise, the skilled and practical, and the wealthy, have different functions in society, so also do different building types. Some buildings are appropriate for society as a whole, some for the foremost citizens, some for the common people. Among leading citizens, those presiding over domestic councils require different buildings than those involved in executing decisions or those occupied with accumulating wealth. More broadly, the type of government determines the architectural decisions of the ruler—tyrants must make different architectural decisions than magistrates.[53]

Architecture is an ethical discipline, and one that also involves the acquisition of knowledge. It is not a perfected set of principles but involves an ongoing process of study. It requires empirical investigation and observation, as well as the collection of advice from the past—"we shall collect, compare, and extract into our own work all the soundest and most useful advice that our learned ancestors have handed down to us in writing, and whatever principles we ourselves have noted in the very execution of their works. We shall go on to report things contrived through our own invention, by careful, painstaking investigation, things we consider to be of some future use." This reporting, Alberti repeatedly emphasizes, must be open and clear, achieved with the use of concise, understandable technical language.[54]

Although Alberti's literary and practical interests were in some ways very different from Cusanus's theological concerns, the two men shared the conviction that knowledge and practical life, including practical mathematics, were closely interrelated.[55] Cusanus's views concerning such connections are strikingly evident in his *Idiota: De sapientia, De mente, De staticis experimentis* (ca. 1450). *Idiota* contains four books, the first two of which, on wisdom, take the form of a dialogue between an unlearned man (the *idiota*) and an educated orator. The *idiota* points the way to wisdom by rejecting the learning of the orator, which is based on the authority of books. He suggests instead that wisdom can be found in the streets and the marketplace, where ordinary weighing and measuring occur. By elaborating

[53] Alberti, *Art of Building*, trans. Rykwert *et al.*, pp. 4–5, 367 n. 12 ("good citizens approve"), 37 ("judicious mind"), 92–94 (building types), 117–125 (types of government and architecture). For a discussion of Alberti's civic and moral positions, as well as his debt to Cicero's *De officiis*, see John Onians, *Bearers of Meaning: The Classical Orders in Antiquity, the Middle Ages, and the Renaissance* (Princeton, N.J.: Princeton Univ. Press, 1988), pp. 147–157.

[54] Alberti, *Art of Building*, trans. Rykwert *et al.*, p. 7; see also p. 24, where Alberti stresses inventiveness rather than slavish imitation as a desirable approach to the architectural orders. On writing about architecture, he remarks: "But since it is our desire to be as limpid, clear, and expeditious as possible in dealing with a subject otherwise knotty, awkward, and for the most part thoroughly obscure, we shall explain, as is our custom, the precise nature of our undertaking. For the very springs of our argument should be laid open, so that the discussion that follows may flow more easily." *Ibid.*, p. 7. Elsewhere (p. 155) Alberti lamented the obscurity and poor Latinity of Vitruvius's text, but, like Vitruvius, he stressed the difficulty of writing on architecture. He preferred that his own "speech seemed lucid [rather] than appeared eloquent" and believed that his writing was "in proper Latin, and in comprehensible form." Openness and clarity are values that Alberti repeatedly emphasizes. See Richard Krautheimer, "Alberti and Vitruvius," *Acts of the Twentieth International Congress of the History of Art* (Princeton, N.J.: Princeton Univ. Press, 1963), Vol. 2, pp. 42–52; and Alberti, "Ludi rerum mathematicarum" (cit. n. 47), p. 173.

[55] For the reciprocal influence of Alberti and Cusanus see Riccardo Gavagna, "Un abbinamento editoriale del '500: Vitruvio e Cusano," *Rivista Critica di Storia della Filosofia*, 1975, *30*:400–410; and Gavagna, "Cusano e Alberti a proposito del 'De architectura' di Vitruvio," *ibid.*, 1979, *34*:162–176. Cusanus's own mathematical writings include the *Transmutationes geometricae* (1445), which treats changing a circle into a square and a straight line and includes many examples from practical geometrical problems.

upon these activities as they see them in the Roman Forum, the unlearned man leads the orator to wisdom, that is, comprehension of the divine, through mathematics. Wisdom, which all men by nature desire to know, is higher than knowledge and utterly unknowable, unspeakable by all language, unintelligible by any understanding, immeasurable by all measure. Cusanus's elaboration involves the notion that the divine involves absolute number, precision, equality, and other mathematical entities that in the human realm can only be relative.[56]

The third book, *De mente,* concerns the human mind; Cusanus argues that the word for it—*mens*—derives from the same word as *measure* (*mensur*). In this book, which is set in the Jubilee year of 1450, a philosopher seeking knowledge about the mind from ancient books in Rome is found by the orator and led to the Forum, where the *idiota* is busy making spoons in the Temple of Eternity. The unlearned man defends his "unworthy" occupation by claiming that the maker of spoons and pots is even closer to the divine than the painter and others who imitate nature. There are no spoons in nature; likewise, the divine creation was original, not imitative.[57]

Cusanus moves from this remarkable vindication of artisanal labor to the fourth book, *De staticis experimentis,* which describes a series of experimentis in practical measurement, including the determination of specific gravities and the assaying of metals. Speaking through the *idiota,* Cusanus points to many situations in which comparative measurement of the weight of various substances, including the blood and urine of sick and well men, water from different fountains, stones, and metals, could lead to a better knowledge of the nature of those substances. The *idiota* insists that the myriad results of such measurements should be written down in charts and that one result would be human benefit—for example, physicians could better diagnose and treat patients.[58] For Cusanus, divine and human mathematics are separated by a conceptual gulf that renders the first perfect and absolute, the second always imperfect and relative. Nevertheless, the two remain linked, and Cusanus demonstrates a continuing interest in both.

Around the time that Cusanus completed *De mente,* Sigismund Malatesta (1417–1468), lord of Rimini, instructed his secretary, Roberto Valturio, to write a treatise on military matters. Sigismund had acquired Rimini through his father, the famous *condottiero* Pandolfo Malatesta, who had bestowed it upon him and his brothers. At the precocious age of fourteen, Sigismund wrested it by military force from his brothers and from an uncle who was contesting the bequest. In 1433, some time after his spectacular adolescent military victory, Sigismund, lord of Rimini, was knighted by the emperor Sigismund during a stop in Rimini on his way north after the imperial coronation. Thereafter, the lord of Rimini consolidated his rule by means of an extensive building campaign that included the construction of the great fort and palace, the Rocca Malatestiana, and the radical

[56] Nicolai de Cusa, *Opera omnia,* Vol. 5: *Idiota: De sapientia, De mente, De staticis experimentis,* rev. ed., ed. Renata Steiger *et al.* from edition of Ludwig Bauer (Hamburg: Felix Meiner, 1983). There is an anonymous seventeenth-century English translation—Cusanus, *The Idiot in Four Books* (London: William Leake, 1650), from which I have paraphrased text on pp. 7–8; and more recent English translations of the third book—Nicholas de Cusa, *Idiota de Mente: The Layman: About Mind,* trans. Clyde Lee Miller (New York: Abaris, 1979)—and of the fourth book—"De staticis experimentis," in Henry Viets, "*De staticis experimentis* of Nicolaus Cusanus," *Annals of Medical History,* 1922, *4*:115–135.

[57] Cusa, *Idiota de Mente,* trans. Miller, pp. 41–45. For the importance of the dialogue for the idea of creativity see Hans Blumenberg, " 'Nachahmung der Natur': Zur Vorgeschichte der Idee des schöpferischen Menschen," *Studium Generale,* 1957, *10*:266–283. *Idiota* involves a complicated commentary on Plato's *Apology* (21a–23b), in which Socrates recognizes the wisdom of understanding his own ignorance in the face of divine wisdom, and on the *Timaeus.*

[58] Cusa, *Opera omnia,* Vol. 5: *Idiota: De staticis experimentis,* ed. Steiger *et al.* (cit. n. 56), pp. 219–242; and Viets, "*De staticis experimentis*" (cit. n. 56).

redesign (primarily by Alberti) of the Church of San Francesco, with a facade in the form of a triumphal arch, that came to be known as the "Tempio Malatestiano" (see Figure 6).[59] The form and decorations of the Tempio Malatestiano bring together Roman antiquity, humanist learning, and Sigismund's power and rulership—affiliations that Valturio also repeatedly invokes in his military treatise.

The *Elenchus et index rerum militarium* (often cited as *De re militari*) is a humanist tour de force in which Valturio compiles a vast amount of material from ancient military literature and discusses Sigismund's own military and architectural feats as if they constituted extensions of Roman magnificence. The most widely disseminated military treatise of the fifteenth century, it exists in at least twenty-two manuscript copies and (after 1472) in numerous printed editions and translations. Yet the favor it found in its own time has not been shared by modern scholars, who have generally ignored or scorned it. Samuel Y. Edgerton, Jr., for example, suggests that the popularity of the treatise discouraged the printed publication of engineering books for fifty years. His judgment seems to be based on a model of technological progress in which the most desirable engineering books are deemed to be those that convey the latest technical information.[60] Such a judgment would not have been shared by Valturio's contemporaries, who seem to have appreciated his erudite explication of ancient writings, his ability to bring together ancient and contemporary weapons, and his brilliant success at representing the power and glory of Sigismund Malatesta, his patron and prince.

Unlike most of his multifarious ancient military sources, Valturio supplements his examination of strategy, generalship, military rituals, order of battle, recruiting, and the like with extensive treatment of weapons technology—ancient, fantastical, and modern. Manuscript copies of the treatise include numerous beautifully hand-painted illustrations of military machines, indebted in part to the illustrations of Kyeser's *Bellifortis*.[61]

[59] For Sigismund Malatesta see esp. Philip J. Jones, *The Malatesta of Rimini and the Papal State: A Political History* (Cambridge: Cambridge Univ. Press, 1974), pp. 176–239; and Mario Tabanelli, *Sigismondo Pandolfo Malatesta: Signore del medioevo e del rinascimento* (Faenza: Stabilimento Grafico Fratelli Lega, 1977). Older studies include Salvatore Sibilia, *Sigismondo Malatesta: Signore di Rimini* (Turin: G. B. Paravia, 1942); and Charles Yriarte, *Un condottiere au xv siècle Rimini: Études sur les lettres et les arts à la cour des Malatesta* (Paris: J. Rothschild, 1882). For Sigismund's building projects see especially Carla Tomasini Pietramellara and Angelo Turchini, eds., *Castel Sismondo e Sigismondo Pandolfo Malatesta* (Rimini: Ghigi, 1985); and Woods-Marsden, "Images of Castles in the Renaissance" (cit. n. 7). For the Tempio Malatestiano see Corrado Ricci, *Il Tempio Malatestiano,* with appendix by Pier Giorgio Passini (1924; rpt., Rimini: Ghigi, 1974); and Charles Hope, "The Early History of the Tempio Malatestiano," *J. Warburg Courtauld Inst.*, 1992, *55*:51–154, esp. pp. 58–59 (for the papal bulls), 91 (for Piero della Francesca), 86–87 and 118–119 (for Roberto Valturio), 95 ff. (for Alberti), and 59–62, 66–68, 94–96, and 151–154 (for Sigismund's personal involvement in the project). For Piero della Francesca's work in Rimini see Marilyn A. Lavin, *Piero della Francesca a Rimini: L'affresco nel Tempio Malatestiano* (Bologna: Nuova Alfa, 1984).

[60] Valturio began the treatise around 1450 and completed it about 1455–1460. See Roberto Valturio, *Elenchus et index rerum militarium quae singulis codicis huius voluminibus continentur ut inpromptu sint universa hoc est ne talim Rerum curiosa perleganti* (Verona: Johannes ex Verona oriundus Nicholai Cyrugie Medici Filius, 1472). For a partial facsimile reproduction of one manuscript at Turin, Archivio Storico Amma, MS. lat., cart., sec. XV, 142 cc., and useful articles concerning the illustrations see Bassignana, ed., *Macchine di Valturio* (cit. n. 44), which includes an (incomplete) list of manuscript copies and further bibliography; and Galluzzi, ed., *Prima di Leonardo* (cit. n. 1), pp. 199–201. For Edgerton's discussion see Samuel Y. Edgerton, Jr., *The Heritage of Giotto's Geometry: Art and Science on the Eve of the Scientific Revolution* (Ithaca, N.Y.: Cornell Univ. Press, 1991), pp. 145–147.

[61] The respect accorded Valturio by the Malatesta is indicated by his burial in one of the exterior niches of the Tempio Malatestiano. Most of the material concerning military technology is in book 10, which concerns military armor, weapons, machines, and other devices. The early printed edition contains eighty-two woodcuts of military machines and devices, possibly made by Matteo de' Pasti, a sculptor in the court of Sigismund; see Bassignana, ed., *Macchine di Valturio,* pp. 13–16.

Figure 6. *Tempio Malatestiano, eastern view, Rimini, Italy. (Courtesy of Alinari/Art Resource, New York.) The lord of Rimini, Sigismund Malatesta, had the Church of San Francesco in Rimini rebuilt to his greater glory. The new facade, which resembles a Roman triumphal arch, was designed by Alberti. Some of the decorations inside were designed with the help of Roberto Valturio. Sigismund's burial niche is decorated with a fresco painted by Piero della Francesca. The fresco depicts Sigismund kneeling before a St. Sigismund who strikingly resembles the emperor of the same name. Valturio is buried in one of the outside niches.*

Valturio suggests that by means of authorship he has placed Sigismund "in the seats of the celestial and terrestrial councils of the immortals." Just as the fame of Hector and Achilles in the Trojan War, and of many other ancients as well, has been transmitted "by the protection of letters," so his writings will serve the same function for the lord of Rimini.[62]

War itself boasts an ancient and honorable history. Emphasizing the civic nature of the military arts, Valturio suggests that military leaders develop by means of three necessary elements—nature, teaching, and practice. Also important are the instruments of war and the "many ornaments of liberal and noble arts" that are part of military science. Following the Vitruvian list of disciplines necessary to the architect, Valturio insists that the military leader must be literate, emulate the precepts of philosophy, understand history, and have good recollection. He is not ignorant of oratory and poetry; he understands music, arithmetic, geometry, the stars of the sky, the laws and rules of diverse people, and, finally, gymnastics and military practice in both peace and war.[63]

The military leader should be both "literate . . . and learned." He should be knowledgeable, not in "the vulgar, crass, and crude" things that are now used in camp, but in "noble and very precise subjects joined with knowledge of many things." Such knowledge enables the captain to read and understand the precepts and examples of those who wrote. Indeed, Valturio counsels Sigismund, when at home "you can be alone with old friends, that is, with books. Before all things, it is necessary to be soaked with ingenuity and almost seized by the work of the best teachers."[64]

Valturio provides a long list of Greek and Roman generals who were both great soldiers and learned and studious men. Primacy belongs to Alexander of Macedon, whose tutor was Aristotle. The list also includes Julius Caesar—"it is said that Julius Caesar never missed a day in the pursuit of letters"—and Augustus, an avid reader and patron of poets and scholars. Valturio moves from his last ancient example, Marcus Aurelius, directly to Sigismund himself, describing the lord of Rimini as one who reads and hears much, who has great knowledge and the seeds of all the arts, who takes pleasure in subtle questions, favors orators and poets, and collects books, which will be useful in the present and the future. Valturio praises Sigismund for his military prowess, his poetry, his buildings (he describes the palace and fort of Rimini in detail), and his growing library, which he compares to the famous libraries of antiquity. Given the vicissitudes of fortune, the impermanence of all monuments, the monument of letters lasts longest.[65]

The writings of Valturio, Cusanus, and Alberti exemplify a strain of humanism that was deeply concerned with the technical and constructive arts. Yet the three men were not artisans but university-educated humanists who stand among the most learned and influential authors of the fifteenth century. Their collective grasp of ancient learning and hu-

[62] Valturio, *Elenchus et index rerum militarium* (cit. n. 60), "Ad magnanimum . . . Sigismundu[m]," n.p.: "concilium caelestium terrestriumq[u]e insedibus i[m]mortalitatis"; "litterar[um] custodiae." Valturio adds that he hopes that he himself might also achieve immortality in future generations through his writings.

[63] *Ibid.*, bk. 1, ch. 2.: "plurima liberaliu[m] et optimarum artiu[m] orname[n]ta." (Book 1 treats the origin of war and its history in ancient times.) For the Roman architect Vitruvius's list of the disciplines that should be known by the architect see *De architectura* 1.1.3.

[64] Valturio, *Elenchus et index rerum militarium,* bk. 1, ch. 3: "litteratum . . . et eruditum"; "Non vulgarem: non crassam: et barbaram"; "sed pr[a]eclaram et accuratam illam cum rerum multarum scientia coniunctam" (what the crude, vulgar camp learning is, he does not elaborate); "cum veteribus amicis idest libris esse seorsu[m] possis: ante omnia tinctu[m] esse oportet ingeniu[m] et quasi iniciatu[m] pr[a]eceptoris optimi opera."

[65] *Ibid.:* "Nullum a studiis litterarum diem intermisisse Iul. Cae. traditur." Valturio describes the Rocca Malatestiana, including its walls, towers, and munitions, as "a true marvel of the magnificence of Italy [italic(a)e magnificenti(a)e vera admiratio]."

manist Latinity was unparalleled in their own time. These men were not stooping to dabble in the low mechanical arts. Rather, they were participating in a reformulation of the value and place of technological culture. Valturio's studious military leader with a technological arsenal at hand; Cusanus's orator leaving his books to learn from transactions in the marketplace and from the unlearned artisan; Alberti's reformulation of the principles of painting, sculpture, and architecture in terms of rational principles, mathematics, and engineering—these were not the delusions of men whose fantasies had taken them outside of their own cultures. Neither is there the slightest reason to believe that any of the three was engaged in combating the hierarchical social structure of his time. The careers and authorship of all three were fundamentally based on patronage, whether clerical or princely. Their writings represent the highest expression of a culture in which political power and certain constructive and mechanical arts had come to form a close alliance.

PRACTITIONER AUTHORSHIP IN LATE QUATTROCENTO ITALY

Just as humanists were influenced by visual and constructive arts in the fifteenth century, artisan practitioners were inspired by humanist writings on such arts. In the second half of the century, artisan-trained practitioners such as Lorenzo Ghiberti, Antonio Averlino, Piero della Francesca, Francesco di Giorgio Martini, and Leonardo da Vinci practiced skilled crafts and also wrote treatises. By means of authorship, they aided in the transformation of certain arts from skilled practice to rational discipline and simultaneously improved their own fortunes as well. Artisans trained in workshops, such as Ghiberti, Francesco di Giorgio, and Leonardo da Vinci, struggled to learn Latin; yet they shared with the humanists a cultural arena of common interests and patronage. Practitioners and humanists together created a middle ground of discursive and constructive practice within the space that had separated learned and artisanal culture in earlier centuries.

The Florentine Lorenzo Ghiberti (1378?–1455) trained as a goldsmith in the workshop of his stepfather Bartolo di Michele, called Bartoluccio. In 1400 he established his reputation by winning a competition to make relief panels for the gilded bronze doors of the Florentine baptistery. The award, which made the Bartoluccio workshop the most important bronze foundry in the city, also provided the basis for Ghiberti's subsequent career (see Figure 7).[66] Much later, around 1447, Ghiberti wrote his tripartite treatise on sculpture known as the *Commentarii,* which includes a history of sculpture that begins in antiquity and ends with a catalogue of his own works. Like Valturio, Ghiberti excerpts extensively from ancient writings, beginning with a citation from the book on siege machines by Athenaios Mechanicos that advocates brevity and clarity over rhetorical polish. Ghiberti believed that in antiquity skilled practitioners routinely wrote books that elucidated the principles of their art but that this practice had lapsed. Only in his own age (and his own work) were ancient heights again reached in matters of art itself and in the elucidation of the principles of art in writing.[67]

[66] For Ghiberti's life and works see Richard Krautheimer with Trude Krautheimer-Hess, *Lorenzo Ghiberti,* 3rd ed. (Princeton, N.J.: Princeton Univ. Press, 1982). See also *Lorenzo Ghiberti: "Materia e rationamenti"* (Firenze, Museo dell'Accademia e Museo di San Marco, 18 Ottobre 1978–31 Gennaio 1979) (Florence: Centro Di, 1978); and *Lorenzo Ghiberti nel suo tempo,* 2 vols. (Atti del Convegno Internazionale di Studi, Firenze, 18–21 Ottobre 1978) (Florence: Leo S. Olschki, 1980).

[67] Julius von Schlosser, *Lorenzo Ghibertis Denkwürdigkeiten (I commentarii),* 2 vols. (Berlin: Julius Bard, 1912). See also Klaus Bergdolt, ed. and trans., *Der dritte Kommentar Lorenzo Ghibertis: Naturwissenschaften und Medizin in der Kunsttheorie der Frührenaissance* (Weinheim: Acta Humaniora, 1988), a commentary on and German translation of Ghiberti's third book on perspective. There is also an Italian edition in which Ghiberti's often difficult Italian has been modernized: Lorenzo Ghiberti, *I commentari,* ed. Ottovio Morisani (Naples: Riccardo Ricciardi, 1947).

Figure 7. *Lorenzo Ghiberti's gilded bronze doors, known as the "Gates of Paradise," baptistery, Florence, Italy. (Courtesy of Alinari/Art Resource, New York.) Ghiberti's spectacular doors display ten large panels of Old Testament scenes rendered in Albertian perspective. Ghiberti included his own self-portrait within one of the medallions of the frame, placing it near his signature at about the eye level of the average viewer. The doors took more than twenty years to complete. The scenes were first modeled in wax, then cast in bronze, then gilded with gold, and finally finished. As he completed the process of construction, Ghiberti also worked on his* Commentarii.

The difficulty and obscurity of the *Commentarii* are the evident result of Ghiberti's attempt to utilize Latin texts, a project for which his command of the language was clearly inadequate. Yet his linguistic difficulties should not obscure his shrewd choice of excerpts or his astute use of them to develop his own notion of learned practice. Ghiberti repeatedly refers to the writings (gleaned primarily from Vitruvius and Pliny) of ancient practitioners. Painting and sculpture "is a knowledge adorned with many disciplines and various teachings, which is the highest invention of all the other arts. It is made with certain meditation which is completed through material and reasoning." Things that are fabricated, Ghiberti insists, "through proportions by astuteness and by reason" can be demonstrably explained. Paraphrasing Vitruvius (*De architectura,* 1.1.2), he points to the weaknesses of both the unlettered practitioner and the unskilled learned man. Both skill (*ingegnio*) and learning (*disciplina*) are necessary to sculpture. Moreover, the sculptor must have good character traits as well. He must learn philosophy because it fills him "with [a] great soul," one that "is not arrogant" and, moreover, is "moderate and humble and faithful and without avarice." A work made without faith or chasteness cannot be perfect.[68]

The author of the first Italian vernacular architectural treatise, Antonio Averlino (ca. 1400–ca. 1462), was trained as a goldsmith in Florence, probably in the Ghiberti workshop. In 1433 Averlino, who called himself Filarete, "lover of virtue," was working in Rome, assisting in the coronation ceremonies of the emperor Sigismund. From 1433 to 1445 he was at work on the bronze relief panels for the great doors of St. Peter's. In the early 1450s, enticed by the patronage of Francesco Sforza, he moved to Milan, where he supervised the design and construction of a number of important buildings over the protests of local guildsmen. He completed his architectural treatise in the early 1460s, dedicating one copy to Francesco Sforza and another to Piero di Medici of Florence.[69]

Francesco Sforza (1401–1466) was one of the most famous *condottiere* of the fifteenth century who in 1450 became the duke of Milan through marriage, diplomatic maneuver, and military conquest. Averlino's treatise, a direct product of Sforza's patronage, is a narrative tale in which a new city, Sforzinda, is designed and constructed. The treatise provides a highly idealized account of the relationship between architect and patron, represented by Averlino and Francesco Sforza themselves. Throughout, the architect is treated as an equal, admired greatly, and paid handsomely.[70]

[68] Schlosser, *Lorenzo Ghibertis Denkwürdigkeiten,* p. 4 (1.2): "L'iscultura et pictura è scientia di più discipline et di varij amaestramenti ornati, la quale di tutte l'altre arti è somma inventione, è fabricata con certa meditatione la quale si compie per materia et ragionamenti." *Ibid.,* pp. 4–6: "per proportione d'astutia et di ragione"; "con magno animo"; "non sia arrogante"; "agevole et humile et fedele et sança avaritia."

[69] There is a modern Italian edition of the treatise and an English translation, both of which contain biographical and bibliographical information. Antonio Averlino detto il Filarete, *Trattato di architettura,* 2 vols., ed. Anna Maria Finoli and Liliana Grassi (Milan: Il Polifilo, 1972), wherein pp. lxxxviii–xci is a useful chronology of Averlino's life; for the English translation see John R. Spencer, ed. and trans., *Filarete's Treatise on Architecture,* 2 vols. (New Haven, Conn.: Yale Univ. Press, 1965). See also Angiola Maria Romanini, "Antonio Averlino (Averulino), detto Filarete," in *Dizionario biografico degli italiani* (Rome: Istituto della Enciclopedia Italiana, 1962), Vol. 4, pp. 662–667.

[70] See, e.g., Spencer, ed. and trans., *Filarete's Treatise on Architecture,* Vol. 1, p. 79, where a dinner conversation between the architect, the patron, and other nobles is described. The architect (Averlino himself) explains the many good omens that have occurred during the building of the city and reports that "my Lord and the others were so taken with love for me that they gave me enough to live honorably." For Francesco Sforza and the court of Milan under his rule see esp. Franco Catalano, *Francesco Sforza* (Milan: dall'Oglio, 1983); Giorgio Chittolini, ed., *Gli Sforza, la chiesa lombarda, la corte di Roma: Strutture e pratiche beneficiarie nel ducato di Milano (1450–1535)* (Naples: Liguori, 1989); *Gli Sforza a Milano e in Lombardia* (cit. n. 11); Gary Ianziti, *Humanistic Historiography under the Sforzas: Politics and Propaganda in Fifteenth-Century Milan* (Oxford: Clarendon, 1988); and Evelyn S. Welch, *Art and Authority in Renaissance Milan* (New Haven, Conn.: Yale Univ. Press,

Architect and patron alike are fascinated both by design and by the technical aspects of construction. Averlino describes a dinner in which the architect, the lord, and the lord's son eat together as equals and discuss the new city. The lord's son is "so pleased by our conversation" that he falls "in love with this science of architecture" and begs his father to let the architect show him everything that had been done. The request granted, the young prince becomes the pupil of the architect, ever gaining in skill and delighted with his instruction. The prince's interests include strictly technological ones. He wants to know how metal is melted, how furnaces for melting bronze are made; he also asks for information about glass furnaces.[71]

The close association of the architect practitioner and the noble patron is paralleled by institutions in the new city in which learned disciplines and the crafts are taught side by side. Averlino proposes a school for impoverished youth that would provide instruction in letters and good habits but also in "every branch of knowledge and every skill." Instructors would include doctors of laws, of medicine, of canon law, and of rhetoric and poetry. The proposed school is unusual, however, because "some manual arts would also be taught here" by craftsmen. These include "a master of painting, a silversmith, a master of carving in marble and one for wood, a turner, an iron smith, a master of embroidery, a tailor, a pharmacist, a glassmaker and a master of clay." All the children could thereby be trained under the same roof in the discipline to which their souls and intelligences were most suited. Filarete's city also contains a "House of Virtue" in which instruction is offered in the seven liberal arts, areas are designated for military exercises, and all the crafts and trades are both practiced and taught. When the students are "judged by good masters, and have been educated in this place," they are "given the degree like the doctors." Three governors rule the temple of virtue—a doctor, a decorated soldier, and an artisan.[72]

Like Averlino, the painter Piero della Francesca (ca. 1415–1492) combined craft practice with authorship. Piero was born in Sansepolcro in central Italy and often worked there, but he also traveled to other Italian courts and cities at the behest of princely patrons. In addition to painting, Piero worked at constructive projects such as fortification and hydraulic engineering. As a painter, he was called to Perugia, Florence, Arezzo, Rome, and Rimini (where he painted a fresco, in the Tempio Malatestiano [see Figure 6], of Sigismund Malatesta kneeling before a St. Sigismund who strikingly resembles the emperor). He also worked in Urbino in the court of Federico II da Montefeltro.[73] Federico was a skilled

---

1996). For Francesco's father, Musio Attendolo Sforza, and his early career as a soldier see Mallett, *Mercenaries and Their Masters* (cit. n. 11), pp. 66–69 and passim; and Georges Peyronnet, "François Sforza: De condottiere à Duc de Milan," in *Gli Sforza a Milano e in Lombardia*, pp. 7–25. For the Sforza court under Francesco's son Galeazzo Maria Sforza (1444–1476) see Gregory Lubkin, *A Renaissance Court: Milan under Galeazzo Maria Sforza* (Berkeley/Los Angeles: Univ. California Press, 1994).

[71] Spencer, ed. and trans., *Filarete's Treatise on Architecture*, Vol. 1, p. 79. Filarete explicates technology as well as design. His treatise contains one of the earliest descriptions of a modern blast furnace and describes techniques such as how to make a plaster that does not stain. He mentions technical books that he is in the process of writing (none of which, if they were completed, are extant) on "agriculture," on "technical matters," and on "engines." See *ibid.*, pp. 232–233 (furnaces), 258–259 (plaster), 277 (treatise on engines), 317 (where he notes that two books of his treatise on agriculture are complete); see books 22, 23, and 24 of his treatise for his discussions of drawing, perspective, and painting. See also John R. Spencer, "Filarete's Description of a Fifteenth-Century Italian Iron Smelter at Ferriere," *Technol. Cult.*, 1963, *4*:201–206.

[72] Spencer, ed. and trans., *Filarete's Treatise on Architecture*, Vol. 1, pp. 228, 254–255.

[73] Piero was the son of a leather merchant whose family became involved in the production of the herb *guado*, which—made into dyer's indigo—was an important ingredient in the production of cloth. He owned property in Sansepolcro and seems to have enjoyed financial security based on his own and his family's holdings. For his life and works, with reference to the extant documents, see esp. Eugenio Battisti, *Piero della Francesca*, 2 vols. (Milan: Istituto Editoriale Italiano, 1971), which includes a summary of documents and sources (Vol. 2,

military leader who reshaped his city-state by means of a massive building program and surrounded himself with courtiers distinguished in letters and the arts. He had a frieze depicting military and other machines taken from the treatises of Valturio and Francesco di Giorgio built into the facade of his palace in Urbino.[74]

In Urbino and elsewhere, Piero della Francesca created luminous, riveting paintings that reveal his fundamental grasp of mathematics and perspective.[75] Through his writings, Piero elaborated his acute understanding of mathematical principles, which was informed by his knowledge of ancient texts, especially those of Euclid. *Trattato d'abaco* (ca. 1450), concerning commercial mathematics and geometry, was written for a member of a merchant family of Sansepolcro.[76] *De prospectiva pingendi,* written for Federico II da Montefeltro in the 1470s, treats painter's perspective; and *De corporibus regularibus,* composed between 1482 and 1492 for Federico's son, Guidobaldo da Montefeltro, is devoted to the

---

pp. 213–246); Carlo Bertelli, *Piero della Francesca,* trans. Edward Farrelly (New Haven, Conn.: Yale Univ. Press, 1992), esp. pp. 7–50; Paolo Dal Poggetto, *Piero e Urbino, Piero e le corti rinascimentali* (Venice: Marsilio, 1992); Marilyn Aronberg Lavin, *Piero della Francesca* (New York: Abrams, 1992); and Antonio Paolucci, *Piero della Francesca: Catalogo completo dei dipinti* (Florence: Cantini, 1990). For Piero's activity as "ingegnere idraulico" see Margaret Daly Davis, *Piero della Francesca's Mathematical Treatises: The "Trattato d'abaco" and "Libellus de quinque corporibus regularibus"* (Ravenna: Longo, 1977), pp. 16–17 n. 49.

[74] Federico (1422–1482), prince of Urbino, was a brilliant soldier who led his army in the service of the pope, Venice, Florence, and the Aragonese of Naples and against his fierce rival Sigismund Malatesta of Rimini. His courtier Vespasiano da Bisticci presents him in terms of the "Valturian" soldier-scholar: "no other united as he did, in his own person the soldier and the man of letters, or knew how to make intellect augment the force of battalions." Vespasiano da Bisticci, *Renaissance Princes, Popes, and Prelates: The Vespasiano Memoirs: Lives of Illustrious Men of the Fifteenth Century,* trans. William George and Emily Waters (1926; rpt., New York: Harper & Row, Harper Torchbooks, 1963), p. 83. For Federico and his court at Urbino see esp. Giorgio Cerboni Baiardi, Giorgio Chittolini, and Piero Floriani, eds., *Federico di Montefeltro: Lo stato, Le arti, La cultura,* 3 vols. (Rome: Bulzoni, 1986); Luciano Cheles, *Lo studiolo di Urbino: Iconografia di un microcosmo principesco* (Ferrara: Franco Cosimo Panini, 1991); Walter Tommasoli, *La vita di Federico da Montefeltro (1422–1482)* (Urbino: Argalia, 1978); and Pasquale Rotondi, *The Ducal Palace of Urbino: Its Architecture and Decoration* (London: Alec Tiranti, 1969). For the frieze on the facade of the palace see esp. Grazia Bernini Pezzini, *Il fregio dell'arte della guerra nel palazzo ducale di Urbino: Catalogo dei Rilievi* (Rome: Istituto Poligrafico e Zecca dello Stato, 1985); and Antonio Manno, "Architettura e arti meccaniche nel fregio del palazzo ducale di Urbino," in *Federico di Montefeltro,* ed. Baiardi *et al.,* Vol. 2: *Le arti,* pp. 89–104. The copy of Valturio's treatise that was in Federico's library is now Biblioteca Apostolica Vaticana, MS. urb. lat. 281.

[75] See Martin Kemp, *The Science of Art: Optical Themes in Western Art from Brunelleschi to Seurat* (New Haven, Conn.: Yale Univ. Press, 1990), pp. 27–35, for a brief, illuminating discussion of Piero's painting, in which he describes Piero as "a man deeply conversant with pure and applied mathematics, capable of writing treatises to match in quality anything produced in the Italy of his day" (p. 27). See also J. V. Field, "Mathematics and the Craft of Painting: Piero della Francesca and Perspective," in *Renaissance and Revolution,* ed. Field and James (cit. n. 5), pp. 73–95; Field, "Piero della Francesca's Treatment of Edge Distortion," *J. Warburg Courtauld Inst.,* 1986, *49*:66–90; and Marilyn Aronberg Lavin, ed., *Piero della Francesca and His Legacy* (Studies in the History of Art, 48) (Washington, D.C.: National Gallery of Art, 1995).

[76] Piero della Francesca, *Trattato d'abaco dal Codice Ashburnhamiano 280 (359*–291*) della Biblioteca Medicea Laurenziana di Firenze,* ed. Gino Arrighi (Pisa: Domus Galilaeana, 1970). This treatise was written for the Picchi family of Sansepolcro; see Battisti, *Piero della Francesca* (cit. n. 73), Vol. 1, p. 281, and Vol. 2, index, s.v. "Picchi famiglia di Sansepolcro." For the abacus tradition within which Piero worked see Richard A. Goldthwaite, "Schools and Teachers of Commercial Arithmetic in Renaissance Florence," *Journal of European Economic History,* 1972, *1*:418–433; Paul F. Grendler, *Schooling in Renaissance Italy: Literacy and Learning, 1300–1600* (Baltimore: Johns Hopkins Univ. Press, 1989), pp. 22–23, 72–73; and Warren van Egmond, *Practical Mathematics in the Italian Renaissance: A Catalogue of Italian Abbacus Manuscripts and Printed Books to 1600* (Monografia no. 4) (Florence: Istituto e Museo di Storia della Scienza, 1980). Yet Piero moves rapidly beyond the needs of commercial mathematics. He introduces innovative methods of algebraic computation, treats the measurement not of real buildings but of abstract polygons and polyhedra, handles both plane and solid geometry, and, following Euclid's *Elements,* discusses the measurements of the five regular solids and irregular solids within a sphere. See Davis, *Piero della Francesca's Mathematical Treatises* (cit. n. 73), pp. 11–20; and S. A. Jayawardene, "The 'Trattato d'Abaco' of Piero della Francesca," in *Cultural Aspects of the Italian Renaissance: Essays in Honor of Paul Oskar Kristeller,* ed. Cecil H. Clough (Manchester: Manchester Univ. Press, 1976), pp. 229–243.

five regular solids. Piero explicated both the mathematical reasoning and the procedures required for the practice of painting, yet in range and complexity his account goes far beyond the needs of workshop practice. His prospective audience included noble patrons and learned men. Both treatises were quickly translated into Latin.[77]

Francesco di Giorgio Martini (1439–1501), from Siena, was a contemporary of Piero's who also found patronage at court in Urbino. Francesco started out as a painter and sculptor and became one of the most respected architects and military engineers of the late fifteenth century. His career was based primarily in three locations: his native city of Siena; Urbino, where he enjoyed the patronage of Federico II da Montefeltre and his successor Guidobaldo; and Naples, where he was employed by Alfonso II, duke of Calabria (1448–1495).[78]

In addition to his work as an architect and engineer, Francesco struggled to learn Latin, translated Vitruvius, and created pictorial notebooks and illustrated treatises on architecture, fortification, and military engineering. An early example of his authorship, *Opusculum de architectura,* is a book of drawings of machines and plans of fortification without descriptive text, dedicated to Federico, duke of Urbino. Reminiscent of Mariano Taccola's *De machinis,* the machine drawings—which number over two hundred—are divided into types, such as mills, pumps, siphons and waterwheels, and military devices. Francesco also authored two major treatises on architecture, engineering, and the military arts. Although the exact dates of the treatises are debated, a comparison of the earlier work to the later one reveals the striking development of his ideas and of his skills as a translator of Vitruvius.[79]

Francesco divided *Trattato I* (probably written around the late 1470s) into untitled

[77] Piero della Francesca, *De prospectiva pingendi,* ed. Giusta Nicco-Fassola, 2nd ed., ed. Eugenio Battisti *et al.* (Florence: Le Lettere, 1984); and Girolamo Mancini, "L'opera 'De corporibus regularibus' di Pietro Franceschi detto Della Francesca, usurpata da Fra Luca Pacioli," *Memorie della Rendiconti della Accademia dei Lincei: Classe di Scienze Morali, Storiche e Filologiche,* Ser. 5, 1916, *14*:441–580. The Latin translation of *De prospectiva pingendi* was made by Matteo da Borgo, who probably translated the *Libellus* on the regular solids as well; see Davis, *Piero della Francesca's Mathematical Treatises,* p. 54. Although extant copies of Piero's perspectival treatise are lacking front matter that would include a dedicatory letter, evidence that it was dedicated to Federico exists in *De corporibus regularibus,* concerning the five regular solids. The dedication letter of this later treatise, addressed to Guidobaldo, duke of Urbino, expresses Piero's hope that *De corporibus regularibus* would find a place on the shelf next to the book on perspective that he had written for Guidobaldo's father, Federico da Montefeltre. New editions of the treatises, based on additional manuscript sources, by Marisa Dalai Emiliani, Cecil Grayson, and Carlo Maccagni are forthcoming.

[78] For a comprehensive introduction to Francesco di Giorgio's career and to an extensive bibliography see Francesco Paolo Fiore and Manfredo Tafuri, eds., *Francesco di Giorgio architetto* (Milan: Electa, 1993); Gustina Scaglia, *Francesco di Giorgio: Checklist and History of Manuscripts and Drawings in Autographs and Copies from ca. 1470 to 1687 and Renewed Copies (1764–1839)* (Bethlehem, Pa.: Lehigh Univ. Press, 1992); Ralph Toledano, *Francesco di Giorgio Martini: Pittore e scultore,* trans. Massimo Parizzi (Milan: Electa, 1987); and Francesco di Giorgio Martini, *Trattati di architettura ingegneria e arte militare,* ed. Corrado Maltese, transcribed by Livia Maltese Degrassi, 2 vols. (Milan: Il Polifilo, 1967), Vol. 1, pp. xi–xxii.

[79] The dates and chronology of Francesco's notebooks and treatises are the focus of ongoing scholarly debate. See esp. Richard J. Betts, "On the Chronology of Francesco di Giorgio's Treatises: New Evidence from an Unpublished Manuscript," *Journal of the Society of Architectural Historians,* 1977, *36*:3–14; Carolyn Kolb, "The Francesco di Giorgio Material in the Zichy Codex," *J. Soc. Architect. Hist.,* 1988, *47*:132–159; and Massimo Mussini, "La trattatistica di Francesco di Giorgio: Un problema critico aperto," in *Francesco di Giorgio architetto,* ed. Fiore and Tafuri, pp. 358–359. For the *Opusculum,* some of the drawings of which derive from the books of Taccola, see Scaglia, *Francesco di Giorgio: Checklist,* pp. 43–50; for the manuscript see London, British Museum, Codex 197 B 21 [MS Harley 3281] (Scaglia, *Francesco di Giorgio: Checklist,* pp. 101–104); for the manuscript that includes the dedication to Federico see Turin, Biblioteca Reale, Codex serie militare 383 [cod. 14856; 14876; 14896 d[otazione] c[orona]; see Scaglia, *Francesco di Giorgio: Checklist,* p. 56, for Francesco's sketches on the autograph manuscript of Taccola's *De ingeneis* in Munich, Bayerische Staatsbibliothek, Codex CLM 197, part II. For Francesco's two major treatises, along with a thorough discussion of the manuscripts, see Francesco di Giorgio, *Trattati,* ed. Maltese (Vol. 1 contains *Trattato I,* Vol. 2, contains *Trattato II*).

sections according to subject matter, such as fortresses, cities, temples, theaters, geometry, waterwheels and water mills, ways of lifting and transporting water, and military machines. *Trattato II,* although lacking a formal dedication, contains a lengthy tribute to Federico II da Montefeltre. Francesco commences this, his most mature treatise, with an introductory essay, based loosely on Pliny, that extols arithmetic and geometry. He maintains that in antiquity drawing was "reputed in the first grade of the liberal arts," rather than—as in his own time—"reputed low and inferior to many other mechanical arts." Drawing is necessary "to every human work," including invention, the explanation of concepts, and operations such as the military arts; and it is related to geometry, arithmetic, and perspective, all of which are necessary for making things "with right reason."[80]

Francesco di Giorgio's career was in many ways parallel to that of Leonardo da Vinci (1452–1519), and indeed the two men knew each other and were consulted together on at least one occasion (concerning the cathedral at Pavia). Leonardo was apprenticed in the Florentine workshop of Verrocchio in the late 1460s, at a time when the shop's numerous projects included the task of placing the heavy gilded copper sphere on top of the lantern that centered Brunelleschi's cupola. Leonardo was fascinated by this operation and by the machines employed to accomplish it, as he was also intrigued by the machines that Brunelleschi had invented to build the cupola itself. Eventually, however, he left Florence, turning to the patronage of princes, most importantly Ludovico Sforza of Milan, and others such as Cesare Borgia, *condottiero* nephew of the pope, and, finally, Francis I, king of France. These patrons valued not only Leonardo's work as a painter but also his abilities as an engineer. Leonardo's lifelong technological interests included hydraulic works, machines of all kinds, fortification, and weaponry.[81]

During most of his working life as a painter, sculptor, and engineer, especially after his arrival in Milan in the early 1480s, Leonardo created voluminous notebooks, of which approximately a third (about six thousand sheets) still exist, including the Madrid Codices, two long-lost volumes discovered in the early 1960s in the Spanish National Library. Taken as a whole, Leonardo's notebooks represent at least two kinds of authorship. Some are private notes and sketches, made in the notebooks that he habitually carried with him—Codex Madrid II is an example. Leonardo clearly intended others, such as Codex Madrid I, to be formal treatises. Leonardo planned treatises on mechanics, painting, anatomy, and hydraulics, among other things. While he was well aware of the technical writings of his contemporaries and predecessors, his own authorship was also, at least in part, motivated

[80] Francesco di Giorgio, *Trattati,* ed. Maltese, Vol. 2, pp. 293–294 (for citations), 293 n. 2 (for a discussion of the relationship of this text to Pliny's): "nel primo grado de le liberali era reputata"; "reputata vile et inferiore a molte altre arti mecaniche"; "in ogni opera umana"; "con dritta ragione." See also Vol. 2, p. 425, for Francesco's tribute to Federico. For Francesco's architectural theory, which involved anthropomorphic, modular methods of design, see esp. Henry Millon, "The Architectural Theory of Francesco di Giorgio," *Art Bull.,* 1958, *40*:257–261.

[81] The best general introduction to Leonardo's life and work is Martin Kemp, *Leonardo da Vinci: The Marvellous Works of Man and Nature* (Cambridge, Mass.: Harvard Univ. Press, 1981). For Francesco di Giorgio and Leonardo in Pavia and Milan see Pietro C. Marani, "Francesco di Giorgio a Milano e a Pavia: Consequenze e ipotesi," in *Prima di Leonardo,* ed. Galluzzi (cit. n. 1), pp. 93–104; and Gustina Scaglia, "Leonardo da Vinci e Francesco di Giorgio a Milano nel 1490," in *Leonardo e l'età della ragione,* ed. Enrico Bellone and Paolo Rossi (Milan: Scientia, 1982), pp. 225–253. For Leonardo as an engineer see esp. Paolo Galluzzi, "The Career of a Technologist," in *Leonardo da Vinci: Engineer and Architect,* ed. Galluzzi (Montreal: Montreal Museum of Fine Arts, 1987), pp. 41–109; for his interest in Brunelleschi's machines see Salvatore Di Pasquale, "Leonardo, Brunelleschi, and the Machinery of the Construction Site," *ibid.,* pp. 163–181. See also Frank D. Prager and Gustina Scaglia, *Brunelleschi: Studies of His Technology and Inventions* (Cambridge, Mass.: MIT Press, 1970), esp. pp. 65–109; and Scaglia, "Drawings of Brunelleschi's Mechanical Inventions for the Construction of the Cupola," *Marsyas,* 1960–1961, *10*:45–67.

by patronage concerns. For example, we know from a sixteenth-century note that he began part of a book on painting (the *paragone* comparing the nobility of painting to that of sculpture) at the request of his patron Ludovico Sforza. Beyond the concerns of patronage, Leonardo was intensely involved in the world of books, as is evidenced by the list of over a hundred codices in his personal library. His own early writings were clearly inspired by Alberti's *Della pittura.*[82]

Leonardo failed to complete his treatises and presented none to patron or publisher, a circumstance that presents problems of evaluation. Unlike the writings of most other fifteenth-century technical authors, his manuscripts exist almost exclusively as autographs. Yet as Ladislao Reti pointed out, Leonardo was well aware that his unique drawings could not be reproduced by scribes, nor could woodcut print technology reproduce their fine detail. The new method of copper engraving was adequate in theory but would have been too expensive and difficult, given the large number of drawings embedded in text that he produced. Leonardo developed a whole set of illustrative techniques that allowed him to demonstrate clearly exactly how a machine, a device, or even a human body was put together. Although his inability to complete his treatises had complex causes, his awkward positioning between the cultures of scribe and print, neither of which could in his time adequately reproduce his books, clearly must have been a factor.[83]

Leonardo's Codex Madrid I constitutes a formal treatise on mechanics with two parts, one on theory and the other on practice (see Figure 8). Concerning practice, he presents often strikingly beautiful drawings of machines and mechanisms and comments on how to build them and how they work. His working methods have been elucidated by Reti, who compared several sheets of Codex Atlanticus (from one of Leonardo's private notebooks) with those of the formal treatise. Machines roughly sketched in Codex Atlanticus are redrawn with care and deliberation in Codex Madrid I. Paragraphs of text in Codex Atlanticus are crossed out and then reappear neatly inscribed (but still in mirror writing) in Codex Madrid I. Leonardo treats machines, including how to make them and how they

[82] For the complex tradition of Leonardo's notebooks see Galluzzi, "Career of a Technologist," pp. 43–47; Augusto Marinoni, "I manoscritti di Leonardo da Vinci e le loro edizioni," in *Leonardo: Saggi e ricerche,* ed. Comitato Nazionale per le Onoranze a Leonardo da Vinci nel Quinto Centenario della Nascità (Rome: Istituto Poligrafico dello Stato, 1954), pp. 231–274; Marinoni, "The Writer: Leonardo's Literary Legacy," in *The Unknown Leonardo,* ed. Ladislao Reti (New York: McGraw-Hill, 1974), pp. 56–85; and Carlo Pedretti, *Leonardo da Vinci: On Painting, a Lost Book (Libro A)* (Berkeley/Los Angeles: Univ. California Press, 1964), pp. 9 (for the note by Giovan Paolo Lomazzo), 252–259. For the Madrid Codices see Leonardo da Vinci, *The Madrid Codices,* 5 vols., ed. Reti (New York: McGraw-Hill, 1974), Vol. 2, fols. 2v–3r, Vol. 3, pp. 91–108, Vol. 5, pp. 5–8 (for Leonardo's library). Despite intense scholarship and the approximate dating of many individual sheets, Leonardo's notebooks as he wrote them cannot be fully reconstructed. Evidence that Leonardo planned formal treatises includes remarks addressed explicitly to readers and experiments with alternative tables of contents scattered throughout his notebooks; see, e.g., Jean Paul Richter, ed., *The Notebooks of Leonardo da Vinci,* 2 vols. (1883; rpt., New York: Dover, 1970), Vol. 1, pp. 13–17, 69, 190, Vol. 2, pp. 107–112, 144, 175–181, 388–389.

[83] For Leonardo's brilliantly innovative illustrative techniques see James Ackerman, "The Involvement of Artists in Renaissance Science," in *Science and the Arts in the Renaissance,* ed. John W. Shirley and F. David Hoeniger (Washington, D.C.: Folger Shakespeare Library, 1985), pp. 94–129, esp. pp. 102–111; Galluzzi, "Career of a Technologist" (cit. n. 81), pp. 96–109; and Kemp, *Leonardo da Vinci* (cit. n. 81), pp. 54–57. Francesco di Giorgio was an important predecessor in such illustration; see Martin Kemp, " 'La diminutione di ciascun piano': La rappresentazione delle forme nello spazio di Francesco di Giorgio," in *Prima di Leonardo,* ed. Galluzzi (cit. n. 1), pp. 105–112. For Leonardo's difficulty with contemporary methods of printing by woodcut and copper engraving and his invention (on paper) of relief etching see Ladislao Reti, "Leonardo da Vinci and the Graphic Arts: The Early Invention of Relief-Etching," *Burlington Magazine,* 1971, *112*:189–195. Leonardo discusses this invention in Codex Madrid II; see Leonardo, *Madrid Codices,* ed. Reti, Vol. 2, fol. 119r, Vol. 5, pp. 255–256.

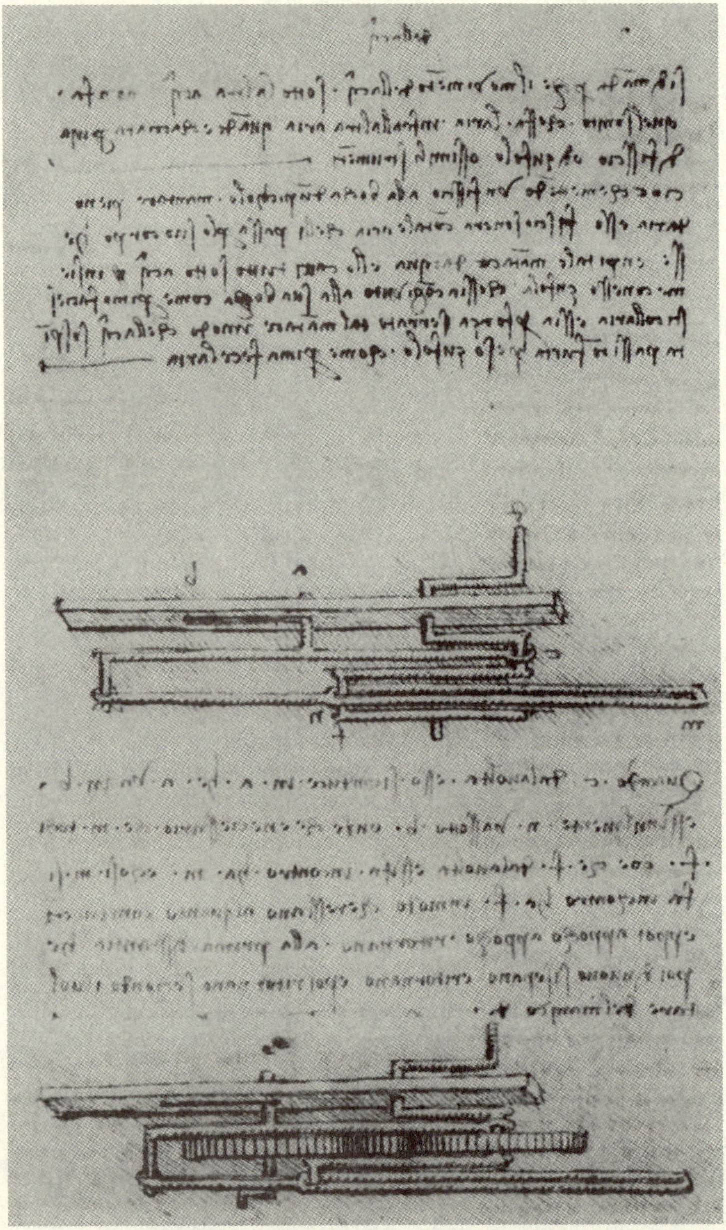

Figure 8. *Leonardo da Vinci's epicycloid gears and the notion of air and water, Madrid Codex I, fol. 111v. (Courtesy of Biblioteca Nacional, Madrid.) This sheet from Madrid Codex I is one of numerous examples in which Leonardo discusses natural and mechanical phenomena as interrelated aspects of reality that can be fruitfully examined together. Here, he asks why the motion of water forced through water (by bellows, for example) does not make a sound, whereas the motion of air forced through air (by a whistle) does make one. He then carefully describes the motion of the epicycloid gears that he has drawn. Although it is not always clear precisely what connections Leonardo is trying to make, it is abundantly evident that for him the world of nature and the world of the machine reflect one another and that analogies between the two realms have explanatory force.*

work; he also discusses theoretical mechanics, which for him primarily concerned weight, force, impact, and motion.[84]

Paolo Galluzzi has argued that Leonardo's technical authorship was clearly different from that of predecessors such as Mariano Taccola and Francesco di Giorgio because of his profound interest in theory and his use of practical mechanics to study theoretical issues. As Galluzzi has noted, Leonardo brought together two separate traditions of the mechanical arts—one based on Aristotle's mechanics and medieval statics and kinetics, the second involving the explication of machines and the mechanical arts of the fifteenth-century engineer. My caveat is that Leonardo's predecessors included not only Mariano Taccola and Francesco di Giorgio but also Fontana, Alberti, Ghiberti, Roberto Valturio, and Piero della Francesca, among others. Leonardo's attempt to integrate theory (especially ancient and medieval theories of mechanics) with practice, as well as his use of a combination of pictorial and textual explication in his treatises, represents a development from the work of his immediate predecessors, one that arose naturally out of the rich tradition of technical authorship that preceded him.[85]

CONCLUSION

The new alliance of *praxis* and *techne* in fifteenth-century Italian and south German regions was a nexus within which authorship concerning the mechanical arts flourished. Men from diverse backgrounds—university-educated humanists and artisan-trained practitioners—took to writing about matters that had previously existed primarily in the arena of skilled practice. These authors usually wrote within the context of patronage, dedicating their books to princes, emperors, and oligarchs. Such books possessed value in relationships between patrons and clients because the arts that they treated themselves had gained cultural significance. Newly associated with learning and the liberal arts, and with the writings and heroes of antiquity, they had come to be considered intrinsic to civic, moral, and political life beyond their technical realization in the material world. The new cultural value of the constructive arts promoted the development and explication of their rational principles in writing.

The rise in status of the mechanical arts is one result of the use of building, urban redesign, and visual ornamentation to represent and promote political power in the fifteenth century. Authorship involving such arts also produced new images of practitioners—from the gunners of the *Feuerwerkbuch* to the goldsmith and architect Averlino—self-described "lover of virtue"—who embodied both skill and virtue. Nevertheless, the new cultural status of the mechanical arts, although it clearly aided the careers of certain practitioners, particularly if they wrote books, should not be taken to indicate a rise in the status of artisans in general. Whereas authorship helped to transform some arts from the arena of skilled know-how to that of discursive knowledge, it did not change artisans into learned men. It is more accurate to say that it prepared certain of the mechanical arts for appropriation by learned culture.

[84] Leonardo, *Madrid Codices,* ed. Reti, Vol. 1, fol. 18r, Vol. 4, pp. 45–48; Leonardo da Vinci, *Il Codice Atlantico di Leonardo da Vinci: Edizione in facsimile dopo il restauro dell' originale conservato nella Biblioteca Ambrosiana di Milano,* 12 vols. (Florence: Giunti Barbèra, 1973; New York: Harcourt Brace Jovanovich, Johnson Reprint, 1973), Vol. 1, fol. 77v; and Augusto Marinoni, *Il Codice Atlantico della Biblioteca Ambrosiana di Milano,* diplomatic and critical transcription, 12 vols. (Florence: Giunti Barbèra, 1975), Vol. 1, p. 57. See also Carlo Pedretti, *The Codex Atlanticus of Leonardo da Vinci: A Catalogue of Its Newly Restored Sheets,* 2 pts. (New York: Harcourt Brace Jovanovich, Johnson Reprint, 1978), Pt. 1, p. 57.

[85] Galluzzi, "Career of a Technologist" (cit. n. 81), pp. 92–95.

The complex history of that appropriation in the sixteenth and seventeenth centuries is beyond the scope of this essay. By the phrase "from mechanical know-how to mechanical knowledge," I do not refer directly to Galilean mechanics or to the "mechanization of the world picture." Rather, I mean to suggest a change in the status of the mechanical arts accomplished in part by means of authorship—from a kind of artisanal know-how transmitted for the most part orally and utilized in the service of material construction to a group of rationalized disciplines informed by principles and explicated in books both pictorially and in writing. Without ignoring occasional prior writings on the mechanical arts, including ancient exemplars, I emphasize the significance of the great expansion of authorship concerning such arts in the fifteenth century and point to the essential context of such authorship—patronage and the use of the constructive arts for the display and legitimization of political power.

The fifteenth century is a forgotten era in the historiography of medieval and early modern science, all the more so with the burgeoning of interest in local context and meaning. Yet if the proper study of mechanics and the mechanical world view as it developed in the late sixteenth and seventeenth centuries involves the meticulous study of key texts and the careful investigation of the local cultures that inform the meanings of those texts, it should also take account of what Hans-Georg Gadamer calls the "traditionary material" that was available. Gadamer's phrase is useful because it suggests not continuous traditions, unbroken threads from the past, but cultural legacies and materials available for appropriation.[86] To put it another way, if it is methodologically sound to study "islands of discourse" in the context of local cultures (as I believe it is), it is also important to assess critically the ways in which we construct the boundaries of those discursive islands and to assess the nature of the materials from the past that are available for appropriation within those boundaries.

Fifteenth-century writings on the mechanical arts can be investigated on their own terms; they can be understood, as well, with regard to their significance for a later period. The substance of this essay concerns authorship on the mechanical arts during the fifteenth century. Yet I also suggest that such authorship had significance for the development of experimentalism and the mechanical philosophy in a later era. Scholars of the later tradition often seem to assume that seventeenth-century developments occurred against the background of a long, relatively uniform tradition of low status assigned to the mechanical arts. I hope that the present study—which, despite important differences, belongs to the tradition of scholarship represented by the late writings of Edgar Zilsel—has demonstrated the extent to which such an assumption involves an oversimplification. As a study based in the fifteenth century, it does not of course tell the complete story of transformation and appropriation that forms the background for the development of experimentalism and the mechanical philosophy. It does, however, demonstrate that the fifteenth-century proliferation of authorship in the mechanical arts forms an essential early chapter in such a story and that the history itself is far more complicated than is often assumed.

Printing greatly increased the number of technical books produced and gave far wider exposure to practitioner authors, as well as to the learned. As books became commodities, hundreds of printed technical books on artillery and fortification, mechanics, mining, metallurgy, machines, and architecture appeared, along with books of secrets and works on many other topics. Yet the positive valuation of the mechanical arts and their role in the

[86] Hans-Georg Gadamer, *Truth and Method,* 2nd rev. ed., trans. Joel Weinsheimer and Donald G. Marshall (New York: Continuum, 1993).

legitimization of rulership were cultural phenomena that developed significantly in the century before printing. Organized textually and pictorially in manuscript books and provided with elaborated rational principles, the mechanical arts became central features of moral and political life, as they also began to take part in the construction of knowledge itself.

Index

CONTRIBUTORS

Martin Bernal is a professor in the departments of Government and of Near Eastern Studies at Cornell University. Previously a scholar of China, he is the author of *Chinese Socialism to 1907* (Cornell University Press, 1976). His *Black Athena: The Afroasiatic Roots of Classical Civilization,* 2 vol. (Rutgers University Press) appeared in 1987 and 1991. Volume 3 is anticipated in 2003. *Black Athena Writes Back,* a response to *Black Athena Revisited,* a compilation of criticisms hostile to *Black Athena,* is due out from Duke University Press, in the spring of 2001.

Alan C. Bowen is Director of the Institute for Research in Classical Philosophy and Science in Princeton. His research concentrates on the exact sciences in Greco-Latin antiquity, especially on astronomy and harmonic science. He has just completed, with Robert B. Todd, *Physics and Astronomy in Later Stoic Philosophy: Cleomedes, "The Heavens"* for the University of California Press and is currently writing a book on Greco-Latin planetary theory before Ptolemy.

Marshall Clagett is Professor Emeritus of the History of Science and Mathematics in the School of Historical Studies, Institute for Advanced Study in Princeton. A recipient of the Sarton Medal of the History of Science Society, his earlier works include extensive publications on medieval natural philosophy and mathematics, most notably *Archimedes in the Middle Ages,* 5 vol. (University of Wisconsin Press and American Philosophical Society, 1964–84). In recent years, he has been working on *Ancient Egyptian Science* (American Philosophical Society, 1989–), of which three volumes have appeared.

Brian Copenhaver is Professor of History and Philosophy and Provost of the College of Letters and Science at UCLA. His books include *Hermetica: The Greek Corpus Hermeticum and the Latin Asclepius in English Translation, with Notes and Introduction* (Cambridge University Press, 1992). His current works-in-progress focus on Giovanni Pico della Mirandola's *Oration* and Polydore Vergil's *De inventoribus rerum.*

Bernard R. Goldstein is University Professor Emeritus at the University of Pittsburgh. His publications include studies of astronomy in the Greek, Arabic, Hebrew, and Latin traditions from antiquity through the early modern period. He has recently completed a book with José Chabás: *Astronomy in the Iberian Peninsula: Abraham Zacut and the Transition from Manuscript to Print* (Philadelphia: American Philosophical Society, to appear in 2000).

Edward Grant is Distinguished Professor Emeritus of History and Philosophy of Science at Indiana University, Bloomington. A recipient of the Sarton Medal of the History of Science Society, he is the author of numerous books and articles on medieval science and natural philosophy. His recent books are *Planets, Stars, and Orbs: The Medieval Cosmos, 1200–1687* (Cambridge, 1994) and *The Foundations of Modern Science in the Middle Ages* (Cambridge, 1996). His latest book, on reason in the Middle Ages, is currently in press.

Alexander Jones is Professor of Classics and the History and Philosophy of Science and Technology at the University of Toronto. His recent book *Astronomical Papyri from Oxyrhynchus,* the most substantial of his many publications on Greco-Roman astronomy, confirms the central speculations of the article in this anthology while modifying the details. With J. L. Berggren, he has produced a new translation of the theoretical part of Ptolemy's *Geography* (Princeton University Press).

Helen Rodnite Lemay is Professor of History at the State University of New York at Stony Brook. She has published a number of studies on medieval women's medicine, including *Women's Secrets: A Translation of Pseudo-Albertus Magnus' "De secretis mulierum" with Commentaries* (SUNY Press, 1992). She is now preparing a critical edition of William of Conches's *Glosae super Macrobium* for the series Corpus Christianorum: Continuatio Mediaevalis.

David C. Lindberg, Hilldale Professor of the History of Science at the University of Wisconsin, is the author of *Theories of Vision from al-Kindi to Kepler* (University of Chicago Press, 1976), *Roger Bacon's Philosophy of Nature* (Oxford University Press, 1983), and *The Beginnings of Western Science* (University of Chicago Press, 1992). He has been a Guggenheim Fellow, and a recipient of the Watson Davis Prize and the Sarton Medal of the History of Science Society.

Sir Geoffrey E. R. Lloyd is Professor of Ancient Philosophy and Science and Master of Darwin College in the University of Cambridge. He is the author or editor of more than a dozen books on various aspects of Greek science and natural philosophy, often in anthropological and cross-cultural perspective. These include *The Revolutions of Wisdom: Studies in the Claims and Practice of*

Ancient Greek Science (University of California Press, 1987), *Demystifying Mentalities* (Cambridge University Press, 1990), *Aristotelian Explorations* (Cambridge University Press, 1996), and *Adversaries and Authorities: Investigations into Ancient Greek and Chinese Science* (Cambridge University Press, 1996). His scholarship has been recognized by knighthood and the Sarton Medal of the History of Science Society.

Pamela O. Long is a historian who has taught most recently at Johns Hopkins University, and is presently (2000–2001) a senior fellow at the Dibner Institute of the History of Science and Technology, M.I.T. Her articles on technical authorship and related topics have appeared in *Technology and Culture, History and Technology, Journal of Medieval and Renaissance Studies,* and elsewhere. She is the editor of *Science and Technology in Medieval Society* (Annals of the New York Academy of Sciences, vol. 441), and the author of *Openness, Secrecy, Authorship: Technical Arts and the Culture of Knowledge from Antiquity to the Renaissance* (Johns Hopkins University Press, in press).

Stephen McCluskey is Professor of History at West Virginia University. His research focuses on the astronomies of early cultures, ranging from the indigenous peoples of Pre-Columbian America to the monastic communities of the early Middle Ages. He is one of the editors of *Archaeoastronomy: The Journal of Astronomy in Culture* (University of Texas Press) and has recently published *Astronomies and Cultures in Early Medieval Europe* (Cambridge University Press, 1998).

Michael McVaugh is Wells Professor of History at the University of North Carolina at Chapel Hill. Among other works, he is the author of *Medicine Before the Plague: Practitioners and their Patients in the Crown of Aragon, 1285–1345* (Cambridge University Press, 1993), the co-editor (with Edith Sylla) of *Texts and Contexts in Ancient and Medieval Science* (Brill, 1997), and has been co-editor of the *Arnaldi de Villanova Opera Medica Omnia* since 1975. He is presently working on a study of surgical learning of the thirteenth and fourteenth centuries.

Otto Neugebauer (1899–1990), a professor in the Department of the History of Mathematics at Brown University since the 1930s, was arguably the most eminent historian of science of the twentieth century. He was the foremost specialist in the comparative history of ancient and medieval mathematical astronomy for the cultures extending from the Atlantic to the Indus valley. He wrote, among other books, *The Exact Sciences in Antiquity,* 2nd ed. (Brown University Press, 1957; Dover, 1969), *A History of Ancient Mathematical Astronomy* (Springer Verlag, 1975), *Astronomy and History: Selected Essays* (Springer Verlag, 1983), and (with Noel M. Swerdlow), *Mathematical Astronomy in Copernicus's* De Revolutionibus (Springer Verlag, 1984).

William R. Newman is Professor in the Department of History and Philosophy of Science at Indiana University. He has published *The Summa Perfectionis of Pseudo-Geber* (Brill, 1991), an edition, translation, and analysis of the major alchemical work of the High Middle Ages, and *Gehennical Fire: The Lives of George Starkey, An American Alchemist in the Scientific Revolution* (Harvard University Press, 1994). He is researching three subjects—the influence of J. B. Van Helmont on Starkey and Robert Boyle, the atomism of the seventeenth-century physician Daniel Sennert, and the pre-modern project to create artificial human life.

David Pingree is Professor and Chair of the Department of the History of Mathematics at Brown University, where he succeeded Neugebauer in 1971. A classicist and Sanskritist who became interested in Akkadian, he is the author and editor of more than twenty-five books, including *From Astral Omens to Astrology: From Babylon to Binaker* (Istituto italiano per l'Africa et l'Oriente, 1997) and, with Hermann Hunger, *Astral Sciences in Mesopotamia* (Brill, 1999). A former John D. and Catherine T. MacArthur Fellow, he is currently involved in research projects involving the edition and interpretation of texts in Akkadian, Greek, Latin, Arabic, and Sanskrit.

Victor Roberts worked with Professor E. S. Kennedy at the American University in Beirut, Lebanon, in the early 1960s. An Iraqi citizen, he was an undergraduate when he wrote and published his two important papers on Ibn al-Shāṭir, the one reprinted here and the earlier "The Planetary Theory Ibn al-Shāṭir: Latitudes for the Planets," *Isis* 57 (1966) 208–209. The North American history of science community has lost touch with him for many years.

Francesca Rochberg is Professor of History at the University of California-Riverside. A John D. and Catherine T. MacArthur Fellow in 1982–87, her publications include first editions of cuneiform texts in her books *Aspects of Babylonian Celestial Divination: The Lunar Eclipse Tablets of Enuma Anu Enlil,* (1988) and *Babylonian Horoscopes* (American Philosophical Society, 1998) as well as historical and philological studies of Babylonian astrology, astronomy, cosmology, and religion.

A(bdelhamid). I. Sabra's first venture in history of science was in seventeenth-century optics, with a London 1955 Ph.D. thesis later published as *Theories of Light from Descartes to Newton* (London, 1967; Cambridge, 1981). In his alma mater, the University of Alexandria, Egypt, later at London's Warburg Institute (1962–72), and at Harvard since 1972, he turned his attention to medieval Arabic science, with studies, editions, and translations mainly in the areas of optics, mathematics, and astronomy. He is currently Research Professor of the History of Arabic Science in Harvard's Department of the History of Science, working on completing his edition and translation of Ibn al-Haytham's *Optics.*

Michael H. Shank is Professor of History of Science and a Senior Member of the Institute for Research in the Humanities at the University of Wisconsin-Madison. He is the author of *"Unless You Believe, You Shall Not Understand:" Logic, University, and Society in Late Medieval Vienna* (Princeton University Press, 1988), and the co-editor, with David Lindberg, of volume two of the forthcoming *Cambridge History of Science.* He is completing a book on Regiomontanus's *Disputationes.*

Nancy Siraisi is Professor of History at Hunter College and the Graduate Center of the City University of New York. Her research interests are in the history of medicine and life sciences in cultural and intellectual context, chiefly in the Italian universities, ca. 1250–1600. She is the author of the prize-winning *Medieval and Early Renaissance medicine: An Introduction to Knowledge and Practice* (University of Chicago Press, 1990) and of *The Clock and the Mirror: Girolamo Cardano and Renaissance Medicine* (Princeton University Press, 1997).

Nathan Sivin is Professor of Chinese Culture and of the History of Science at the University of Pennsylvania. He has written widely on ancient Chinese alchemy, mathematical astronomy, and medicine. His most recent books are *Science in Ancient China* (Variorum, 1995), *Medicine, Philosophy, and Religion in Ancient China* (Variorum, 1995), and *Science and Civilisation in China,* vol. 6, pt. 6: Medicine (Cambridge University Press, 2000).

A. Mark Smith is currently a member of the Department of History at the University of Missouri, Columbia. He has recently published *Ptolemy's Theory of Visual Perception* (American Philosophical Society, 1996) and *Ptolemy and the Foundations of Ancient Mathematical Optics: A Source Based Guided Study* (American Philosophical Society, 1999), and just finished a critical edition of the first three books of Alhacen's "De aspectibus" (American Philosophical Society, forthcoming).

Owsei Temkin is Welch Professor Emeritus in the Department of History of Science, Medicine and Technology at Johns Hopkins University. Since the 1930s, he has published a wide variety of works in the history of ancient, medieval, and modern medicine. He is he author, editor, and translator of a dozen books, including such classics as *Soranus of Ephesus, Gynecology* (Johns Hopkins University Press, 1956), *Galenism: Rise and Decline of a Medical Philosophy* (1973), *Hippocrates in a World of Pagans and Christians* (Johns Hopkins University Press, 1991).

Lynn Thorndike (1882–1965) spent most of his career as Professor of History at Columbia University. The third recipient of the Sarton Medal of the History of Science Society, he was a pioneer in the study of the occult sciences in the Latin Middle Ages. He published seven books on topics ranging from astronomy to natural history through biographies, textbooks, and reference works, but is best-known for his 8-volume *History of Magic and Experimental Science* (Columbia University Press, 1923–1958).

Linda Ehrsam Voigts is Curators' Professor of English at the University of Missouri, Kansas City. A specialist in medieval vernacular science and medicine, she has published, with Michael McVaugh, *A Latin technical phlebotomy and its Middle English translation* (American Philosophical Society, 1984). Her most recent contributions include *Scientific and Medical Writings in Old and Middle English: An Electronic Reference* (information on more than 9,000 texts).

Heinrich von Staden is Professor of Classics and History of Science at the Institute for Advanced Study in Princeton. His publications include *Herophilus: The Art of Medicine in Early Alexandria* (Cambridge University Press, 1989; 2nd ed. 1994), numerous articles on Greek and Roman medicine, biology, philosophy, history, and literature, and several essays on the nineteenth-century reception of classical antiquity.